Contents

Introduction ... xi
A Note to the Reader ... xvii

1 Sets and Spaces ... 1
1.1 Sets ... 1
1.2 Ordered Sets ... 9
   1.2.1 Relations ... 10
   1.2.2 Equivalence Relations and Partitions ... 14
   1.2.3 Order Relations ... 16
   1.2.4 Partially Ordered Sets and Lattices ... 23
   1.2.5 Weakly Ordered Sets ... 32
   1.2.6 Aggregation and the Pareto Order ... 33
1.3 Metric Spaces ... 45
   1.3.1 Open and Closed Sets ... 49
   1.3.2 Convergence: Completeness and Compactness ... 56
1.4 Linear Spaces ... 66
   1.4.1 Subspaces ... 72
   1.4.2 Basis and Dimension ... 77
   1.4.3 Affine Sets ... 83
   1.4.4 Convex Sets ... 88
   1.4.5 Convex Cones ... 104
   1.4.6 Sperner's Lemma ... 110
   1.4.7 Conclusion ... 114
1.5 Normed Linear Spaces ... 114
   1.5.1 Convexity in Normed Linear Spaces ... 125
1.6 Preference Relations ... 130
   1.6.1 Monotonicity and Nonsatiation ... 131
   1.6.2 Continuity ... 132
   1.6.3 Convexity ... 136
   1.6.4 Interactions ... 137
1.7 Conclusion ... 141
1.8 Notes ... 142

2 Functions ... 145
2.1 Functions as Mappings ... 145
   2.1.1 The Vocabulary of Functions ... 145

viii   Contents

	2.1.2	Examples of Functions	156
	2.1.3	Decomposing Functions	171
	2.1.4	Illustrating Functions	174
	2.1.5	Correspondences	177
	2.1.6	Classes of Functions	186
	2.2	Monotone Functions	186
	2.2.1	Monotone Correspondences	195
	2.2.2	Supermodular Functions	198
	2.2.3	The Monotone Maximum Theorem	205
	2.3	Continuous Functions	210
	2.3.1	Continuous Functionals	213
	2.3.2	Semicontinuity	216
	2.3.3	Uniform Continuity	217
	2.3.4	Continuity of Correspondences	221
	2.3.5	The Continuous Maximum Theorem	229
	2.4	Fixed Point Theorems	232
	2.4.1	Intuition	232
	2.4.2	Tarski Fixed Point Theorem	233
	2.4.3	Banach Fixed Point Theorem	238
	2.4.4	Brouwer Fixed Point Theorem	245
	2.4.5	Concluding Remarks	259
	2.5	Notes	259
3		**Linear Functions**	263
	3.1	Properties of Linear Functions	269
	3.1.1	Continuity of Linear Functions	273
	3.2	Affine Functions	276
	3.3	Linear Functionals	277
	3.3.1	The Dual Space	280
	3.3.2	Hyperplanes	284
	3.4	Bilinear Functions	287
	3.4.1	Inner Products	290
	3.5	Linear Operators	295
	3.5.1	The Determinant	296
	3.5.2	Eigenvalues and Eigenvectors	299
	3.5.3	Quadratic Forms	302

FOUNDATIONS OF MATHEMATICAL ECONOMICS

FOUNDATIONS OF MATHEMATICAL ECONOMICS

Michael Carter

The MIT Press
Cambridge, Massachusetts
London, England

© 2001 Massachusetts Institute of Technology

All rights reserved. No part of this book may be reproduced in any form by any electronic or mechanical means (including photocopying, recording, or information storage and retrieval) without permission in writing from the publisher.

This book was set in Times New Roman in '3B2' by Asco Typesetters, Hong Kong.

Printed and bound in the United States of America.

Library of Congress Cataloging-in-Publication Data

Carter, Michael, 1950–
 Foundations of mathematical economics / Michael Carter.
   p. cm.
 Includes bibliographical references and index.
 ISBN 0-262-03289-9 (hc. : alk. paper) — ISBN 0-262-53192-5 (pbk. : alk. paper)
 1. Economics, Mathematical. I. Title.
HB135 .C295 2001
330′.01′51—dc21   2001030482

To my parents, Merle and Maurice Carter, who provided a firm foundation for life

Contents

3.6	Systems of Linear Equations and Inequalities	306
3.6.1	Equations	308
3.6.2	Inequalities	314
3.6.3	Input–Output Models	319
3.6.4	Markov Chains	320
3.7	Convex Functions	323
3.7.1	Properties of Convex Functions	332
3.7.2	Quasiconcave Functions	336
3.7.3	Convex Maximum Theorems	342
3.7.4	Minimax Theorems	349
3.8	Homogeneous Functions	351
3.8.1	Homothetic Functions	356
3.9	Separation Theorems	358
3.9.1	Hahn-Banach Theorem	371
3.9.2	Duality	377
3.9.3	Theorems of the Alternative	388
3.9.4	Further Applications	398
3.9.5	Concluding Remarks	415
3.10	Notes	415

4 Smooth Functions — 417

4.1	Linear Approximation and the Derivative	417
4.2	Partial Derivatives and the Jacobian	429
4.3	Properties of Differentiable Functions	441
4.3.1	Basic Properties and the Derivatives of Elementary Functions	441
4.3.2	Mean Value Theorem	447
4.4	Polynomial Approximation	457
4.4.1	Higher-Order Derivatives	460
4.4.2	Second-Order Partial Derivatives and the Hessian	461
4.4.3	Taylor's Theorem	467
4.5	Systems of Nonlinear Equations	476
4.5.1	The Inverse Function Theorem	477
4.5.2	The Implicit Function Theorem	479
4.6	Convex and Homogeneous Functions	483
4.6.1	Convex Functions	483

	4.6.2	Homogeneous Functions	491
	4.7	Notes	496
5		**Optimization**	497
	5.1	Introduction	497
	5.2	Unconstrained Optimization	503
	5.3	Equality Constraints	516
	5.3.1	The Perturbation Approach	516
	5.3.2	The Geometric Approach	525
	5.3.3	The Implicit Function Theorem Approach	529
	5.3.4	The Lagrangean	532
	5.3.5	Shadow Prices and the Value Function	542
	5.3.6	The Net Benefit Approach	545
	5.3.7	Summary	548
	5.4	Inequality Constraints	549
	5.4.1	Necessary Conditions	550
	5.4.2	Constraint Qualification	568
	5.4.3	Sufficient Conditions	581
	5.4.4	Linear Programming	587
	5.4.5	Concave Programming	592
	5.5	Notes	598
6		**Comparative Statics**	601
	6.1	The Envelope Theorem	603
	6.2	Optimization Models	609
	6.2.1	Revealed Preference Approach	610
	6.2.2	Value Function Approach	614
	6.2.3	The Monotonicity Approach	620
	6.3	Equilibrium Models	622
	6.4	Conclusion	632
	6.5	Notes	632
		References	635
		Index of Symbols	641
		General Index	643

Introduction

Economics made progress without mathematics, but has made faster progress with it. Mathematics has brought transparency to many hundreds of economic arguments.
—Deirdre N. McCloskey (1994)

Economists rely on models to obtain insight into a complex world. Economic analysis is primarily an exercise in building and analyzing models. An economic model strips away unnecessary detail and focuses attention on the essential details of an economic problem. Economic models come in various forms. Adam Smith used a verbal description of a pin factory to portray the principles of division of labor and specialization. Irving Fisher built a hydraulic model (comprising floating cisterns, tubes, and levers) to illustrate general equilibrium. Bill Phillips used a different hydraulic model (comprising pipes and colored water) to portray the circular flow of income in the national economy. Sir John Hicks developed a simple mathematical model (IS-LM) to reveal the essential differences between Keynes's *General Theory* and the "classics." In modern economic analysis, verbal and physical models are seen to be inadequate. Today's economic models are almost exclusively mathematical.

Formal mathematical modeling in economics has two key advantages. First, formal modeling makes the assumptions explicit. It clarifies intuition and makes arguments transparent. Most important, it uncovers the limitations of our intuition, delineating the boundaries and uncovering the occasional counterintuitive special case. Second, the formal modeling aids communication. Once the assumptions are explicit, participants spend less time arguing about what they really meant, leaving more time to explore conclusions, applications, and extensions.

Compare the aftermath of the publication of Keynes's *General Theory* with that of von Neumann and Morgenstern's *Theory of Games and Economic Behavior*. The absence of formal mathematical modeling in the *General Theory* meant that subsequent scholars spent considerable energy debating "what Keynes really meant." In contrast, the rapid development of game theory in recent years owes much to the advantages of formal modeling. Game theory has attracted a predominance of practitioners who are skilled formal modelers. As their assumptions are very explicit, practitioners have had to spend little time debating the meaning of others' writings. Their efforts have been devoted to exploring ramifications and applications. Undoubtedly, formal modeling has enhanced the pace of innovation in game-theoretic analysis in economics.

Economic models are not like replica cars, scaled down versions of the real thing admired for their verisimilitude. A good economic model strips away all the unnecessary and distracting detail and focuses attention on the essentials of a problem or issue. This process of stripping away unnecessary detail is called abstraction. Abstraction serves the same role in mathematics. The aim of abstraction is not greater generality but greater simplicity. Abstraction reveals the logical structure of the mathematical framework in the same way as it reveals the logical structure of an economic model.

Chapter 1 establishes the framework by surveying the three basic sources of structure in mathematics. First, the order, geometric and algebraic structures of sets are considered independently. Then their interaction is studied in subsequent sections dealing with normed linear spaces and preference relations.

Building on this foundation, we study mappings between sets or functions in chapters 2 and 3. In particular, we study functions that preserve the structure of the sets which they relate, treating in turn monotone, continuous, and linear functions. In these chapters we meet the three fundamental theorems of mathematical economics—the (continuous) maximum theorem, the Brouwer fixed point theorem, and the separating hyperplane theorem, and outline many of their important applications in economics, finance, and game theory.

A key tool in the analysis of economic models is the approximation of smooth functions by linear and quadratic functions. This tool is developed in chapter 4, which presents a modern treatment of what is traditionally called multivariate calculus.

Since economics is the study of rational choice, most economic models involve optimization by one or more economic agents. Building and analyzing an economic model involves a typical sequence of steps. First, the model builder identifies the key decision makers involved in the economic phenomenon to be studied. For each decision maker, the model builder must postulate an objective or criterion, and identify the tools or instruments that she can use in pursuit of that objective. Next, the model builder must formulate the constraints on the decision maker's choice. These constraints normally take the form of a system of equations and inequalities linking the decision variables and defining the feasible set. The model therefore portrays the decision maker's problem as an exercise in constrained optimization, selecting the best alternative from a feasible set.

Typically analysis of an optimization model has two stages. In the first stage, the constrained optimization problem is solved. That is, the optimal choice is characterized in terms of the key parameters of the model. After a general introduction, chapter 5 first discusses necessary and sufficient conditions for unconstrained optimization. Then four different perspectives on the Lagrangean multiplier technique for equality constrained problems are presented. Each perspective adds a different insight contributing to a complete understanding. In the second part of the chapter, the analysis is extended to inequality constraints, including coverage of constraint qualification, sufficient conditions, and the practically important cases of linear and concave programming.

In the second stage of analysis, the sensitivity of the optimal solution to changes in the parameters of the problem is explored. This second stage is traditionally (in economics) called comparative statics. Chapter 6 outlines four different approaches to the comparative static analysis of optimization models, including the traditional approaches based on the implicit function theorem or the envelope theorem. It also introduces a promising new approach based on order properties and monotonicity, which often gives strong conclusions with minimal assumptions. Chapter 6 concludes with a brief outline of the comparative static analysis of equilibrium (rather than optimization) models.

The book includes a thorough treatment of some material often omitted from introductory texts, such as correspondences, fixed point theorems, and constraint qualification conditions. It also includes some recent developments such as supermodularity and monotone comparative statics. We have made a conscious effort to illustrate the discussion throughout with economic examples and where possible to introduce mathematical concepts with economic ideas. Many illustrative examples are drawn from game theory.

The completeness of the real numbers is assumed, every other result is derived within the book. The most important results are stated as theorems or propositions, which are proved explicitly in the text. However, to enhance readability and promote learning, lesser results are stated as exercises, answers for which will be available on the internet (see the note to the reader). In this sense the book is comprehensive and entirely self-contained, suitable to be used as a reference, a text, or a resource for self-study.

The sequence of the book, preceding from sets to functions to smooth functions, has been deliberately chosen to emphasize the structure of the underlying mathematical ideas. However, for instructional purposes or for self-study, an alternative sequence might be preferable and easier to motivate. For example, the first two sections of chapter 1 (sets and ordered sets) could be immediately followed by the first two sections of chapter 2 (functions and monotone functions). This would enable the student to achieve some powerful results with a minimum of fuss. A second theme could then follow the treatment of metric spaces (and the topological part of section 1.6) with continuous functions culminating in the continuous maximum theorem and perhaps the Banach fixed point theorem. Finally the course could turn to linear spaces, linear functions, convexity, and linear functionals, culminating in the separating hyperplane theorem and its applications. A review of fixed point theorems would then highlight the interplay of linear and topological structure in the Brouwer fixed point theorem and its generalizations. Perhaps it would then be advantageous to proceed through chapters 4, 5, and 6 in the given sequence. Even if chapter 4 is not explicitly studied, it should be reviewed to understand the notation used for the derivative in the following chapters.

The book can also be used for a course emphasizing microeconomic theory rather than mathematical methods. In this case the course would follow a sequence of topics, such as monotonicity, continuity, convexity, and homogeneity, interspersed with analytical tools such as constrained optimization, the maximum, fixed point, and separating hyperplane theorems, and comparative statics. Each topic would be introduced and illustrated via its role in the theory of the consumer and the producer.

Achieving consistency in notation is a taxing task for any author of a mathematical text. Wherever I could discern a standard notation in the economics literature, I followed that trend. Where diversity ruled, I have tended to follow the notation in Hal Varian's *Microeconomic Analysis*, since it has been widely used for many years. A few significant exceptions to these rules are explicitly noted.

Many people have left their mark on this book, and I take great pleasure in acknowledging their contribution. Foremost among my creditors is Graeme Guthrie whose support, encouragement, and patient exposition of mathematical subtleties has been invaluable. Richard Edlin and Mark Pilbrow drafted most of the diagrams. Martin Osborne and Carolyn Pitchik made detailed comments on an early draft of the manuscript and Martin patiently helped me understand intricacies of TEX and LATEX.

Other colleagues who have made important comments include Thomas Cool, John Fountain, Peter Kennedy, David Miller, Peter Morgan, Mike Peters, Uli Schwalbe, David Starrett, Dolf Talman, Paul Walker, Richard Watt, and Peyton Young. I am also very grateful for the generous hospitality of Eric van Damme and CentER at the University of Tilburg and Uli Schwalbe and the University of Hohenheim in providing a productive haven in which to complete the manuscript during my sabbatical leave. Finally, I acknowledge the editorial team at The MIT Press, for their proficiency in converting my manuscript into a book. I thank them all.

A Note to the Reader

Few people rely solely on any social science for their pleasures, and attaining a suitable level of ecstasy involves work.... It is a nuisance, but God has chosen to give the easy problems to the physicists.
—Lave and March (1975)

Some people read mathematics books for pleasure. I assume that you are not one of this breed, but are studying this book to enhance your understanding of economics. While I hope this process will be enjoyable, to make the most of it will require some effort on your part. Your reward will be a comprehension of the foundations of mathematical economics, you will appreciate the elegant interplay between economic and mathematical ideas, you will know why as well as how to use particular tools and techniques.

One of the most important requirements for understanding mathematics is to build up an appropriate mental framework or structure to relate and integrate the various components and pieces of information. I have endeavored to portray a suitable framework in the structure of this book, in the way it is divided into chapters, sections, and so on. This is especially true of the early mathematical chapters, whose structure is illustrated in the following table:

Sets	Functions
Ordered sets	Monotone functions
Metric spaces	Continuous functions
Linear spaces	Linear functions
Convex sets	Convex functions
Cones	Homogeneous functions

This is the framework to keep in mind as you proceed through the book.

You will also observe that there is a hierarchy of results. The most important results are stated as theorems. You need to be become familiar with these, their assumptions and their applications. Important but more specialized results are stated as propositions. Most of the results, however, are given as exercises. Consequently exercise has a slightly different meaning here than in many texts. Most of the 820 exercises in the book are not "finger exercises," but substantive propositions forming an integral part of the text. Similarly examples contain many of the key ideas and warrant careful attention.

There are two reasons for this structure. First, the exercises and examples break up the text, highlighting important ideas. Second, the exercises provide the potential for deeper learning. It is an unfortunate fact of life that for most of us, mathematical skills (like physical skills) cannot be obtained by osmosis through reading and listening. They have to be acquired through practice. You will learn a great deal by attempting to do these exercises. In many cases elaborate hints or outlines are given, leaving you to fill in the detail. Then you can check your understanding by consulting the comprehensive answers, which are available on the Internet at *http://mitpress.mit.edu/carter-foundations*.

FOUNDATIONS OF MATHEMATICAL ECONOMICS

1 Sets and Spaces

All is number.
—Pythagoras

God created the integers; everything else is the work of man
—L. Kronecker

One of the most important steps in understanding mathematics is to build a framework to relate and integrate the various components and pieces of information. The principal function of this introductory chapter is to start building this framework, reviewing some basic concepts and introducing our notation. The first section reviews the necessary elements of set theory. These basics are developed in the next three sections, in which we study sets that have a specific structure. First, we consider ordered sets (section 1.2), whose elements can be ranked by some criterion. A set that has a certain form or structure is often called *a space*. In the following two sections, we tour in turn the two most important examples: metric spaces and linear spaces. Metric spaces (section 1.3) generalize the familiar properties of Euclidean geometry, while linear spaces (section 1.4) obey many of the usual rules of arithmetic while. Almost all the sets that populate this book will inhabit a linear, metric space (section 1.5), so a thorough understanding of these sections is fundamental to the remainder of the book. The chapter ends with an extended example (section 1.6) in which we integrate the order, algebraic, and geometric perspectives to study preference relations that are central to the theory of the consumer and other areas of economics.

1.1 Sets

A *set* is a collection of objects (called *elements*) such as the set of people in the world, books in the library, students in the class, weekdays, or commodities available for trade. Sometimes we denote a set by listing all its members between braces { }, for example,

Weekdays = {Monday, Tuesday, Wednesday, Thursday, Friday}

Some of the elements may be omitted from the list when the meaning is clear, as in the following example:

alphabet = {A, B, C, ..., Z}

More frequently we denote a set by specifying a rule determining membership, for example,

ECON301 = {students : who are studying Economics 301}

The elements of a set may themselves be sets. Such sets of sets are often called *classes, collections,* or *families*. We write $x \in X$ to denote that x is an element or member of the set X, while $x \notin X$ indicates that x is not in X. The most fundamental set in economics is $\Re$, the set of real numbers. Another important set is $\mathfrak{N} = \{1, 2, 3, \ldots\}$, the set of positive integers.

Exercise 1.1
Denote the set of odd positive integers in two different ways.

A *subset* S of a set T (denoted $S \subseteq T$) is a set containing some (possibly all, possibly none) of the elements of T. For example, the vowels form a subset of the alphabet

vowels = {A, E, I, O, U}

An important example in economics is the set of nonnegative real numbers

$\Re_+ = \{x \in \Re : x \geq 0\} \subseteq \Re$

since economic quantities such as prices and incomes are usually nonnegative. $\Re_+$ and $\mathfrak{N}$ are different subsets of $\Re$. If $S \subseteq T$, then T is called a *superset* of S. We sometimes emphasize the inclusive role of T by using the notation $T \supseteq S$.

Two sets S and T are said to be *equal* ($S = T$) if they comprise exactly the same elements. S is a *proper subset* of T if $S \subseteq T$ but $S \neq T$. We will use the notation $S \subset T$ to denote that S is a proper subset of T. Note that every set is a subset of itself. It is important to clearly distinguish the notions *belonging* ($\in$) and *inclusion* ($\subseteq$). If $x \in X$ is an element of the set X, then x belongs to X, while the set $\{x\}$ is a subset of X.

Exercise 1.2
Show that $A \subseteq B$ and $B \subseteq A$ implies that $A = B$.

For any set S, we use $|S|$ to denote the number of elements in S. A set is *finite* if it contains a finite number of elements, otherwise it is an *infinite* set. The set of all subsets of a finite set S is called the *power set* of S and is denoted $\mathscr{P}(S)$. The *empty* or *null set* is a special set which contains no elements. Denoted $\emptyset$, the empty set is a subset of every set.

Exercise 1.3
Give examples of finite and infinite sets.

Example 1.1 (Sample space) In a random experiment, the set S of all possible outcomes is called the *sample space*. An event is a subset of the possible outcomes, that is, a subset of S.

Exercise 1.4
Describe the sample space for the experiment of tossing a single die. What is the event E that the result is even?

Example 1.2 (A game) A game is a mathematical model of strategic decision making combining elements of both conflict and cooperation. It specifies a finite set N of participants, called the *players*. Each player $i \in N$ has a set of possible actions A_i, which is called her *action space*. A game is finite if A_i is finite for every $i \in N$. The outcome depends on the action chosen by each of the players.

Exercise 1.5 (Rock–Scissors–Paper)
To decide whose turn it is to wash the dishes, Jenny and Chris play the following game. Each player simultaneously holds up two fingers (scissors), an open palm (paper), or a closed fist (rock). The winner is determined by the following rules:

- Scissors beats (cuts) paper
- Paper beats (covers) rock
- Rock beats (blunts) scissors

The loser does the dishes. Specify the set of players and the action space for each player.

Exercise 1.6 (Oligopoly)
An electricity grid connects n hydroelectric dams. Each dam i has a fixed capacity Q_i. Assuming that the dams are operated independently, the production decision can be modeled as a game with n players. Specify the set of players and the action space of each player.

Example 1.3 (Coalitions) In a game, subsets of the set of players N are called coalitions. The set of all coalitions is the power set of N, denoted $\mathscr{P}(N)$. It includes the set of all players N (called the grand coalition) and the empty coalition $\emptyset$. The set of *proper coalitions* excludes the trivial coalitions $\emptyset$ and N.

Remark 1.1 There is a subtle distinction in the usage of the word *proper* between set theory and game theory. In conventional usage, $\emptyset$ is a proper subset of a nonempty set N, but it is not a proper coalition.

Exercise 1.7
List all the coalitions in a game played by players named 1, 2, and 3. How many coalitions are there in a ten player game?

If S is a subset of X, the *complement* of S (with respect to X), denoted S^c, consists of all elements of X that are *not in* S, that is,

$$S^c = \{x \in X : x \notin S\}$$

If both S and T are subsets of X, their *difference* $S \setminus T$ is the set of all elements in S which do not belong to T, that is,

$$S \setminus T = \{x \in X : x \in S, x \notin T\}$$

This is sometimes known as the *relative complement* of T in S. The *union* of the two sets S and T is the set of all elements which belong to either S or T or both, that is,

$$S \cup T = \{x : x \in S, \text{ or } x \in T, \text{ or both}\}$$

The *intersection* of two sets S and T is set of all elements that simultaneously belong to both S and T,

$$S \cap T = \{x : x \in S \text{ and } x \in T\}$$

These set operations are illustrated in figure 1.1 by means of *Venn diagrams*, where the shaded areas represent the derived set.

Exercise 1.8 (DeMorgan's laws)
Show that

$$(S \cup T)^c = S^c \cap T^c$$
$$(S \cap T)^c = S^c \cup T^c$$

Set union and intersection have straightforward extensions to collections of sets. The union of a collection $\mathscr{C}$ of sets

$$\bigcup_{S \in \mathscr{C}} S = \{x : x \in S \text{ for some } S \in \mathscr{C}\}$$

1.1 Sets

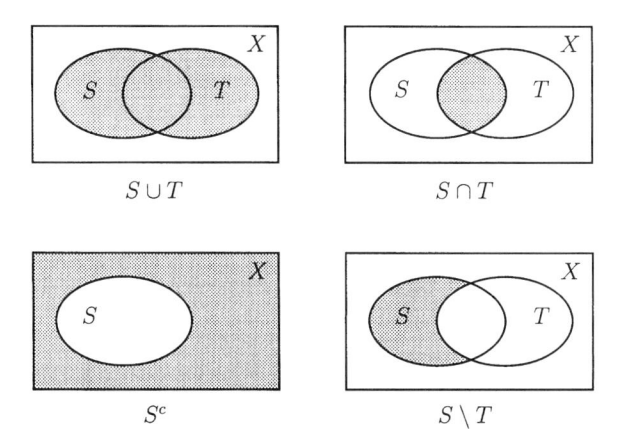

Figure 1.1
Venn diagrams

is the set of all elements that belong to a least one of the sets in $\mathscr{C}$. The intersection of a collection $\mathscr{C}$ of sets

$$\bigcap_{S \in \mathscr{C}} S = \{x : x \in S \text{ for every } S \in \mathscr{C}\}$$

is the set of all elements that belong to each of the sets in $\mathscr{C}$. If the sets in a collection $\mathscr{C}$ have no elements in common, then their intersection is the empty set.

Exercise 1.9
Let $\mathscr{C}$ be the collection of coalitions in a five-player game ($N = \{1, 2, 3, 4, 5\}$). What is the union and the intersection of the sets in $\mathscr{C}$?

Union and intersection are one way of generating new sets from old. Another way of generating new sets is by welding together sets of disparate objects into another set called their product. The *product* of two sets X and Y is the set of ordered pairs

$$X \times Y = \{(x, y) : x \in X, y \in Y\}$$

A familiar example is the coordinate plane $\Re \times \Re$ which is denoted $\Re^2$ (figure 1.2). This correspondence between points in the plane and ordered pairs (x, y) of real numbers is the foundation of analytic geometry. Notice how the order matters. $(1, 2)$ and $(2, 1)$ are different elements of $\Re^2$.

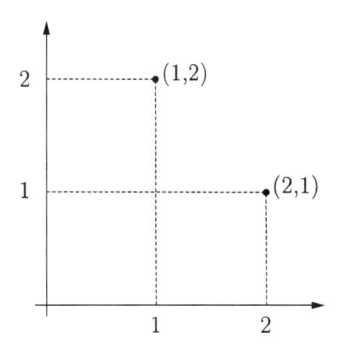

Figure 1.2
The coordinate plane $\mathfrak{R}^2$

The product readily generalizes to many sets, so that

$$X_1 \times X_2 \times \cdots \times X_n = \{(x_1, x_2, \ldots, x_n) : x_i \in X_i\}$$

is the set of all ordered lists of elements of X_i, and $\mathfrak{R}^n = \{(x_1, x_2, \ldots, x_n) : x_i \in \mathfrak{R}\}$ is the set of all ordered lists of n real numbers. An ordered list of n elements is called an *n-tuple*. $\mathfrak{R}^n$ and its nonnegative subset $\mathfrak{R}^n_+$ provide the domain of most economic quantities, such as commodity bundles and price lists. To remind ourselves when we are dealing with a product space, we will utilize boldface to distinguish the elements of a product space from the elements of the constituent sets, as in

$$\mathbf{x} = (x_1, x_2, \ldots, x_n) \in X$$

where $X = X_1 \times X_2 \times \cdots \times X_n$.

Example 1.4 (Action space) The outcome of a game depends on the action chosen by each of the players. If there are n players each of whom chooses an action a_i from a set A_i, the combined choice is the n-tuple $(a_1, a_2, \ldots, a_n)$. The set of all possible outcomes A is the product of the individual action spaces

$$A = A_1 \times A_2 \times \cdots \times A_n = \{(a_1, a_2, \ldots, a_n) : a_1 \in A_1, a_2 \in A_2, \ldots, a_n \in A_n\}$$

A is called the *action space* of the game. A typical element $\mathbf{a} = (a_1, a_2, \ldots, a_n) \in A$, called an *action profile*, lists a particular choice of action for each player and determines the outcome of the game.

Exercise 1.10
Let the two possible outcomes of coin toss be denoted H and T. What is the sample space for a random experiment in which a coin is tossed three times?

Given any collection of sets $X_1, X_2, \ldots, X_n$, we use X_{-i} to denote the product of all but the ith set, that is,

$$X_{-i} = X_1 \times X_2 \times \cdots \times X_{i-1} \times X_{i+1} \times \cdots \times X_n$$

An element $\mathbf{x}_{-i}$ of X_{-i} is a list containing one element from each of the sets except X_i:

$$\mathbf{x}_{-i} = (x_1, x_2, \ldots, x_{i-1}, x_{i+1}, \ldots, x_n)$$

By convention, the ordered pair $(x_i, \mathbf{x}_{-i})$ denotes the list of elements $(x_1, x_2, \ldots, x_n)$ with x_i restored to its rightful place in the order, that is,

$$(x_i, \mathbf{x}_{-i}) = \mathbf{x} = (x_1, x_2, \ldots, x_{i-1}, \underset{\Uparrow}{x_i}, x_{i+1}, \ldots, x_n)$$

Example 1.5 The preceding notation is used regularly in game theory, when we want to explore the consequences of changing actions one player at time. For example, if $\mathbf{a}^* = (a_1^*, a_2^*, \ldots, a_n^*)$ is a list of actions in a game (an action profile), then $\mathbf{a}_{-i}^*$ denotes the actions of all players except player i. $(a_i, \mathbf{a}_{-i}^*)$ denotes the outcome in which player i takes action a_i, while all the other players j take action $a_j^*, j \neq i$ (see example 1.51).

Next we introduce two examples of set products that form the basis for consumer and producer theory. We will use these sets regularly to illustrate further concepts.

Example 1.6 (Consumption set) The arena of consumer theory is the *consumption set*, the set of all feasible consumption bundles. Suppose that there are n commodities. The behavior of a consumer can be described by a list of purchases $(x_1, x_2, \ldots, x_n)$, where x_i is the quantity of the ith commodity. For example, x_1 might be pounds of cheese and x_2 bottles of wine. Since purchases cannot be negative, each quantity x_i belongs to $\Re_+$. A particular consumption bundle $\mathbf{x} = (x_1, x_2, \ldots, x_n)$ is a list of nonnegative real numbers. The consumption set X is a subset of $\Re_+^n$, the product of n copies of $\Re_+$, which is known as the *nonnegative orthant* of

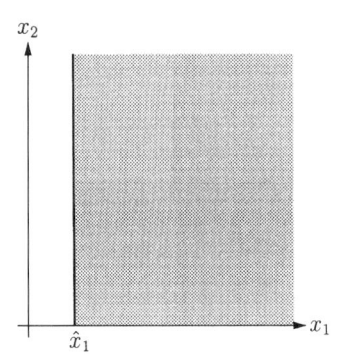

Figure 1.3
A consumption set with two commodities

$\Re^n$. The consumption set may be a proper subset, since not all consumption bundles will necessarily be feasible. For example, we may wish to preclude from consideration any bundles that do not ensure subsistence for the consumer. Figure 1.3 illustrates a possible consumption set for two commodities, where a minimum quantity of $\hat{x}_1$ is required for subsistence.

Example 1.7 (Production possibility set) A producer combines various goods and services (called inputs) to produce one or more products (called outputs). A particular commodity may be both an input and an output. The *net output* y_i of a commodity is the output produced minus any input required. The net output is positive if output exceeds input, and negative otherwise. A production plan is a list of the net outputs of the various goods and services $y = (y_1, y_2, \ldots, y_n)$. That is, a production plan is an element of the product set

$$\Re^n = \{(y_1, y_2, \ldots, y_n) : y_i \in \Re\}$$

The *production possibility set* Y is set of all technologically feasible production plans,

$$Y = \{(y_1, y_2, \ldots, y_n) \in \Re^n : y \text{ is technologically feasible}\}$$

It is a proper subset of the product set $\Re^n$. The precise composition of Y depends on the production technology. Producer theory begins by assuming some properties for Y. We meet some of these in subsequent sections.

Exercise 1.11
Assume that $Y \subset \Re^n$ is a production possibility set as defined in the previous example. What is $Y \cap \Re^n_+$?

Example 1.8 (Input requirement set) In classical producer theory a firm produces a single output using n different inputs. If we let y denote the quantity of output $\mathbf{x}$ denote the quantities of the various inputs, we can represent a production plan as the pair $(y, -\mathbf{x})$ where $\mathbf{x} \in \Re^n_+$. The production possibility set is

$$Y = \{(y, -\mathbf{x}) \in \Re^{n+1}_+ : (y, -\mathbf{x}) \text{ is technologically feasible}\}$$

It is often more convenient to work with the *input requirement set*

$$V(y) = \{\mathbf{x} \in \Re^n_+ : (y, -\mathbf{x}) \in Y\}$$

which is the set of all input bundles sufficient to produce y units of output. It details all the technologically feasible ways of producing y units of output. One of the tasks of economic analysis is to identify the least costly method of producing a given level of output. In this representation of the technology, both inputs and outputs are measured by positive quantities.

Exercise 1.12 (Free disposal and monotonicity)
A conventional assumption in production theory is *free disposal*, namely

$$\mathbf{y} \in Y \Rightarrow \mathbf{y}' \in Y \quad \text{for every } \mathbf{y}' \leq \mathbf{y}$$

A technology is said to be *monotonic* if

$$\mathbf{x} \in V(y) \Rightarrow \mathbf{x}' \in V(y) \quad \text{for every } \mathbf{x}' \geq \mathbf{x}$$

where $\mathbf{x}' \geq \mathbf{x}$ means that $x_i \geq x'_i$ for every i (see example 1.26). Show that free disposal implies that

1. the technology is monotonic and
2. the input requirement sets are nested, that is, $V(y') \supseteq V(y)$ for every $y' \leq y$.

1.2 Ordered Sets

Economics is the study of rational choice. Economic analysis presumes that economic agents seek the best element in an appropriate set of feasible alternatives. Consumers are assumed to choose the best consumption bundle among those that are affordable. Each producer is assumed to choose the most profitable production plan in its production possibility set. Each player in a game is assumed to choose her best alternative given her predictions of the choices of the other players. Consequently economic analysis requires that the analyst can rank alternatives and identify the best element in various sets of choices. Sets whose elements can be ranked are called *ordered sets*. They are the subject of this section.

An ordered set is a set on which is defined an order relation, which ranks the elements of the set. Various types of ordered sets arise in economics. They differ in the specific properties assumed by the associated relation. This section starts with an outline of relations in general and a discussion of their common properties. This leads to a discussion of the two most common types of relations—equivalence relations and order relations. Next we discuss in turn the two main types of ordered sets—partially ordered sets and weakly ordered sets. Finally we consider the extension of orders to the product of sets. Figure 1.4 illustrates the relationship of the various types of relations. It also serves as a road map for the section.

1.2.1 Relations

Given two sets X and Y, any subset R of their product $X \times Y$ is called a *binary relation*. For any pair of elements $(x, y) \in R \subseteq X \times Y$, we say that x is related to y and write $x \, R \, y$. Although formally expressed as a subset of the product, a relation is usually thought of in terms of the rule expressing the relationship between the elements.

Example 1.9 Let

$X = \{\text{Berlin, London, Tokyo, Washington}\}$

and

$Y = \{\text{Germany, Japan, United Kingdom, United States}\}$

The relation

$R = \{(\text{Berlin, Germany}), (\text{London, United Kingdom}),$
$\quad (\text{Tokyo, Japan}), (\text{Washington, United States})\}$

expresses the relation "x is the capital of y."

1.2 Ordered Sets

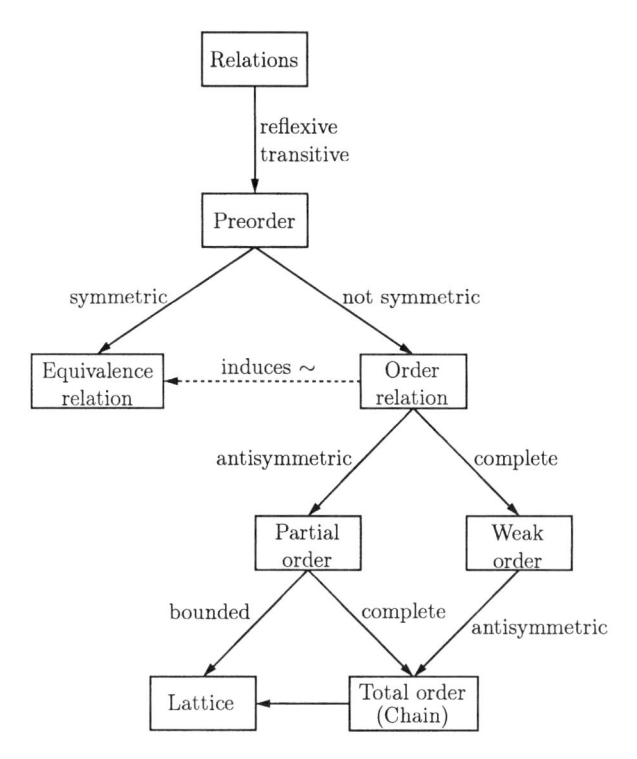

Figure 1.4
Types of relations

Example 1.10 Let $X = Y = \{1, 2, 3\}$. The set $X \times Y$ is the set of all ordered pairs

$$X \times Y = \{(1,1), (1,2), (1,3), (2,1), (2,2), (2,3), (3,1), (3,2), (3,3)\}$$

The relation "less than" between X and Y is the set of ordered pairs "$<$" $= \{(1,2), (1,3), (2,3)\}$ which expresses the ranking that $1 < 2$, $1 < 3$, and $2 < 3$. When X and Y are subsets of $\Re$, we can illustrate the relation by means of its "graph." Figure 1.5 illustrates the product $X \times Y$, where the elements of the relation "$<$" are circled.

Any relation $R \subseteq X \times Y$ has an inverse relation $R^{-1} \subseteq Y \times X$ defined by

$$R^{-1} = \{(y, x) \in Y \times X : (x, y) \in R\}$$

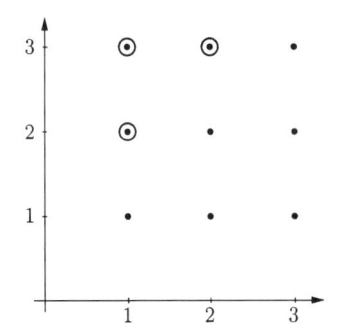

Figure 1.5
A relation

For example, the inverse of the relation "is the capital of" is the relation "the capital of y is x." The inverse of "$<$" is "$>$." It is sometimes useful to identify the set of elements which are involved in a relation. For any relation $R \subseteq X \times Y$, the *domain* of R is set of all $x \in X$ that are related to some $y \in Y$, that is,

domain $R = \{x \in X : (x, y) \in R\}$

The *range* is the corresponding subset of Y, that is,

range $R = \{y \in Y : (x, y) \in R\}$

Exercise 1.13
What are the domain and range of the relation "$<$" in example 1.10?

Most relations encountered in economics are defined on the elements of a single set, with $X = Y$. We then speak of relation on X.

Exercise 1.14
Depict graphically the relation $\{(x, y) : x^2 + y^2 = 1\}$ on $\Re$.

Any relation R can be characterized by the properties that it exhibits. The following properties of binary relations have been found to be important in a variety of contexts. A relation R on X is

reflexive if $x \, R \, x$
transitive if $x \, R \, y$ and $y \, R \, z \Rightarrow x \, R \, z$
symmetric if $x \, R \, y \Rightarrow y \, R \, x$

antisymmetric if $x\,R\,y$ and $y\,R\,x \Rightarrow x = y$
asymmetric if $x\,R\,y \Rightarrow$ not $(y\,R\,x)$
complete if either $x\,R\,y$ or $y\,R\,x$ or both

for all x, y, and z in X.

Example 1.11 Let R be the relation "at least as high as" applied to the set of all mountain peaks. R is reflexive, since every mountain is at least as high as itself. It is complete, since all mountains can be compared. It is transitive, since if A is at least has high as B and B is at least as high as C, then A is at least as high as C. However, it is not symmetric, asymmetric, nor antisymmetric.

It is not symmetric, since if A is higher than B, A is at least as high as B, but B is not at least as high as A. It is not antisymmetric, since if two distinct mountains A and B are of the same height, we have $A\,R\,B$ and $B\,R\,A$ but without $A = B$. Neither is it asymmetric, since if A and B have the same height, then $A\,R\,B$ and $B\,R\,A$.

Exercise 1.15
What properties does the relation "is strictly higher than" exhibit when applied to the set of mountains?

Exercise 1.16
Consider the relations $<$, $\leq$, $=$ on $\Re$. Which of the above properties do they satisfy?

Example 1.12 (Preference relation) The most important relation in economics is the consumer preference relation $\succsim$ on the consumption set X. The statement $\mathbf{x} \succsim \mathbf{y}$ means that the consumer rates consumption bundle $\mathbf{x}$ at least as good as consumption bundle $\mathbf{y}$. The consumer's preference relation is usually assumed to be complete and transitive. We explore the consumer preference relation in some detail in section 1.6, where we will introduce some further assumptions.

Any relation which is reflexive and transitive is called a *preorder* or *quasi-order*. A set on which is defined a preorder is called *preordered set*. Preorders fall into two fundamental categories, depending on whether or not the relation is symmetric. A symmetric preorder is called an *equivalence relation*, while any preorder that is not symmetric is called an *order relation*. Both classes of relations are important in economics, and we deal

Table 1.1
Classes of relations

	Reflexive	Transitive	Symmetric	Antisymmetric	Asymmetric	Complete
Equivalence	Y	Y	Y			
Order Quasi-ordering Preordering	Y	Y	N			
Partial ordering	Y	Y	N	Y		
Total ordering Linear ordering Chain	Y	Y	N	Y		Y
Weak ordering	Y	Y	N			Y

with each in turn. Table 1.1 summarizes the properties of the common classes of relations.

1.2.2 Equivalence Relations and Partitions

An *equivalence relation* R on a set X is a relation that is reflexive, transitive, and symmetric. Given an equivalence relation $\sim$, the set of elements that are related to a given element a,

$$\sim(a) = \{x \in X : x \sim a\}$$

is called the *equivalence class* of a.

There is intimate connection between equivalence relations on a set and partitions of that set. A partition is a decomposition of a set into subsets. More formally, a *partition* of a set X is a collection of disjoint subsets of X whose union is the full set X. Given an equivalence relation on a set X, every element of X belongs to one and only one equivalence class. Thus the collection of equivalence classes partitions X. Conversely, every partition of X induces some equivalence relation on X.

The simplest possible partition of a set X comprises a subset S and its complement S^c. The collection $\{S, S^c\}$ form a partition of X since $S \cup S^c = X$ and $S \cap S^c = \emptyset$. At the other extreme, all one element subsets of X comprise another partition. Less trivial examples are given in the following examples.

Example 1.13 (Mutually exclusive events) Recall that an event E in a random experiment is a subset of the sample space S, the set of all possible outcomes. Two events E_1 and E_2 are mutually exclusive if they cannot occur together, that is, if $E_1 \cap E_2 = \emptyset$. If we decompose the possible outcomes into a collection of mutually exclusive events $\{E_1, E_2, \ldots, E_n\}$ with $S = \bigcup_i E_i$, then the events E_i form a partition of the sample space S.

Example 1.14 (Teams) Suppose that a game of n players is played in teams, with each player belonging to one and only one team. Suppose that there are k teams $\{T_1, T_2, \ldots, T_k\}$. Let R be the relation "belongs to the same team as." Then R is an equivalence relation, since it is reflexive, transitive, and symmetric. The teams are coalitions that partition the set of players.

Example 1.15 (Rational numbers) A fraction is the ratio of two integers. The fractions 1/2 and 2/4 both represent the same real number. We say that two fractions p/q and r/s are equal if $ps = qr$. Thus defined, equality of fraction is an equivalence relation in the set of fractions of integers. Each rational number is identified with an equivalence class in the set of fractions.

Example 1.16 (Indifference classes) The consumer preference relation $\succsim$ is not symmetric and hence is not an equivalence relation. However, it induces a symmetric relation $\sim$ on the consumption set which is called *indifference*. For any two consumption bundles **x** and **y** in X, the statement **x** $\sim$ **y** means that the consumer is indifferent between the two consumption bundles **x** and **y**; that is, **x** is at least as good as **y**, but also **y** is at least as good as **y**. More precisely

$$\mathbf{x} \sim \mathbf{y} \Leftrightarrow \mathbf{x} \succsim \mathbf{y} \text{ and } \mathbf{y} \succsim \mathbf{x}$$

Indifference is an equivalence relation. The equivalence classes of the indifference relation are called indifference classes, and they form a partition of the consumption set, which is sometimes called an indifference map. The indifference map is often depicted graphically by a set of indifference curves. Each indifference curve represents one indifference class.

Exercise 1.17
Show that any equivalence relation on a set X partitions X.

Exercise 1.18 (Coalitions)
In a game played by members of the set N, is the set of proper coalitions (example 1.3) a partition of N?

1.2.3 Order Relations

A relation that is reflexive and transitive but *not* symmetric is called an *order relation*. We denote a general order relation $x \succsim y$ and say that "x follows y" or "x dominates y." Every order relation $\succsim$ on a set X induces two additional relations $\succ$ and $\sim$. We say that "x strictly dominates y," denoted $x \succ y$, if x dominates y but y does not dominate x, that is,

$$x \succ y \Leftrightarrow x \succsim y \text{ and } y \not\succsim x$$

The relation $\succ$ is transitive but not reflexive. Every order relation $\succsim$ also induces an equivalence relation $\sim$ defined by

$$x \sim y \Leftrightarrow x \succsim y \text{ and } y \succsim x$$

for all x, y in X. An *ordered set* $(X, \succsim)$ consists of a set X together with an order relation $\succsim$ defined on X.

It is sometimes useful to use the inverse relations $\precsim$ and $\prec$. We say that y precedes x if x follows y,

$$y \precsim x \Leftrightarrow x \succsim y$$

or y strictly precedes x if x strictly follows y,

$$y \prec x \Leftrightarrow x \succ y$$

Remark 1.2 (Weak and strong orders) The reflexive relation $\succsim$ is often called a *weak* order, while its nonreflexive counterpart $\succ$ is called a *strong* or *strict* order. For example, in the consumer's preference relation, $\succsim$ is called weak preference and $\succ$ strong preference. Note however that the adjective "weak" is also applied to a completely ordered set (section 1.2.5).

The following interactions between these orderings are often used in practice.

Exercise 1.19

$x \succ y$ and $y \sim z \Rightarrow x \succ z$

$x \sim y$ and $y \succ z \Rightarrow x \succ z$

Exercise 1.20
Show that $\succ$ is asymmetric and transitive.

1.2 Ordered Sets

Exercise 1.21
Show that $\sim$ is reflexive, transitive, and symmetric, that is, an equivalence relation.

Remark 1.3 (Acyclicity) In consumer and social choice theory, a weaker condition than transitivity is sometimes invoked. A binary relation $\succsim$ on X is acyclical if for every list $x_1, x_2, \ldots, x_k \in X$,

$$x_1 \succ x_2, x_2 \succ x_3, \ldots, x_{k-1} \succ x_k \Rightarrow x_1 \succsim x_k$$

This is a minimal requirement for a consistent theory of choice.

Example 1.17 (Natural order on $\Re$) The natural order on $\Re$ is $\geq$ with the inverse $\leq$. It induces the strict orders $>$ and $<$ and the equivalence relation $=$. All order relations are generalizations of aspects of the natural order on $\Re$.

Example 1.18 (Integer multiples) For the set $\Re$ of positive integers, the relation "m is a multiple of n" is an order relation. For example, $4 \succsim 2$ and $15 \succsim 3$, while $2 \not\succsim 4$ and $5 \not\succsim 2$. Figure 1.6 illustrates the implied strict relation $\succ$ on $\{1, 2, \ldots, 9\}$, where the arrows indicate that m is a proper multiple of n. The two pathways connecting the integers 8 and 2 illustrate the property of transitivity.

Exercise 1.22
Show that the relation in example 1.18 is an order relation. That is, show that it is reflexive and transitive, but not symmetric.

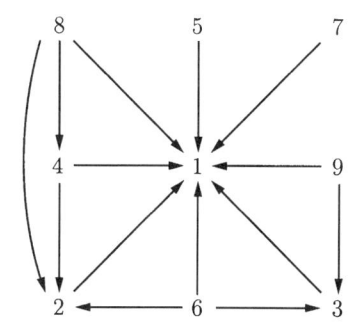

Figure 1.6
Integer multiples

Intervals

Given a set X ordered by $\succsim$ and two elements $a, b \in X$ with $a \precsim b$, the *closed interval* $[a, b]$ is the set of all elements between a and b, that is,

$$[a, b] = \{x \in X : a \precsim x \precsim b\}$$

With $a \prec b$, the *open interval* (a, b) is the set of all elements strictly between a and b, that is,

$$(a, b) = \{x \in X : a \prec x \prec b\}$$

Note that $a, b \in [a, b]$ while (a, b) may be empty. We also encounter hybrid intervals

$$[a, b) = \{x \in X : a \precsim x \prec b\} \quad \text{and} \quad (a, b] = \{x \in X : a \prec x \precsim b\}$$

Example 1.19 In Example 1.18, the elements of the intervals $[2, 8]$ and $(2, 8)$ are $\{2, 4, 8\}$ and $\{4\}$ respectively.

Exercise 1.23
Assume that the set $X = \{a, b, x, y, z\}$ is ordered as follows:

$$x \prec a \prec y \prec b \sim z$$

Specify the closed interval $[a, b]$ and the open interval (a, b).

Example 1.20 (Intervals in $\Re$) Intervals are especially common subsets of $\Re$. For example,

$$[0, 1] = \{x \in \Re : 0 \leq x \leq 1\}$$
$$(-1, 1) = \{x \in \Re : -1 < x < 1\}$$

Upper and Lower Contour Sets

Analogous to intervals are the upper and lower contour sets. Given a set X ordered by $\succsim$, the set

$$\succsim(a) = \{x \in X : x \succsim a\}$$

of all elements that follow or dominate a is called the upper contour set of $\succsim$ at a. The set of elements that strictly dominate a is

$$\succ(a) = \{x \in X : x \succ a\}$$

Similarly the lower contour set at a,

$\precsim(a) = \{x \in X : x \precsim a\}$

contains all elements which precede a in the order $\succsim$. The set of elements which strictly precede a is

$\prec(a) = \{x \in X : x \prec a\}$

Note that $a \in \succsim(a)$ but that $a \notin \succ(a)$.

Example 1.21 In example 1.18, the upper contour set of 7 is the set of all multiples of 7, namely

$\succsim(7) = \{7, 14, 21, \ldots\}$

Similarly

$\succ(7) = \{14, 21, \ldots\}$

$\precsim(7) = \{1, 7\}$

$\prec(7) = \{1\}$

Exercise 1.24
Assume that the set $X = \{a, b, x, y, z\}$ is ordered as follows:

$x \prec a \prec y \prec b \sim z$

Specify the upper and lower contour sets of y.

Example 1.22 (Upper and lower contour sets in $\Re$) A special notation is used for upper and lower contour sets in $\Re$. For any $a \in \Re$ the upper contour set at a (in the natural order) is denoted $[a, \infty)$. That is,

$[a, \infty) = \{x \in \Re : x \geq a\}$

Similarly

$(a, \infty) = \{x \in \Re : x > a\}$

$(-\infty, a] = \{x \in \Re : x \leq a\}$

$(-\infty, a) = \{x \in \Re : x < a\}$

The set of nonnegative real numbers $\Re_+^n$ is the upper contour set at 0, that is $\Re_+ = [0, \infty)$. The set of positive real numbers is $(0, \infty)$, which is often denoted $\Re_{++}$.

Maximal and Best Elements

Given any order relation $\succsim$ on a set X, an element x is a *maximal element* if there is no element that strictly dominates it; that is, there is no element $y \in X$ such that $y \succ x$. $x \in X$ is called the *last* or *best* element in X if it dominates every other element, that is, $x \succsim y$ for all $y \in X$. In general, there can be multiple maximal and best elements.

Exercise 1.25
Formulate analogous definitions for minimal and first or worst elements.

Example 1.23 Let X be the set of positive integers $\{1, 2, \ldots, 9\}$, ordered by the relation m is a multiple of n (example 1.18). The numbers $5, 6, 7, 8, 9$ are all maximal elements (they have no arrowheads pointing at them). There is no best element. The number 1 is the first number and the only minimal number.

Exercise 1.26
Find the maximal and minimal elements of the set $X = \{a, b, x, y, z\}$ when ordered $x \prec a \prec y \prec b \sim z$. Are they also best and worst elements respectively?

Exercise 1.27
Every best element is a maximal element, and not vice versa.

Exercise 1.28
Every finite ordered set has a least one maximal element.

The following characterization of maximal and best elements in terms of upper contour sets is often useful. (See, for example, proposition 1.5.) Analogous results hold for minimal and first elements.

Exercise 1.29
Let X be ordered by $\succsim$.

x^* is maximal $\Leftrightarrow \succ(x^*) = \emptyset$

x^* is best $\Leftrightarrow \precsim(x^*) = X$

Upper and Lower Bounds

To delineate sets that have no maximal or best elements, such as the interval $(0, 1)$, we often identify upper and lower bounds. Let A be a nonempty subset of an ordered set X. An element $x \in X$ is called an *upper bound* for A if x dominates every element in A, that is, $x \succsim a$ for every $a \in A$. $x \in X$ is called a *least upper bound* for A if it precedes every upper bound for A.

Exercise 1.30
Formulate analogous definitions for lower bound and greatest lower bound.

Example 1.24 Consider again the set of positive integers $\mathfrak{N}$ ordered by m is a multiple of n (example 1.18). It has a unique minimal element 1 and no maximal element. Any finite subset $\{n_1, n_2, \ldots, n_k\} \subseteq \mathfrak{N}$ has a least upper bound that is called the *least common multiple*. It has a greatest lower bound that is called the *greatest common divisor*.

Exercise 1.31
For the set of positive integers $\mathfrak{N}$ ordered by m is a multiple of n (example 1.18), specify upper and lower bounds for the set $A = \{2, 3, 4, 5\}$. Find the least upper bound and greatest lower bound.

Example 1.25 (Intervals) b is the least upper bound of the closed interval $[a, b]$. b is an upper bound of the open interval (a, b) but not necessarily the least upper bound. Similarly a is a lower bound of (a, b) and the greatest lower bound of $[a, b]$.

Exercise 1.32
Assume that the set $X = \{a, b, x, y, z\}$ is ordered as follows:

$x \prec a \prec y \prec b \sim z$

Find the least upper bounds of the intervals $[a, b]$ and (a, b).

Exercise 1.33
Let X be ordered by $\succsim$.

x is an upper bound of $A \Leftrightarrow A \subseteq \precsim(x)$

x is a lower bound of $A \Leftrightarrow A \subseteq \succsim(x)$

Product Orders

Recall that we generate a product of sets by welding together individual sets. If we take the product $X = X_1 \times X_2 \times \cdots \times X_n$ of a collection of ordered sets X_i, there is a natural order induced on the product by

$$\mathbf{x} \succsim \mathbf{y} \Leftrightarrow x_i \succsim_i y_i \quad \text{for all } i = 1, 2, \ldots, n$$

$\mathbf{x}, \mathbf{y} \in X$. We often want to distinguish between the cases in which $x_i \succ_i y_i$ for all i and those in which $x_i \sim_i y_i$ for some i. We do this by means of notational convention, reserving $\succ$ for the case

$$\mathbf{x} \succ \mathbf{y} \Leftrightarrow x_i \succ_i y_i \quad \text{for all } i = 1, 2, \ldots, n$$

using $\succneqq$ to indicate the possibility that $x_i = y_i$ for some i, that is,

$$\mathbf{x} \succneqq \mathbf{y} \Leftrightarrow x_i \succsim y_i \quad \text{for all } i = 1, 2, \ldots, \text{ and } \mathbf{x} \neq \mathbf{y}$$

Even if all the order $\succsim_i$ are complete, the natural product order $\succsim$ is only a partial order on the product space $X_1 \times X_2 \times \cdots \times X_n$. When $x_i \succ y_i$ for some i while $x_i \prec y_i$ for others, $\mathbf{x}$ and $\mathbf{y}$ are not comparable.

Example 1.26 (Natural order on $\Re^n$) Elements of $\Re^n$ inherit a natural order from $\geq$ on $\Re$. Thus for any $\mathbf{x}, \mathbf{y} \in \Re^n$,

$$\mathbf{x} \geq \mathbf{y} \Leftrightarrow x_i \geq y_i \quad \text{for all } i = 1, 2, \ldots, n$$

Readers of the literature need to be alert to what various authors mean by $\mathbf{x} > \mathbf{y}$ in $\Re^n$. We adopt the convention that

$$\mathbf{x} > \mathbf{y} \Leftrightarrow x_i > y_i \quad \text{for all } i = 1, 2, \ldots, n$$

using $\mathbf{x} \gneqq \mathbf{y}$ for the possibility the $\mathbf{x}$ and $\mathbf{y}$ are equal in some components

$$\mathbf{x} \gneqq \mathbf{y} \Leftrightarrow x_i \geq y_i \quad \text{for all } i = 1, 2, \ldots, n \text{ and } \mathbf{x} \neq \mathbf{y}$$

Some authors use $>$ where we use $\gneqq$, and use $\gg$ in place of $>$. Other conventions are also found.

The natural order is not the only way in which to order the product of weakly ordered sets. An example of a complete order on a product space $X = X_1 \times X_2 \times \cdots \times X_n$ is the *lexicographic order*, in which $\mathbf{x} \succ^L \mathbf{y}$ if $x_k \succ y_k$ in the first component in which they differ. That is,

$$\mathbf{x} \succ^L \mathbf{y} \Leftrightarrow x_k \succ y_k \quad \text{and} \quad x_i = y_i \quad \text{for all } i = 1, 2, \ldots, k-1$$

A dictionary is ordered lexicographically, which is the origin of the name. (Lexicography is the process or profession of compiling dictionaries.) Lexicographic orders are used occasionally in economics and game theory (see example 1.49).

Exercise 1.34
Using the natural order $\geq$ on $\Re$, order the plane $\Re^2$ by the lexicographic order. It is a total order?

Exercise 1.35
Let X be the product of n sets X_i each of which is ordered by $\succsim_i$. Show that the lexicographic order $\succsim^L$ is complete if and only if the component orders $\succsim_i$ are complete.

1.2.4 Partially Ordered Sets and Lattices

In general, an ordered set may have many maximal elements and its subsets may have multiple least upper bounds. Uniqueness may be achieved by imposing the additional requirement of antisymmetry. The result is called a partially ordered set.

A *partial order* is a relation that is reflexive, transitive, and antisymmetric. The most common example of an antisymmetric order relation is the numerical order $\leq$ on $\Re$, where $x \geq y$ and $y \geq x$ implies that $x = y$. A set X on which is defined a partial order $\succsim$ is called a *partially ordered set* or *poset*. Partially ordered sets have numerous applications in economics.

Example 1.27 The set $\Re$ of positive integers is partially ordered by the relation "m is a multiple of n" (example 1.18). The ordering is only partial, since not all integers are comparable under this relation.

Example 1.28 ($\Re^n$) The natural order on $\Re^n$ (example 1.26) is only partial, although $\geq$ is complete on $\Re$. In $\Re^2$, for example, $(2, 1) \geq (1, 1)$, but the elements $(2, 1)$ and $(1, 2)$ are not comparable. Therefore $\Re^n$ with the natural order is a partially ordered set.

Example 1.29 (Set inclusion) The set of subsets of any set is partially ordered by set inclusion $\subseteq$.

Exercise 1.36
Show that set inclusion is a partial order on the power set of a set X.

Example 1.30 Let X be the set of steps necessary to complete a project (e.g., a building or a computer program). For any $x, y \in X$, let $x \prec y$ denote that task x has to be completed before y and define $x \precsim y$ if $x \prec y$ or $x = y$. The set of tasks X is partially ordered by $\precsim$.

The significance of antisymmetry is that, if they exist, the least upper bound and greatest lower bound of any subset of a poset are unique. The least upper bound of a set S is called the *supremum* of S and denoted sup S. The greatest lower bound is called the *infimum* of S and denoted inf S.

Exercise 1.37
Let A be a nonempty subset of X that is partially ordered by $\succsim$. If A has a least upper bound, then it is unique. Similarly A has at most one greatest lower bound.

Exercise 1.38
Characterize the equivalence classes of the relation $\sim$ induced by a partial order $\succsim$.

Remark 1.4 (Best versus supremum) The best element in a set is an element of the set, whereas the supremum of a set may not necessarily belong to the set. For example, 1 is the supremum of the interval $(0, 1)$, which has no best element. Another example is given in exercise 1.31. This distinction is of practical importance in optimization, where the search for the *best* alternative may identify the supremum of the choice set, which may not be a feasible alternative.

When the supremum of a partially ordered set X belongs to X, it is necessarily the best element of X. In this case, the supremum is called the *maximum* of X. Similarly, when the infimum of a partially ordered set X belongs to X, it is called the *minimum* of X.

Exercise 1.39
The set of subsets of a set X is partially ordered by inclusion. What is the maximum and minimum of $\mathcal{P}(X)$.

Chains

A partial ordering is "partial" in the sense that not all elements are necessarily comparable. For example, in the partial order $\subseteq$ if S and T are disjoint nonempty sets, then neither $S \subseteq T$ nor $T \subseteq S$. If all elements in a partially ordered set are comparable, so that the ordering $\succsim$ is *complete*, it is called a *total* or *linear ordering*. A totally ordered set is called a *chain*.

Example 1.31 The set of real numbers with the usual order $\leq$ is a chain.

1.2 Ordered Sets

Exercise 1.40
In the multiple ordering of $\mathfrak{N}$ (example 1.18), find a subset which is a chain.

Exercise 1.41
A chain has at most one maximal element.

Example 1.32 (Game tree) In many strategic games (example 1.2), the temporal order of moves is vital. Economists model such dynamic interactions by means of the *extensive form*; an essential ingredient is a game tree. A *game tree* is a partially ordered finite set $(T, \succ)$ in which the predecessors $\prec(t)$ of every element t are totally ordered, that is, $\prec(t)$ is a chain for every t.

The elements of T are called *nodes*. Nodes that have no successors are called *terminal nodes*. Thus the set of terminal nodes Z is defined by

$$Z = \{t \in T : \succ(t) = \varnothing\}$$

Terminal nodes are the maximal elements of $(T, \succ)$. The remaining nodes $X = T \setminus Z$ are called *decision nodes*. Similarly nodes that have no predecessors are called *initial nodes*. The set of initial nodes W is

$$W = \{t \in T : \prec(t) = \varnothing\}$$

Initial nodes are the minimal elements of $(T, \succ)$.

As a partial order, $\succ$ is asymmetric, transitive, and antisymmetric. The additional requirement that $\prec(t)$ is a chain for every t implies that there is a unique path to every node from some initial node (exercise 1.42). A partially ordered set with this additional property is called *an arborescence*.

Exercise 1.42
Let $(T, \succ)$ be a game tree (arborescence). For every noninitial node, call

$$p(t) = \sup \prec(t)$$

the immediate predecessor of t. Show that

1. $p(t)$ is unique for every $t \in T \setminus W$.

2. There is a unique path between any node and an initial node in a game tree.

Remark 1.5 (Zorn's lemma) Zorn's lemma asserts that if X is a partially ordered set in which every chain has an upper bound, then X has a maximal element. Zorn's lemma is the fundamental existence theorem of advanced mathematics. It is not possible to prove Zorn's lemma in the usual sense of deducing it from more primitive propositions. It can be shown that Zorn's lemma is equivalent to the seemingly obvious *axiom of choice* that states: Given any nonempty class of disjoint nonempty sets, it is possible to select precisely one element from each set. The axiom of choice or one of its equivalents is usually taken as an axiom in any mathematical system.

Lattices

A *lattice* is a partially ordered set (poset) in which every pair of elements have a least upper bound and a greatest lower bound. If x and y are any two elements in a lattice L, their least upper bound, denoted $x \vee y$, is an element of L which is called the *join* of x and y. Their greatest lower bound, denoted $x \wedge y$ is called their *meet*. These notations are analogous to set union and intersection, which provides a useful example of a lattice.

Example 1.33 The real numbers $\Re$ with $x \vee y = \max\{x,y\}$ and $x \wedge y = \min\{x,y\}$ form a lattice.

Example 1.34 Let $X = \{1,2,3\}$. The partially ordered set $X \times X$ is a lattice, where

$(x_1, x_2) \vee (y_1, y_2) = (\max\{x_1, y_1\}, \max\{x_2, y_2\})$

$(x_1, x_2) \wedge (y_1, y_2) = (\min\{x_1, y_1\}, \min\{x_2, y_2\})$.

See figure 1.7. Although the points $(2,1)$ and $(1,2)$ are not comparable under the natural order, they have a least upper bound of $(2,2)$ and a greatest lower bound of $(1,1)$. Therefore $(2,1) \vee (1,2) = (2,2)$ and $(2,1) \wedge (1,2) = (1,1)$.

Exercise 1.43
In example 1.34, what is $(1,2) \vee (3,1)$? $(1,2) \wedge (3,2)$?

Example 1.35 The positive integers $\Re$ ordered by "m is a multiple of n" (example 1.18) constitute a lattice. $m \wedge n$ is the least common multiple of m and n, while $m \vee n$ is the greatest common divisor of m and n.

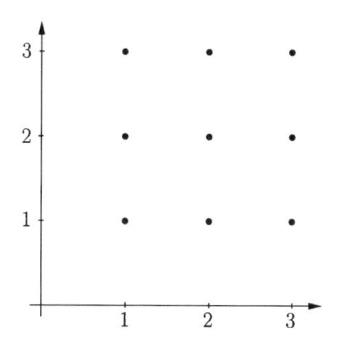

Figure 1.7
A simple lattice in $\mathscr{R}^2$

Example 1.36 (Set inclusion) For any set X, the poset $(\mathscr{P}(X), \subseteq)$ is a lattice. For any two sets S and T, their join is $S \vee T = S \cup T$ and their meet is $S \wedge T = S \cap T$.

The lattice of subsets of the four element set $\{a, b, c, d\}$ is illustrated in figure 1.8. Its regular structure justifies the term "lattice."

Example 1.37 (Information partitions) Let S denote the sample space of random experiment. An *information partition* P is a partition of S into mutually exclusive events with interpretation that the decision maker knows which event takes place. The information partition captures the decision makers information about the random experiment.

Let $\mathscr{P}$ be the set of all partitions of S. We say that a partition P_1 is finer than P_2 if each set in P_2 can be written as the union of sets in P_1. A finer partition provides better information about the outcome. Let $P_1 \succsim P_2$ denote the relation P_1 is finer than P_2. Then the ordered set $(\mathscr{P}, \succsim)$ is a lattice, where $P_1 \vee P_2$ is the coarsest partition that is finer than both and $P_1 \wedge P_2$ is the finest partition that is coarser than both.

Exercise 1.44
The operations $\vee$ and $\wedge$ have the following consistency properties. For every x, y in a lattice $(X, \succsim)$,

1. $x \vee y \succsim x \succsim x \wedge y$
2. $x \succsim y \Leftrightarrow x \vee y = x$ and $x \wedge y = y$
3. $x \vee (x \wedge y) = x = x \wedge (x \vee y)$

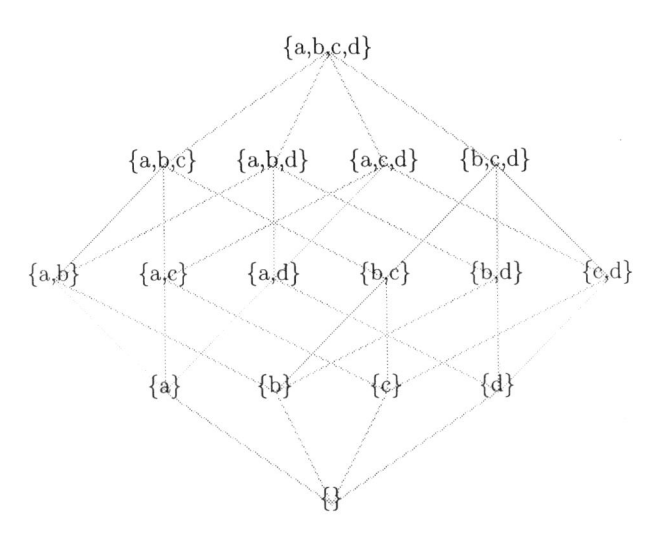

Figure 1.8
The lattice of subsets of a four-element set

Exercise 1.45
Any chain is a lattice.

Exercise 1.46
The product of two lattices is a lattice.

The previous exercise implies that the product of *n* lattices is a lattice, with $\vee$ and $\wedge$ defined componentwise, that is,

$$\mathbf{x} \vee \mathbf{y} = (x_1 \vee y_1, x_2 \vee y_2, \ldots, x_n \vee y_n)$$

$$\mathbf{x} \wedge \mathbf{y} = (x_1 \wedge y_1, x_2 \wedge y_2, \ldots, x_n \wedge y_n)$$

Example 1.38 $\Re^n$ is a lattice with $\vee$ and $\wedge$ defined componentwise.

A lattice *L* is *complete* if every nonempty subset $S \subseteq L$ has a least upper bound and a greatest lower bound in *L*. Set inclusion is the only complete lattice in the above examples. A *sublattice* is a subset $S \subseteq L$ that is a lattice in its own right, that is, $x \wedge y \in S$ and $x \vee y \in S$ for every $x, y \in S$.

Example 1.39 The set of real numbers $\Re$ is an example of set that is completely ordered but not a complete lattice. There are sets that do not have upper (e.g., $\Re$) or lower bounds.

Remark 1.6 (Extended real numbers) The fact that the set $\Re$ of real numbers is not a *complete* lattice (example 1.39) often causes technical difficulties. Therefore a useful analytical device is to extend the set $\Re$ so that it is always complete. To do this, we add to new elements $+\infty$ and $-\infty$, with $-\infty < x < \infty$ for every $x \in \Re$. The set

$$\Re^* = \Re \cup \{-\infty\} \cup \{+\infty\}$$

is called the set of *extended real numbers*.

Since $-\infty < x < \infty$ for every $x \in \Re$, every subset $S \subseteq \Re^*$ has an upper bound ($+\infty$) and a lower bound ($-\infty$). Consequently every nonempty set has a least upper bound and a greatest lower bound. $\Re^*$ is a complete lattice. If moreover we adopt the convention that $\sup \emptyset = -\infty$ and $\inf \emptyset = +\infty$, then every subset of $\Re^*$ has a least upper bound and greatest lower bound.

In fact the definition of a complete lattice is partially redundant. For completeness it suffices that every subset has a greatest lower bound, since this implies the existence of a least upper bound. This result (exercise 1.47) is valuable in establishing completeness. (Similarly, if every subset of a partially ordered set has a least upper bound, the poset is a complete lattice.)

Exercise 1.47

Let X be a partially ordered set which has a best element x^*. If every nonempty subset S of a X has a greatest lower bound, then X is a complete lattice.

Example 1.40 The set of points $\{1, 2, 3\} \times \{1, 2, 3\}$ (example 1.34) is a sublattice of $\Re^2$.

However, note that the requirement of being a sublattice is more stringent than being a complete lattice in its own right.

Example 1.41 In the previous example, let X be the set of points illustrated in figure 1.9.

$$X = \{(1, 1), (2, 1), (3, 1), (1, 2), (1, 3), (3, 3)\}$$

X is a complete lattice but is it not a sublattice of $\{1, 2, 3\} \times \{1, 2, 3\}$ or $\Re^2$. The point $(2, 2)$ is the least upper bound of $\{(2, 1), (1, 2)\}$ in $\Re^2$, while $(3, 3)$ is the least upper bound in X.

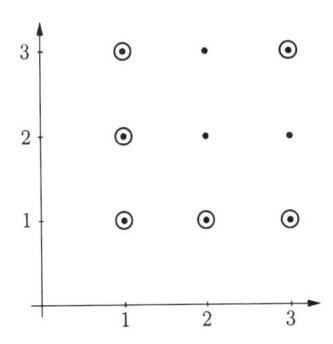

Figure 1.9
A lattice that is not a sublattice

Exercise 1.48
Let a, b be elements in a lattice L with $a \precsim b$. Then the subsets $\succsim(b)$, $\precsim(a)$ and $[a, b]$ are sublattices. The sublattices are complete if L is complete.

Remark 1.7 It is worth noting the role of successive assumptions. Antisymmetry ensures the uniqueness of the least upper bound and greatest lower bound of any set *if they exist*. A poset is a lattice if it contains the least upper bound and greatest lower bound of every *pair* of elements in the set. A lattice is complete if furthermore it contains the least upper bound and greatest lower bound *for every set*. Note that "completeness" in reference to a lattice is used in a slightly different sense to completeness of the underlying relation. A complete lattice is a complete ordering, although the converse is not necessarily true.

Strong Set Order

Any lattice $(X, \succsim)$ induces a relation on the subsets of X that is called the *strong set order*, denoted $\succsim_S$. Given $S_1, S_2 \subseteq X$,

$$S_2 \succsim_S S_1 \Leftrightarrow x_1 \wedge x_2 \in S_1 \quad \text{and} \quad x_1 \vee x_2 \in S_2$$

for every x_1 in S_1 and $x_2 \in S_2$. This order will play an important role in section 2.2.

Example 1.42 The strong set order is quite different to set inclusion. For example, consider example 1.34. Let $S_1 = \{(1, 1), (2, 1), (3, 1)\}$ and $S_2 = \{(1, 2), (2, 2), (3, 2)\}$ (figure 1.10). Then $S_2 \succsim_S S_1$ although S_1 and S_2 are disjoint. Further consider a proper subset of S_1 such as $S_3 = \{(2, 1), (3, 1)\}$. $S_2 \not\succsim_S S_3$, since $(1, 1) = (2, 1) \wedge (1, 2) \notin S_3$.

1.2 Ordered Sets

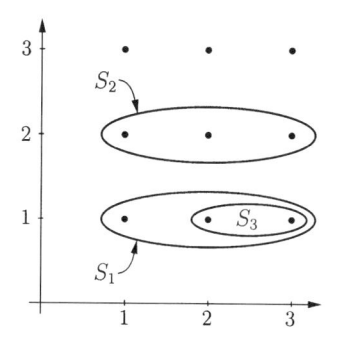

Figure 1.10
The strong set order

The strong set order is not reflexive, and hence it is not an order on $\mathscr{P}(X)$. However, it is a partial order on the set of all sublattices of X. The details are given in the following exercise.

Exercise 1.49

1. For any lattice X, the strong set order $\succsim_S$ is antisymmetric and transitive. [*Hint*: Use exercise 1.44.]
2. $S \succsim_S S$ if and only S is a sublattice.
3. $\succsim_S$ is a partial order on the set of all sublattices of X.

The nature of the strong set order is characterized by the following result, which says (roughly) that $S_2 \succsim_S S_1$ implies that the lowest element of S_2 dominates the lowest element of S_1. Similarly the best element of S_2 is greater than the best element of S_1.

Exercise 1.50

If S_1, S_2 are subsets of a complete lattice,

$$S_1 \succsim_S S_2 \Rightarrow \inf S_1 \succsim \inf S_2 \quad \text{and} \quad \sup S_1 \succsim \sup S_2$$

For closed intervals of a chain, such as $\Re$, this characterization can be strengthened.

Exercise 1.51

If S_1, S_2 are intervals of a chain,

$$S_1 \succsim_S S_2 \Leftrightarrow \inf S_1 \succsim \inf S_2 \quad \text{and} \quad \sup S_1 \succsim \sup S_2$$

1.2.5 Weakly Ordered Sets

The second class of order relations important in economics is obtained by imposing completeness rather than antisymmetry on the preorder. A *weak order* is a relation that is complete and transitive. It is sometimes called simply an *ordering*. The most important example in economics is the consumer's preference relation (example 1.12), which is considered in detail in section 1.6.

Exercise 1.52
Many economics texts list three assumptions—complete, transitive, and reflexive—in defining the consumer's preference relation. Show that reflexivity is implied by completeness, and so the third assumption is redundant.

Exercise 1.53
Why would antisymmetry be an inappropriate assumption for the consumer's preference relation?

Exercise 1.54
In a weakly ordered set, maximal and best elements coincide. That is,

x is maximal $\Leftrightarrow x$ is best

Exercise 1.55
A weakly ordered set has a most one best element. True or false?

In a weakly ordered set, every element is related to every other element. Given any element y, any other element $x \in X$ belongs to either the upper or lower contour set. Together with the indifference sets, the upper and lower contour sets partition the set X in various ways. Furthermore the upper and lower contour sets are nested. The details are given in the following exercises.

Exercise 1.56
If $\succsim$ is a weak order on X, then for every $y \in X$,

1. $\succsim(y) \cup \precsim(y) = X$ and $\succsim(y) \cap \precsim(y) = I_y$
2. $\succsim(y) \cup \prec(y) = X$ and $\succsim(y) \cap \prec(y) = \emptyset$
3. $\succ(y)$, I_y and $\prec(y)$ together partition X

Exercise 1.57
If $\succsim$ is a weak order on X,

$$x \gtrsim y \Rightarrow \gtrsim(x) \subseteq \gtrsim(y)$$

$$x \succ y \Rightarrow \succ(x) \subsetneqq \succ(y)$$

The principal task of optimization theory and practice is identify the best element(s) in a choice set X, which is usually weakly ordered by some criterion. To identify the best element, optimization theory draws on other properties (linear and metric) of the choice set. Techniques of optimization are explored in chapter 5. To prepare the ground, we next investigate the metric and linear properties of sets in sections 1.3 and 1.4. Before leaving order relations, we touch on the problem of aggregating different orders on a common set.

1.2.6 Aggregation and the Pareto Order

The product order defines a natural order on the product of ordered sets. Economists frequently confront an analogous situation, involving *different* orders over a *common* set. Specifically, suppose that there is a set X on which is defined a profile of distinct orderings $(\gtrsim_1, \gtrsim_2, \ldots, \gtrsim_n)$. These different orders might correspond to different individuals or groups or to different objectives. The problem is to aggregate the separate orders into a common or social order.

Analogous to the product order, a natural way in which to aggregate the individual preferences is to define the social preference by

$$x \gtrsim^P y \Leftrightarrow x \gtrsim_i y \quad \text{for all } i = 1, 2, \ldots, n$$

$x, y \in X$, and

$$x \succ^P y \Leftrightarrow x \succ_i y \quad \text{for all } i = 1, 2, \ldots, n \tag{1}$$

This is known as the *Pareto order*. For state x to strictly preferred to state y in the Pareto order requires that x strictly dominate y in every individual order.

The outcome x is said to *Pareto dominate* y if $x \succ^P y$ in the Pareto order. It is called *Pareto efficient* or *Pareto optimal* if it is maximal in the weak Pareto order, that is if there is no outcome y such that all individuals strictly prefer y to x. The set of states that is maximal in the Pareto ordering is called the *Pareto optimal set*.

Pareto $= \{x \in X : \text{there is no } y \in X \text{ such that } y \succ^P x\}$

Exercise 1.58
There always at least one Pareto optimal outcome in any finite set, that is

X finite $\Rightarrow$ Pareto $\neq \emptyset$

Remark 1.8 (Weak versus strong Pareto order) Two distinct Pareto orders are commonly used in economics, which is a potential source of confusion. The order defined by (1) is called the *weak* Pareto order. Some authors replace (1) with

$$x \succ^P y \Leftrightarrow x \succsim_i y \quad \text{for all } i = 1, 2, \ldots, \text{ and } x \succ_j y \text{ for some } j$$

This is called the *strong Pareto order*, since it ranks more alternatives. Alternative x dominates y in the strong Pareto order provided that at least one individual strictly prefers x to y (and no one strictly prefers y to x). Similarly x is called *strongly Pareto optimal* if it is maximal in the strong Pareto order.

The weak and strong Pareto orders are distinct, and they can lead to different answers in some situations [although the distinction is immaterial in one domain of prime economic interest (exercise 1.249)]. The strong Pareto order is more commonly applied in welfare economics, whereas the weak order is usually adopted in general equilibrium theory and game theory.

Exercise 1.59
Investigate which definition of the Pareto order is used in some leading texts, such as Kreps (1990), Mas-Colell et al. (1995), Varian (1992), and Osborne and Rubinstein (1994).

Even if all the constituent orders $\succsim_i$ are complete, the Pareto order $\succsim^P$ is only a partial order. Where $x \succ_i y$ for some i while $x \prec_i y$ for others, x and y are not comparable. This deficiency provides scope for two fertile areas of economic analysis, social choice theory and game theory. Since they also provide good illustrations for the material of this book, we briefly describe each of these areas.

Social Choice

A social choice problem comprises

- a finite set N of *individuals* or *agents*

- a set X of outcomes or *social states*
- for each individual $i \in N$ a preference relation $\succsim_i$ on the set of outcomes X

The theory of social choice is concerned with aggregating individuals orderings over a set of *social states* X into a social preference. While non–Pareto orders have been considered, the Pareto criterion is so compelling that the central problem of social choice can be regarded as completing the Pareto order in a way that respects the individual preference orderings. Unfortunately, the principal results are essentially negative, as exemplified by the famous impossibility theorem of Arrow.

Example 1.43 (Arrow's impossibility theorem) One way to complete the Pareto order would be to define

$$x \succsim y \Leftrightarrow x \succsim_i y$$

for some specific individual i. In effect, individual i is made a dictator. While the dictatorial ordering satisfies the Pareto principle, it is not currently regarded as politically correct!

Another property that reflects sympathy between the social ordering and individual orders requires that the social order does not distinguish between alternatives that are indistinguishable to individuals. We say that the product order $\succsim$ satisfies *independence of irrelevant alternatives* (IIA) if for every set of social states $A \subset X$, given two sets of individual preferences $\succsim_i$ and $\succsim_i'$ which are identical over the set A, that is,

$$x \succsim_i y \Leftrightarrow x \succsim_i' y \quad \text{for every } x, y \in A$$

then the corresponding social orders $\succsim$ and $\succsim'$ also order A identically, that is,

$$x \succsim y \Leftrightarrow x \succsim' y \quad \text{for every } x, y \in A$$

In 1950 Nobel laureate Kenneth Arrow (1963) showed that it is impossible to complete the weak Pareto ordering in a way that is independent of irrelevant alternatives but not dictatorial. The following three exercises provide a straightforward proof of Arrow's theorem.

Exercise 1.60 (Field expansion lemma)
A group S of individuals is *decisive* over a pair of alternatives $x, y \in X$ if

$$x \succ_i y \quad \text{for every } i \in S \Rightarrow x \succ y$$

Assume that the social order is consistent with the Pareto order and satisfies the IIA condition. Show that, if a group is decisive over any pair of states, it is decisive over every pair of alternatives. (Assume that there are at least four distinct states, that is, $|X| \geq 4$.)

Exercise 1.61 (Group contraction lemma)
Assume that the social order is consistent with the Pareto order and satisfies the IIA condition, and that $|X| \geq 3$. If any group S with $|S| > 1$ is decisive, then so is a proper subset of that group.

Exercise 1.62
Using the previous exercises, prove Arrow's impossibility theorem.

Exercise 1.63 (The Liberal Paradox)
Liberal values suggest that there are some choices that are purely personal and should be the perogative of the individual concerned. We say that a social order exhibits *liberalism* if for each individual i, there is a pair of alternatives $x, y \in X$ over which she is decisive, that is, for which

$$x \succ_i y \Rightarrow x \succ y$$

(A dictator is decisive over all alternatives.) Show that is impossible to complete a Pareto order in a way that respects liberalism. This inconsistency between liberalism and the Pareto principle is known the Liberal Paradox. [*Hint*: It suffices that show that there are not even two persons who are decisive over personal choices. Consider a pair of alternatives for each person, and show that the implied Pareto order is intransitive.]

Remark 1.9 (Rawlsian social choice) A criterion of social justice first advocated by the philosopher John Rawls (1971) has attracted a lot of attention from economists. Effectively, the Rawlsian maximin criterion is analogous to completing the Pareto ordering lexicographically, assigning priority to the preferences of the least well off individual, then the next least favored, and so on. The analogy is inexact, since the identity of the least well off individual varies with the social state.

Coalitional Games

A *coalitional game* comprises

- a finite set N of *players*
- a set X of *outcomes*

- for each player $i \in N$ a preference relation $\succsim_i$ on the set of outcomes X
- for every proper coalition $S \subseteq N$, a set $W(S) \subseteq X$ of outcomes that it can obtain by its own actions

In effect each coalition S is decisive for the outcomes in $W(S)$. Although the structure of a coalitional game is similar to that of a social choice problem, the analytical focus is different. Rather than attempting to produce a complete social order over X, coalitional game theory aims to isolate a subset of X which is maximal with respect to some partial order.

Example 1.44 (Cost allocation) The Southern Electricity Region of India comprises four states: Andra Pradesh (AP), Kerala, Mysore, and Tamil Nadu (TN). In the past each state had tended to be self-sufficient in the generation of electricity. This led to suboptimal development for the region as a whole, with reliance on less economic alternative sources in Andra Pradesh and Tamil Nadu, instead of exploiting the excellent hydro resources in Mysore and Kerala.

The costs of developing the electric power system in the region, under various assumptions about the degree of cooperation between states, are summarized in the following table. To simplify the calculations, Kerala and Mysore have been amalgamated into a hybrid state (KM), since they are essentially similar in their hydro resources and power requirements. (The cost estimates were derived from a general investment planning and system operation model comprising 800 variables and 300 constraints.)

Coalition structure	TN	AP	KM	Total cost in region
Self-sufficiency for each area	5,330	1,870	860	8,060
Cooperation between TN and AP, self-sufficiency for KM	5,520	1,470	860	7,850
Cooperation between TN and KM, self-sufficiency for AP	2,600	1,870	2,420	6,890
Cooperation between AP and KM, self-sufficiency for TN	5,330	480	1,480	7,290
Full cooperation	3,010	1,010	2,510	6,530

Clearly, the region as a whole benefits from cooperation, since total costs are minimized by exploiting the rich hydro resources in Kerala and Mysore. However, this increases the costs incurred in Kerala and Mysore to 2,510 million rupees, whereas they can provide for the own needs at a much lower cost of 860 million rupees. To induce the Kerala and Mysore to cooperate in a joint development, the other states must contribute to the development of their hydro resources.

We can model this problem as a coalitional game in which the players are Andra Pradesh (AP), Tamil Nadu (TN), and the hybrid state Kerala Mysore (KM). The outcomes are their respective cost shares (x_{AP}, x_{TN}, x_{KM}), where x_{AP} is the cost borne by Andra Pradesh. The set of outcomes $X = \Re_+^3$. Each player prefers a lower cost share, that is,

$$x_i' \succsim_i x_i \Leftrightarrow x_i' \leq x_i$$

By being self-sufficient, each state can ensure that it pays no more than its own costs, so that

$$W(AP) = \{(x_{AP}, x_{TN}, x_{KM}) : x_{AP} \leq 1{,}870\}$$

$$W(TN) = \{(x_{AP}, x_{TN}, x_{KM}) : x_{TN} \leq 5{,}300\}$$

$$W(KM) = \{(x_{AP}, x_{TN}, x_{KM}) : x_{KM} \leq 860\}$$

Alternatively, Andra Pradesh and Tamil Nadu could undertake a joint development, sharing the total cost of 6,990 between them. Thus

$$W(AP, TN) = \{(x_{AP}, x_{TN}, x_{KM}) : x_{AP} + x_{TN} \leq 6{,}990\}$$

Similarly

$$W(AP, KM) = \{(x_{AP}, x_{TN}, x_{KM}) : x_{AP} + x_{KM} \leq 1{,}960\}$$

$$W(TN, KM) = \{(x_{AP}, x_{TN}, x_{KM}) : x_{TN} + x_{KM} \leq 5{,}020\}$$

Finally the three states could cooperate sharing the total costs 6,530

$$W(AP, TN, KM) = \{(x_{AP}, x_{TN}, x_{KM}) : x_{AP} + x_{TN} + x_{KM} = 6{,}530\}$$

Coalitional game theory typically respects the Pareto order, attempting to extend it by recognizing the decisiveness of coalitions over certain subsets of the feasible outcomes. The primary example of such a *solution concept* is the core, which extends the Pareto order to coalitions as well as the group as a whole.

1.2 Ordered Sets

Example 1.45 (Core) The *core* of a coalitional game is the set of outcomes for which no coalition can do better by unilateral action, that is,

Core = $\{x \in X :$ there does not exist S and $y \in W(S)$ such that
$\quad\quad y \succ_i x$ for every $i \in S\}$

For some games there may be no such unimprovable allocation, in which case we say that the core is *empty*.

Exercise 1.64
Show that every core allocation is Pareto optimal, that is,

core $\subseteq$ Pareto

Exercise 1.65
Find the core of the cost allocation game (example 1.44).

Example 1.46 (Coalitional game with transferable payoff) In an important class of coalitional games, the set of possible outcomes X comprises allocations of fixed sum of money or other good, denoted $w(N)$, among the players. That is,

$$X = \left\{ \mathbf{x} \in \Re^n : \sum_{i \in N} x_i = w(N) \right\}$$

Individual coalitions can allocate smaller sums, denoted $w(S)$, among their members, so that

$$w(S) = \left\{ \mathbf{x} \in \Re^n : \sum_{i \in S} x_i \leq w(S) \right\}$$

Individual players rank allocations on the basis of their own shares, that is,

$\mathbf{x}' \succsim_i \mathbf{x} \Leftrightarrow x'_i \geq x_i$

Concisely, *a coalitional game with transferable payoff* comprises

- a finite set of players N
- for every coalition $S \subseteq N$, a real number $w(S)$ that is called the *worth* of the coalition S

Conventionally $w(\emptyset) = 0$. A TP-coalitional game (N, w) is called *essential* if there is some surplus to distribute, that is,

$$w(N) > \sum_{i \in N} w(\{i\})$$

The label *transferable payoff* reflects the fact that potential worth of a coalition $w(S)$ can be freely allocated among the members of the coalition. For convenience, we will refer to such games as *TP-coalitional games*. As well as being practically important (exercise 1.66), TP-coalitional games provide an excellent illustration of many of the concepts introduced in this chapter and also in chapter 3.

Exercise 1.66 (Cost allocation)
Formulate the cost allocation problem in example 1.44 as a TP-coalitional game. [Hint: Regard the potential cost savings from cooperation as the sum to be allocated.]

Exercise 1.67
Show that the core of coalitional game with transferable payoff is

$$\text{core} = \left\{ \mathbf{x} \in X : \sum_{i \in S} x_i \geq w(S) \text{ for every } S \subseteq N \right\}$$

Example 1.47 (Simple games) A *simple* game is a TP-coalitional game in which the worth of each coalition $w(S)$ is either 0 or 1 ($w(N) = 1$). A coalition for which $w(S) = 1$ is called a *winning coalition*. Simple games often provide a suitable model for situations involving the exercise of power.

Example 1.48 (Unanimity games) In some simple games a particular coalition T is necessary and sufficient to form a winning coalition, so that

$$w(S) = \begin{cases} 1 & \text{if } S \supseteq T \\ 0 & \text{otherwise} \end{cases}$$

Each member i of the essential coalition T is called a *veto player*, since no winning coalition can be formed without i. The game is called a *unanimity game*, since winning requires the collaboration of all the veto players.

For a given set of players N, each coalition T defines a different unanimity game u_T given by

$$u_T(S) = \begin{cases} 1 & \text{if } S \supseteq T \\ 0 & \text{otherwise} \end{cases}$$

Unanimity games play a fundamental role in the theory of TP-coalitional games.

Exercise 1.68
Specify the set of unanimity games for the player set $N = \{1, 2, 3\}$.

Exercise 1.69
Show that the core of a simple game is nonempty if and only if it is a unanimity game.

Example 1.49 (Nucleolus) We can measure the potential dissatisfaction of a coalition S with a particular outcome $\mathbf{x} \in X$ by the difference between its worth and its total share, defining

$$d(S, \mathbf{x}) = w(S) - \sum_{i \in S} x_i$$

The amount $d(S, \mathbf{x})$ is called the *deficit* of the coalition S at the outcome $\mathbf{x}$. If $d(S, \mathbf{x}) \geq 0$, the coalition is receiving less than its worth. The larger its deficit $d(S, \mathbf{x})$, the greater is its potential dissatisfaction with the outcome $\mathbf{x}$.

For any outcome $\mathbf{x}$, let $\mathbf{d}(\mathbf{x})$ denote a list of deficits for each proper coalition arranged in decreasing order. That is, the first element of the list $\mathbf{d}(\mathbf{x})$ is the deficit of the most dissatisfied coalition at the outcome $\mathbf{x}$. Since there are 2^n coalitions, the list $\mathbf{d}(\mathbf{x})$ has 2^n components. It is an element of the space $\Re^{2^n}$.

We can order these lists lexicographically (section 1.2.3). $\mathbf{d}(\mathbf{x})$ precedes $\mathbf{d}(\mathbf{y})$ in the lexicographic order on $\Re^{2^n}$ if the coalition which is most dissatisfied with $\mathbf{x}$ has a smaller deficit than the coalition which is most dissatisfied at $\mathbf{y}$.

The lexicographic order $\succsim^L$ on $\Re^{2^n}$ induces a preorder $\succsim^d$ on the set of outcomes X defined by

$$\mathbf{x} \succsim^d \mathbf{y} \Leftrightarrow \mathbf{d}(\mathbf{x}) \precsim^L \mathbf{d}(\mathbf{y})$$

and

$$\mathbf{x} \succ^d \mathbf{y} \Leftrightarrow \mathbf{d}(\mathbf{x}) \prec^L \mathbf{d}(\mathbf{y})$$

Outcome $\mathbf{x}$ is preferred to outcome $\mathbf{y}$ in the deficit order $\succsim^d$ if $\mathbf{d}(\mathbf{x})$ precedes $\mathbf{d}(\mathbf{y})$ in the lexicographic order; that is, the maximum deficit at $\mathbf{x}$ is less than the maximum deficit at $\mathbf{y}$.

The *nucleolus* of a TP-coalitional game is the set of most preferred or best elements in the deficit order $\succsim^d$, that is,

$$\text{Nu} = \{\mathbf{x} \in X : \mathbf{x} \succsim^d \mathbf{y} \text{ for every } \mathbf{y} \in X\}$$

In section 1.6 we will show that there is a unique best element for every TP-coalitional game. Consequently the nucleolus is a useful solution concept for such games, with many desirable properties.

Exercise 1.70
In the cost allocation game, find $\mathbf{d}(\mathbf{x})$ for

$$\mathbf{x}^1 = (180, 955, 395) \quad \text{and} \quad \mathbf{x}^2 = (200, 950, 380)$$

Show that $\mathbf{d}(\mathbf{x}^1) <^L \mathbf{d}(\mathbf{x}^2)$ and therefore $\mathbf{x}^1 \succ^d \mathbf{x}^2$.

Exercise 1.71
Is the deficit order $\succsim^d$ defined in example 1.49

- a partial order?
- a weak order?

on the set X.

Exercise 1.72
$\mathbf{x}$ belongs to the core if and only if no coalition has a positive deficit, that is,

$$\text{core} = \{\mathbf{x} \in X : d(S, \mathbf{x}) \leq 0 \text{ for every } S \subseteq N\}$$

Exercise 1.73
Show that $\text{Nu} \subseteq \text{core}$ assuming that $\text{core} \neq \emptyset$.

Strategic Games

Our earlier description of a strategic game in example 1.2 was incomplete in that it lacked any specification of the preferences of the players. A full description of a *strategic game* comprises:

- A finite set N of *players*.
- For each player $i \in N$ a nonempty set A_i of *actions*.
- For each player $i \in N$ a *preference relation* $\succsim_i$ on the action space $A = A_1 \times A_2 \times \cdots \times A_n$.

We can summarize a particular game by the ordered triple $(N, A, (\succsim_1, \succsim_2, \ldots, \succsim_n))$, comprising the set of players, their strategies, and their preferences.

Example 1.50 (Strictly competitive game) A strategic game between two players is *strictly competitive* if the preferences of the players are strictly opposed, that is for every $\mathbf{a}^1, \mathbf{a}^2 \in A$

$$\mathbf{a}^1 \succsim_1 \mathbf{a}^2 \Leftrightarrow \mathbf{a}^2 \succsim_2 \mathbf{a}^1$$

In other words, $\succsim_2$ is the inverse of $\succsim_1$.

A strategic game is analogous to the problem of social choice, in that it involves different orderings over a common space $A = A_1 \times A_2 \times \cdots \times A_n$. However, strategic game theory adopts yet another way of resolving (in an analytical sense) the conflict between competing orderings. Rather than attempting to combine the individual preference orderings $\succsim_i$ into a complete order $\succsim$, the game theorist attempts to identify certain action profiles $\mathbf{a} \in A$ as likely outcomes of independent play. The primary criterion is Nash equilibrium.

Example 1.51 (Nash equilibrium) In a strategic game, a *Nash equilibrium* is a choice of action for each player $\mathbf{a}^* = (a_1^*, a_2^*, \ldots, a_n^*)$ such that for every player $i \in N$,

$$(a_i^*, \mathbf{a}_{-i}^*) \succsim_i (a_i, \mathbf{a}_{-i}^*) \quad \text{for every } a_i \in A_i$$

Each player's chosen action a_i^* is at least as preferred as any other action $a_i \in A_i$ *given the choices of the other players*. These choices are made simultaneously, and a Nash equilibrium results when each player's action is an optimal response to those of the other players. No player will regret her action when the actions of the other players are revealed. A Nash equilibrium is called *strict* if

$$(a_i^*, \mathbf{a}_{-i}^*) \succ_i (a_i, \mathbf{a}_{-i}^*) \quad \text{for every } a_i \in A_i \setminus \{a_i^*\}$$

Example 1.52 (The Prisoner's Dilemma) Two suspects are arrested and held in separate cells. Each is independently offered the option of turning "state's evidence" by confessing the crime and appearing as a witness against the other. He will be freed while his partner receives a sentence of four years. However, if both confess, they can expect a sentence of three years each. If neither confess, the police only have sufficient evidence to charge them with a lesser offense, which carries a penalty of one year's imprisonment.

Each player has just two actions: Confess (denoted C) and not confess (N). The following table summarizes actions available to the players and their consequences (expected years of imprisonment).

		Player 2	
		C	N
Player 1	C	3, 3	0, 4
	N	4, 0	1, 1

Assuming that the suspects do not like prison, the players preferences are

$(C, N) \succ_1 (N, N) \succ_1 (C, C) \succ_1 (N, C)$

$(N, C) \succ_2 (N, N) \succ_2 (C, C) \succ_2 (C, N)$

where (C, N) denotes the action profile in which player 1 confesses and player 2 does not. Note that each player would prefer to confess irrespective of the choice of the other player. The Nash equilibrium outcome of this game is (C, C) is which both suspects confess, receiving a sentence of three years each.

Note that the Nash equilibrium is inconsistent with the Pareto order, since both players prefer (N, N) to the Nash equilibrium outcome (C, C). The Prisoner's Dilemma game is a model for many social phenomena, in which independent action does not achieve Pareto optimal outcomes.

Exercise 1.74
Show formally that the action profile (C, C) is a Nash equilibrium.

Example 1.53 (Dominance) In a strategic game each player's complete preference ordering $\succsim_i$ over outcomes (action profiles A) defines a partial ordering over the player i's own actions. We say that action a_i^2 *weakly dominates* a_i^1 for player i if

$(a_i^2, \mathbf{a}_{-i}) \succsim_i (a_i^1, \mathbf{a}_{-i})$ for every $\mathbf{a}_{-i} \in A_{-i}$

that a_i^2 *strictly dominates* a_i^1 if

$(a_i^2, \mathbf{a}_{-i}) \succ_i (a_i^1, \mathbf{a}_{-i})$ for every $\mathbf{a}_{-i} \in A_{-i}$

In the Prisoner's Dilemma, C strictly dominates N for both players.

Exercise 1.75
Let $\succsim_i'$ denote the partial order induced on player i's action space by her preferences over A. That is,

$$a_i^2 \succsim_i' a_i^1 \Leftrightarrow (a_i^2, \mathbf{a}_{-i}) \succsim_i (a_i^1, \mathbf{a}_{-i}) \qquad \text{for every } \mathbf{a}_{-i} \in A_{-i}$$

Show that if there exists an action profile $\mathbf{a}^*$ such that a_i^* is the unique maximal element in $(A_i, \succsim_i')$ for every player i, then $\mathbf{a}^*$ is the unique Nash equilibrium of the game.

1.3 Metric Spaces

In a metric space, attention is focused on the spatial relationships between the elements. A *metric space* is a set X on which is defined a measure of distance between the elements. To conform with our conventional notion of distance, the distance measure must satisfy certain properties. The distance between distinct elements should be positive. It should be symmetric so that it does not matter in which direction it is measured. Last, the shortest route between two distinct elements is the direct route (the triangle inequality). See figure 1.11. A distance measure with these properties is called a metric.

Formally, a *metric* on a set X is a measure that associates with every pair of points $x, y \in X$ a real number $\rho(x, y)$ satisfying the following properties:

1. $\rho(x, y) \geq 0$
2. $\rho(x, y) = 0$ if and only if $x = y$

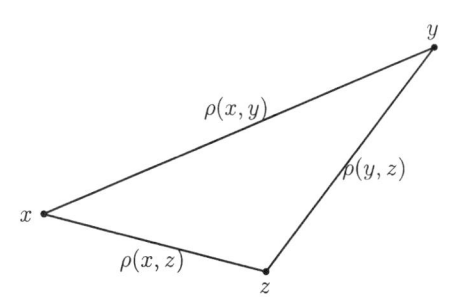

Figure 1.11
The triangle inequality

3. $\rho(x, y) = \rho(y, x)$ (symmetry)
4. $\rho(x, y) \leq \rho(x, z) + \rho(z, y)$ (triangle inequality)

A metric space is a set X together with its metric ρ and is denoted by the ordered pair (X, ρ). If the metric is understood, it may be referred to as the metric space X. The elements of a metric space are usually called *points*.

The most familiar metric space is the set $\Re$ of real numbers, where the distance between any two elements x and y is naturally measured by their difference $|x - y|$, where we take absolute values to ensure nonnegativity. There are various ways in which we can generalize this to other sets. Some of these are explored in the following example.

Example 1.54 (Consumption bundles) Consider how we might define the distance between consumption bundles. Recall that a consumption bundle **x** is a list $(x_1, x_2, \ldots, x_n)$ of quantities of different commodities, where x_i is the quantity of good i. Given two consumption bundles **x** and **y**, one way to measure the distance is to consider the difference in consumption of each commodity in turn and sum them, giving

$$\rho_1(\mathbf{x}, \mathbf{y}) = \sum_{i=1}^{n} |x_i - y_i|$$

Instead of taking the absolute value of the differences, an alternative measure would be to square the differences and take their square root

$$\rho_2(\mathbf{x}, \mathbf{y}) = \sqrt{\sum_{i=1}^{n} (x_i - y_i)^2}$$

Finally, we might consider that the commodity whose quantity has changed most should determine the distance between the commodity bundles, giving

$$\rho_\infty(\mathbf{x}, \mathbf{y}) = \max_{i=1}^{n} |x_i - y_i|$$

Each of these measures is a metric, an appropriate measure of the distance between consumption bundles.

The preceding example introduced three ways in which we might generalize the notion of distance between real numbers to the n-dimensional space $\Re^n$. Each of the three metrics for $\Re^n$ discussed in the preceding example is used in mathematical analysis. Most familiar is the *Euclidean metric*

$$\rho_2(\mathbf{x},\mathbf{y}) = \sqrt{\sum_{i=1}^n (x_i - y_i)^2}$$

which generalizes the usual notion of distance in two and three dimensional space. The third metric

$$\rho_\infty(\mathbf{x},\mathbf{y}) = \max_{i=1}^n |x_i - y_i|$$

is known as the *sup* metric. It is often more tractable in computations. We will see later that the distinctions between these three metrics are often immaterial, since the most important properties are independent of the particular metric.

Exercise 1.76
Show that $\rho(x,y) = |x - y|$ is a metric for $\Re$.

Exercise 1.77
Show that $\rho_\infty(\mathbf{x},\mathbf{y}) = \max_{i=1}^n |x_i - y_i|$ is a metric for $\Re^n$.

Analogous to linear spaces, a *subspace* of a metric space (X,ρ) is a subset $S \subset X$ in which distance is defined by the metric inherited from the space (X,ρ). For example, the consumption set $X \subseteq \Re^n_+$ (example 1.6) can be thought of as a subspace of the metric space $\Re^n$, with one of its associated metrics $\rho_1, \rho_2, \rho_\infty$.

Two further illustrations of metric spaces are given in the following examples. Other interesting examples involving sets of functions will be met in chapter 2.

Example 1.55 (Discrete metric space) Any set X can be converted into a metric space by equipping it with the discrete metric

$$\rho(x,y) = \begin{cases} 0 & \text{if } x = y \\ 1 & \text{otherwise} \end{cases}$$

Such a metric space is not very useful, except as a source of possible counterexamples.

Example 1.56 (Hamming distance) Let X be the set of all n-tuples of zeros and ones, that is,

$$\mathbf{x} = \{(x_1, x_2, \ldots, x_n) : x_i \in \{0, 1\}\}$$

The elements of **x** can be regarded as binary strings or messages of length n. The Hamming distance ρ^H between any two elements **x** and **y** in X is the number of places in which **x** and **y** differ. (X, ρ^H) is a metric space used in coding theory, where the distance between two points (strings or messages) is the number of locations in which they differ. It is also used in the theory of automata.

In any metric space the distance of a point from a set is defined to be its distance from the nearest point, that is,

$$\rho(x, S) = \inf\{\rho(x, y) : y \in S\}$$

and the distance between sets is the minimum distance between points in the sets

$$\rho(S, T) = \inf\{\rho(x, y) : x \in S, y \in T\}$$

The *diameter* of a set is the maximum distance between any points in the set

$$d(S) = \sup\{\rho(x, y) : x, y \in S\}$$

A set S is *bounded* if it has a finite diameter, that is, $d(S) < \infty$.

A thorough understanding of the structure of metric spaces requires careful study and attention to detail, which can be somewhat tedious. To understand the rest of this book, the reader needs to be able to distinguish the interior and boundary points, to know the difference between an open and a closed set, and to have some familiarity with the convergence of sequences. The following subsections outline the important properties of metric spaces. Many of these properties will be used in the book, but their use is seldom fundamental in the same way as linearity and convexity. Most of the properties are given as exercises, leaving to the reader the choice of depth in which they are studied. Some readers will be content to note the terminology and the major results. For those who want to go further, much can be learned by attempting all the exercises.

1.3.1 Open and Closed Sets

Proximity and neighborhood are fundamental to the theory of metric spaces. The set of points in close proximity to a given point x_0 is called a ball about x_0. Specifically, given any point x_0 in a metric space (X, ρ) and a distance $r > 0$, the *open ball* about x_0 of radius r is the set of points

$$B_r(x_0) = \{x \in X : \rho(x, x_0) < r\}$$

It is the set of all points that are less than r distant from x_0 (figure 1.12).

Example 1.57 (Unit balls in $\Re^2$) Open balls are not necessarily spherical, and their shape depends on the particular metric. Figure 1.13 illustrates

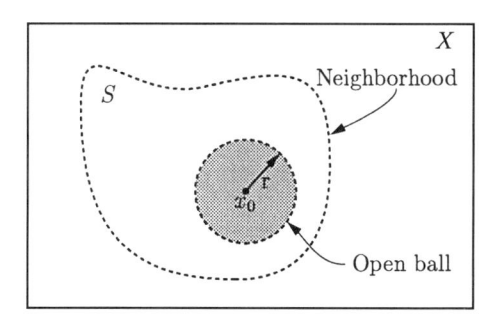

Figure 1.12
An open ball and its neighborhood

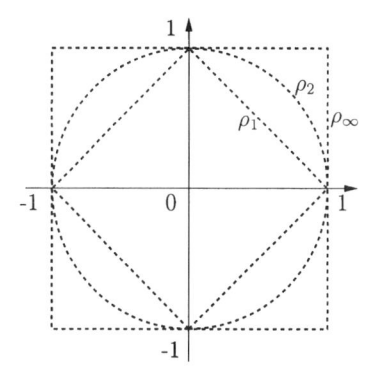

Figure 1.13
Unit balls in $\mathscr{R}^2$

the unit balls $B_r(\mathbf{0})$ associated with different metrics in the plane ($\Re^2$). The unit ball in the Euclidean metric ρ_2 is indeed circular—it is a spherical disk centered at the origin. In the sup metric, ρ_∞, the unit ball is a square. For this reason, the sup metric is sometimes called the taxicab metric. Similarly the metric ρ_1 is sometimes called the "diamond metric."

An open ball is a symmetrical neighborhood. However, we note that symmetry is not essential to the idea of proximity. The important characteristic of the neighborhood of a particular point is that *no nearby points are excluded*. Formally, any set $S \subseteq X$ is a *neighborhood* of x_0 (and x_0 is an *interior point* of S) if S contains an open ball about x_0.

The set of all interior points of a set S is called the *interior* of S, which is denoted int S. A set S is *open* if all its points are interior points, that is, $S = \text{int } S$. In an open set, every point has a neighborhood that is entirely contained in the set, so it is possible to move a little in any direction and remain within the set. We can also define interior points in terms of neighborhoods. A point $x_0 \in X$ is an interior point of $S \subseteq X$ if S contains a neighborhood of x_0. That is, x_0 is an interior point of S if S contains *all* nearby points of x_0. On the other hand, a point $x_0 \in X$ is a *boundary point* of $S \subseteq X$ if every neighborhood of x_0 contains points of S and also contains points of S^c. Each boundary point is arbitrarily close to points in S and to points outside. The *boundary* $\text{b}(S)$ of S is the set of all its boundary points. In line with common usage, the boundary delineates a set from its complement.

The *closure* $\bar{S}$ of a set S is the union of S with its boundary, that is,

$$\bar{S} = S \cup \text{b}(S)$$

Any $x \in \bar{S}$ is called a *closure point* of S. A set S is *closed* if it is equal to its closure, that is, if $S = \bar{S}$.

Remark 1.10 These concepts—balls, neighborhoods, interiors, and boundaries—generalize everyday concepts in familiar three-dimensional Euclidean space. Indeed, the theory of metric spaces is an abstraction of familiar geometry. A thorough understanding of the geometry of more abstract spaces requires an ability to "visualize" these spaces in the mind. Two- and three-dimensional analogues and diagrams like figure 1.14 are very helpful for this purpose. However, we need to bear in mind that these are only aids to understanding, and learn to rely on the definitions. There

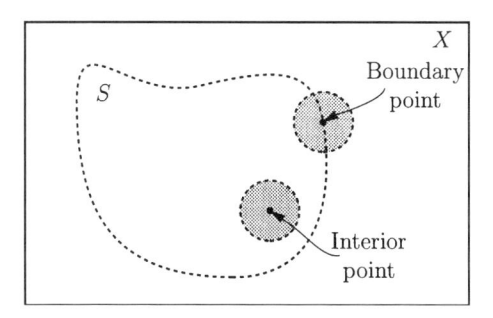

Figure 1.14
Interior and boundary points

are some obvious properties of three-dimensional Euclidean space that do not carry over to all metric spaces. These distinctions are usually explored by studying pathological examples. Since these distinctions will not bother us in this book, we will not pursue them here. However, readers should be wary of leaping to unwarranted conclusions when they encounter more general spaces.

To summarize, every set S in a metric space has two associated sets, int S and $\bar{S}$ with

int $S \subseteq S \subseteq \bar{S}$

In general, S is neither open nor closed, and both inclusions will be proper. However, if equality holds in the left hand inclusion, S is open. If equality holds in the right-hand side, S is closed. Furthermore every point $x \in S$ is *either* an interior point *or* a boundary point. A set is open if it contains no boundary points; it is closed if it contains all its boundary points. If S is open, then its complement is closed. If it is closed, its complement is open. The closure of S is the union of S with its boundary. The interior of S is comprises S minus its boundary. These important properties of open and closed sets are detailed in the following exercises.

Example 1.58 (Closed ball) Given any point x_0 in a metric space (X, ρ), the *closed ball* about x_0 of radius r,

$C_r(x_0) = \{x \in X : \rho(x, x_0) \leq r\}$

is a closed set.

Example 1.59 (Unit sphere) The boundary of the unit ball $B_1(0)$ is the set

$$S_1(0) = \{x \in X : \rho(x, 0) = 1\}$$

is called the *unit sphere*. In $\Re^2$ the unit sphere is $S_1(0) = \{x \in \Re^2 : x_1^2 + x_2^2 = 1\}$, which is the boundary of the set $B_1(0) = \{x \in \Re^2 : x_1^2 + x_2^2 < 1\}$.

Exercise 1.78
What is the boundary of the set $S = \{1/n : n = 1, 2, \ldots\}$?

Exercise 1.79
For any $S \subseteq T$,

1. int $S \subseteq$ int T
2. $\bar{S} \subseteq \bar{T}$

Exercise 1.80
A set is open if and only if its complement is closed.

Exercise 1.81
In any metric space X, the empty set $\emptyset$ and the full space X are both open and closed.

A metric space is *connected* if it cannot be represented as the union of two disjoint open sets. In a connected space the only sets that are both open and closed are X and $\emptyset$. This is case for $\Re$, which is connected. Also the product of connected spaces is connected. Hence $\Re^n$ and $\emptyset$ are the only sets in $\Re^n$ that are both open and closed.

Exercise 1.82
A metric space is connected if and only it cannot be represented as the union of two disjoint *closed* sets.

Exercise 1.83
A metric space X is connected if and only if X and $\emptyset$ are the only sets that are both open and closed.

Exercise 1.84
A subset S of a metric space is both open and closed if and only if it has an empty boundary.

1.3 Metric Spaces

Exercise 1.85
1. Any union of any collection of open sets is open. The intersection of a *finite* collection of open sets is open.
2. The union of a finite collection of closed sets is closed. The intersection of *any* collection of closed sets is closed.

Exercise 1.86
For any set S in a metric space

1. int S is open. It is the largest open set in S.
2. $\bar{S}$ is closed. It is the smallest closed set containing S.

Exercise 1.87
The interior of a set S comprises the set minus its boundary, that is,

int $S = S \backslash b(S)$

Exercise 1.88
A set is closed if and only if it contains its boundary.

Exercise 1.89
A set is bounded if and only it is contained in some open ball.

Exercise 1.90
Given an open ball $B_r(x_0)$ in a metric space, let S be a subset of diameter less than r that intersects $B_r(x_0)$. Then $S \subseteq B_{2r}(x_0)$.

Example 1.60 (Rational approximation) One concept that arises in more advanced work is the notion of a dense set. A set S is *dense* in the metric space X if $\bar{S} = X$. This means that every point in S^c is a boundary point of S. The classic example is the set of rational numbers, which is dense in the set of real numbers. Therefore there are rational numbers that are arbitrarily close to any real number. This is fundamental for computation, since it implies that any real number can be approximated to any degree of accuracy by rational numbers.

Example 1.61 (Efficient production) A production plan $\mathbf{y} \in Y$ is efficient if and only if there is no feasible plan $\mathbf{y}' \in Y$ with $\mathbf{y}' \gneq \mathbf{y}$. $\mathbf{y} \in Y$ is efficient if it impossible to produce the same output with less input, or to produce more output with the same input. Let Eff(Y) denote the set of all efficient production plans. Then

$$\text{Eff}(Y) = \{ \mathbf{y} \in Y : \mathbf{y}' \gneq \mathbf{y} \Rightarrow \mathbf{y}' \notin Y \}$$

Every interior point of the production possibility set Y is inefficient. Assume that $\mathbf{y}^0$ is a production plan in int Y. Then there exists an open ball $B_r(\mathbf{y}^0)$ about $\mathbf{y}^0$ that is contained in int Y. This ball contains a plan $\mathbf{y}' \gneq \mathbf{y}$ in $B_r(\mathbf{y}^0)$, which is feasible. Therefore $\mathbf{y} \notin \text{Eff}(Y)$. Consequently efficient production plans belong to the boundary of the production possibility set, that is, $\text{Eff}(Y) \subseteq \text{b}(Y)$. In general, $\text{Eff}(Y)$ is a proper subset of $\text{b}(Y)$. Not all boundary points are efficient.

Exercise 1.91
Show that *free disposal* (example 1.12) implies that the production possibility set has a nonempty interior.

Remark 1.11 (Topological spaces) The student of mathematical economics will sooner or later encounter a topological space. This is a generalization of a metric space, which can be explained as follows.

We remarked earlier that an open ball is a neighborhood that is symmetrical. Careful study of the preceding exercises reveals that symmetry is irrelevant to distinguishing interior from boundary points, open from closed sets. The fundamental idea is that of a neighborhood. A topological space dispenses with the measure of distance or metric. It *starts* by selecting certain subsets as neighborhoods or open sets, which is known as a *topology* for the set. This suffices to identify interior points, boundary points, closed sets, and so on, with all the properties outlined in the preceding exercises.

Any metric on a space identifies certain subsets as neighborhoods, and hence induces a topology on the space. Furthermore different metrics on a given set may lead to the same topology. For example, we will show in section 1.5 that the three metrics which we proposed for $\Re^n$ all identify the same open sets. We say they generate the same topology. Any property that does not depend on the particular metric, but on the fact that certain sets are open and others are not, is called a topological property (exercise 1.92). Continuity (section 2.3) is the most important topological property.

Exercise 1.92 (Normal space)
A topological space is said to be *normal* if, for any pair of disjoint closed sets S_1 and S_2, there exist open sets (neighborhoods) $T_1 \supseteq S_1$ and $T_2 \supseteq S_2$ such that $T_1 \cap T_2 = \varnothing$. Show that any metric space is normal.

Exercise 1.93
Let S_1 and S_2 be disjoint closed sets in a metric space. Show that there exists an open set T such that

$$S_1 \subseteq T \quad \text{and} \quad S_2 \cap \overline{T} = \emptyset$$

Remark 1.12 (Separation theorems) Many results in economic analysis are based on separation theorems in linear spaces, which will be explored extensively in section 3.9. Exercises 1.92 and 1.93 are topological separation theorems.

Relative Interiors, Open and Closed Sets

Recall that any subset X of a metric space Y is a metric space in its own right (a subspace). When dealing with subspaces, it is important to be clear to which space the topology is relative. A set $S \subseteq X \subseteq Y$ might be open as a subset of the subspace X but not open when viewed as a subset of Y. The following is a typical example from economics.

Example 1.62 In a world of two commodities, suppose that the consumer's consumption set is $X = \Re^2_+$, which is a subspace of $\Re^2$. The consumption set is a metric space in its own right with any of the metrics from $\Re^2$.

Moreover the consumption set X is open in the metric space X, but it is not open in the underlying metric space $\Re^2$. The set $S = \{x \in X : x_1 + x_2 > 1\}$ is also open in X but not in $\Re^2$ (see figure 1.15). To specify the underlying metric space, we say that S is open *relative* to X.

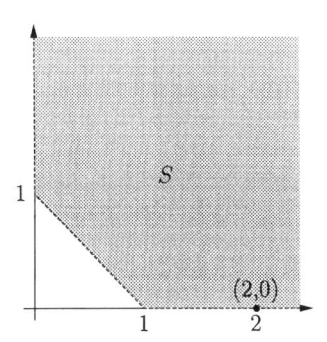

Figure 1.15
The relative topology of the consumption set

Similarly the point $(2,0)$ is an interior point *not* a boundary point of S (relative to X), since it is some distance from any point in $X \setminus S$.

Let X be a subspace of a metric space Y. $x_0 \in S$ is a boundary point of S *relative to X* if every neighborhood of x_0 contains points of S and points of $X \setminus S$. $x_0 \in S$ is an interior point of S *relative to X* if there exists some open ball $B_r(x_0) \subseteq X$ about x_0 such that $B_r(x_0) \subseteq S$. In other words, x_0 is an interior point relative to X if all points $x \in X$ that are less than r distant from x_0 are also in S. S is open relative to X if every $x \in S$ is an interior point relative to X. S is closed relative to X if it contains all its boundary points relative to X.

Exercise 1.94
In the previous example, illustrate the open ball of radius $\frac{1}{2}$ about $(2,0)$. Use the Euclidean metric ρ_2.

The following result links the metric and order structures of real numbers (see exercise 1.20).

Exercise 1.95
A set $S \subseteq \Re$ is connected if and only if it is an interval.

1.3.2 Convergence: Completeness and Compactness

Before the advent of modern calculators, high school students were taught to find square roots by a process of successive approximation. Today the pocket calculator uses a similar algorithm to provide the answer almost instantaneously, and students are no longer required to master the algorithm. This nearly forgotten algorithm is an example of iteration, which is absolutely fundamental to the practice of computation. Most practical computation today is carried out by digital computers, whose comparative advantage lies in iteration.

Pocket calculators use iteration to compute the special functions (roots, exponents, sine, cosine, and their inverses). Computers use iteration to solve equations, whether algebraic or differential. Most practical optimization procedures are based on iteration, the simplex algorithm for linear programming being a classic example. Similarly many dynamic processes are modeled as iterative processes, in which the state at each period is determined by the state in the previous period or periods. The outcome of an iterative process in a metric space is an example of a sequence. One of the most important questions to be asked of any iterative computational process is whether or not it converges to the desired answer. A necessary requirement for converging on the right answer is converging on any answer. In this section we study the general theory of convergence of sequences in a metric space.

A *sequence* in a metric space X is a list of particular elements $x^1, x^2, x^3, \ldots$ of X. We will use the notation (x^n) to denote a sequence. A sequence is finite if it is a finite list; otherwise, it is infinite. It is important to note that a sequence of points in a set is *not* a subset of the set, since it does not necessarily contain distinct elements. The set of elements in an infinite sequence may be a finite set. For example, $\{0, 1\} \subset \Re$ is the set of elements in the sequence $1, 0, 1, 0, \ldots$. The sequence $0, 1, 0, 1, \ldots$ is different sequence containing the same elements. (We will be able to give a more robust definition of a sequence in the next chapter.) Typical instances of sequences in economics include a series of observations on some economic variable (a time series), the outputs of an iterative optimization process, or moves in a game. Each of the elements x^n in a sequence (x^n) is a element in the metric space X—it may a single number (a measure of some economic quantity), an n-tuple (a consumption bundle), a set (a production possibility set), a function (a statistical estimator), or something more complicated like a whole economy or game.

Example 1.63 (Repeated game) Suppose that a set of n players repeatedly play the same game. At each stage each player i chooses from an action a_i from a set A_i. Let a_i^t denote the choice of player i at time t, and let $\mathbf{a}^t = (a_1^t, a_2^t, \ldots, a_n^t)$ denote the combined choice. The outcome of the repeated game is a sequence of actions $(\mathbf{a}^0, \mathbf{a}^1, \mathbf{a}^2, \ldots)$. If there are a finite number of stages T, the game is called a finitely repeated game, and the outcome $(\mathbf{a}^0, \mathbf{a}^1, \mathbf{a}^2, \ldots, \mathbf{a}^T)$ is a finite sequence. Otherwise, it is called an infinitely repeated game. (It is conventional to label the first stage "period 0.")

At any time t the finite sequence of past actions $(\mathbf{a}^0, \mathbf{a}^1, \mathbf{a}^2, \ldots, \mathbf{a}^{t-1})$ is called the *history* of the game to time t. The set of possible histories at time t is the product

$$H^t = \underbrace{A \times A \times \cdots \times A}_{t \text{ times}}$$

where $A = A_1 \times A_2 \times \cdots \times A_n$.

A sequence (x^n) in x converges to point $x \in X$ if the elements of (x^n) get arbitrarily close to x, so every neighborhood of x eventually contains all subsequent elements in the sequence. More precisely, the sequence (x^n) converges to x if there exists a stage N such that x^n belongs to the open ball $B_r(x)$ for all $n \geq N$. Every neighborhood of x contains all but a finite number of terms in the sequence. The point x is called the *limit* of the sequence (x^n). Convergence of sequence is often denoted by

$$x^n \to x \quad \text{or} \quad x = \lim_{n \to \infty} x^n$$

Example 1.64 The formula

$$x^0 = 2, \quad x^{n+1} = \frac{1}{2}\left(x^n + \frac{2}{x^n}\right), \qquad n = 0, 1, 2, \ldots$$

defines an infinite sequence, whose first five terms are

$(2, 1.5, 1.416666666666667, 1.41421568627451, 1.41421356237469)$

The sequence converges to $\sqrt{2}$ (example 1.103).

Exercise 1.96
If a sequence converges, its limit is unique. Therefore we are justified in talking about *the* limit of a convergent sequence.

Exercise 1.97
Every convergent sequence is bounded; that is, the set of elements of a convergent sequence is a bounded set. [Hint: If $x^n \to x$, show that there exists some r such that $\rho(x^n, x) < r$ for all n.]

Exercise 1.98
At a birthday party the guests are invited to cut their own piece of cake. The first guest cuts the cake in half and takes one of the halves. Then, each guest in turn cuts the remainder of the cake in half and eats one portion. How many guests will get a share of the cake?

Remark 1.13 (Consistent estimators) One of the principal topics of advanced econometrics concerns the asymptotic properties of estimators, that is, their behavior as the sample size becomes large. Often it is easier to analyze the limiting behavior of some econometric estimator than it is to derive its properties for any finite sample. For example, suppose that $\hat{\theta}^n$ is an estimate of some population parameter θ that is based on a sample of size n. The estimator $\hat{\theta}$ is said to be *consistent* if the sequence $(\hat{\theta}^n)$ converges to the true value θ. However, the estimator $\hat{\theta}$ is a random variable (example 2.19), which requires an appropriate measure of distance and convergence, called convergence in probability (Theil 1971, pp. 357–62).

Exercise 1.99 (Cauchy sequence)
Let (x^n) be a sequence that converges to x. Show that the points of (x^n) become arbitrarily close to one another in the sense that for every $\varepsilon > 0$ there exists an N such that

$$\rho(x^m, x^n) < \varepsilon \qquad \text{for all } m, n \geq N$$

A sequence with this property is called a Cauchy sequence.

Exercise 1.100
Any Cauchy sequence is bounded.

Exercise 1.99 showed that every convergent sequence is a Cauchy sequence; that is, the terms of the sequence become arbitrarily close to one another. The converse is not always true. There are metric spaces in which a Cauchy sequence does not converge to *an element of the space*. A *complete metric space* is one in which every Cauchy sequence is convergent. Roughly speaking, a metric space is complete if every sequence that tries to converge is successful, in the sense that it finds its limit in the space. It is a fundamental result of elementary analysis that the set $\Re$ is complete; that is, every Cauchy sequence of real numbers converges. This implies that $\Re^n$ is complete (exercise 1.211).

Basic Fact $\Re$ *is complete*.

Remark 1.14 (Cauchy convergence criterion)   The practical importance of completeness is as follows: To demonstrate that a sequence in a complete metric space is convergent, it is sufficient to demonstrate that it is a Cauchy sequence. This does not require prior knowledge of the limit. Hence we can show that an iterative process converges without knowing its limiting outcome. This is called the Cauchy convergence criterion.

Example 1.65 (Monotone sequences)   Another useful convergence criterion (in $\Re$) is monotonicity. A sequence (x^n) of real numbers is *increasing* if $x^{n+1} \geq x^n$ for all n. It is *decreasing* if $x^{n+1} \leq x^n$ for all n. A sequence

(x^n) is *monotone* if it is either increasing or decreasing. Every bounded monotone sequence of real numbers converges. This fact links the order and metric properties of $\Re$.

Exercise 1.101 (Bounded monotone sequence)
A monotone sequence in $\Re$ converges if and only if it is bounded. [Hint: If x^n is a bounded monotone sequence, show that $x^n \to \sup\{x^n\}$.]

Exercise 1.102
For every $\beta \in \Re_+$, the sequence $\beta, \beta^2, \beta^3, \ldots$ converges if and only if $\beta \leq 1$ with

$$\beta^n \to 0 \Leftrightarrow \beta < 1$$

Exercise 1.103
Show that

1. $\frac{1}{2}\left(x + \frac{2}{x}\right) \geq \sqrt{2}$ for every $x \in \Re_+$. [Hint: Consider $(x - \sqrt{2})^2 \geq 0$.]

2. the sequence in example 1.64 converges to $\sqrt{2}$.

Exercise 1.104
Extend example 1.64 to develop an algorithm for approximating the square root of any positive number.

The following exercises establish the links between convergence of sequences and geometry of sets. First, we establish that the boundary of a set corresponds to the limits of sequences of elements in the set. This leads to an alternative characterization of closed sets which is useful in applications.

Exercise 1.105
Let S be a nonempty set in a metric space. $x \in \bar{S}$ if and only if it is the limit of a sequence of points in S.

Exercise 1.106
A set S is closed if and only if the limit of every convergent sequence belongs to S.

Exercise 1.107
A closed subset of a complete metric space is complete.

A sequence (S^n) of subsets of a metric space X is nested if $S^1 \supseteq S^2 \supseteq S^2 \supseteq \cdots$.

Exercise 1.108 (Cantor intersection theorem)
Let (S^n) be a nested sequence of nonempty closed subsets of a complete metric space with $d(S^n) \to 0$. Their intersection $S = \bigcap_{n=1}^{\infty} S^n$ contains exactly one point.

Exercise 1.109 (A topological duel)
Let C be the set of all subsets of a metric space X with nonempty interior. Consider the following game with two players. Each player in turn selects a set S^n from C such that

$$S^1 \supseteq S^2 \supseteq S^2 \supseteq \cdots$$

Player 1 wins if $\bigcap_{n=1}^{\infty} S^n \neq \emptyset$. Otherwise, player S wins. Show that player 1 has a winning strategy if X is complete.

One of the most important questions that we can ask of any iterative process is whether or not it converges. It is impossible to ensure that any arbitrary sequence converges. For example, neither of the real sequences $(1, 2, 3, \ldots)$ and $(1, 0, 1, 0, \ldots)$ converges. However, the behavior of the second sequence is fundamentally different from the first. The second sequence $(1, 0, 1, 0, \ldots)$ has a convergent subsequence $(0, 0, 0, \ldots)$ consisting of every second term. (The remaining terms $(1, 1, 1, \ldots)$ form another convergent subsequence.) A metric space X is *compact* if every sequence has a convergent subsequence. A subset S of a metric space is compact if it is a compact subspace, that is, if every sequence in S has a subsequence that converges to a limit in S. Compactness is related to the earlier properties of closedness and boundedness, as detailed in the following proposition.

Proposition 1.1 *In any metric space, a compact set is closed and bounded.*

Proof Assume that S is compact. To show that S is closed, let x be any point in $\overline{S}$. There exists a sequence (x^n) in S which converges to x (exercise 1.105). Since S is compact, the sequence (x^n) converges to an element of S. Since the limit of a sequence is unique, this implies that $x \in S$, and therefore that $\overline{S} \subseteq S$. Therefore S is closed.

To show that S is bounded, we assume the contrary. Choose some $x^0 \in S$ and consider the sequence of open balls $B(x^0, n)$ for $n = 1, 2, 3, \ldots$.

If S is unbounded, no ball contains S. Therefore, for every n, there exists some point $x^n \notin B(x^0, n)$. The sequence (x^n) cannot have a convergent subsequence, contradicting the assumption that S is compact. □

In general, the converse of this proposition is false, that is, a closed and bounded set is not necessarily compact. However, the converse is true in the space which economists normally inhabit, $\Re^n$ (proposition 1.4). Also a closed subset of a compact set is compact.

Exercise 1.110
A closed subset of a compact set is compact.

Exercise 1.111
A Cauchy sequence is convergent ⇔ it has a convergent subsequence.

Actually compact spaces have a much stronger property than boundedness. A metric space X is *totally bounded* if, for every $r > 0$, it is contained in a finite number of open ball $B_r(x_i)$ of radius r, that is,

$$X = \bigcup_{i=1}^{n} B_r(x_i)$$

The open balls are said to *cover X*.

Exercise 1.112
A compact metric space is totally bounded.

Exercise 1.113
A metric space if compact if and only if it is complete and totally bounded.

This leads us toward an equivalent formulation of compactness, which is useful in many applications (e.g., in the proof of proposition 1.5). A collection $\mathscr{C}$ of subsets of a metric space X is said to *cover X* if X is contained in their union, that is,

$$X = \bigcup_{S \in \mathscr{C}} S$$

$\mathscr{C}$ is an *open cover* if all the sets S are open and a *finite cover* if the number of sets in $\mathscr{C}$ is finite. Exercise 1.112 showed that every compact set has a finite cover of open balls of a given size. In the next two exercises we show that if X is compact, every open cover has a finite subcover; that is, if

$$X = \bigcup_{S \in \mathscr{C}} S, \quad S \text{ open}$$

there exists sets $S_1, S_2, \ldots, S_n \in \mathscr{C}$ such that

$$X = \bigcup_{i=1}^{n} S_i$$

Exercise 1.114 (Lebesgue number lemma)
Let $\mathscr{C}$ be an open cover for a compact metric space X. Call a subset T "big" if is not contained in a single $S \in \mathscr{C}$, that is, if it requires more than one open set $S \in \mathscr{C}$ to cover it. Let $\mathscr{B}$ be the collection of all big subsets of X, and define $\delta = \inf_{T \in \mathscr{B}} d(T)$. Use the following steps to show that $\delta > 0$:

Step 1. $d(T) > 0$ for every $T \in \mathscr{B}$.

Step 2. Suppose, however, $\delta = \inf_{T \in \mathscr{B}} d(T) = 0$. Then, for every $n = 1, 2 \ldots$, there exists some big set T_n with $0 < d(T_n) < 1/n$.

Step 3. Construct a sequence $(x^n : x^n \in T_n)$. This sequence has a convergent subsequence $x^m \to x_0$.

Step 4. Show that there exists some $S^0 \in \mathscr{C}$ and r such that $B_r(x_0) \subseteq S_0$.

Step 5. Consider the concentric ball $B_{r/2}(x)$. There exists some N such that $x^n \in B_{r/2}(x)$ for every $n \geq N$.

Step 6. Choose some $n \geq \min\{N, 2/r\}$. Show that $T_n \subseteq B_r(x) \subseteq S^0$.

This contradicts the assumption that T_n is a big set. Therefore we conclude that $\delta > 0$.

In the previous exercise we showed that for every open covering there exists a diameter δ such that every set of smaller diameter than δ is wholly contained in at least one S. The critical diameter δ is known as a *Lebesgue number* for the $\mathscr{C}$. Thus, in a compact metric space, every open cover has a Lebesgue number. In the next exercise we use this fact to show that every compact space has a finite cover.

Exercise 1.115 (Finite cover)
Let $\mathscr{C}$ be an open cover of a compact metric space, with Lebesgue number δ. Let $r = \delta/3$.

Step 1. There exists a finite number of open balls $B_r(x_n)$ such that $X = \bigcup_{i=1}^{n} B_r(x_i)$.

Step 2. For each i, there exists some $S_i \in \mathscr{C}$ such that $B_r(x_i) \subseteq S_i$.

Step 3. $\{S_1, S_2, \ldots, S_n\}$ is a finite cover for X.

Yet another useful characterization of compactness is given in the following exercise. A collection C of subsets of a set has the *finite intersection property* if every finite subcollection has a nonempty intersection.

Exercise 1.116 (Finite intersection property)
A metric space X is compact if and only if every collection $\mathscr{C}$ of closed sets with the finite intersection property has a nonempty intersection.

We will used this property in the following form (see exercise 1.108).

Exercise 1.117 (Nested intersection theorem)
Let $S_1 \supseteq S_2 \supseteq S_3 \ldots$ be a nested sequence of nonempty compact subsets of a metric space X. Then

$$S = \bigcap_{i=1}^{\infty} S_i \neq \varnothing$$

Exercise 1.118
In any metric space the following three definitions of compactness are equivalent:

1. Every sequence has a convergent subsequence.

2. Every open cover has a finite subcover.

3. Every collection of closed subsets with the finite intersection property has a nonempty intersection.

Remark 1.15 Completeness and compactness are the fundamental properties of metric spaces. Their names are suggestive. Completeness relates to richness of the space. An incomplete space lacks certain necessary elements. On the other hand, compactness is a generalization of finiteness. Many properties, which are trivially true of finite sets, generalize readily to compact sets, and fail without compactness. A good example of the role of compactness can be found in the proof of proposition 1.5.

In the most common metric space $\Re$, the properties of completeness and compactness are closely related. Completeness of $\Re$ implies another fundamental theorem of analysis, the Bolzano-Weierstrass theorem. This theorem, which will be used in section 1.5, states that every bounded sequence of real numbers has a convergent subsequence. In turn, the Bolzano-Weierstrass theorem implies that $\Re$ is complete. The details are provided in the following exercises.

Exercise 1.119 (Bolzano-Weierstrass theorem)
Every bounded sequence of real numbers has a convergent subsequence. [Hint: Construct a Cauchy sequence by successively dividing the interval containing the bounded sequence. Then use the completeness of $\Re$.]

Exercise 1.120
Use the Bolzano-Weierstrass theorem to show that $\Re$ is complete.

The following proposition is regarded as the most important theorem in topology. We give a simplified version for the product of two metric spaces. By induction, it generalizes to any finite product. In fact the theorem is also true of an infinite product of compact spaces.

Proposition 1.2 (Tychonoff's theorem) *The product of two compact metric spaces is compact.*

Proof Let $X = X_1 \times X_2$, where X_1 and X_2 are compact. Let (x^n) be a sequence in X. Each term x^n is an ordered pair (x_1^n, x_2^n). Focusing on the first component, the sequence of elements (x_1^n) in x_1 has a convergent subsequence, with limit x_1 since X_1 is compact. Let (x^m) be the subsequence in which the first component converges. Now, focusing on the second component in the subsequence (x^m), the sequence of elements (x_2^m) has a convergent subsequence, with limit x_2. Thus (x^n) has a subsequence that converges to (x_1, x_2). □

Remark 1.16 A similar induction argument could be used to show that $\Re^n$ is complete. However, we will give a slightly more general result below, showing that any finite-dimensional linear metric space is complete.

Example 1.66 (Compact strategy space) Consider a game of n players each of whom has a strategy space S_i. The strategy space of the game is the product of the individual strategy spaces

$$S = S_1 \times S_2 \times \cdots \times S_n$$

If each of the individual player's strategy spaces S_i is compact, then the combined strategy space is compact. This is an essential component of the Nash theorem establishing the existence of an equilibrium in a noncooperative game (example 2.96).

1.4 Linear Spaces

It is a fundamental property of economic quantities that they can be added and scaled in a natural way. For example, if firm A produces y_1^A units of good 1 while firm B produces y_1^B units of the same good, the aggregate output of the two firms is $y_1^A + y_1^B$. If firm A then doubles it output while firm B reduces its output by 50 percent, their respective outputs are $2y_1^A$ and $\frac{1}{2}y_1^B$, and their combined output is $2y_1^A + \frac{1}{2}y_1^B$.

Similarly lists of economic quantities can be added and scaled item by item. For example, if $\mathbf{y} = (y_1, y_2, \ldots, y_n)$ is a production plan with net outputs y_i, $2\mathbf{y} = (2y_1, 2y_2, \ldots, 2y_n)$ is another production plan in which all the inputs and outputs have been doubled. The production plan $\frac{1}{2}\mathbf{y}$ produces half the outputs (of $\mathbf{y}$) with half the inputs. Similarly, if $\mathbf{x} = (x_1, x_2, \ldots, x_n)$ and $\mathbf{y} = (y_1, y_2, \ldots, y_n)$ are two consumption bundles, $\mathbf{x} + \mathbf{y}$ is another consumption bundle containing $x_1 + y_1$ units of good 1, $x_2 + y_2$ units of good 2 and so on. We can also combine adding and scaling. The consumption bundle $\frac{1}{2}(\mathbf{x} + \mathbf{y})$ is the average of the two bundles $\mathbf{x}$ and $\mathbf{y}$. It contains $\frac{1}{2}(x_1 + y_1)$ units of good 1. The important point is that adding, scaling, and averaging consumption bundles and production plans does not change their fundamental nature. The consequence of these arithmetic operations is simply another consumption bundle or production plan.

A set whose elements can be added and scaled in this way is called a linear space. Formally, a *linear space* is a set X whose elements have the following properties:

Additivity

For every pair of elements $\mathbf{x}$ and $\mathbf{y}$ in X, there exists another element $\mathbf{x} + \mathbf{y} \in X$, called the sum of $\mathbf{x}$ and $\mathbf{y}$ such that

1. $\mathbf{x} + \mathbf{y} = \mathbf{y} + \mathbf{x}$ (commutativity)
2. $(\mathbf{x} + \mathbf{y}) + \mathbf{z} = \mathbf{x} + (\mathbf{y} + \mathbf{z})$ (associativity)
3. there is a null element $\mathbf{0}$ in X such that $\mathbf{x} + \mathbf{0} = \mathbf{x}$

4. to every $\mathbf{x} \in X$ there exists a unique element $-\mathbf{x} \in X$ such that $\mathbf{x} + (-\mathbf{x}) = \mathbf{0}$

Homogeneity

For every element $\mathbf{x} \in X$ and number $\alpha \in \Re$, there exists an element $\alpha \mathbf{x} \in X$, called the *scalar multiple* of $\mathbf{x}$ such that

5. $(\alpha \beta) \mathbf{x} = \alpha (\beta \mathbf{x})$ (associativity)
6. $1 \mathbf{x} = \mathbf{x}$

Moreover the two operations of addition and scalar multiplication obey the usual distributive rules of arithmetic, namely

7. $\alpha (\mathbf{x} + \mathbf{y}) = \alpha \mathbf{x} + \alpha \mathbf{y}$
8. $(\alpha + \beta) \mathbf{x} = \alpha \mathbf{x} + \beta \mathbf{x}$

for all $\mathbf{x}, \mathbf{y} \in X$ and $\alpha, \beta \in \Re$.

We say that a linear space is "closed" under addition and scalar multiplication. A linear space is sometimes called a *vector space*, and the elements are called *vectors*.

This long list of requirements does not mean that a linear space is complicated. On the contrary, linear spaces are beautifully simple and possess one of the most complete and satisfying theories in mathematics. Linear spaces are also immensely useful providing one of the principal foundations of mathematical economics. The most important examples of linear spaces are $\Re$ and $\Re^n$. Indeed, the abstract notion of linear space generalizes the algebraic behavior of $\Re$ and $\Re^n$. The important requirements are additivity and homogeneity. The additional requirements such as associativity and commutativity merely ensure that the arithmetic in a linear space adheres to the usual conventions of arithmetic in $\Re$, in which the order of addition or scaling is irrelevant. More subtle examples of linear spaces include sets of functions and sets of games.

Example 1.67 ($\Re^n$) The set of all lists of n quantities, $\Re^n$, is a linear space. Each element $\mathbf{x} \in \Re^n$ is an n-tuple of real numbers, that is, $\mathbf{x} = (x_1, x_2, \ldots, x_n)$ where each $x_i \in \Re$. Clearly, if $\mathbf{y} = (y_1, y_2, \ldots, y_n)$ is another n-tuple, then

$$\mathbf{x} + \mathbf{y} = (x_1 + y_1, x_2 + y_2, \ldots, x_n + y_n)$$

is also a *n*-tuple, another list of numbers $x_i + y_i$ in $\Re^n$. Similarly

$$\alpha \mathbf{x} = (\alpha x_1, \alpha x_2, \ldots, \alpha x_n)$$

is also a *n*-tuple in $\Re^n$. To verify that $\Re^n$ satisfies the above rules of arithmetic is straightforward but tedious. For example, to verify the commutative law (rule (1)), we note that

$$\begin{aligned}\mathbf{x} + \mathbf{y} &= (x_1 + y_1, x_2 + y_2, \ldots, x_n + y_n) \\ &= (y_1 + x_1, y_2 + x_2, \ldots, y_n + x_n) \\ &= \mathbf{y} + \mathbf{x}\end{aligned}$$

Example 1.68 (Sequences) The set of all sequences of real numbers $\{x_1, x_2, \ldots\}$ is also a linear space.

Example 1.69 (Polynomials) An expression of the form $5 + 3t^2 - 220t^4 + t^7$, where $t \in \Re$ is called a *polynomial*. A general polynomial can be expressed as

$$\mathbf{x} = a_0 + a_1 t + a_2 t^2 + \cdots + a_n t^n$$

The *degree* of a polynomial is the highest power of t in its expression. $\mathbf{x}$ is of degree n, and $5 + 3t^2 - 220t^4 + t^7$ is a polynomial of degree 7. We add polynomials by adding the coefficients of corresponding terms. For example, if y is another polynomial

$$\mathbf{y} = b_0 + b_1 t + b_2 t^2 + \cdots + b_m t^m$$

their sum (supposing that $m \leq n$) is

$$\begin{aligned}\mathbf{x} + \mathbf{y} &= (a_0 + b_0) + (a_1 + b_1)t + (a_2 + b_2)t^2 + \cdots + (a_m + b_m)t^m \\ &\quad + a_{m+1} t^{m+1} + \cdots + a_n t^n\end{aligned}$$

Similarly scalar multiplication is done term by term, so

$$\alpha \mathbf{x} = \alpha a_0 + \alpha a_1 t + \alpha a_2 t^2 + \cdots + \alpha a_n t^n$$

The set of all polynomials is a linear space. Polynomials are often used in economics to provide tractable functional forms for analysis and estimation.

Example 1.70 (The space of TP-coalitional games) Recall that a TP-coalitional game (example 1.46) comprises

- a finite set of players N
- for each coalition $S \subseteq N$, a real number $w(S)$ that is called its *worth*

We use the notation $G = (N, w)$ to denote an arbitrary game among players N with coalitional worths $\{w(S) : S \subseteq N\}$.

Given any specific game $G = (N, w)$, if the worth of each coalition is multiplied by some number $\alpha \in \Re$, we obtain another coalitional game among the same set of players. We can denote this game $\alpha G = (N, \alpha w)$. Similarly, given two specific games $G_1 = (N, w_1)$ and $G_2 = (N, w_2)$, we can conceive another game among the same players in which the worth of each coalition is the sum of its worth in G_1 and G_2. That is, in the new game, the worth of each coalition is given by

$$w(S) = w_1(S) + w_2(S) \qquad \text{for every } S \subseteq N$$

We denote the construction of the new game by $G_1 + G_2 = (N, w_1 + w_2)$. We see that TP-coalitional games can be added and scaled in a natural way, so that the set of all coalitional games among a fixed set of players forms a linear space, which we denote $\mathscr{G}^N$. The null vector in this space is the null game in $(N, 0)$ in which the worth of each coalition (including the grand coalition) is zero. It is straightforward, though tedious, to verify that the space of TP-coalitional games satisfies the other requirements of a linear space.

One of the most common ways of making new linear spaces is by welding together existing spaces by taking their product. In this way we can think of $\Re^n$ as being the product of n copies of $\Re$.

Exercise 1.121
If X_1 and X_2 are linear spaces, then their product

$$X = X_1 \times X_2$$

is a linear space with addition and multiplication defined as follows:

$$(x_1, x_2) + (y_1, y_2) = (x_1 + y_1, x_2 + y_2)$$

$$\alpha(x_1, x_2) = (\alpha x_1, \alpha x_2)$$

The following standard rules of arithmetic are often used in computing with linear spaces.

Exercise 1.122
Use the definition of a linear space to show that

1. $\mathbf{x} + \mathbf{y} = \mathbf{x} + \mathbf{z} \Rightarrow \mathbf{y} = \mathbf{z}$
2. $\alpha \mathbf{x} = \alpha \mathbf{y}$ and $\alpha \neq 0 \Rightarrow \mathbf{x} = \mathbf{y}$
3. $\alpha \mathbf{x} = \beta \mathbf{x}$ and $\mathbf{x} \neq \mathbf{0} \Rightarrow \alpha = \beta$
4. $(\alpha - \beta)\mathbf{x} = \alpha \mathbf{x} - \beta \mathbf{x}$
5. $\alpha(\mathbf{x} - \mathbf{y}) = \alpha \mathbf{x} - \alpha \mathbf{y}$
6. $\alpha \mathbf{0} = \mathbf{0}$

for all $\mathbf{x}, \mathbf{y}, \mathbf{z} \in X$ and $\alpha, \beta \in \Re$.

Remark 1.17 (Real and complex linear spaces) We implicitly assumed in the preceding discussion that the "scalars" relevant to a linear space were real numbers $\alpha \in \Re$. This corresponds to physical reality of scaling consumption bundles and production plans, so it is appropriate for most applications in economics. A linear space with real scalars is called a *real linear space*. For some purposes it is necessary to extend the set of scalars to include complex numbers, giving rise to a *complex linear space*. We will encounter only real linear spaces in this book.

The consumption set and the production possibility set are not linear spaces in their own right. A linear space is symmetrical in the sense that $-\mathbf{x} \in X$ for $\mathbf{x} \in X$. Therefore a linear space must include negative quantities, which precludes the consumption set. Although the production possibility set Y includes negative (inputs) as well as positive (outputs) quantities, it is usually the case that production is irreversible. Consequently, if $\mathbf{y} \in Y$ is a feasible production plan, $-\mathbf{y}$ (which involves recovering the inputs from the outputs) is not feasible, and hence $-\mathbf{y} \notin Y$. Neither the consumption nor the production possibility set is a linear space in its own right. However, both are subsets of the linear space $\Re^n$, and they inherit many of the attributes of linearity from their parent space. The next example illustrates some aspects of linearity in the production possibility set. Some further examples of linearity in economics follow.

Example 1.71 (Production plans) We can illustrate some of the consequences of the conventional rules of arithmetic in the context of production plans. Let $\mathbf{x}$, $\mathbf{y}$, and $\mathbf{z}$ be production plans. The first rule (commutativity) states that the order of addition is irrelevant, $\mathbf{x} + \mathbf{y}$ and $\mathbf{y} + \mathbf{x}$ are the same production plan—they produce the same outputs using the same inputs. Similarly adding $\mathbf{z}$ to $\mathbf{x} + \mathbf{y}$ produces the same net outputs as adding $\mathbf{x}$ to $\mathbf{y} + \mathbf{z}$ (associativity). The null vector is zero production plan, in which all net outputs are zero (rule 3). Rule 7 states that scaling the combined production plan $\mathbf{x} + \mathbf{y}$ generates the same result as combining the scaled production plans $\alpha \mathbf{x}$ and $\alpha \mathbf{y}$.

Example 1.72 (Aggregate demand and supply) Consider an economy consisting of k consumers. Suppose that each consumer i purchases the consumption bundle $\mathbf{x}^i$. Aggregate demand $\mathbf{x}$ is the sum of the individual purchases

$$\mathbf{x} = \mathbf{x}^1 + \mathbf{x}^2 + \cdots + \mathbf{x}^k$$

where for each commodity j, the total demand x_j is the sum of the individual demands

$$x_j = x_j^1 + x_j^2 + \cdots + x_j^k$$

and x_j^i is the demand of consumer i for good j.

Suppose that there are n producers. Each produces the net output vector $\mathbf{y}^i$. Aggregate supply is the sum of the supplies of the separate firms

$$\mathbf{y} = \mathbf{y}^1 + \mathbf{y}^2 + \cdots + \mathbf{y}^n$$

Equilibrium requires that aggregate demand equal aggregate supply, that is,

$$\mathbf{x} = \mathbf{y}$$

This simple equation implies that for every commodity j, the quantity demanded by all consumers is equal to the total quantity produced, that is,

$$x_j^1 + x_j^2 + \cdots + x_j^k = y_j^1 + y_j^2 + \cdots + y_j^n$$

or $x_j = y_j$.

Example 1.73 (Constant returns to scale) It is conventionally assumed that it is possible to replicate any production process. That is, if $\mathbf{y}$ is a feasible production plan, we assume that $n \times \mathbf{y}$ is also feasible for every $n \in N$. That is, replication implies that for every $n \in \mathcal{N}$

$ny \in Y$ for every $y \in Y$

It is often further assumed that it is feasible to scale up or down, so that any positive scaling is feasible, that is,

$\alpha y \in Y$ for every $y \in Y$ and for every $\alpha > 0$ \hfill (2)

A technology is said to exhibit constant returns to scale if the production possibility set satisfies (2). Note that constant returns to scale is a *restricted form of homogeneity*, since it is limited to positive multiples.

Example 1.74 (Inflation and average prices)   Let $\mathbf{p}^t = (p_1^t, p_2^t, \ldots, p_n^t)$ be a list of the prices of the n commodities in an economy. If the economy experiences 10% inflation, the prices at time $t+1$ are 1.1 times the prices at time t, that is,

$$\mathbf{p}^{t+1} = 1.1\mathbf{p}^t = (1.1p_1^t, 1.1p_2^t, \ldots, 1.1p_n^t)$$

Comparing the prices prevailing at two different times, if $\mathbf{p}^2$ can be obtained from $\mathbf{p}^1$ merely by scaling so that the prices of all goods change at the same rate, we say that it is a general price change; that is, there is some $\alpha \in \Re$ such that $\mathbf{p}^2 = \alpha\mathbf{p}^1$. On the other hand, if the prices of different commodities change at different rates, so that $\mathbf{p}^2 \neq \alpha\mathbf{p}^1$ for any $\alpha \in \Re$, we say that relative prices have changed. A pure inflation is an example of a linear operation on the set of prices.

Even when relative prices change, we can summarize the prices prevailing at two distinct times by computing their average

$$\bar{\mathbf{p}} = \tfrac{1}{2}(\mathbf{p}^1 + \mathbf{p}^2)$$

where $\bar{p}_j = (p_j^1 + p_j^2)/2$ is the average price of good j.

1.4.1 Subspaces

A *linear combination* of elements in a set $S \subseteq X$ is a finite sum of the form

$$\alpha_1 \mathbf{x}_1 + \alpha_2 \mathbf{x}_2 + \cdots + \alpha_n \mathbf{x}_n$$

where $\mathbf{x}_1, \mathbf{x}_2, \ldots, \mathbf{x}_n \in S$ and $\alpha_1, \alpha_2, \ldots, \alpha_n \in \Re$. The *span* or *linear hull* of a set of elements S, denoted lin S, is the set of all linear combinations of elements in S, that is,

$$\text{lin } S = \{\alpha_1 \mathbf{x}_1 + \alpha_2 \mathbf{x}_2 + \cdots + \alpha_n \mathbf{x}_n : \mathbf{x}_1, \mathbf{x}_2, \ldots, \mathbf{x}_n \in S,\ \alpha_1, \alpha_2, \ldots, \alpha_n \in \Re\}$$

Exercise 1.123
What is the linear hull of the vectors $\{(1,0),(0,2)\}$ in $\Re^2$?

Example 1.75 (Coalitional games) The characteristic function w of a TP-coalitional game $(N,w) \in \mathscr{G}^N$ (example 1.70) is a linear combination of unanimity games u_T (example 1.48), that is,

$$w(S) = \sum_T \alpha_T u_T(S)$$

for every coalition $S \subseteq N$ (exercise 1.124).

Exercise 1.124
Given a fixed set of players N, each coalition $T \subseteq N$ determines a unanimity game u_T (example 1.48) defined by

$$u_T(S) = \begin{cases} 1 & \text{if } S \supseteq T \\ 0 & \text{otherwise} \end{cases}$$

1. For each coalition $S \subset N$, recursively define the *marginal value of a coalition* by

$$\alpha_i = w(i)$$

$$\alpha_S = w(S) - \sum_{T \subset S} \alpha_T$$

(Recall that $T \subset S$ means that T is a proper subset of S, i.e., $T \subseteq S$ but $T \neq S$.) Show that

$$\sum_{T \subseteq S} \alpha_T = w(S) \qquad \text{for every } S \subseteq N$$

2. Show that

$$w(S) = \sum_{T \subseteq N} \alpha_T u_T(S)$$

for every coalition $S \subseteq N$.

A subset S of a linear space X is a *subspace* of X if for every $\mathbf{x}$ and $\mathbf{y}$ in S, the combination $\alpha \mathbf{x} + \beta \mathbf{y}$ belongs to S, that is,

$$\alpha \mathbf{x} + \beta \mathbf{y} \in S \qquad \text{for every } \mathbf{x}, \mathbf{y} \in S \text{ and } \alpha, \beta \in \Re \tag{3}$$

Condition (3) combines the two principal requirements of linearity, namely

additivity $\mathbf{x} + \mathbf{y} \in S$ for every $\mathbf{x}, \mathbf{y} \in S$.

homogeneity $\alpha \mathbf{x} \in S$ for every $\mathbf{x} \in S$ and $\alpha \in \Re$.

Every subspace of a linear space is a linear space in its own right. By definition, it satisfies the principal requirements of additivity and homogeneity. These in turn imply the existence of the null vector (rule 3) and an inverse for every vector (rule 4; exercise 1.125). Furthermore, any subset of linear space will inherit the conventional arithmetic properties of its parent space, thus satisfying rules 1, 2, 5, 6, 7, and 8. Therefore, to verify that the subset S is in fact a subspace, it suffices to confirm that it satisfies the two properties of additivity and homogeneity; that is, it is closed under addition and scalar multiplication.

Exercise 1.125
If $S \subseteq X$ is a subspace of a linear space X, then

1. S contains the null vector $\mathbf{0}$
2. for every $\mathbf{x} \in S$, the inverse $-\mathbf{x}$ belongs to S

Example 1.76 (Subspaces of $\Re^3$) The subspaces of $\Re^3$ are

- the origin $\{\mathbf{0}\}$
- all lines through the origin
- all planes through the origin
- $\Re^3$ itself

Exercise 1.126
Give some examples of subspaces in $\Re^n$.

Exercise 1.127
Is $\Re^n_+$ a subspace of $\Re^n$?

Example 1.77 (Polynomials of degree less than n) Let $\mathscr{P}_n$ denote the set of all polynomials of degree less than n. Since addition and scalar multiplication cannot increase the degree of a polynomial, the set $\mathscr{P}_n$ for any n is a subspace of the set of all polynomials $\mathscr{P}$.

1.4 Linear Spaces

Exercise 1.128
The linear hull of a set of vectors S is the smallest subspace of X containing S.

Exercise 1.129
A subset S of a linear space is a subspace if and only if $S = \text{lin } S$.

Exercise 1.130
If S_1 and S_2 are subspaces of linear space X, then their intersection $S_1 \cap S_2$ is also a subspace of X.

Example 1.78 In $\Re^3$, the intersection of two distinct planes through the origin is a line through the origin. The intersection of two distinct lines through the origin is the subspace $\{\mathbf{0}\}$.

The previous exercise regarding the intersection of two subspaces can be easily generalized to any arbitrary collection of subspaces (see exercises 1.152 and 1.162). On the other hand, the union of two subspaces is not in general a subspace. However, two subspaces of a linear space can be joined to form a larger subspace by taking their sum. The *sum* of two subsets S_1 and S_2 of a linear space X is the set of all element $\mathbf{x}_1 + \mathbf{x}_2$ where $\mathbf{x}_1 \in S_1$ and $\mathbf{x}_2 \in S_2$, that is,

$$S_1 + S_2 = \{\mathbf{x} \in X : \mathbf{x} = \mathbf{x}_1 + \mathbf{x}_2, \mathbf{x}_1 \in S_1, \mathbf{x}_2 \in S_2\}$$

The sum of two sets is illustrated in figure 1.16.

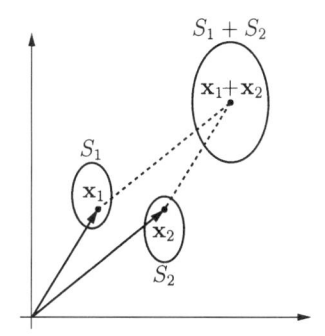

Figure 1.16
The sum of two sets

Exercise 1.131
If S_1 and S_2 are subspaces of linear space X, their sum $S_1 + S_2$ is also a subspace of X.

Exercise 1.132
Give an example of two subspaces in $\Re^2$ whose union is not a subspace. What is the subspace formed by their sum?

Linear Dependence and Independence

A element $\mathbf{x} \in X$ is *linearly dependent* on a set S of vectors if $\mathbf{x} \in \text{lin } S$, that is, if $\mathbf{x}$ can be expressed as a linear combination of vectors from S. This means that there exist vectors $\mathbf{x}_1, \mathbf{x}_2, \ldots, \mathbf{x}_n \in S$ and numbers $\alpha_1, \alpha_2, \ldots, \alpha_n \in \Re$ such that

$$\mathbf{x} = \alpha_1 \mathbf{x}_1 + \alpha_2 \mathbf{x}_2 + \cdots + \alpha_n \mathbf{x}_n \tag{4}$$

Otherwise, $\mathbf{x}$ is *linearly independent* of S. We say that a set S is linearly dependent if some vector $\mathbf{x} \in S$ is linearly dependent on the other elements of S, that is $\mathbf{x} \in \text{lin}(S \setminus \{\mathbf{x}\})$. Otherwise, the set is said to be linearly independent.

Exercise 1.133
Show that a set of vectors $S \subseteq X$ is linearly dependent if and only if there exists distinct vectors $\mathbf{x}_1, \mathbf{x}_2, \ldots, \mathbf{x}_n \in S$ and numbers $\alpha_1, \alpha_2, \ldots, \alpha_n$, not all zero, such that

$$\alpha_1 \mathbf{x}_1 + \alpha_2 \mathbf{x}_2 + \cdots + \alpha_n \mathbf{x}_n = \mathbf{0} \tag{5}$$

The null vector therefore is a nontrivial linear combination of other vectors. This is an alternative characterization of linear dependence found in some texts.

Exercise 1.134
Is the set of vectors $\{(1,1,1), (0,1,1), (0,0,1)\} \subseteq \Re^3$ linearly dependent?

Exercise 1.135 (Unanimity games)
Let $U = \{u_T : T \subseteq N, T \neq \emptyset\}$ denote the set of all unanimity games (example 1.48) playable by a given set of players N. Show that U is linearly independent.

Exercise 1.136
Every subspace S of a linear space is linearly dependent.

1.4.2 Basis and Dimension

In an arbitrary set, the elements may be completely unrelated to one another, for example, as in the set {Tonga, blue, wood, bread, 1, Pascal}. In a linear space the elements are related to one another in a precise manner, so any element can be "represented" by other elements. It is this structure that makes linear spaces especially useful.

A *basis* for a linear space X is a linearly independent subset S that spans X, that is, lin $S = X$. Since S spans X, every $\mathbf{x} \in X$ can be represented as a linear combination of elements in S. That is, for every $\mathbf{x} \in X$ there exist elements $\mathbf{x}_1, \mathbf{x}_2, \ldots, \mathbf{x}_n \in S$ and numbers $\alpha_1, \alpha_2, \ldots, \alpha_n \in \Re$ such that

$$\mathbf{x} = \alpha_1 \mathbf{x}_1 + \alpha_2 \mathbf{x}_2 + \cdots + \alpha_n \mathbf{x}_n \tag{6}$$

Furthermore, since S is linearly independent, this representation is unique (for the basis S). In this sense a basis encapsulates the whole vector space. It is a minimal spanning set.

Exercise 1.137 (Unique representation)
Show that the representation in equation (6) is unique, that is, if

$$\mathbf{x} = \alpha_1 \mathbf{x}_1 + \alpha_2 \mathbf{x}_2 + \cdots + \alpha_n \mathbf{x}_n$$

and also if

$$\mathbf{x} = \beta_1 \mathbf{x}_1 + \beta_2 \mathbf{x}_2 + \cdots + \beta_n \mathbf{x}_n$$

then $\alpha_i = \beta_i$ for all i.

Exercise 1.138
Every linear space has a basis. [Hint: Let P be the set of all linearly independent subsets of a linear space X. P is partially ordered by inclusion. Use Zorn's lemma (remark 1.5) to show that P has a maximal element B. Show that B is a basis of X.]

Example 1.79 (Standard basis for $\Re^n$) The set of *unit vectors*

$$\mathbf{e}_1 = (1, 0, 0, \ldots, 0)$$
$$\mathbf{e}_2 = (0, 1, 0, \ldots, 0)$$
$$\mathbf{e}_3 = (0, 0, 1, \ldots, 0)$$
$$\mathbf{e}_n = (0, 0, 0, \ldots, 1)$$

is called the *standard basis* for $\Re^n$. Every list $\mathbf{x} = (x_1, x_2, \ldots, x_n)$ has a unique representation in terms of the standard basis

$$\mathbf{x} = \alpha_1 \mathbf{e}_1 + \alpha_2 \mathbf{e}_2 + \cdots + \alpha_n \mathbf{e}_n$$

Expanding this representation

$$\begin{pmatrix} x_1 \\ x_2 \\ \vdots \\ x_n \end{pmatrix} = \alpha_1 \begin{pmatrix} 1 \\ 0 \\ \vdots \\ 0 \end{pmatrix} + \alpha_2 \begin{pmatrix} 0 \\ 1 \\ \vdots \\ 0 \end{pmatrix} + \cdots \alpha_n \begin{pmatrix} 0 \\ 0 \\ \vdots \\ 1 \end{pmatrix}$$

we see that

$$\alpha_1 = x_1, \alpha_2 = x_2, \ldots, \alpha_n = x_n$$

Example 1.80 (Standard basis for P) Let P denote the set of all polynomials and $B \subset P$ be the set of polynomials $\{1, t, t^2, t^3, \ldots\}$ (example 1.69). Since every polynomial

$$\mathbf{x} = a_0 + a_1 t + a_2 t^2 + a_3 t^3 + \cdots \in P$$

is a linear combination of polynomials in B, B spans P. Furthermore B is linearly independent. Therefore B is a basis for P.

Example 1.81 (Standard basis for $\mathscr{G}^N$) The set $U = \{u_T : T \subseteq N, T \neq \emptyset\}$ of unanimity game u_T (example 1.48) defined by

$$u_T(S) = \begin{cases} 1 & \text{if } S \supseteq T \\ 0 & \text{otherwise} \end{cases}$$

form a basis for the linear space $\mathscr{G}^N$ of all TP-coalitional games amongst a fixed set of players N (exercise 1.146).

Example 1.82 (Arrow-Debreu securities) The burgeoning field of financial economics (Duffie 1992; Luenberger 1997; Varian 1987) is founded on a simple linear model of financial assets. The model has two periods. In the first period ("today"), assets are bought and sold. In the second period ("tomorrow"), exactly one of a finite number S *states of the world* eventuate and the assets are realized, when their value depends on the state of the world. Formally, an *asset* or *security* is a title to receive a return or payoff r_s "tomorrow" if state s occurs. Any asset is therefore fully described by its return vector $\mathbf{r} = (r_1, r_2, \ldots, r_S)$, which details its prospective return of payoff in each state. Negative returns are allowed in certain states, in which case the holder is obligated to pay r_s. Consequently the return vectors $\mathbf{r}$ belong to the linear space $\Re^S$.

A special role is accorded *Arrow-Debreu securities*. These are (hypothetical) financial assets that pay $1 if and only if a particular state of the world occurs. The return vector of the s Arrow-Debreu security is $\mathbf{e}_s = (0, \ldots, 1, \ldots, 0)$, where the 1 occurs in the location s. Arrow-Debreu securities form a basis for the linear space $\Re^S$ of all securities. Consequently any actual financial asset $\mathbf{r}$ is equivalent to a portfolio of Arrow-Debreu assets, since the return vector $\mathbf{r}$ can be constructed from a linear combination of elementary (Arrow-Debreu) assets $\mathbf{e}_s$. For example, if there are three states of the world, the asset with return vector $(3, 4, 5)$ is equivalent to a portfolio containing 3, 4, and 5 units respectively of the Arrow-Debreu securities $(1, 0, 0)$, $(0, 1, 0)$, and $(0, 0, 1)$.

Remark 1.18 (Primary colors and the spectrum) Around 1800 the physicist Thomas Young observed that all the colors of visible spectrum could be generated by mixing three, but not less than three, pure colors. The ability to recreate the spectrum from just three colors explains human color vision and underlies the technology of color photography and television. Red, green, and blue are the usually chosen as the three *primary colors*. However, it is well known that other combinations also serve to generate the spectrum. For example, Young initially chose red, yellow, and blue as the primary colours.

Mixing colors is analogous to the linear combination of vectors in a linear space. A set of primary colors represents the spectrum in the same sense in which a basis represents a linear space. Any color can be obtained as a linear combination of the primary colors, while fewer than three primary colors is insufficient to generate the whole spectrum. Other colors can be substituted for one of the primary colors to provide a different but equally adequate spanning set.

Exercise 1.139
Is $\{(1, 1, 1), (0, 1, 1), (0, 0, 1)\}$ a basis for $\Re^3$? Is $\{(1, 0, 0), (0, 1, 0), (0, 0, 1)\}$?

A linear space which has a basis with a finite number of elements is said to be *finite dimensional*. Otherwise, the linear space is called *infinite dimensional*. In a finite-dimensional space X, every basis has the same number of elements, which is called the *dimension* of X and denoted dim X.

Exercise 1.140
Any two bases for a finite-dimensional linear space contain the same number of elements.

Exercise 1.141 (Coalitional games $\mathscr{G}^N$)
The linear space $\mathscr{G}^N$ of TP-coalitional games has dimension $2^n - 1$ where n is the number of players.

The following facts about bases and dimension are often used in practice.

Exercise 1.142
A linearly independent set in a linear space can be extended to a basis.

Exercise 1.143
Any set of $n+1$ elements in an n-dimensional linear space is linearly dependent.

The next two results highlight the dual features of a basis, namely that a basis is both

- a maximal linearly independent set
- a minimal spanning set

Exercise 1.144
A set of n elements in an n-dimensional linear space is a basis if and only if it is linearly independent.

Exercise 1.145
A set of n elements in an n-dimensional linear space X is a basis if and only if it spans X.

Exercise 1.146 (Standard basis for $\mathscr{G}^N$)
Show that the set of unanimity games $U = \{u_T : T \subseteq N, T \neq \emptyset\}$ forms a basis for the space of TP-coalitional games $\mathscr{G}^N$.

As a linear space in its own right, a subspace has a unique dimension. The dimension of a subspace cannot exceed that of it parent space. Furthermore a proper subspace of a finite-dimensional space necessarily has a lower dimension (and a smaller basis) than its parent space.

Exercise 1.147 (Dimension of a subspace)
A proper subspace $S \subset X$ of an n-dimensional linear space X has dimension less than n.

1.4 Linear Spaces

Finite-dimensional linear spaces are somewhat easier to analyze, since our intuition is contradicted less often and there are fewer pathological cases. Furthermore there are some results that only hold in finite-dimensional spaces. For the most part, however, finite- and infinite-dimensional spaces are completely analogous. Although infinite-dimensional spaces play an important role in more advanced analysis, the linear spaces encountered in this book will usually be finite dimensional.

Coordinates

The unique numbers $\alpha_1, \alpha_2, \ldots, \alpha_n$ that represent a vector **x** with respect to a given basis are called the *coordinates* of **x** relative to the basis. It is important to note that the coordinates vary with the chosen basis (exercise 1.148). Some bases offer more convenient representations than others. For example, in $\Re^n$, the coordinates of any *n*-tuple $\mathbf{x} = (x_1, x_2, \ldots, x_n)$ *with respect to the standard basis* is simply the components of **x** (example 1.79). Similarly the coordinates of a polynomial with respect to the standard basis for P (example 1.80) are the coefficients of the polynomial, since every polynomial $\mathbf{x} \in P$,

$$\mathbf{x} = a_0 + a_1 t + a_2 t^2 + a_3 t^3 + \cdots$$

Despite the simplicity of these representations, it is important to remember the distinction between an element of a linear space and its coordinates with respect to a particular basis.

Exercise 1.148
What are the coordinates of the vector $(1, 1, 1)$ with respect to the basis $\{(1, 1, 1), (0, 1, 1), (0, 0, 1)\}$? What are its coordinates with respect to the standard basis $\{(1, 0, 0), (0, 1, 0), (0, 0, 1)\}$?

Remark 1.19 (Notation) Choice of notation involves a trade-off between consistency and flexibility. We will consistently use a boldface, for example **x** and **y**, to denote elements of a linear space. We will use subscripts to denote their coordinates with respect to a particular basis (which will almost always be the standard basis), as in

$$\mathbf{x} = (x_1, x_2, \ldots, x_n)$$

The coordinates are always numbers (scalars) and will be in the ordinary face.

We will use both subscripts and superscripts to label particular elements of a linear space. Therefore $\mathbf{x}_1$ and $\mathbf{y}^A$ are particular vectors, while y_1^A is the first coordinate of $\mathbf{y}^A$ with respect to a particular basis. The alternative convention of reserving subscripts for coordinates, and using superscripts to distinguish vectors, is too inflexible for our purposes. In economic models we will often have two or more sources of labels for vectors, and the use of both subscripts and superscripts will enhance clarity. For example, we might need to label strategy choices or consumption bundles by player or agent (subscript) and also by time period (superscript).

Weighted Sums and Averages

Subspaces of a linear space X are those sets that contain arbitrary weighted sums of their elements. That is, $S \subseteq X$ is a subspace if

$$\mathbf{x} = \alpha_1 \mathbf{x}_1 + \alpha_2 \mathbf{x}_2 + \cdots + \alpha_n \mathbf{x}_n \in S$$

for all $\mathbf{x}_1, \mathbf{x}_2, \ldots, \mathbf{x}_n \in S$, and $\alpha_i \in \Re$. We have already seen how permitting arbitrary weighted sums is too general for some important sets in economics, such as consumption and production sets.

However, restricted classes of weighted sums occur frequently in economics. In production theory, it is natural to consider *nonnegative* weighted sums of production plans. In the theory of the consumer, it is appropriate to average different consumption bundles. The *weighted average* a set of elements $\{\mathbf{x}_1, \mathbf{x}_2, \ldots, \mathbf{x}_n\}$ in a linear space is a weighted sum

$$\mathbf{x} = \alpha_1 \mathbf{x}_1 + \alpha_2 \mathbf{x}_2 + \cdots + \alpha_n \mathbf{x}_n$$

in which the weights are nonnegative and sum to one, that is, $\alpha_i \geq 0$ and $\alpha_1 + \alpha_2 + \cdots + \alpha_n = 1$. Each of these restricted weighted sums characterizes a class of sets with special properties. These include affine sets, convex sets, and convex cones, whose relationships are detailed in table 1.2. We now consider each class in turn. Convex sets and cones are absolutely

Table 1.2
Classes of subset in a linear space

	$\sum \alpha_i$ unrestricted	$\sum \alpha_i = 1$
$\alpha_i \gtreqless 0$	Subspace	Affine set
$\alpha_i \geq 0$	Convex cone	Convex set

fundamental in mathematical economics and game theory, and these sections should be studied carefully. Affine sets lie midway between convex sets and subspaces. They play a less prominent role and are included here for completeness.

1.4.3 Affine Sets

Subspaces are the *n*-dimensional analogues of straight lines and planes passing through the origin. When translated so that they do not pass through the origin, straight lines, planes, and their analogues are called affine sets. The theory of affine sets closely parallels the theory of subspaces, for which they are a slight generalization. The solutions of a system of linear equations form an affine set (exercise 3.101). Affine sets also occur in the theory of simplices, which are used in general equilibrium theory and game theory.

A subset S of a linear space X is called an *affine set* if for every **x** and **y** in S the combination $\alpha \mathbf{x} + (1 - \alpha)\mathbf{y}$ belongs to S, that is,

$$\alpha \mathbf{x} + (1 - \alpha)\mathbf{y} \in S \quad \text{for every } \mathbf{x}, \mathbf{y} \in S \text{ and } \alpha \in \Re$$

For distinct **x** and **y** in X, the set of all points

$$\{\alpha \mathbf{x} + (1 - \alpha)\mathbf{y} : \alpha \in \Re\}$$

is called the *line through* **x** *and* **y** (figure 1.17). It is the straight line through **x** and **y** and extending beyond the endpoints in both directions. A set is an affine if the straight line *through* any two points remains entirely within the set. Affine sets have many synonyms, including linear manifolds, linear varieties, and flats.

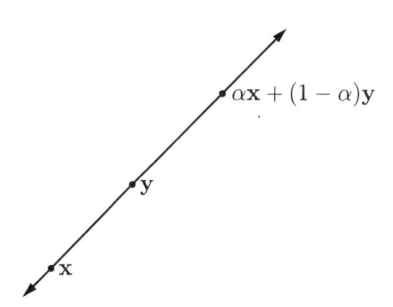

Figure 1.17
The line through **x** and **y**

For every affine set S that does not necessarily pass through the origin, there is a corresponding subspace that does. That is, there is a unique subspace V such that

$S = \mathbf{x} + V$

for some $\mathbf{x} \in S$. We say that S is parallel to V. In $\Re^3$, affine sets include planes, straight lines, and points. The following exercises formalize the relationship between affine sets and subspaces.

Exercise 1.149
In any linear space every subspace is an affine set, and every affine set containing $\mathbf{0}$ is a subspace.

Exercise 1.150
For every affine set S there is a unique subspace V such that $S = \mathbf{x} + V$ for some $\mathbf{x} \in S$.

Exercise 1.151
Let X be a linear space. Two affine subsets S and T are *parallel* if one is a translate of the other, that is,

$S = T + \mathbf{x}$ for some $\mathbf{x} \in X$

Show that the relation S is parallel to T is an equivalence relation in the set of affine subsets of X.

Example 1.83 ($\Re^2$) Let $\mathbf{x}$ and $\mathbf{y}$ be two points in $\Re^2$. The straight line through $\mathbf{x}$ and $\mathbf{y}$ is an affine set. It is a subspace if and only if the straight line passes through the origin (figure 1.18).

Example 1.84 ($\Re^3$) In $\Re^3$ the affine sets are

- $\emptyset$
- all points $\mathbf{x} \in \Re^3$
- all straight lines
- all planes
- $\Re^3$

Exercise 1.152
The intersection of any collection of affine sets is affine.

1.4 Linear Spaces

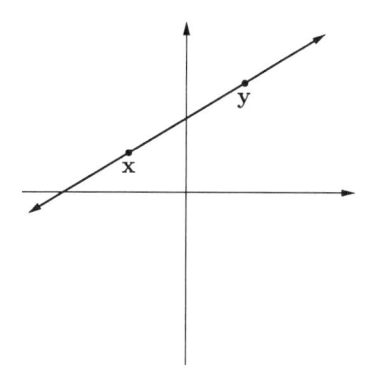

Figure 1.18
An affine set in the plane

The *dimension* of an affine set is defined as the dimension of the subspace to which it is parallel (exercise 1.150). Affine sets of dimension 0, 1, and 2 are called points, lines, and planes respectively.

The proper affine subsets of a linear space X are partially ordered by inclusion. Any maximal element of this partially ordered set is called a hyperplane. That is, a *hyperplane* is a maximal proper affine subset, the biggest possible affine set that is not the whole space. In an n-dimensional space, every $(n-1)$-dimensional affine set is a hyperplane. Lines and planes are hyperplanes in $\Re^2$ and $\Re^3$ respectively.

Exercise 1.153
Let H be a hyperplane in a linear space X. Then H is parallel to unique subspace V such that

1. $H = \mathbf{x}_0 + V$ for some $\mathbf{x}_0 \in H$
2. $\mathbf{x}_0 \in V \Leftrightarrow H = V$
3. $V \subsetneqq X$
4. $X = \text{lin}\{V, \mathbf{x}_1\}$ for every $\mathbf{x}_1 \notin V$
5. for every $\mathbf{x} \in X$ and $\mathbf{x}_1 \notin V$, there exists a unique $\alpha \in \Re$ such that $\mathbf{x} = \alpha x_1 + \mathbf{v}$ for some $\mathbf{v} \in V$

Example 1.85 (Preimputations: feasible outcomes in a coalitional game)
The set of feasible outcomes in a TP-coalitional game with transferable payoff (example 1.46)

$$X = \left\{ \mathbf{x} \in \Re^n : \sum_{i \in N} x_i = w(N) \right\}$$

is a hyperplane in $\Re^n$. In other words, it is an affine subset of dimension $n - 1$. Elements of X are sometimes called *preimputations*.

Exercise 1.154
Show that

$$X = \left\{ \mathbf{x} \in \Re^n : \sum_{i \in N} x_i = w(N) \right\}$$

is an affine subset of $\Re^n$.

Affine Combinations and Affine Hulls

Linear combinations of vectors in a linear space allowed arbitrary sums. Slightly more restrictive, an *affine combination* of vectors in a set $S \subseteq X$ is a finite sum of the form

$$\alpha_1 \mathbf{x}_1 + \alpha_2 \mathbf{x}_2 + \cdots + \alpha_n \mathbf{x}_n$$

where $\mathbf{x}_1, \mathbf{x}_2, \ldots, \mathbf{x}_n \in S$, $\alpha_1, \alpha_2, \ldots, \alpha_n \in \Re$ and $\alpha_1 + \alpha_2 + \cdots + \alpha_n = 1$.

Analogous to the linear hull, the *affine hull* of a set of vectors S, denoted aff S, is the set of all affine combinations of vectors in S, that is,

$$\begin{aligned} \text{aff } S = \{ &\alpha_1 \mathbf{x}_1 + \alpha_2 \mathbf{x}_2 + \cdots + \alpha_n \mathbf{x}_n : \\ &\mathbf{x}_1, \mathbf{x}_2, \ldots, \mathbf{x}_n \in S, \\ &\alpha_1, \alpha_2, \ldots, \alpha_n \in \Re \\ &\alpha_1 + \alpha_2 + \cdots + \alpha_n = 1 \} \end{aligned} \quad (7)$$

The affine hull of a set S is the smallest affine set containing S.

Example 1.86 The affine hull of the standard basis $\{\mathbf{e}_1, \mathbf{e}_2, \mathbf{e}_3\}$ for $\Re^3$ is the plane through the points $(1, 0, 0), (0, 1, 0), (0, 0, 1)$. It has dimension 2. By contrast, the linear hull of the three vectors $\{\mathbf{e}_1, \mathbf{e}_2, \mathbf{e}_3\}$ is the whole space $\Re^3$.

Exercise 1.155
A set S in a linear space is affine if and only if $S = \text{aff } S$.

1.4 Linear Spaces

Exercise 1.156
Is $\Re^n_+$ an affine subset of $\Re^n$?

Affine Dependence and Independence

A vector $\mathbf{x} \in X$ is *affinely dependent* on a set S of vectors if $\mathbf{x} \in \text{aff } S$, that is, if $\mathbf{x}$ can be expressed as an affine combination of vectors from S. This means that there exist vectors $\mathbf{x}_1, \mathbf{x}_2, \ldots, \mathbf{x}_n \in S$ and numbers $\alpha_1, \alpha_2, \ldots, \alpha_n \in \Re$ with $\alpha_1 + \alpha_2 + \cdots + \alpha_n = 1$ such that

$$\mathbf{x} = \alpha_1 \mathbf{x}_1 + \alpha_2 \mathbf{x}_2 + \cdots + \alpha_n \mathbf{x}_n \tag{8}$$

Otherwise, $\mathbf{x}$ is *affinely independent* of S. We say that a set S is affinely dependent if some vector $\mathbf{x} \in S$ is affinely dependent on the other elements of S, that is, $\mathbf{x} \in \text{aff}(S \setminus \{\mathbf{x}\})$. Otherwise, the set is affinely independent.

Exercise 1.157
The set $S = \{\mathbf{x}_1, \mathbf{x}_2, \ldots, \mathbf{x}_n\}$ is affinely dependent if and only if the set $\{\mathbf{x}_2 - \mathbf{x}_1, \mathbf{x}_3 - \mathbf{x}_1, \ldots, \mathbf{x}_n - \mathbf{x}_1\}$ is linearly dependent.

Exercise 1.157 implies that the maximum number of affinely independent elements in an *n*-dimensional space is $n + 1$. Moreover the maximum dimension of a proper affine subset is *n*. Analogous to exercise 1.133, we have the following alternative characterization of affine dependence.

Exercise 1.158
The set $S = \{\mathbf{x}_1, \mathbf{x}_2, \ldots, \mathbf{x}_n\}$ is affinely dependent if and only if there exist numbers $\alpha_1, \alpha_2, \ldots, \alpha_n$, not all zero, such that

$$\alpha_1 \mathbf{x}_1 + \alpha_2 \mathbf{x}_2 + \cdots + \alpha_n \mathbf{x}_n = \mathbf{0}$$

with $\alpha_1 + \alpha_2 + \cdots + \alpha_n = 0$.

Analogous to a basis, every vector in the affine hull of a set has a unique representation as an affine combination of the elements of the set.

Exercise 1.159 (Barycentric coordinates)
If $S = \{\mathbf{x}_1, \mathbf{x}_2, \ldots, \mathbf{x}_n\}$ is affinely independent, every $\mathbf{x} \in \text{aff } S$ has a unique representation as an affine combination of the elements of S; that is, there are unique scalars $\alpha_1, \alpha_2, \ldots, \alpha_n$ such that

$$\mathbf{x} = \alpha_1 \mathbf{x}_1 + \alpha_2 \mathbf{x}_2 + \cdots + \alpha_n \mathbf{x}_n$$

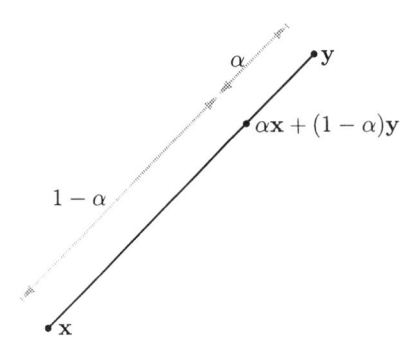

Figure 1.19
The line joining two points

with $\alpha_1 + \alpha_2 + \cdots + \alpha_n = 1$. The numbers $\alpha_1, \alpha_2, \ldots, \alpha_n$ are called the *barycentric coordinates* of **x** with respect to S.

1.4.4 Convex Sets

A subset S of a linear space X is a *convex set* if for every **x** and **y** in S, the weighted average $\alpha\mathbf{x} + (1-\alpha)\mathbf{y}$ with $0 \leq \alpha \leq 1$ belongs to S, that is,

$$\alpha\mathbf{x} + (1-\alpha)\mathbf{y} \in S \quad \text{for every } \mathbf{x}, \mathbf{y} \in S, \text{ and } 0 \leq \alpha \leq 1 \tag{9}$$

For distinct **x** and **y** in X, the set of weighted averages or convex combinations

$$\{\alpha\mathbf{x} + (1-\alpha)\mathbf{y} : 0 \leq \alpha \leq 1\}$$

is the straight line joining the two points (figure 1.19). A set is convex if the line joining any two points remains entirely within the set (figure 1.20). Note that X and $\emptyset$ are trivially convex.

In an obvious extension of the notation for an interval, we will let $[\mathbf{x}, \mathbf{y}]$ denote the line joining two points **x** and **y**, that is,

$$[\mathbf{x}, \mathbf{y}] = \{\bar{\mathbf{x}} \in X : \bar{\mathbf{x}} = \alpha\mathbf{x} + (1-\alpha)\mathbf{y}, 0 \leq \alpha \leq 1\}$$

Similarly $(\mathbf{x}, \mathbf{y})$ denotes the line joining two points, but excluding the end points, that is,

$$(\mathbf{x}, \mathbf{y}) = \{\bar{\mathbf{x}} \in X : \bar{\mathbf{x}} = \alpha\mathbf{x} + (1-\alpha)\mathbf{y}, 0 < \alpha < 1\}$$

1.4 Linear Spaces

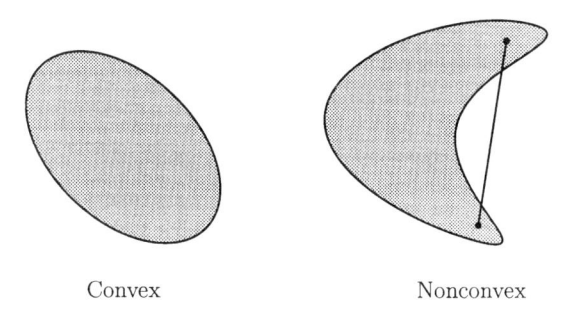

Figure 1.20
Convex and nonconvex sets

Exercise 1.160 (Intervals)
Show that the open interval (a, b) and the closed interval $[a, b]$ are both convex sets of $\Re$ with the natural order (example 1.20). The hybrid intervals $[a, b)$ and $(a, b]$ are also convex. Show that intervals are the only convex sets in $\Re$.

Example 1.87 (Consumption set)   If **x** and **y** are two consumption bundles, their weighted average $\alpha\mathbf{x} + (1 - \alpha)\mathbf{y}$ is another consumption bundle containing a weighted average of the amount of each commodity in **x** and **y**. More specifically, the consumption bundle $\frac{1}{2}\mathbf{x} + \frac{1}{2}\mathbf{y}$ contains the average of each commodity in **x** and **y**, that is,

$$\tfrac{1}{2}\mathbf{x} + \tfrac{1}{2}\mathbf{y} = (\tfrac{1}{2}x_1 + \tfrac{1}{2}y_1, \tfrac{1}{2}x_2 + \tfrac{1}{2}y_2, \ldots, \tfrac{1}{2}x_n + \tfrac{1}{2}y_n)$$

The consumption set X is a convex subset of $\Re^n_+$.

Example 1.88 (Input requirement set)   Recall that the input requirement set $V(y) \subseteq \Re^n_+$ details the inputs necessary to produce y units of a single output. Assume that $\mathbf{x}_1$ and $\mathbf{x}_2$ are two different ways of producing y. For example, $\mathbf{x}_1$ might be a capital intensive production process, whereas $\mathbf{x}_2$ might use less capital and relatively more labor. A natural question is whether it is possible to combine these two production processes and still produce y, that is does $\alpha\mathbf{x}_1 + (1 - \alpha)\mathbf{x}_2$ belong to $V(y)$. The answer is yes if $V(y)$ is a convex set. In producer theory, it is conventional to assume that $V(y)$ is convex for every output level y, in which case we say that the technology is convex.

Exercise 1.161
The core of a TP-coalitional game is convex.

Exercise 1.162
The intersection of any collection of convex sets is convex.

Example 1.89 (Slicing an egg) As an illustration of the preceding result, consider slicing an egg. An egg is a good example of a convex set. Observe that no matter in what direction we slice an egg, provided that the slices are parallel, the slices are also convex. With just a little license, we can think of slices as resulting from the intersection of two convex sets, the egg and the plane (affine set) containing the knife. Provided that the knife does not deviate from a single plane, we are guaranteed a convex slice. A banana illustrates that the converse is not true. A banana will also produces convex slices, but the banana itself is not a convex set.

As the preceding example illustrates, a set may have convex cross sections without itself being convex. This is an important distinction in producer theory.

Example 1.90 (Convex technology) The input requirement set $V(y)$ is a cross section of the production possibility set Y. It is conventional to assume that the input requirement set $V(y)$ is convex for every y. This is less restrictive than assuming that the production possibility set Y is convex. Exercise 1.162 demonstrates that

$$Y \text{ convex} \Rightarrow V(y) \text{ convex} \quad \text{for every } y$$

but the converse is not generally true. If the technology exhibits increasing returns to scale, Y is not convex although $V(y)$ may be. Technology emulates the banana rather than the egg.

Exercise 1.163
Devise a formal proof of

$$Y \text{ convex} \Rightarrow V(y) \text{ convex} \quad \text{for every } y$$

Sums and products of convex sets are also convex, as detailed in the following exercises. Convexity of a sum is used in establishing the existence of a general equilibrium in an exchange economy, while convexity of the product is used in establishing the existence of a noncooperative equilibrium of a game.

Exercise 1.164 (Sum of convex sets)
If $\{S_1, S_2, \ldots, S_n\}$ is a collection of convex subsets of a linear space X, their sum $S_1 + S_2 + \cdots + S_n$ is also a convex set.

Exercise 1.165 (Product of convex sets)
If $S_1, S_2, \ldots, S_n$ are convex subsets of the linear spaces $X_1, X_2, \ldots, X_n$, their product $S_1 \times S_2 \times \cdots \times S_n$ is a convex subset of the product space $X_1 \times X_2 \times \cdots \times X_n$.

Example 1.91 (Aggregate production possibility set) Suppose that an economy contains n producers dealing in m commodities. The technology of each producer is summarized by its production possibility set $Y^j \subset \Re^m$. Aggregate production $\mathbf{y}$ is the sum of the net outputs of each of the producers $\mathbf{y}^j$, that is,

$$\mathbf{y} = \mathbf{y}^1 + \mathbf{y}^2 + \cdots + \mathbf{y}^m$$

The set of feasible aggregate production plans, the *aggregate production possibility set*, is the sum of the individual production sets

$$Y = Y^1 + Y^2 + \cdots + Y^m \subset \Re^m$$

The aggregate net output $\mathbf{y}$ is feasible if and only if $\mathbf{y} = \mathbf{y}^1 + \mathbf{y}^2 + \cdots + \mathbf{y}^m$ and $\mathbf{y}^j \in Y^j$ for every j. A *sufficient* condition for the aggregate production possibility set Y to be convex is that each firm has a convex production set (exercise 1.164).

Exercise 1.166
S convex $\Rightarrow \alpha S$ convex for every $\alpha \in \Re$.

Exercise 1.167
If $\{S_1, S_2, \ldots, S_n\}$ is a collection of convex subsets of a linear space X, any linear combination $\alpha_1 S_1 + \alpha_2 S_2 + \cdots + \alpha_n S_n$, $\alpha_i \in \Re$ is also a convex set.

Exercise 1.168 is a useful characterization of convex sets.

Exercise 1.168
A set S is convex if and only if $S = \alpha S + (1 - \alpha)S$ for every $0 \leq \alpha \leq 1$.

Exercise 1.169
The collection of all convex subsets of a linear space ordered by inclusion forms a complete lattice.

Convex Combinations and Convex Hulls

A *convex combination* of elements in a set $S \subseteq X$ is a finite sum of the form

$$\alpha_1 \mathbf{x}_1 + \alpha_2 \mathbf{x}_2 + \cdots + \alpha_n \mathbf{x}_n$$

where $\mathbf{x}_1, \mathbf{x}_2, \ldots, \mathbf{x}_n \in S$ and $\alpha_1, \alpha_2, \ldots, \alpha_n \in \Re_+$ with $\alpha_1 + \alpha_2 + \cdots + \alpha_n = 1$. The weights α_i are nonnegative fractions between 0 and 1. In many applications the weights have a natural interpretation as proportions or probabilities. The *convex hull* of a set of vectors S, denoted conv S, is the set of all convex combinations of vectors in S, that is,

$$\text{conv } S = \{\alpha_1 \mathbf{x}_1 + \alpha_2 \mathbf{x}_2 + \cdots + \alpha_n \mathbf{x}_n :$$

$$\mathbf{x}_1, \mathbf{x}_2, \ldots, \mathbf{x}_n \in S,$$

$$\alpha_1, \alpha_2, \ldots, \alpha_n \in \Re_+$$

$$\alpha_1 + \alpha_2 + \cdots + \alpha_n = 1\}$$

See figure 1.21.

The definition of a convex set (9) requires that it contain the convex combination of any *two* elements. An equivalent criterion, which is often used in practice, requires that a convex set contains the convex combination of an arbitrary number of elements.

Exercise 1.170
A set is convex if and only if it contains all convex combinations of its elements.

Exercise 1.171
The convex hull of a set of vectors S is the smallest convex subset of X containing S.

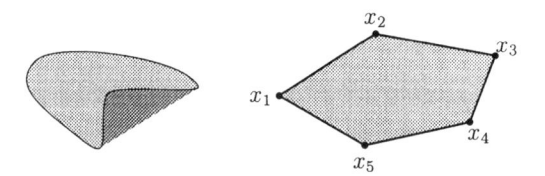

Figure 1.21
Two convex hulls

1.4 Linear Spaces

Exercise 1.172
A set S is convex if and only if $S = \text{conv } S$

A useful fact is that taking sums and convex hulls commutes. That is, the convex hull of the sum of a collection of sets is equal to the sum of their convex hulls.

Exercise 1.173
For any finite collection of sets $\{S_1, S_2, \ldots, S_n\}$,

$$\text{conv} \sum_{i=1}^{n} S_i = \sum_{i=1}^{n} \text{conv } S_i$$

[Hint: Establish the result for $n = 2$. The generalization to any finite n is immediate.]

Remark 1.20 (Shapley-Folkman theorem) Let $\{S_1, S_2, \ldots, S_n\}$ be a collection of nonempty (possibly nonconvex) subsets of an m-dimensional linear space, and let **x** belong to $\text{conv} \sum_{i=1}^{n} S_i$. Then by the previous exercise

$$\mathbf{x} = \sum_{i=1}^{n} \mathbf{x}_i$$

where $\mathbf{x}_i \in \text{conv } S_i$. The Shapley-Folkman theorem shows that all but at most m of the x_i actually belong to S_i. In this sense

$$\text{conv} \sum_{i=1}^{n} S_i \approx \sum_{i=1}^{n} S_i$$

The sum of a large number of sets is approximately convex.

This is relevant in economic models, where convexity is a common assumption. It is comforting to know that aggregation tends to convexify. Even if convexity is not appropriate for individual economic agents, convexity in the aggregate may be a reasonable approximation. For example, suppose that the sets S_i are the production possibility sets of the n producers in an economy with m commodities. $S = \sum_{i=1}^{n} S_i$ is the aggregate production possibility set (example 1.91). In a large economy with many more producers n than commodities m, the aggregate production possibility set may be reasonably convex even if the technology of individual firms is nonconvex. Proof of the Shapley-Folkman theorem requires additional tools. We will give two different proofs in chapter 3 (exercises 3.112 and 3.210).

The following example demonstrates the convexifying effect of aggregation.

Example 1.92 Let $S_i = \{0, 1\}$, $i = 1, 2, \ldots, n$ be a collection of subsets of $\Re$. Then

- $S = \sum_{i=1}^{n} S_i = \{1, 2, \ldots, n\}$
- conv S_i = the closed interval $[0, 1]$
- conv $S = [0, n]$

Any real number $\mathbf{x} \in$ conv S can be written in many ways as the sum

$$\mathbf{x} = \sum_{i=1}^{n} x_i$$

where $x_i \in [0, 1]$. Among these representations, there is a least one in which every x_i *except one* is an integer, either 0 or 1. In this sense conv $S \approx S$. For example, with $n = 3$, the number 2.25 can be written as $2.25 = 0.80 + 0.75 + 0.7$. It can also be represented as $2.25 = 1 + 1 + 0.25$.

So far our treatment of convex sets has paralleled exactly our presentation of subspaces and affine sets, which are both particular examples of convex sets. In general, however, convex sets are less regularly structured than affine sets and subspaces. Consequently there is no direct counterpart of a basis for a convex set. However, there is an analogous representation theory that we discuss in the next subsection. We then turn to some new concepts which arise in general convex sets, and some special classes of convex sets.

Dimension and Carathéodory's Theorem

The dimension of a convex set is measured in terms of its affine hull. Specifically, the *dimension* of a convex set S is defined to be the dimension of its affine hull (exercise 1.150).

Example 1.93 The affine hull of an egg is the entire three-dimensional space $\Re^3$. Hence an egg is a three-dimensional convex set. If we contemplate an egg slice of negligible dimension, its affine hull is the two-dimensional plane $\Re^2$. Hence a planar egg slice is a two-dimensional convex set.

Exercise 1.174
Suppose a producer requires n inputs to produce a single output. Assume that the technology is convex. What is the dimension of the input requirement set $V(y)$?

By definition, any element in the convex hull of a set S can be represented as a convex combination of a finite number of elements of S. In fact it is sufficient to take dim $S + 1$ distinct points. This is analogous to the representation of elements of a subspace by a basis.

Exercise 1.175 (Carathéodory's theorem)
Let S be a nonempty subset of a linear space, and let $m = \dim S = \dim \text{aff } S$. Suppose that $\mathbf{x}$ belongs to conv S so that there exist $\mathbf{x}_1, \mathbf{x}_2, \ldots, \mathbf{x}_n \in S$ and $\alpha_1, \alpha_2, \ldots, \alpha_n \in \Re_+$ with $\alpha_1 + \alpha_2 + \cdots + \alpha_n = 1$ such that

$$\mathbf{x} = \alpha_1 \mathbf{x}_1 + \alpha_2 \mathbf{x}_2 + \cdots + \alpha_n \mathbf{x}_n \tag{10}$$

1. If $n > \dim S + 1$, show that the elements $\mathbf{x}_1, \mathbf{x}_2, \ldots, \mathbf{x}_n \in S$ are affinely dependent, and therefore there exist numbers $\beta_1, \beta_2, \ldots, \beta_n$, not all zero, such that

$$\beta_1 \mathbf{x}_1 + \beta_2 \mathbf{x}_2 + \cdots + \beta_n \mathbf{x}_n = \mathbf{0} \tag{11}$$

and

$$\beta_1 + \beta_2 + \cdots + \beta_n = 0$$

2. Show that for any number t, $\mathbf{x}$ can be represented as

$$\mathbf{x} = \sum_{i=1}^{n} (\alpha_i - t\beta_i) \mathbf{x}_i \tag{12}$$

3. Let $t = \min_i \{\alpha_i / \beta_i : \beta_i > 0\}$. Show that $\alpha_i - t\beta_i \geq 0$ for every t and $\alpha_i - t\beta_i = 0$ for at least one t. For this particular t, (12) is a convex representation of $\mathbf{x}$ using only $n - 1$ elements.

4. Conclude that every $\mathbf{x} \in \text{conv } S$ can be expressed as a convex combination of at most dim $S + 1$ elements.

None   Finite   Infinite

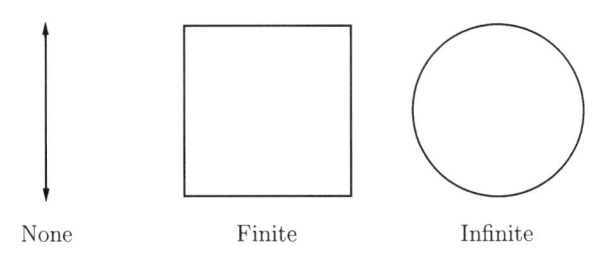

Figure 1.22
Sets with and without extreme points

Extreme Points and Faces

An element **x** in a convex set S is an *extreme point* or *vertex* of S if it does not lie on any line segment in S; that is, there are no two distinct points $\mathbf{x}_1$ and $\mathbf{x}_2$ in S such that

$$\mathbf{x} = \alpha \mathbf{x}_1 + (1 - \alpha) \mathbf{x}_2$$

for some $\alpha \in (0, 1)$. In other words, an extreme point cannot be written as the convex combination of other points in the set. A set may have no extreme points, a finite number or an infinite number of extreme points. Figure 1.22 illustrates the three cases. We use $\text{ext}(S)$ to denote the set of extreme points of S.

Exercise 1.176
If **x** is not an extreme point of the convex set $S \subseteq X$, then there exists $\mathbf{y} \in X$ such $\mathbf{x} + \mathbf{y} \in S$ and $\mathbf{x} - \mathbf{y} \in S$.

A convex subset F of a convex set S is called a *face* of S if no point of F is an interior point of a line segment whose end points are in S but not in F. Formally, if for any $\mathbf{x}, \mathbf{y} \in S$, the point $\bar{\mathbf{x}} = \alpha \mathbf{x} + (1 - \alpha) \mathbf{y} \in F$ for any $\alpha \in (0, 1)$, then **x** and **y** are also in F. An extreme point is a face containing a single point (figure 1.23).

Example 1.94   For any $c > 0$, consider the "cube"

$$C = \{\mathbf{x} = (x_1, x_2, \ldots, x_n) \in \Re^n : -c \leq x_i \leq c, i = 1, 2, \ldots, n\}$$

Each point of the form $(\pm c, \pm c, \ldots \pm c)$ is a vertex of the cube. Each point of the form $(x_1, \ldots, x_{i-1}, \pm c, x_{i+1}, \ldots, x_n)$, where x_i is fixed at $\pm c$, lies on a face of cube. A three-dimensional cube is illustrated in figure 1.24.

1.4 Linear Spaces

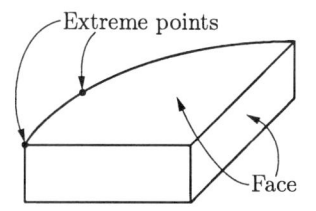

Figure 1.23
Faces and extreme points

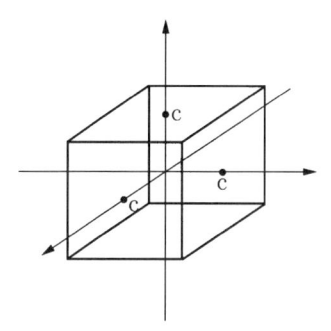

Figure 1.24
A three-dimensional cube

Exercise 1.177
1. Show that the cube $C_2 = \{\mathbf{x} \in \Re^2 : -c \leq x_1 \leq c, -c \leq x_2 \leq c\}$ in $\Re^2$ lies in the convex hull of the points $(\pm c, \pm c)$, that is,

$$C_2 \subseteq \text{conv}\left\{\begin{pmatrix} c \\ c \end{pmatrix}, \begin{pmatrix} -c \\ c \end{pmatrix}, \begin{pmatrix} c \\ -c \end{pmatrix}, \begin{pmatrix} -c \\ -c \end{pmatrix}\right\}$$

2. Suppose for any $n = 2, 3, \ldots$, that the cube $C_{n-1} \subseteq \text{conv}\{(\pm c, \pm c, \ldots, \pm c)\} \subset \Re^{n-1}$. Show that n-dimensional cube $C_n \subseteq \text{conv}\{(\pm c, \pm c, \ldots, \pm c)\} \subset \Re^n$.

3. Conclude that the only extreme points of the cube

$$C_n = \{\mathbf{x} = (x_1, x_2, \ldots, x_n) \in \Re^n : -c \leq x_i \leq c, \ i = 1, 2, \ldots, n\}$$

are the points of the form $(\pm c, \pm c, \ldots \pm c)$.

4. Show that C_n is the convex hull of its extreme points.

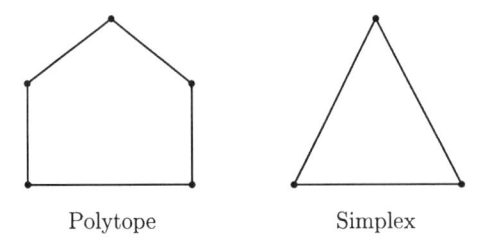

Figure 1.25
A polytope and a simplex

Exercise 1.178
If F is a face of a convex set S, then $S \setminus F$ is convex.

Exercise 1.179 will be used in chapter 3.

Exercise 1.179
Let S be a convex set in a linear space:

1. S and $\emptyset$ are faces of S.
2. The union of a collection of faces of S is a face.
3. The intersection of any nested collection of faces of S is a face.
4. The collection of all faces of S (partially ordered by inclusion) is a complete lattice.

Polytopes and Simplices

The simplest of all convex sets are convex polytopes and simplices. The convex hull of a finite set of points $E = \{\mathbf{x}_1, \mathbf{x}_2, \ldots, \mathbf{x}_n\}$ is called a *polytope* (figure 1.25). If in addition the points $\mathbf{x}_1, \mathbf{x}_2, \ldots, \mathbf{x}_n$ are affinely independent, conv E is called a *simplex* with vertices $\mathbf{x}_1, \mathbf{x}_2, \ldots, \mathbf{x}_n$. Polytopes and simplices figure prominently in optimization theory, general equilibrium theory, and game theory.

The following exercise shows that polytopes have a convenient representation in terms of their extreme points. This result is a generalization of exercise 1.177. It will be further generalized in chapter 2 to compact convex sets, a result known as the Krein-Milman theorem (exercise 3.209).

Exercise 1.180
Let E be the set of extreme points of a polytope S. Then $S = \text{conv } E$.

Exercise 1.181
Let S be the simplex generated by the finite set of points $E = \{\mathbf{x}_1, \mathbf{x}_2, \ldots, \mathbf{x}_n\}$. Show that each of the vertices $\mathbf{x}_i$ is an extreme point of the simplex.

Simplices are the most elementary of convex sets and every convex set is the union of simplices. For this reason results are often established for simplices and then extended to more general sets (example 1.100, exercise 1.229). The dimension of a simplex with n vertices is $n - 1$. Since the vertices of a simplex are affinely independent, each element in a simplex has a *unique* representation as a convex combination of the vertices (exercise 1.159). The coefficients in this representation are called the *barycentric coordinates* of the point.

Example 1.95 (Standard simplex in $\Re^n$) The *standard* or *unit simplex* in $\Re^n$ is the $(n-1)$-dimensional convex hull of the unit vectors $\mathbf{e}_1, \mathbf{e}_2, \ldots, \mathbf{e}_n$, that is,

$$\Delta^{n-1} = \text{conv}\{\mathbf{e}_1, \mathbf{e}_2, \ldots, \mathbf{e}_n\}$$

Elements $\mathbf{x}$ of Δ^{n-1} are nonnegative vectors in $\Re^n$ whose components sum to one, that is,

$$\Delta^{n-1} = \left\{ \mathbf{x} \in \Re^n : x_i \geq 0 \text{ and } \sum_{i=1}^{n} x_i = 1 \right\}$$

Each component x_i is a fraction between 0 and 1. Standard simplices provide a natural space for the weights in convex combinations and for probability distributions (example 1.98). The one-dimensional simplex is a line, the two-dimensional simplex is a triangle, and the three-dimensional simplex is a tetrahedron (figure 1.26).

Exercise 1.182
Every n-dimensional convex set contains an n-dimensional simplex.

The following examples give some impression of the utility of simplices in economics and game theory.

Example 1.96 (Sectoral shares) One of the most striking features of economic development is the changing sectoral composition of output and employment, where the predominance of economic activity shifts

100   Chapter 1   Sets and Spaces

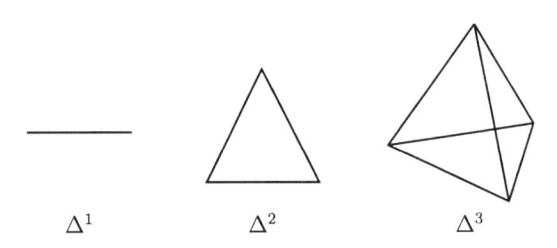

Δ^1 Δ^2 Δ^3

Figure 1.26
Some simplices

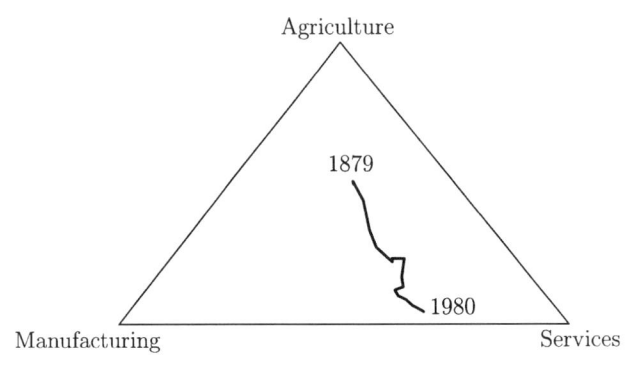

Figure 1.27
Illustrating the changing sectoral distribution of employment in the United States

from agriculture to manufacturing, and then to services. This transition can be illustrated graphically by plotting sectoral shares in the two-dimensional unit simplex, where each point represents the respective shares of agriculture, manufacturing and services.

Figure 1.27 illustrates this structural change in the United States over the period 1879 to 1980. In 1879, 50 percent of the workforce were engaged in agriculture and mining. By 1953, this had declined to 12 percent, falling to 4.5 percent in 1980. For nearly a century, employment growth was shared by both manufacturing and services. Recently, however, manufacturing employment has also declined as the United States moves inexorably toward a "service economy." By 1980 two out of every three workers were employed in the service sector. Both the rapid decline in agriculture and the subsequent emergence of the service economy are graphically evident in figure 1.27.

Example 1.97 (Imputations: reasonable outcomes in a coalitional game) The outcome $\mathbf{x} \in X$ of a TP-coalitional game is an allocation of the available sum $w(N)$ among the players so that

$$X = \left\{ \mathbf{x} \in \Re^n : \sum_{i \in N} x_i = w(N) \right\}$$

Each player i receives x_i. Assuming that the players are rational, it seems reasonable to assume that no player will agree to an outcome that is inferior to that which she can obtain acting alone. Each player will insist that $x_i \geq v(\{i\})$. The presumption of *individual rationality* requires that

$$x_i \geq v(\{i\}) \quad \text{for every } i \in N$$

Any feasible, individually rational outcome in a coalitional game is called an *imputation*. Typically the set of imputations

$$I = \{\mathbf{x} \in X : x_i \geq v(\{i\})\} \quad \text{for every } i \in N$$

is an $(n-1)$-dimensional simplex in $\Re^n$.

Consider the three-player game

$w(\{1\}) = 10 \quad w(\{1,2\}) = 50$
$w(\{2\}) = 20 \quad w(\{1,3\}) = 60 \quad w(\{1,2,3\}) = 100$
$w(\{3\}) = 30 \quad w(\{2,3\}) = 70$

The set of imputations is

$$I = \{\mathbf{x} \in \Re^3 : x_1 \geq 10, x_2 \geq 20, x_3 \geq 30, x_1 + x_2 + x_3 = 100\} \tag{13}$$

This is illustrated by the dark shaded area in the left-hand panel of figure 1.28, which is a two-dimensional simplex. The larger lightly shaded area comprises all nonnegative allocations.

A more concise pictorial representation of the set of imputations can be obtained by projecting the two-dimensional simplex onto the plane from a suitable viewpoint. This gives us a planar representation of the set imputations, which is illustrated by the dark shaded area in the right-hand panel of figure 1.28. Each of the vertices is labeled with one of the players. The payoff to each player is measured from the baseline opposite the vertex corresponding to the player. Each point in the simplex has the property that the sum of its coordinates is a constant $v(N)$, which is the sum available for distribution.

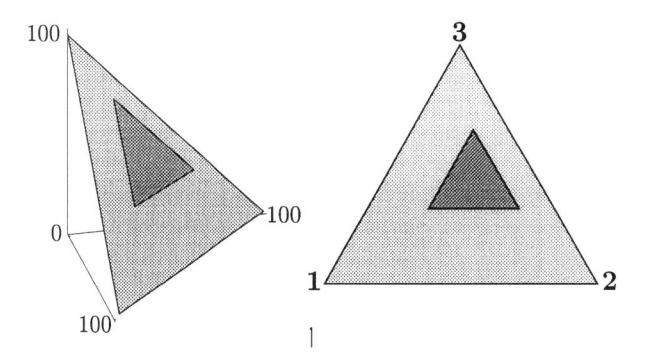

Figure 1.28
Outcomes in a three-player cooperative game

Exercise 1.183
The set of imputations of an essential TP-coalitional game (N, w) is an $(n-1)$-dimensional simplex in $\Re^n$.

Example 1.98 (Mixed strategies) In a finite strategic game, each player i has a set $S_i = (s_1, s_2, \ldots, s_m)$ of possible strategies. Each element of $s_j \in S_i$ is called a *pure strategy*. In a static strategic game (example 1.2 and section 1.2.6), each pure strategy corresponds to an action so that $S_i = A_i$. In a dynamic game (example 1.63), a pure strategy s_j may be a sequence of actions to be carried out by player i. It is often advantageous for a player to choose her strategy randomly in order to keep her opponent guessing. Such a random choice is called a mixed strategy. Formally, a *mixed strategy* for player i is a probability distribution over her set of pure strategies. That is, a mixed strategy is a set of probability weights $\mathbf{p} = (p_1, p_2, \ldots, p_m)$, where p_j is the probability attached to pure strategy s_j. Since $\mathbf{p}$ is a probability distribution

- $0 \leq p_j \leq 1, j = 1, 2, \ldots, m$
- $\sum_{j=1}^{m} p_j = 1$

Each mixed strategy $\mathbf{p}$ corresponds to a point in the unit simplex Δ^{m-1}. Therefore the set of mixed strategies is the $(m-1)$-dimensional unit simplex.

Example 1.99 (Rock–Scissors–Paper) If one player plays a pure strategy in *Rock–Scissors–Paper* (exercise 1.5), this can always be exploited by the

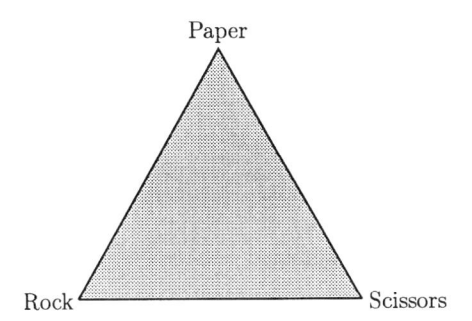

Figure 1.29
Mixed strategies in Rock–Scissors–Paper

other player. Therefore it is necessary to play a mixed strategy. One particular mixed strategy is $\sigma = (\frac{1}{2}, \frac{1}{3}, \frac{1}{6})$ which involves playing "Rock" with probability $\frac{1}{2}$, "Scissors" with probability $\frac{1}{3}$, and "Paper" with probability $\frac{1}{6}$. The set Σ of all mixed strategies is the two-dimensional unit simplex (figure 1.29). The vertices represent the pure strategies, while the edges represent mixtures of two of the three strategies. Any point in the interior of the simplex involves a mixture of all three strategies—it is called a *completely mixed strategy*. We show later that the mixed strategy $(\frac{1}{3}, \frac{1}{3}, \frac{1}{3})$ in which the player chooses each action with equal probability is the unique equilibrium of the game.

Example 1.100 (The price simplex) Let $(\hat{p}_1, \hat{p}_2, \ldots, \hat{p}_m)$ denote the prices of the m goods in a general equilibrium model. Sometimes it is convenient to *normalize* the prices by dividing each price by the sum of all prices, defining the normalized price

$$p_i = \frac{\hat{p}_i}{\sum_{j=1}^{m} \hat{p}_j}$$

This normalization preserves relative prices and has the consequence that the normalized prices p_i always sum to one. Therefore the normalized price vectors are contained in the $(m-1)$-dimensional unit simplex

$$\Delta^{m-1} = \left\{ \mathbf{p} \in \Re_+^m : \sum_{i=1}^{m} p_i = 1 \right\}$$

We will use this normalization in proving the existence of a general equilibrium in an exchange economy (example 2.95).

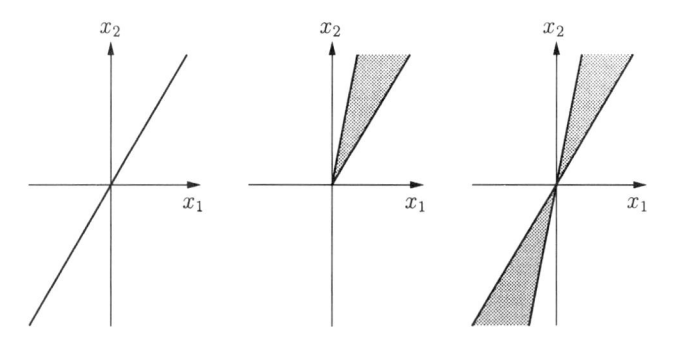

Figure 1.30
Some cones in $\mathscr{R}^2$

1.4.5 Convex Cones

A subset S of a linear space X is a *cone* if $\alpha S \subseteq S$ for every $\alpha \geq 0$, that is,

$$\alpha \mathbf{x} \in S \quad \text{for every } \mathbf{x} \in S \text{ and } \alpha \geq 0$$

This a slight relaxation of the homogeneity requirement of a linear space. If, in addition, S is convex, it is called a *convex cone* (figure 1.30). Note that every cone contains $\mathbf{0}$, which is called the vertex.

Exercise 1.184
Give examples of

1. a cone that is not convex
2. a convex set that is not a cone
3. a convex cone

Example 1.101 (Constant returns to scale) A production technology exhibits *constant returns to scale* if any feasible production plan y remains feasible when it is scaled up or down, that is,

$$\alpha \mathbf{y} \in Y \quad \text{for every } \mathbf{y} \in Y \text{ and } \alpha \geq 0$$

In other words, the technology exhibits constant returns to scale if the production possibility set Y is a cone.

Convex cones provide a slight generalization of subspaces. A subspace of linear space is a subset that is closed under addition and scalar multiplication. A convex cone is a slightly broader class of a set that is closed under addition and *nonnegative* multiplication. That is, S is a convex cone if

$$\alpha \mathbf{x} + \beta \mathbf{y} \in S \quad \text{for every } \mathbf{x}, \mathbf{y} \in S \text{ and } \alpha, \beta \in \Re_+$$

Compare this with equation (3) defining a subspace. This alternative characterization of a convex cone is established in exercise 1.186.

Exercise 1.185
Show that set $\Re_+^n$ is a cone in $\Re^n$.

Exercise 1.186
A subset S of a linear space is a convex cone if and only if

$$\alpha \mathbf{x} + \beta \mathbf{y} \in S \quad \text{for every } \mathbf{x}, \mathbf{y} \in S \text{ and } \alpha, \beta \in \Re_+$$

Exercise 1.187
A set S is a convex cone if and only if

1. $\alpha S \subseteq S$ for every $\alpha \geq 0$
2. $S + S \subseteq S$

Convex cones arise naturally in economics, where quantities are required to be nonnegative. The set of nonnegative prices vectors is a convex cone ($\Re_+^n$) and the production possibility set is often assumed to be a convex cone (example 1.102).

Example 1.102 (Convex technology) Among the typical assumptions on technology cited by Debreu (1959, pp. 41–42) are

additivity $Y + Y \subseteq Y$

constant returns to scale $\alpha Y \subseteq Y$ for every $\alpha \geq 0$

Additivity requires that production processes be independent. Together, these conventional assumptions imply that the production possibility set Y is a *convex* cone. In general, convexity is too stringent a requirement to demand of the technology.

Exercise 1.188
Another conventional (and trivial) assumption on technology cited by Debreu (1959, p. 41) is $\mathbf{0} \in Y$, which he calls the *possibility of inaction*. Show that the three assumptions

convexity Y is convex
additivity $Y + Y \subseteq Y$
possibility of inaction $\mathbf{0} \in Y$

together imply that the technology exhibits constant returns to scale.

Exercise 1.189 (Superadditive games)
A natural assumption for TP-coalitional games is superadditivity, which requires that coalitions cannot lose through cooperation. Specifically, a TP-coalitional game is *superadditive* if

$$w(S \cup T) \geq w(S) + w(T)$$

for all distinct coalitions S, T, $S \cap T = \emptyset$. Show that the set of superadditive games forms a convex cone in $\mathscr{G}^N$ (example 1.70).

Analogous to convex sets exercises 1.162 and 1.164, cones are preserved through intersection and addition.

Exercise 1.190
If $\{S_1, S_2, \ldots, S_n\}$ is a collection of cones in a linear space X, then

- their intersection $\bigcap_{i=1}^{n} S_i$
- their sum $S_1 + S_2 + \cdots + S_n$

are also cones in X.

Nonnegative Linear Combinations and Conic Hulls

A *nonnegative linear combination* of elements in a set $S \subseteq X$ is a finite sum of the form

$$\alpha_1 \mathbf{x}_1 + \alpha_2 \mathbf{x}_2 + \cdots + \alpha_n \mathbf{x}_n$$

where $\mathbf{x}_1, \mathbf{x}_2, \ldots, \mathbf{x}_n \in S$ and $\alpha_1, \alpha_2, \ldots, \alpha_n \in \Re_+$. The *conic hull* of a set of vectors S, denoted cone S, is the set of all nonnegative combinations of vectors in S, that is,

$$\text{cone } S = \{\alpha_1 \mathbf{x}_1 + \alpha_2 \mathbf{x}_2 + \cdots + \alpha_n \mathbf{x}_n :$$
$$\mathbf{x}_1, \mathbf{x}_2, \ldots, \mathbf{x}_n \in S,$$
$$\alpha_1, \alpha_2, \ldots, \alpha_n \in \Re_+\}$$

See figure 1.31.

1.4 Linear Spaces

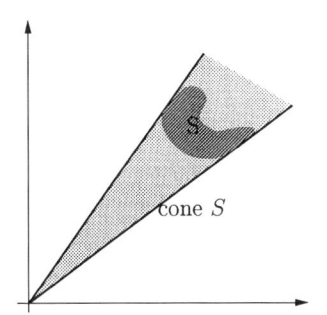

Figure 1.31
The conic hull of S

Example 1.103 (Linear production model) One of simplest models of production begins with the assumption that there are a finite number of basic *activities* or production plans $\mathbf{y}_1, \mathbf{y}_2, \ldots, \mathbf{y}_m$. These basic activities can be operated independently at constant returns to scale, that is,

- $\mathbf{y}_i + \mathbf{y}_j \in Y$ for all $i, j = 1, 2, \ldots, m$
- $\alpha \mathbf{y}_i \in Y$ for all $i = 1, 2, \ldots, m$ and $\alpha > 0$

so that the production possibility set Y is a convex cone (example 1.102). In fact the production possibility set Y is precisely the conic hull of the basic activities, that is,

$$Y = \text{cone}\{\mathbf{y}_1, \mathbf{y}_2, \ldots, \mathbf{y}_m\}$$

Exercise 1.191
Suppose that a firm's technology is based on the following eight basic activities:

$\mathbf{y}_1 = (-3, -6, 4, 0)$

$\mathbf{y}_2 = (-7, -9, 3, 2)$

$\mathbf{y}_3 = (-1, -2, 3, -1)$

$\mathbf{y}_4 = (-8, -13, 3, 1)$

$\mathbf{y}_5 = (-11, -19, 12, 0)$

$\mathbf{y}_6 = (-4, -3, -2, 5)$

$\mathbf{y}_7 = (-8, -5, 0, 10)$

$\mathbf{y}_8 = (-2, -4, 5, -2)$

which can be operated independently at any scale. The aggregate production possibility set is

$Y = \text{cone}\{\mathbf{y}_1, \mathbf{y}_2, \mathbf{y}_3, \mathbf{y}_4, \mathbf{y}_5, \mathbf{y}_6, \mathbf{y}_7, \mathbf{y}_8\}$

1. Show that it is impossible to produce output without using any inputs, that is,

$\mathbf{y} \in Y, \mathbf{y} \geq \mathbf{0} \Rightarrow \mathbf{y} = \mathbf{0}$

This is called the *no-free-lunch property*.

2. Show that Y does not exhibit *free disposal* (exercise 1.12).

3. Show that activities $\mathbf{y}_4$, $\mathbf{y}_5$, $\mathbf{y}_6$, and $\mathbf{y}_8$ are inefficient. (Compare with $\mathbf{y}_2$, $\mathbf{y}_1$, $\mathbf{y}_7$, and $\mathbf{y}_3$ respectively.)

4. Show that activities $\mathbf{y}_1$ and $\mathbf{y}_2$ are inefficient. (Compare with a combination of $\mathbf{y}_3$ and $\mathbf{y}_7$.)

5. Specify the set of efficient production plans.

The following results are analogous to those for convex hulls.

Exercise 1.192
The conic hull of a set of vectors S is the smallest convex cone in X containing S.

Exercise 1.193
A set S is a convex cone if and only if $S = \text{cone } S$.

Carathéodory's Theorem Again

As a convex set, the dimension of a convex cone S is defined to be the dimension of its affine hull. However, since every cone contains $\mathbf{0}$, its affine hull is in fact a subspace. Hence the dimension of a convex cone is the dimension of its linear hull, the maximum number of linearly independent elements which it contains. By Carathéodory's theorem, any point in a convex cone S can be represented as a convex combination of $\dim S + 1$ distinct points in S. In fact, because of its tighter structure, $\dim S$ points suffices for a convex cone. This has an interesting implication for the linear production model (example 1.104).

Exercise 1.194 (Carathéodory's theorem for cones)
Let S be a nonempty subset of a linear space and let $m = \dim \text{cone } S$. For every $\mathbf{x} \in \text{cone } S$, there exist $\mathbf{x}_1, \mathbf{x}_2, \ldots, \mathbf{x}_n \in S$ and $\alpha_1, \alpha_2, \ldots, \alpha_n \in \Re_+$ such that

$$\mathbf{x} = \alpha_1 \mathbf{x}_1 + \alpha_2 \mathbf{x}_2 + \cdots + \alpha_n \mathbf{x}_n \tag{14}$$

1. If $n > m = \dim \text{cone } S$, show that the elements $\mathbf{x}_1, \mathbf{x}_2, \ldots, \mathbf{x}_n \in S$ are linearly dependent and therefore there exist numbers $\beta_1, \beta_2, \ldots, \beta_n$, not all zero, such that

$$\beta_1 \mathbf{x}_1 + \beta_2 \mathbf{x}_2 + \cdots + \beta_n \mathbf{x}_n = \mathbf{0}$$

2. Show that for any number t, $\mathbf{x}$ can be represented as

$$\mathbf{x} = \sum_{i=1}^{n} (\alpha_i - t\beta_i) \mathbf{x}_i$$

3. Let $t = \min_i \{\alpha_i / \beta_i : \beta_i > 0\}$. Show that $\alpha_i - t\beta_i \geq 0$ for every t and $\alpha_i - t\beta_i = 0$ for at least one t. For this particular t, (14) is a nonnegative representation of $\mathbf{x}$ using only $n-1$ elements.

4. Conclude that every $\mathbf{x} \in \text{cone } S$ can be expressed as a nonnegative combination of at most $\dim S$ elements.

Example 1.104 In the linear production model (example 1.103), the production possibility set $Y = \text{cone}\{\mathbf{y}_1, \mathbf{y}_2, \ldots, \mathbf{y}_m\}$ is a subset of $\Re^n$ where is n is the number of commodities. Assume that $m > n$. Exercise 1.194 implies that every feasible production plan $\mathbf{y} \in Y$ can be obtained with at most n basic processes.

The preceding exercise can be extended to arbitrary convex sets, providing an alternative proof of exercise 1.175. This illustrates a common technique called *homogenization*, in which a result is first established for convex cones (which are easier) and then extended to arbitrary convex sets.

Exercise 1.195
Let S be a nonempty subset of a linear space, and let $m = \dim S = \dim \text{aff } S$. Consider the set

$$\tilde{S} = \left\{ \begin{pmatrix} \mathbf{x} \\ 1 \end{pmatrix} : \mathbf{x} \in S \right\}$$

illustrated in figure 1.32.

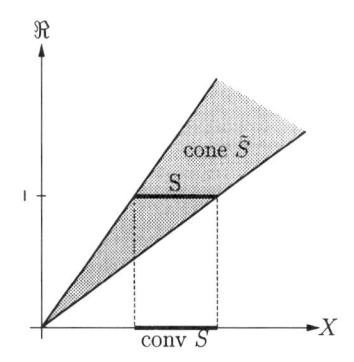

Figure 1.32
Carathéodory's theorem for cones

1. Show that dim cone $\tilde{S} = \dim S + 1$.

2. For every $\mathbf{x} \in \text{conv } S$, there exists $m+1$ points $\mathbf{x}_1, \mathbf{x}_2, \ldots, \mathbf{x}_{m+1} \in S$ such that

$$\mathbf{x} \in \text{conv}\{x_1, x_2, \ldots, x_{m+1}\}$$

1.4.6 Sperner's Lemma

Suppose that a simplex S is partitioned into a finite collection of subsimplices $S_1, S_2, \ldots, S_k \subseteq S$ so that $S = \bigcup S_i$. If no further restriction is placed on this collection, the subsimplices may overlap or intersect in the middle of a face, as illustrated in figure 1.33. A simplicial partition precludes arbitrary intersections. That is, a *simplicial partition* of a simplex is a partition into finitely many simplices such that either any two simplices are disjoint or they have a common face as their intersection (figure 1.34).

Let S be a simplex with vertices $\{\mathbf{x}_1, \mathbf{x}_2, \ldots, \mathbf{x}_n\}$. Suppose that S is simplicially partitioned, and let V denote the set of all vertices of the subsimplices. Assign to each vertex $\mathbf{x} \in V$ a label $1, 2, \ldots, n$. Such an

Figure 1.33
Invalid intersections of subsimplices

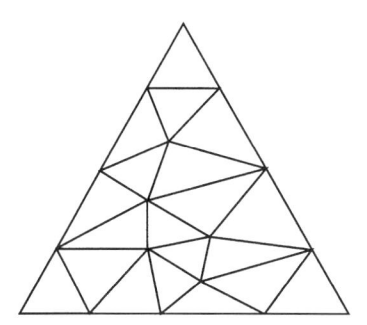

Figure 1.34
A simplicial partition

assignment is called an *admissible labeling* provided that

- each vertex of the original simplex S retains its own label and
- each vertex on a face of S receives a label corresponding to one of the vertices of that face

If one of the subsimplices has a complete set of labels $1, 2, \ldots, n$, then we say that the subsimplex is completely labeled. Surprisingly, every admissibly labeled simplicial partition has at least one completely labeled subsimplex, a profound result known as Sperner's lemma.

An admissibly labeled simplicial partition of a two-dimensional simplex is illustrated in figure 1.36. Each of the original vertices retains its own label. Each vertex along a face of the original simplex is assigned the label of one of the vertices of the face, while the labels assigned to the interior points are quite arbitrary. The shaded subsimplex has a complete set of labels 1, 2, and 3. Since the labels in the interior were assigned arbitrarily, the existence of a completely labeled subsimplex seems implausible in general. Suppose that the label of the vertex 2 is changed to 3. Then another subsimplex becomes completely labeled. On the other hand, suppose that it is changed to 1. Then three subsimplices are completely labeled. Sperner's lemma asserts that there is always an odd number of completely labeled subsimplices. It is will be used to prove the Brouwer fixed point theorem (theorem 2.6) in chapter 2.

Proposition 1.3 (Sperner's lemma) *An admissibly labeled simplicial partition of a simplex always contains at least one subsimplex that is completely labeled.*

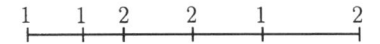

Figure 1.35
An admissibly labeled partition of a one-dimensional simplex

Proof Let S be an n-dimensional simplex. We proceed by induction on the dimension of the simplex, with one- and two-dimensional cases serving as a model for the general case.

For $n = 1$, let the two vertices of the one-dimensional simplex be labeled 1 and 2. An admissibly labeled simplicial partition divides the line joining $\mathbf{x}_1$ and $\mathbf{x}_2$ into segments (figure 1.35) with vertices labeled 1 or 2. A segment may have no, one, or two vertices labeled 1. Let c denote the number of segments with just one vertex labeled 1, and let d denote the number of segments with both vertices labeled 1. The total number of 1 vertices, *counted segment by segment*, is $c + 2d$. But, interior vertices have been counted twice in this total, since each interior vertex is shared by two segments. Let a denote the number of interior 1 vertices. There is a single boundary 1 vertex, $\mathbf{x}_1$. Therefore the previous count must be equal to $2a + 1$. That is,

$$2a + 1 = c + 2d$$

which implies that c is necessarily odd. If a segment has just one vertex labeled 1, the other vertex must be labeled 2—such a segment is completely labeled. We conclude that there are an odd number of completely labeled segments.

For $n = 2$, let S be the two-dimensional simplex generated by the points $\mathbf{x}_1, \mathbf{x}_2, \mathbf{x}_3$. Create an admissibly labeled simplicial partition (figure 1.36). Call a side of a subsimplex *distinguished* if it carries both the labels 1 and 2. A subsimplex may have none, one, or two distinguished sides. (Why are three distinguished sides impossible? See exercise 1.196.) Let c denote the number of subsimplices with one distinguished side and d denote the number of subsimplices with two distinguished sides. The total number of distinguished sides, counted simplex by simplex, is $c + 2d$. But every interior distinguished side is shared by two subsimplices, and therefore has been included twice in preceding total. Let a denote the number of interior distinguished sides and b the number of distinguished sides on the boundary. The previous count must be equal to $2a + b$, that is,

$$2a + b = c + 2d$$

1.4 Linear Spaces

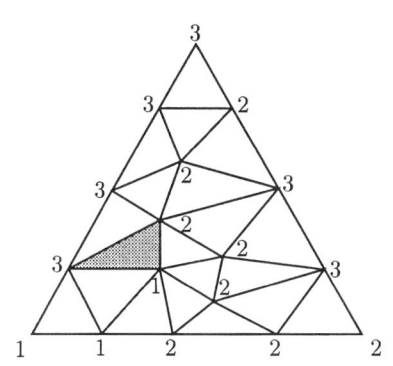

Figure 1.36
An admissibly labeled simplicial partition

Every distinguished side on the boundary is a completely labeled subsimplex of one-dimensional simplex. We have just shown that b is odd, and therefore c must also be odd. A subsimplex with precisely one distinguished side is completely labeled. We conclude that there are an odd number of completely labeled subsimplices.

For $n > 2$, assume every admissibly labeled simplicial partition of an $(n-1)$-dimensional simplex contains an odd number of completely labeled subsimplices. Let Λ be an admissibly labeled simplicial subdivision of an n-dimensional simplex S. Call an $(n-1)$-dimensional face of a subsimplex *distinguished* if it carries all the labels $1, 2, \ldots, n-1$. For each n-dimensional subsimplex $T \in \Lambda$, there are three possibilities (exercise 1.196):

- T has no distinguished faces.
- T has one distinguished face.
- T has two distinguished faces.

Let c denote the number of subsimplices with just one distinguished face and d denote the number of subsimplices with two distinguished faces. The total number of distinguished faces, counted simplex by simplex, is $c + 2d$. But, in a simplicial partition, every interior distinguished face is shared by two subsimplices, and therefore has been included twice in preceding total. Let a denote the number of interior distinguished faces and b the number of distinguished faces on the boundary. The previous count must be equal to $2a + b$, that is,

$$2a + b = c + 2d$$

Every distinguished face on the boundary is a completely labeled subsimplex of an $(n-1)$-dimensional simplex. By assumption, there are an odd number of completely labeled subsimplices on the boundary. That is, b is odd and therefore c must also be odd. A subsimplex with precisely one distinguished face is completely labeled. We conclude that there are an odd number of completely labeled subsimplices.

Since we have established the result for $n = 1$ and $n = 2$, we conclude that every admissibly labeled simplicial partition of an n-dimensional simplex has an odd number of completely labeled subsimplices. In particular, since zero is not an odd number, there is at least one completely labeled subsimplex. □

Exercise 1.196
Why can a subsimplex have no more than two distinguished faces?

1.4.7 Conclusion

Linear spaces and their subsets—affine sets, convex sets, and cones—are the natural domain of the typical objects of economic analysis, such as consumption bundles, production plans and financial portfolios, and TP-coalitional games. Linearity reflects our physical ability to combine and scale these basic objects into new objects of the same type.

A subspace is a subset that is a linear space in its own right, meeting the twin requirements of linearity, namely additivity and homogeneity. Affine sets, convex sets, and cones are subsets that retain some (but not all) of the properties of their underlying spaces. Affine and convex sets satisfy relaxed additivity requirements but not homogeneity. On the other hand, a cone satisfies a relaxed homogeneity condition (without additivity). A convex cone therefore satisfies relaxed forms of both additivity and homogeneity, and is therefore almost but not quite a subspace. In chapter 3 we will note a similar relationship among linear, convex, and homogeneous functions. Another way of distinguishing among subspaces, affine sets, convex sets, and cones is to consider the different types of weighted sum which they embrace, as detailed in table 1.2.

1.5 Normed Linear Spaces

We now explore sets that are simultaneously linear spaces and metric spaces. Any linear space can be made into a metric space by equipping it with a metric. However, to permit the fruitful interaction of the algebraic and geometric structure, it is desirable that the metric or distance function respect the linearity of the space. To achieve the necessary consistency between the algebraic and geometric structure of the space, we derive the metric from another measure, the norm, which respects the linearity of the space.

For any linear space X, a *norm* (denoted $\|\mathbf{x}\|$) is a measure of the size of the elements satisfying the following properties:

1. $\|\mathbf{x}\| \geq 0$
2. $\|\mathbf{x}\| = 0$ if and only if $\mathbf{x} = 0$
3. $\|\alpha \mathbf{x}\| = |\alpha| \|\mathbf{x}\|$ for all $\alpha \in \Re$
4. $\|\mathbf{x} + \mathbf{y}\| \leq \|\mathbf{x}\| + \|\mathbf{y}\|$ (triangle inequality)

A norm on a linear space X induces a metric on X in which the distance between any two elements is given by the norm of their difference

$$\rho(\mathbf{x}, \mathbf{y}) = \|\mathbf{x} - \mathbf{y}\|$$

Note how linearity is used in defining the metric. A linear space together with a norm is called a *normed linear space*. It is a special metric space with a rich interaction of the algebraic and geometric structures. In this section, we highlight some of the features of this interaction which will be useful later in the book.

Exercise 1.197
Show that the metric $\rho(\mathbf{x}, \mathbf{y}) = \|\mathbf{x} - \mathbf{y}\|$ satisfies the properties of a metric, and hence that a normed linear space is a metric space.

Example 1.105 (Production plans) A production plan $\mathbf{y}$ is a list of the net outputs of various goods and services $(y_1, y_2, \ldots, y_n)$, where y_i is the net output of commodity i. How could we measure the "size" of a production plan?

One suggestion would be to sum (or average) the net outputs of all the goods and services, as in $\sum_{i=1}^{n} y_i$ or $(\sum_{i=1}^{n} y_i)/n$. However, recognizing that some of the components will be negative (inputs), it would be more appropriate to take their absolute values. Therefore one possible measure of size is

$$\|\mathbf{y}\|_1 = \sum_{i=1}^{n} |y_i|$$

Another way to compensate for the negativity of some components (inputs) would be to square the individual measures, as in

$$\|\mathbf{y}\|_2 = \sqrt{\sum_{i=1}^{n} y_i^2}$$

This measure, which is analogous to the standard deviation, gives greater weight to larger quantities. Both of these measures qualify as norms, although verifying the triangle inequality for $\|\mathbf{y}\|_2$ is a nontrivial exercise.

Another candidate for a measure of size would be to focus on one particular component, "the output," and measure the size of the production plan by the quantity of this output produced. For example, assume that good n is regarded as the principal output. Could we measure the size of the production plan $\mathbf{y}$ by the quantity y_n? I am afraid not. The measure $\|\mathbf{y}\| = |y_i|$ does not satisfy the requirements of a norm, since $\|\mathbf{y}\| = 0$ does not imply that $\mathbf{y} = \mathbf{0}$. Unfortunately, as researchers are only too aware, it is possible to consume inputs and produce no outputs. This measure does not induce a metric on the production possibility set.

There is a related measure which does qualify as a norm (exercise 1.198) and which induces an appropriate metric. This measure uses the size of the largest component (input or output) as the measure of the size of the production plan, as in

$$\|\mathbf{y}\|_\infty = \max_{i=1}^{n} |y_i|$$

Each of these norms $\|\mathbf{y}\|_1$, $\|\mathbf{y}\|_2$ and $\|\mathbf{y}\|_\infty$ induces one of the standard metrics on $\Re^n$.

Exercise 1.198
Show that $\|\mathbf{y}\|_\infty$ satisfies the requirements of a norm on $\Re^n$.

Exercise 1.199
Show that the average of the net outputs $(\sum_{i=1}^{n} y_i)/n$ does not satisfy the requirements of a norm on the production possibility set.

Example 1.106 (Euclidean space) The Euclidean norm $\|\mathbf{x}\|_2$ generalizes the conventional notion of the length of a vector in two and three dimensional space. In the plane ($\Re^2$), the Euclidean norm is an expression of the theorem of Pythagoras that in a right angle triangle, the square of the

1.5 Normed Linear Spaces

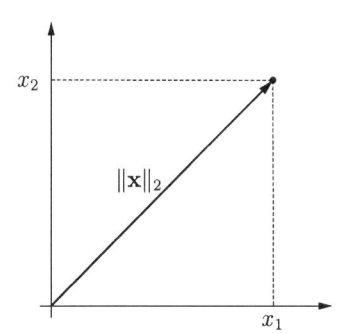

Figure 1.37
The theorem of Pythagorus

length of the hypotenuse is equal to the sum of the squares of the other two sides

$$\|\mathbf{x}\|^2 = |x_1|^2 + |x_2|^2$$

as illustrated figure 1.37. In $\Re^3$, the length of the vector $\mathbf{x} = (x_1, x_2, x_3)$ is

$$\|\mathbf{x}\| = \sqrt{x_1^2 + x_2^2 + x_3^2}$$

Example 1.107 (The space l_∞) Instead of the static choice of a consumption or production plan at a single point in time, consider the problem of choosing a path of consumption over a lifetime. For simplicity, assume that there is a single commodity and let $x_t \in \Re$ denote the consumption of the commodity in period t. Moreover, to avoid the problem of uncertainty regarding the time of death, let us assume that the consumer lives forever. (Alternatively, assume that the decision maker is a social planner concerned with future as well as current generations.) A consumption plan is an infinite sequence of instantaneous consumptions $\mathbf{x} = (x_1, x_2, \ldots)$. The consumption set X is the set of all such infinite sequences of real numbers, that is,

$$X = \{(x_1, x_2, \ldots, x_t, \ldots) : x_t \in \Re\}$$

which is a linear space (example 1.68).

For the moment it is convenient not to exclude negative consumption in particular periods. However, consumption in any period is typically bounded by the available resources; that is, there exists some K such that $|x_t| \le K$ for every t.

The set of all bounded sequences of real numbers $\mathbf{x} = (x_1, x_2, \ldots)$ is a natural setting for the study of simple dynamic models in economics. Equipped with the norm

$$\|\mathbf{x}\| = \sup_i |x_i|$$

it is a normed linear space, which is denoted l_∞. In this norm the magnitude of any consumption plan is the absolute size of the largest consumption planned at any time. l_∞ and related normed linear spaces are now commonplace in dynamic economic models (e.g., Sargent 1987; Stokey and Lucas 1989).

Exercise 1.200
Prove the following useful corollary of the triangle inequality: for any $\mathbf{x}, \mathbf{y}$ in a normed linear space

$$\|\mathbf{x}\| - \|\mathbf{y}\| \le \|\mathbf{x} - \mathbf{y}\|$$

The preceding corollary of the triangle inequality implies that the norm converges along with a sequence, as detailed in the following exercise.

Exercise 1.201
Let $\mathbf{x}_n \to \mathbf{x}$ be a convergent sequence in a normed linear space. Then

$$\|\mathbf{x}_n\| \to \|\mathbf{x}\|$$

Furthermore the norm respects the linearity of the underlying space.

Exercise 1.202
Let $\mathbf{x}_n \to \mathbf{x}$ and $\mathbf{y}_n \to \mathbf{y}$ be convergent sequences in a normed linear space X. The sequence $(\mathbf{x}_n + \mathbf{y}_n)$ converges to $\mathbf{x} + \mathbf{y}$, and $\alpha \mathbf{x}_n$ converges to $\alpha \mathbf{x}$. [Hint: Use the triangle inequality.]

The following corollary will be used in chapter 3.

Exercise 1.203
If S and T are subsets of a normed linear space with

- S closed and
- T compact

then their sum $S + T$ is closed.

Exercise 1.204

Let c_0 be the set of all sequences of real numbers converging to zero, that is,

$$c_0 = \{(x_n) : x_n \in \Re \text{ and } x_n \to 0\}$$

Is c_0 a subspace of l_∞?

Example 1.108 (Geometric series) Given a sequence of $\mathbf{x}_1, \mathbf{x}_2, \mathbf{x}_3, \ldots$ of elements in a normed linear space, their sum

$$\mathbf{x}_1 + \mathbf{x}_2 + \mathbf{x}_3 + \cdots$$

is called a *series*. If the sequence has only finite number of elements, then the series has a finite number of terms and is called a finite series. Otherwise, it is an infinite series with an infinite number of terms. What meaning can we attach to such an infinite sum?

Given an infinite series, we can define the sequence of partial sums whose nth term $\mathbf{s}_n$ is the sum of the first n terms of the series

$$\mathbf{s}_n = \mathbf{x}_1 + \mathbf{x}_2 + \cdots + \mathbf{x}_n$$

If the sequence $\mathbf{s}_1, \mathbf{s}_2, \ldots$ converges to some $\mathbf{s}$, we say that the series $\mathbf{x}_1 + \mathbf{x}_2 + \mathbf{x}_3 + \cdots$ converges, and we call $\mathbf{s}$ the sum of the infinite series, that is,

$$\mathbf{s} = \mathbf{x}_1 + \mathbf{x}_2 + \mathbf{x}_3 + \cdots$$

In the special case where each term $\mathbf{x}_n$ in the sequence is a constant multiple of the previous term ($\mathbf{x}_n = \beta \mathbf{x}_{n-1}$), their sum is called a *geometric series*, which can be written as

$$\mathbf{x} + \beta \mathbf{x} + \beta^2 \mathbf{x} + \cdots$$

where $\mathbf{x} = \mathbf{x}_1$. A geometric series converges if and only if $|\beta| < 1$ (exercise 1.205), and the limit (infinite sum) is

$$\mathbf{s} = \mathbf{x} + \beta \mathbf{x} + \beta^2 \mathbf{x} + \cdots = \frac{\mathbf{x}}{1-\beta}$$

Exercise 1.205

Show that the infinite geometric series $\mathbf{x} + \beta \mathbf{x} + \beta^2 \mathbf{x} + \cdots$ converges provided that $|\beta| < 1$ with

$$\mathbf{x} + \beta \mathbf{x} + \beta^2 \mathbf{x} + \cdots = \frac{\mathbf{x}}{1-\beta}$$

Exercise 1.206
Show that
$$1 + \frac{1}{2} + \frac{1}{4} + \frac{1}{8} + \frac{1}{16} + \cdots = 2$$

Example 1.109 (Present value) Frequently in economics we have to evaluate future income streams recognizing that future income is worth less than current income. For example, in a repeated game (example 1.63) a particular strategy profile will give rise to a sequence of payoffs to each of the players. Typically we evaluate this sequence by its present value, discounting future payoffs to compensate for the delay in their receipt. To be specific, suppose that a particular strategy will generate a constant payoff of x per round for some player. Suppose further that the player discounts future payments at the rate of β per period, so that x dollars to be received in the next period is worth as much as βx dollars in the current period. Then the *present value* of the income stream is a geometric series

$$\text{present value} = x + \beta x + \beta^2 x + \beta^3 x + \cdots$$

Provided that the player discounts future payoffs ($\beta < 1$), the present value is finite and equal to the sum of the series, that is,

$$\text{present value} = \frac{x}{1 - \beta}$$

Exercise 1.207
What is the present value of n periodic payments of x dollars discounted at β per period?

A special feature of a normed linear space is that its structure or geometry is uniform throughout the space. This can be seen in the special form taken by the open balls in a normed linear space. Recall that the open ball about $\mathbf{x}_0$ of radius r is the set

$$B_r(\mathbf{x}_0) = \{\mathbf{x} \in X : \|\mathbf{x} - \mathbf{x}_0\| < r\}$$

By linearity, this can be expressed as

$$B_r(\mathbf{x}_0) = \{\mathbf{x}_0 + \mathbf{x} : \|\mathbf{x}\| < r\}$$

The *unit ball* B is the open ball about 0 of radius 1

$$B = \{\mathbf{x} : \|\mathbf{x}\| < 1\}$$

It is the set of all elements of norm less than 1. Any open ball can be expressed in terms of the unit ball as follows:

$B_r(\mathbf{x}_0) = \mathbf{x}_0 + rB$

That is, any open ball in a normed linear space is simply a translation and scaling of the unit ball. Therefore many important properties of a normed linear space are related to the shape of its unit ball. Figure 1.13 illustrates the unit ball in the plane ($\Re^2$) for some different norms.

The uniform structure of a normed linear space enables the following refinement of exercise 1.93.

Exercise 1.208
Let S_1 and S_2 be disjoint closed sets in a normed linear space with S_1 compact. There exists a neighborhood U of $\mathbf{0}$ such that

$(S_1 + U) \cap S_2 = \emptyset$

Completeness is one of the most desirable properties of a metric space. A complete normed linear space is called a *Banach space*. Almost all the spaces encountered in mathematical economics are Banach spaces.

Exercise 1.209
Let X, Y be Banach spaces. Their product $X \times Y$ with norm

$\|(\mathbf{x}, \mathbf{y})\| = \max\{\|\mathbf{x}\|, \|\mathbf{y}\|\}$

is also a Banach space.

The natural space of economic models is $\Re^n$, the home space of consumption and production sets, which is a typical finite-dimensional normed linear space. In these spaces the interaction between linearity and topology is most acute, and many of the results obtained above can be sharpened. The most important results are summarized in the following proposition.

Proposition 1.4 *Any finite-dimensional normed linear space has the following properties:*

- *It is complete.*
- *All norms are equivalent.*
- *Any subset is compact if and only if it is closed and bounded.*

Let us examine each of these properties in turn. Given the fundamental convenience of completeness, it is very comforting to know that every finite-dimensional normed linear space is complete, in other words, a Banach space. Some of the analytical difficulties of infinite-dimensional spaces arises from the fact that they may be incomplete.

Two norms are equivalent if they generate the same topology, that is if they have the same open and closed sets. In a finite-dimensional normed linear space, the identity of neighborhoods and limits transcends any specific norm associated with the space. In particular, this means that convergence of a sequence is invariant to the choice of norm. Essentially the geometry of all finite-dimensional linear spaces is the same. In this sense, there is only *one* finite-dimensional normed linear space, and $\Re^n$ is a suitable incarnation for this space.

In the previous section, we established (proposition 1.1) that every compact set in a metric space is closed and bounded. Proposition 1.4 shows that the converse is true in a finite-dimensional normed linear space. This is extremely useful in practice, since it provides two simple criteria for identifying compact sets. Typically it is straightforward to show that a set is closed and bounded and hence to conclude that it is compact.

These three important properties of finite-dimensional normed linear spaces (proposition 1.4) are established in the following exercises (1.211, 1.213, and 1.215). All three properties rest on the interplay of two fundamental ideas:

- The spanning of a finite-dimensional linear space by a basis.
- The completeness of the real numbers $\Re$.

These exercises highlight the powerful interaction of algebra (linearity) and geometry in a normed linear space.

One implication of linear independence for geometry is summarized in the following key lemma, which is used in each of the exercises 1.211, 1.213, and 1.215 and also in chapter 3. Roughly speaking, this lemma states that it is impossible to represent arbitrarily small vectors as large linear combinations of linearly independent vectors.

Lemma 1.1 *Let* $S = \{\mathbf{x}_1, \mathbf{x}_2, \ldots, \mathbf{x}_n\}$ *be a linearly independent set of vectors in a normed linear space (of any dimension). There exists some constant $c > 0$ such that for every* $\mathbf{x} \in \text{lin } S$

$$\|\mathbf{x}\| \geq c(|\alpha_1| + |\alpha_2| + \cdots + |\alpha_n|)$$

where $\mathbf{x} = \alpha_1 \mathbf{x}_1 + \alpha_2 \mathbf{x}_2 + \cdots + \alpha_n \mathbf{x}_n$.

Exercise 1.210
To prove lemma 1.1, assume, to the contrary, that for every $c > 0$ there exists $\mathbf{x} \in \text{lin}\{\mathbf{x}_1, \mathbf{x}_2, \ldots, \mathbf{x}_n\}$ such that

$$\|\mathbf{x}\| < c \left(\sum_{i=1}^{n} |\alpha_i| \right)$$

where $\mathbf{x} = \alpha_1 \mathbf{x}_1 + \alpha_2 \mathbf{x}_2 + \cdots + \alpha_n \mathbf{x}_n$. Show that this implies that

1. there exists a sequence $(\mathbf{x}^m)$ with $\|\mathbf{x}^m\| \to 0$
2. there exists a subsequence converging to some $\mathbf{x} \in \text{lin}\{\mathbf{x}_1, \mathbf{x}_2, \ldots, \mathbf{x}_n\}$
3. $\mathbf{x} \neq \mathbf{0}$ contradicting the conclusion that $\|\mathbf{x}^m\| \to 0$

This contradiction proves the existence of a constant $c > 0$ such that

$$\|\mathbf{x}\| \geq c(|\alpha_1| + |\alpha_2| + \cdots + |\alpha_n|)$$

for every $\mathbf{x} \in \text{lin } S$.

Exercise 1.211 (Every finite-dimensional space is complete)
Let $(\mathbf{x}^m)$ be a Cauchy sequence in a normed linear space X of dimension n. Let $\{\mathbf{x}_1, \mathbf{x}_2, \ldots, \mathbf{x}_n\}$ be a basis for X. Each term $\mathbf{x}^m$ has a unique representation

$$\mathbf{x}^m = \alpha_1^m \mathbf{x}_1 + \alpha_2^m \mathbf{x}_2 + \cdots + \alpha_n^m \mathbf{x}_n$$

1. Using lemma 1.1, show that each sequence of scalars α_i^m is a Cauchy sequence in $\Re$ and hence converges to some $\alpha_i \in \Re$.
2. Define $\mathbf{x} = \alpha_1 \mathbf{x}_1 + \alpha_2 \mathbf{x}_2 + \cdots + \alpha_n \mathbf{x}_n$. Show that $\mathbf{x} \in X$ and that $\mathbf{x}^m \to \mathbf{x}$.
3. Conclude that every finite-dimensional normed linear space is complete.

Exercise 1.212 (Equivalent norms)
Two norms $\|\mathbf{x}\|_a$ and $\|\mathbf{x}\|_b$ on a linear space are equivalent if there are positive numbers A and B such that for all $\mathbf{x} \in X$,

$$A\|\mathbf{x}\|_a \leq \|\mathbf{x}\|_b \leq B\|\mathbf{x}\|_a \tag{15}$$

The following exercise shows that there essentially only one finite-dimensional normed linear space.

Exercise 1.213
In a finite-dimensional normed linear space, any two norms are equivalent.

One implication of the equivalence of norms in a normed linear space is that if a sequence converges with respect to one norm, it will converge in every norm. Therefore convergence in a finite-dimensional normed linear space is intrinsic to the sequence, and it does not depend on any particular norm. A useful corollary of this fact is given in the following exercise.

Exercise 1.214
A sequence $(\mathbf{x}^n)$ in $\Re^n$ converges if and only if each of its components x_i^n converges in $\Re$.

Exercise 1.215 (Closed and bounded equals compact)
Let $S \subseteq X$ be a closed and bounded subset of a finite-dimensional normed linear space X with basis $\{\mathbf{x}_1, \mathbf{x}_2, \ldots, \mathbf{x}_n\}$, and let $\mathbf{x}^m$ be a sequence in S. Every term $\mathbf{x}^m$ has a unique representation

$$\mathbf{x}^m = \sum_{i=1}^{n} \alpha_i^m \mathbf{x}_i$$

1. Using lemma 1.1, show that for every i the sequence of scalars (α_i^m) is bounded.

2. Show that $(\mathbf{x}^m)$ has a subsequence $(x_{(1)}^m)$ for which the coordinates of the first coordinate α_1^m converge to α.

3. Repeating this argument n times, show that $(\mathbf{x}^m)$ has a subsequence whose scalars converge to $(\alpha_1, \alpha_2, \ldots, \alpha_n)$.

4. Define $\mathbf{x} = \sum_{i=1}^{n} \alpha_i \mathbf{x}_i$. Show that $\mathbf{x}^m \to \mathbf{x}$.

5. Show that $\mathbf{x} \in S$.

6. Conclude that S is compact.

An immediate corollary is that the closed unit ball in a finite-dimensional space

$$C = \{\mathbf{x} : \|\mathbf{x}\| \leq 1\}$$

is compact (since it is closed and bounded). This is not the case in an infinite-dimensional space, so a linear space is finite-dimensional if and only if its closed unit ball is compact.

1.5.1 Convexity in Normed Linear Spaces

Because of the interaction of algebraic and geometric structure, the topological properties of convex sets are notably simpler than arbitrary sets. The results outlined in the following exercises are often fruitful in economic analysis.

Recall first that many of the important properties of a normed linear space are related to the shape of its unit ball. This is always convex.

Exercise 1.216
In any normed linear space, the unit ball is convex.

Exercise 1.217
Let S be a convex set in a normed linear space. Then int S and $\bar{S}$ are convex.

Similarly it can be shown that closure preserves subspaces, cones and linear varieties. For any convex set the line segment joining an interior point to a boundary point lies in the interior (except for the endpoints).

Exercise 1.218 (Accessibility lemma)
Let S be a convex set, with $\mathbf{x}_1 \in \bar{S}$ and $\mathbf{y}_2 \in $ int S. Then $\alpha \mathbf{x}_1 + (1-\alpha)\mathbf{x}_2 \in$ int S for all $0 < \alpha < 1$.

Exercise 1.219
Let S_i, $i \in I$ be a collection of open convex sets.

$$S = \bigcap_{i \in I} S_i \neq \emptyset \Rightarrow \bar{S} = \bigcap_{i \in I} \bar{S_i}$$

We have encountered two distinct notions of the extremity of a set: *boundary points* and *extreme points*. Boundary points, which demark a set from its complement, are determined by the geometry of a space. Extreme points, on the other hand, are an algebraic rather than a topological concept; they are determined solely by the linear structure of the space. However, in a normed linear space, these two notions of extremity overlap. All extreme points of a convex set are found on the boundary.

Exercise 1.220
If S is a convex set in a normed linear space, $\text{ext}(S) \subseteq \text{b}(S)$.

The converse is not true in general; not all boundary points are extreme points. However, boundary points and extreme points coincide when a set is strictly convex. A set S in a normed linear space is called *strictly convex* if the straight line between any two points lies in the interior of the set. More precisely, S is strictly convex if for every $\mathbf{x}$, $\mathbf{y}$ in X with $\mathbf{x} \neq \mathbf{y}$,

$$\alpha \mathbf{x} + (1 - \alpha)\mathbf{y} \in \text{int } S \quad \text{for every } 0 < \alpha < 1$$

Note that the interior of a convex set is always strictly convex (exercise 1.217). Therefore the additional requirement of strict convexity applies only to boundary points, implying that the straight line between any two boundary points lies in the interior of the set. Hence the boundary of a strictly convex set contains no line segment and every boundary point is an extreme point.

Exercise 1.221
If S is a strictly convex set in a normed linear space, every boundary point is an extreme point, that is, $\text{ext}(S) = \text{b}(S)$.

Exercise 1.222
If S is a convex set in a normed linear space,

S open $\Rightarrow$ S strictly convex

Exercise 1.223
S open $\Rightarrow$ conv S open.

Exercise 1.224
Let $S = \{(x_1, x_2) \in \Re^2 : x_2 \geq 1/|x_1|\}$ which is closed in $\Re^2$. Find conv S. Show that it is open (not closed) in $\Re^2$.

The convex hull of a closed set is not necessarily closed (exercise 1.224). However, if the set is compact, then so is its convex hull. This important result is established in the following exercise as an application of Carathéodory's theorem.

Exercise 1.225
Let S be a compact subset of a finite-dimensional linear space X of dimension n.

1. Show that conv S is bounded.
2. For every $\mathbf{x} \in \overline{\operatorname{conv} S}$, there exists a sequence $(\mathbf{x}^k)$ in conv S that converges to $\mathbf{x}$ (exercise 1.105). By Carathéodory's theorem (exercise 1.175), each term $\mathbf{x}^k$ is a convex combination of at most $n+1$ points, that is,

$$\mathbf{x}^k = \sum_{i=1}^{n+1} \alpha_i^k \mathbf{x}_i^k$$

where $\mathbf{x}_i^k \in S$. Show that we can construct convergent subsequences $\alpha_i^k \to \alpha_i$ and $\mathbf{x}_i^k \to \mathbf{x}_i$. [Hint: See exercise 1.215.]
3. Define $\mathbf{x} = \sum_{i=1}^{n+1} \alpha_i \mathbf{x}_i$. Show that $\mathbf{x}^k \to \mathbf{x}$.
4. Show that $\mathbf{x} \in \operatorname{conv} S$.
5. Show that conv S is closed.
6. Show that conv S is compact.

Remark 1.21 Finite dimensionality is not essential to the preceding result that the convex hull of a compact set is compact, which in fact holds in any Banach space (Pryce 1973, p. 55). However, finite dimensionality is essential to the proof given here, which relies on proposition 1.4 and especially exercise 1.215. Conversely, exercise 1.225 can be used to provide an alternative proof of exercise 1.215, as in the following exercise.

Exercise 1.226
Let S be a closed bounded subset of $\Re^n$. Show that

1. S is a closed subset of some cube

$$C = \{\mathbf{x} = (x_1, x_2, \ldots, x_n) \in \Re^n : -c \le x_i \le c, i = 1, 2, \ldots n\};$$

see example 1.94
2. C is the convex hull of the 2^n points $(\pm c, \pm c, \ldots, \pm c)$
3. C is compact
4. S is compact

The following corollary of exercise 1.225 is important in the theory of optimization and also in game theory.

Exercise 1.227
Any polytope is compact.

Example 1.110 (Mixed strategy space) In a finite strategic game, each player's mixed strategy space Σ_i is a $(m-1)$-dimensional simplex (example 1.98), where $m = |S_i|$ is the number of pure strategies of player i. Exercise 1.225 implies that every Σ_i is compact. Consequently the mixed strategy space of the game

$$\Sigma = \Sigma_1 \times \Sigma_2 \times \cdots \times \Sigma_n$$

is also compact (example 1.66).

In section 1.3.1 we touched briefly on the notion of a relative topology. The distinction is especially pertinent when dealing with convex sets. For example, the situation illustrated in figure 1.28 arises in the theory of TP-coalitional games (section 1.2.6), where the dark shaded triangle (the set of imputations) is a subset of the light shaded triangle (the 2-dimensional simplex). As a subset of Euclidean space $\Re^3$, the set of imputations has an empty interior. Every point in the dark shaded triangle (imputation) is arbitrarily close to points which lie off the hyperplane containing the triangle. Hence every imputation is a boundary point of the set of imputations. Similarly any line in a space of dimension 2 or more has no interior. Generalizing, any set of dimension $n-1$ in an n-dimensional space has an empty interior.

Given a line in space, our intuition would be to refer to any points except the endpoints as interior points. Similarly, in the left panel of figure 1.28, we would like to be able to refer to the interior of the dark shaded triangle as the interior of the set of imputations. Our intuition is to visualize the geometry of a set relative to its affine hull. To give effect to this intuition, we define the topology of a convex set relative to its affine hull. A point $\mathbf{x}$ in a convex set S is a *relative interior point* of S if it is an interior point of S with respect to the relative topology induced by aff S. Similarly the *relative interior* of a subset S of a normed linear space X, denoted ri S, is interior of S regarded as a subset of its affine hull. That is,

$$\text{ri } S = \{\mathbf{x} \in \text{aff } S : \mathbf{x} + rB \subset S \text{ for some } r \in \Re_+\}$$

Of course,

$$\text{ri } S \subseteq S \subseteq \overline{S}$$

1.5 Normed Linear Spaces

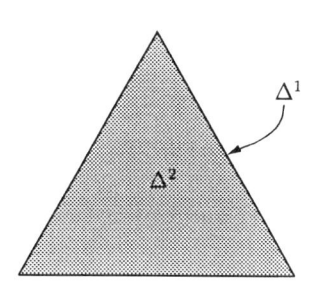

Figure 1.38
The relative interiors of Δ^1 and Δ^2

The set difference $\overline{S}\setminus\text{ri } S$ is called the *relative boundary* of S. S is said to be relatively open if ri $S = S$. For an n-dimensional convex set S in an n-dimensional space, aff $S = X$ and ri $S = $ int S.

For a finite line in space, its affine hull is the straight line extending beyond its endpoints in both direction. Relative to this set, the interior of the finite line is the line minus its endpoints. Similarly, in the game illustrated in figure 1.28, the affine hull of the dark shaded triangle is the plane in which it lies. Relative to this plane, the interior of the shaded triangle is the triangle minus its boundary.

Example 1.111 Each side of the two-dimensional simplex Δ^2 is a one-dimensional simplex Δ^1. The relative interior of Δ^2 is the interior of the triangle (figure 1.38), while ri Δ^1 is the side minus its endpoints. Note that while $\Delta^1 \subset \Delta^2$, ri $\Delta^1 \not\subset$ ri Δ^2. In fact ri Δ^1 and ri Δ^2 are nonempty disjoint sets.

Example 1.112 (Completely mixed strategies and trembles) In a finite game in which each player i has a set S_i of pure strategies, her set of mixed strategies Δ_i is the $(m-1)$-dimensional unit simplex (example 1.98). A mixed strategy σ is called *completely mixed* if every component is strictly positive, $\sigma_i > 0$, so that there is a nonzero probability of every pure strategy being chosen. The set of completely mixed strategies is the relative interior of Δ.

For every pure strategy $s_i \in S_i$, there exists a sequence of completely mixed strategies $\sigma_i^n \in$ ri Σ_i converging to s_i (see exercise 1.105). Sequences of completely mixed strategies, called *trembles*, are used in refining equilibrium concepts (Fudenberg and Tirole 1991, pp. 338–339; Osborne and Rubinstein 1994, pp. 246–253).

Exercise 1.228
The unit simplex in $\Re^n$ has a nonempty relative interior.

Exercise 1.229
If S is a convex set in a finite-dimensional normed linear space

$$S \neq \emptyset \Rightarrow \text{ri } S \neq \emptyset$$

Exercise 1.230
Let S be a nonempty convex set in a finite-dimensional normed linear space.

$$\text{ri } S = \text{int } S \Leftrightarrow \text{int } S \neq \emptyset$$

1.6 Preference Relations

In economics a preference relation (example 1.12) is simply a weak order, that is a relation $\succsim$ on a set X that is complete and transitive. The basic properties of weak orders have been explored in section 1.2. However, the sets on which a preference relation is defined (e.g., the consumption set or strategy space) typically also have algebraic (example 1.87) and geometric structure (example 1.54). All three aspects contribute to economic analysis. In this section we integrate the order, linear, and geometric aspects of preference relation defined on a subset of a normed linear space. We use the consumer's problem to illustrate the usefulness of this interaction.

Example 1.113 (The consumer's problem) Assume that there are n commodities. The consumer's problem is to choose an affordable consumption bundle **x** in the consumption set $X \subset \Re^n$ (example 1.6) that yields the most satisfaction. The consumer's preferences over consumption bundles are assumed to be represented by a preference relation $\succsim$ on X.

The consumer's choice is constrained by her income, m. If the n commodities have prices $p_1, p_2, \ldots, p_n$, the set of affordable commodity bundles

$$X(\mathbf{p}, m) = \{\mathbf{x} \in X : p_1 x_1 + p_2 x_2 + \cdots p_n x_n \leq m\}$$

is called her *budget set*, where $\mathbf{p} = (p_1, p_2, \ldots, p_n)$ is the list of prices. The consumer's problem is to choose a best element in the budget set $X(\mathbf{p}, m)$, that is, to choose $\mathbf{x}^* \in X(\mathbf{p}, m)$ such that $\mathbf{x}^* \succsim \mathbf{x}$ for every $\mathbf{x} \in X(\mathbf{p}, m)$. A best element of $X(\mathbf{p}, m)$ is called the *consumer's optimal choice* given prices $\mathbf{p}$ and income m. Note that there may be more than one optimal choice for any $\mathbf{p}$ and m.

Exercise 1.231
Assume that all prices and income are positive ($\mathbf{p} > \mathbf{0}, m > 0$) and that the consumer can afford some feasible consumption bundle, that is,

$$m > \inf_{\mathbf{x} \in X} \sum_{i=1}^{n} p_i x_i$$

Then the consumer's budget set $X(\mathbf{p}, m)$ is nonempty and compact.

Exercise 1.232
The budget set is convex.

Remark 1.22 In establishing that the budget set is compact (exercise 1.231), we relied on the assumption that the choice was over n distinct commodities so that the consumption set is finite dimensional, $X \subset \Re^n$. In more general formulations involving intertemporal choice or uncertainty, it is not appropriate to assume that the consumption set is finite dimensional. Then, compactness of the budget set is more problematic. Note, however, that finite dimensionality is not required to establish that the budget set is convex (exercise 1.232).

1.6.1 Monotonicity and Nonsatiation

Recall that the natural order on $\Re^n$ (example 1.26) is only a partial order, whereas a preference relation is complete. Therefore the natural order "$\geq$" (example 1.26) cannot represent a preference relation on $X \subseteq \Re^n$. However, an obvious requirement to impose on a preference relation on any $X \subseteq \Re^n$ is that it be consistent with the natural order. This property is usually called monotonicity. A preference relation $\succsim$ on $X \subseteq \Re^n$ is *weakly monotonic* if $\mathbf{x} \geq \mathbf{y}$ implies that $\mathbf{x} \succsim \mathbf{y}$. It is *strongly monotonic* if $\mathbf{x} \gneq \mathbf{y}$ implies $\mathbf{x} \succ \mathbf{y}$. Monotonicity is a natural assumption for the consumer preference relation, embodying the presumption that "more is better." It implies that the consumer is never fully satisfied or *sated*.

Nonsatiation is a weaker assumption on preferences. A best element $\mathbf{x}^*$ in a set X weakly ordered by a preference relation $\succsim$ is called a *bliss point*.

($\mathbf{x}^* \in X$ is a best element if $\mathbf{x}^* \succsim \mathbf{x}$ for every $\mathbf{x} \in X$.) Typically the set X has no best element, in which case we say that the preference relation $\succsim$ is *nonsatiated*. A stronger assumption, which relies on the geometric structure of X, is often imposed in practice. A preference relation is *locally nonsatiated* if given any element $\mathbf{x} \in X$ and neighborhood S around $\mathbf{x}$, there always exists some neighboring element $\mathbf{y} \in S$ that is preferred, that is, $\mathbf{y} \succ \mathbf{x}$. The relationships between these various notions are established in the following exercise.

Exercise 1.233

1. Strong monotonicity $\Rightarrow$ weak monotonicity.
2. Strong monotonicity $\Rightarrow$ local nonsatiation.
3. Local nonsatiation $\Rightarrow$ nonsatiation.

A useful implication of strong monotonicity or local nonsatiation in consumer choice is that the consumer will spend all her income, so every optimal choice lies on the boundary of the budget set. Note that neither weak monotonicity nor nonsatiation is sufficient to provide this result (but see exercise 1.248).

Exercise 1.234
Assume that the consumer's preference relation is strongly monotonic. Then any optimal choice $\mathbf{x}^* \succsim \mathbf{x}$ for every $\mathbf{x} \in X(\mathbf{p}, m)$ exhausts her income, that is, $\sum_{i=1}^{n} p_i x_i = m$.

Exercise 1.235
Extend the previous exercise to encompass the weaker assumption of local nonsatiation.

1.6.2 Continuity

The principal geometric property of a preference relation is continuity. A preference relation $\succsim$ on a metric space is *continuous* if, whenever $\mathbf{x}_0 \succ \mathbf{y}_0$, neighboring points of $\mathbf{x}_0$ are also preferred to $\mathbf{y}_0$. More formally, a preference relation $\succsim$ on a metric space X is *continuous* if, whenever $\mathbf{x}_0 \succ \mathbf{y}_0$, there exist neighborhoods $S(\mathbf{x}_0)$ and $S(\mathbf{y}_0)$ of $\mathbf{x}_0$ and $\mathbf{y}_0$ such that $\mathbf{x} \succ \mathbf{y}$ for every $\mathbf{x} \in S(\mathbf{x}_0)$ and $\mathbf{y} \in S(\mathbf{y}_0)$. In effect, $\succsim$ is continuous provided that small changes in $\mathbf{x}$ and $\mathbf{y}$ do not lead to a reversal of preference.

An alternative definition of continuity is often found in textbooks. A preference relation $\succsim$ on a metric space X is continuous if and only if the upper $\succsim(\mathbf{y})$ and lower $\precsim(\mathbf{y})$ preference sets are closed in X. The equivalence of these definitions is established in the following exercise.

Exercise 1.236

1. Assume that the preference relation $\succsim$ on a metric space X is continuous. Show that this implies that the sets $\succ(\mathbf{y}) = \{\mathbf{x} : \mathbf{x} \succ \mathbf{y}\}$ and $\prec(\mathbf{y}) = \{\mathbf{x} : \mathbf{x} \prec \mathbf{y}\}$ are open for every $\mathbf{y}$ in X.

2. Conversely, assume that the sets $\succ(\mathbf{y}) = \{\mathbf{x} : \mathbf{x} \succ \mathbf{y}\}$ and $\prec(\mathbf{y}) = \{\mathbf{x} : \mathbf{x} \prec \mathbf{y}\}$ are open for every $\mathbf{y}$ in X. Choose any $\mathbf{x}_0$, $\mathbf{z}_0$ in X with $\mathbf{x}_0 \succ \mathbf{z}_0$.

   a. Suppose there exists some $\mathbf{y} \in X$ such that $\mathbf{x}_0 \succ \mathbf{y} \succ \mathbf{z}_0$. Show that there exist neighborhoods $S(\mathbf{x}_0)$ and $S(\mathbf{z}_0)$ of $\mathbf{x}_0$ and $\mathbf{z}_0$ such that $\mathbf{x} \succ \mathbf{z}$ for every $\mathbf{x} \in S(\mathbf{x}_0)$ and $\mathbf{z} \in S(\mathbf{z}_0)$.

   b. Now suppose that there is no such $\mathbf{y}$ with $\mathbf{x}_0 \succ \mathbf{y} \succ \mathbf{z}_0$. Show that

   i. $\succ(\mathbf{z}_0)$ is an open neighborhood of $\mathbf{x}_0$

   ii. $\succ(\mathbf{z}_0) = \succsim(\mathbf{x}_0)$

   iii. $\mathbf{x} \succ \mathbf{z}_0$ for every $\mathbf{x} \in \succ(\mathbf{z}_0)$

   iv. There exist neighborhoods $S(\mathbf{x}_0)$ and $S(\mathbf{z}_0)$ of $\mathbf{x}_0$ and $\mathbf{z}_0$ such that $\mathbf{x} \succ \mathbf{z}$ for every $\mathbf{x} \in S(\mathbf{x}_0)$ and $\mathbf{z} \in S(\mathbf{y}_0)$

   This establishes that a preference relation $\succsim$ on a metric space X is continuous if and only if the sets $\succ(\mathbf{y}) = \{\mathbf{x} : \mathbf{x} \succ \mathbf{y}\}$ and $\prec(\mathbf{y}) = \{\mathbf{x} : \mathbf{x} \prec \mathbf{y}\}$ are open for every $\mathbf{y}$ in X.

3. Show that a preference relation $\succsim$ on a metric space X is continuous if and only if the sets $\succsim(\mathbf{y}) = \{\mathbf{x} : \mathbf{x} \succsim \mathbf{y}\}$ and $\precsim(\mathbf{y}) = \{\mathbf{x} : \mathbf{x} \precsim \mathbf{y}\}$ are closed for every $\mathbf{y}$ in X.

Exercise 1.237
Mas-Colell et al. (1995, p. 46) define continuity of preferences as follows:

The preference relation $\succsim$ on X is continuous if it is preserved under limits. That is, for any sequence of pairs $((\mathbf{x}^n, \mathbf{y}^n))$ with $\mathbf{x}^n \succsim \mathbf{y}^n$ for all n, with $\mathbf{x} = \lim_{n \to \infty} \mathbf{x}^n$, and $\mathbf{y} = \lim_{n \to \infty} \mathbf{y}^n$, we have $\mathbf{x} \succsim \mathbf{y}$.

1. Show that this definition in effect requires that the set $\{(\mathbf{x}, \mathbf{y}) : \mathbf{x} \succsim \mathbf{y}\}$ be a closed subset of $X \times X$.

2. Is this equivalent to the definition given above?

Exercise 1.238
Assume that $\succsim$ is continuous preference on a connected metric space X. For every pair **x**, **z** in X with $\mathbf{x} \succ \mathbf{z}$, there exists **y** such that $\mathbf{x} \succ \mathbf{y} \succ \mathbf{z}$.

Remark 1.23 (Order topology) Any weak order $\succsim$ on a set X induces a natural topology (geometry) in which the sets $\{\mathbf{x} : \mathbf{x} \succ \mathbf{y}\}$ and $\{\mathbf{x} : \mathbf{x} \prec \mathbf{y}\}$ are open. This is called the *order topology* on X. A preference relation (weak order) on a metric space is continuous if the order topology is consistent with the metric topology of space. A preference relation on $\Re^n$ is continuous if the order topology is consistent with the usual topology on $\Re^n$.

Example 1.114 (Lexicographic preferences) The standard example of a noncontinuous preference relation is the lexicographic ordering. Assuming two commodities, the lexicographic ordering on $\Re^2$ is

$$\mathbf{x} \succ \mathbf{y} \Leftrightarrow \left\{ \begin{array}{l} x_1 > y_1 \text{ or} \\ x_1 = y_1 \text{ and } x_2 > y_2 \end{array} \right\}$$

To show that the lexicographic ordering is not continuous, let **x** and **y** be two commodity bundles with the same quantity of good 1 ($x_1 = y_1$) (figure 1.39). Assume that $x_2 > y_2$, and let $r = (x_2 - y_2)/2$. Under the lexicographic ordering, $\mathbf{x} \succ \mathbf{y}$. However, **y** is strictly preferred to some bundles in the neighborhood of **x**. In particular, $\mathbf{y} \succ \mathbf{z} = (x_1 - \varepsilon, x_2)$ for every $\varepsilon < r$.

Continuity is sufficient to ensure the existence of a best element in a compact ordered set. This is essential for a well-defined formulation of the

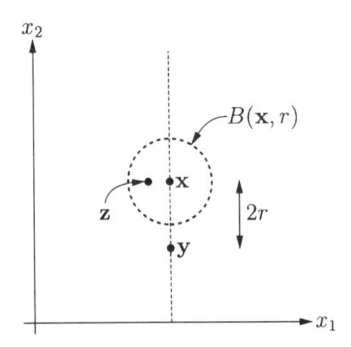

Figure 1.39
Lexicographic preferences are not continuous

consumer's problem (example 1.113). In the next chapter we will see that continuity also ensures the existence of a utility function that represents the preferences.

Proposition 1.5 *A weakly ordered set $(X, \succsim)$ has a best element if X is compact and $\succsim$ is continuous.*

Proof Every finite set $\{\mathbf{y}_1, \mathbf{y}_2, \ldots, \mathbf{y}_n\}$ has a best element (exercise 1.29). Without loss of generality, suppose that this is $\mathbf{y}_1$, so that $\mathbf{y}_1 \succsim \mathbf{y}_i$, $i = 1, 2, \ldots, n$. That is, $\mathbf{y}_1 \in \succsim(\mathbf{y}_i)$, $i = 1, 2, \ldots, n$. Thus we have established that the collection of all upper preference sets $\succsim(\mathbf{y})_\mathbf{y} \in X$ has the finite intersection property; that is, for every finite subcollection $\{\succsim(\mathbf{y}_1), \succsim(\mathbf{y}_2), \ldots, \succsim(\mathbf{y}_n)\}$,

$$\bigcap_{i=1}^{n} \succsim(\mathbf{y}_i) \neq \emptyset$$

Since $\succsim$ is continuous, every $\succsim(\mathbf{y})$ is closed. Since X is compact,

$$\bigcap_{\mathbf{y} \in X} \succsim(\mathbf{y}_i) \neq \emptyset$$

by exercise 1.116. Let $\mathbf{x}^*$ be a point in $\bigcap_{\mathbf{y} \in X} \succsim(\mathbf{y}_i)$. Then $\mathbf{x}^* \succsim \mathbf{y}$ for every $\mathbf{y} \in X$. $\mathbf{x}^*$ is a best element. □

Exercise 1.239
Let $\succsim$ be a continuous preference relation on a compact set X. The set of best elements is nonempty and compact.

Example 1.115 (Existence of an optimal choice) Provided that all prices and income are positive, the budget set is nonempty and compact (exercise 1.231). By proposition 1.5, $X(\mathbf{p}, m)$ contains a best element $\mathbf{x}^* \succsim \mathbf{x}$ for every $\mathbf{x} \in X(\mathbf{p}, m)$.

Exercise 1.240
Assume that a consumer with lexicographic preferences over two commodities requires a positive amount of both commodities so that consumption set $X = \Re^2_{++}$. Show that no optimal choice exists.

Exercise 1.241
Why is the existence of an optimal choice essential for a well-defined formulation of the consumer's problem?

Exercise 1.242 (Nucleolus is nonempty and compact)
Let (N, w) be a TP-coalitional game with a compact set of outcomes X. For every outcome $\mathbf{x} \in X$, let $\mathbf{d}(\mathbf{x})$ be a list of coalitional deficits arranged in decreasing order (example 1.49). Let $d_i(\mathbf{x})$ denote the ith element of $\mathbf{d}(\mathbf{x})$.

1. Show that $X^1 = \{\mathbf{x} \in X : d_1(\mathbf{x}) \leq d_1(\mathbf{y}) \text{ for every } \mathbf{y} \in X\}$ is nonempty and compact.
2. For $k = 2, 3, \ldots, 2^n$, define $X^k = \{\mathbf{x} \in X^{k-1} : \mathbf{d}_k(\mathbf{x}) \leq \mathbf{d}_k(\mathbf{y}) \text{ for every } \mathbf{y} \in X^{k-1}\}$. Show that X^k is nonempty and compact.
3. Show that $\text{Nu} = X^{2^n}$, which is nonempty and compact.

1.6.3 Convexity

The most useful algebraic property of a preference relation is convexity. A preference relation is *convex* if averages are preferred to extremes. Formally, the preference relation $\succsim$ is *convex* if for every $\mathbf{x}, \mathbf{y} \in X$ with $\mathbf{x} \succsim \mathbf{y}$,

$$\alpha \mathbf{x} + (1 - \alpha)\mathbf{y} \succsim \mathbf{y} \qquad \text{for every } 0 \leq \alpha \leq 1$$

The link between convexity of the preference relation and convex sets is given in the following exercise. The method of proof should be carefully noted, since it is widely used in economics.

Exercise 1.243
The preference relation $\succsim$ is convex if and only if the upper preference sets $\succsim(\mathbf{y})$ are convex for every $\mathbf{y}$.

Exercise 1.244
Let $\succsim$ be a convex preference relation on a linear space X. The set of best elements $X^* = \{\mathbf{x} : \mathbf{x} \succsim \mathbf{y} \text{ for every } \mathbf{y} \in X\}$ is convex.

A slightly stronger notion of convexity is often convenient (example 1.116). A preference relation is strictly convex if averages are strictly preferred to extremes. Formally the preference relation $\succsim$ is *strictly convex* if for every $\mathbf{x}, \mathbf{y} \in X$ with $\mathbf{x} \succsim \mathbf{y}$ but $\mathbf{x} \neq \mathbf{y}$,

$$\alpha \mathbf{x} + (1 - \alpha)\mathbf{y} \succ \mathbf{y} \qquad \text{for every } 0 < \alpha < 1$$

Example 1.116 (Unique optimal choice) If the consumer's preference relation is strictly convex, the consumer's optimal choice $\mathbf{x}^*$ (if it exists) is unique. To see this, assume the contrary; that is, assume that there are two distinct best elements $\mathbf{x}^*$ and $\mathbf{y}^*$ in the budget set $X(\mathbf{p}, m)$. That is,

$$\mathbf{x}^* \gtrsim \mathbf{x} \quad \text{and} \quad \mathbf{y}^* \gtrsim \mathbf{x} \quad \text{for every } \mathbf{x} \in X(\mathbf{p}, m)$$

In particular, note that $\mathbf{x}^* \gtrsim \mathbf{y}^*$ and $\mathbf{y}^* \gtrsim \mathbf{x}^*$. By strict convexity, the average of these two bundles $\mathbf{z} = \frac{1}{2}\mathbf{x}^* + \frac{1}{2}\mathbf{y}^*$ is strictly preferred to either $\mathbf{x}^*$ or $\mathbf{y}^*$, that is, $\mathbf{z} \succ \mathbf{x}^*$ and $\mathbf{z} \succ \mathbf{y}^*$. Furthermore, since the budget set is convex (exercise 1.232), $\mathbf{z} \in X(\mathbf{p}, m)$. The consumer can afford the preferred bundle $\mathbf{z}$. We have shown that if the optimal choice were nonunique, we could find another affordable bundle that was strictly preferred. Therefore the optimal choice must be unique if preferences are strictly convex.

Recall that the nucleolus of a TP-coalitional game is the set of outcomes that are maximal in the deficit order $\gtrsim^d$ (example 1.49). Exercise 1.242 showed that this set is always nonempty. In the next exercise we show that the nucleolus contains just one outcome, $\text{Nu} = \{\mathbf{x}^N\}$. In a slight abuse of language, it is conventional to identify the set Nu with its only element. We call the maximal element $\mathbf{x}^N$ the nucleolus and say that "the nucleolus is unique."

Exercise 1.245 (*Nucleolus is unique*)
In a TP-coalitional game the deficit order $\gtrsim^d$ (example 1.49) is strictly convex. Consequently the nucleolus contains a single outcome.

1.6.4 Interactions

To complete this section, we indicate some of the substitutability among algebraic, geometric, and order structures. Throughout this section we have assumed a weak order that is a complete, transitive relation $\gtrsim$ on a set X. Exercise 1.246 shows that the assumption of completeness is redundant provided that $\gtrsim$ is continuous and X is connected (see exercise 1.238). Exercise 1.247 shows how continuity strengthens convexity, while exercise 1.248 establishes a link between nonsatiation and strict convexity. We then introduce the standard model of an exchange economy, and show (exercise 1.249) that weak and strong Pareto optimality (section 1.2.6) coincide in an exchange economy in which the participants have continuous and monotone preferences. After reformulating the exchange economy as an example of a coalitional game, we finish with the first fundamental theorem of welfare economics, which underlies the economist's faith in competitive markets.

Exercise 1.246
Assume that $\succsim$ is a continuous order relation on a connected metric space with $\mathbf{x}_0 \succ \mathbf{y}_0$ for at least one pair $\mathbf{x}_0, \mathbf{y}_0 \in X$.

1. Show that for any $\mathbf{x}_0, \mathbf{y}_0 \in X$ such that $\mathbf{x}_0 \succ \mathbf{y}_0$,
   a. $\prec(\mathbf{x}_0) \cup \succ(\mathbf{y}_0) = \precsim(\mathbf{x}_0) \cup \succsim(\mathbf{y}_0)$
   b. $\prec(\mathbf{x}_0) \cup \succ(\mathbf{y}_0) = X$
2. Suppose that $\succsim$ is not complete. That is, there exists $\mathbf{x}, \mathbf{y} \in X$ such that neither $\mathbf{x} \succsim \mathbf{y}$ nor $\mathbf{x} \precsim \mathbf{y}$. Then show that
   a. $\prec(\mathbf{x}) \cap \prec(\mathbf{y}) \neq X$
   b. $\prec(\mathbf{x}) \cap \prec(\mathbf{y}) \neq \emptyset$
   c. $\prec(\mathbf{x}) \cap \prec(\mathbf{y}) = \precsim(\mathbf{x}) \cap \precsim(\mathbf{y})$
3. Show that X connected implies that $\succsim$ is complete.

Exercise 1.247
If the convex preference relation $\succsim$ is continuous,

$$\mathbf{x} \succ \mathbf{y} \Rightarrow \alpha\mathbf{x} + (1-\alpha)\mathbf{y} \succ \mathbf{y} \quad \text{for every } 0 < \alpha < 1$$

[Hint: Use the accessibility lemma (exercise 1.218).]

Exercise 1.248
If $\succsim$ is strictly convex, nonsatiation is equivalent to local nonsatiation.

Example 1.117 (Exchange economy) In studying aggregate economic interaction, a fruitful simplification is the pure exchange economy in which there is no production. Consumers are endowed with an initial allocation of goods. The only possible economic activity is trade in which consumers exchange their endowments at given prices to obtain preferred consumption bundles. Formally an *exchange economy* comprises

- a set of l commodities
- a set $N = \{1, 2, \ldots, n\}$ of consumers
- for every consumer $i \in N$

  a feasible consumption set $X_i \subseteq \Re^l_+$

  a preference ordering $\succsim_i$ over X_i

  an endowment $\omega_i \in \Re^l_+$

An *allocation* $\underline{x} \in \Re^{ln}$ is a list of commodity bundles assigned to each consumer. That is, $\underline{x} = (x_1, x_2, \ldots, x_n)$, where $x_i \in \Re^l_+$ is the commodity bundle assigned to the ith consumer. An allocation is feasible if

- $x_i \in X_i$ for every consumer i and
- aggregate demand is less than or equal to available supply; that is,

$$\sum_{i \in N} x_i \leq \sum_{i \in N} \omega_i$$

In a *competitive* exchange economy, trade take place at fixed commodity prices $\mathbf{p} = (p_1, p_2, \ldots, p_l)$. Each consumer's income (or wealth) m_i is equal to the value of her endowment, that is,

$$m_i = \sum_{j=1}^{l} p_i \omega_{ij}$$

Each consumer endeavors to exchange commodities to achieve her most preferred bundle, which is affordable given the value of her endowment m_i. A competitive equilibrium is attained when all consumers achieve this goal simultaneously. That is, a *competitive equilibrium* $(\mathbf{p}^*, \underline{x}^*)$ is a set of prices $\mathbf{p}^*$ and a feasible allocation $\underline{x}^*$ such that for every $i \in N$,

- $x_i^* \in X_i$
- $x_i^* \succsim x_i$ for every $x_i \in X(\mathbf{p}, m_i)$
- $\sum_{i \in N} x_i \leq \sum_{i \in N} \omega_i$

Provided that the individual preferences are continuous, proposition 1.5 guarantees the existence of best allocations x^* for every set of prices $\mathbf{p}$ (example 1.115). A deep theorem to be presented in the next chapter guarantees the existence of a set of prices $\mathbf{p}^*$ at which the desired trades are all feasible.

The following exercise is another illustration of the usefulness of the interaction of order and topological structures.

Exercise 1.249
Remark 1.8 distinguished the strong and weak Pareto orders. Show that the distinction is innocuous in an exchange economy in which the agents preferences are monotone and continuous. Specifically, show that an

allocation is weakly Pareto efficient if and only if it is strongly Pareto efficient.

Example 1.118 (Market game) We can model an exchange economy as a coalitional game, thereby establishing a profound link between traditional economic theory and game theory. The set of outcomes X is the set of all allocations

$$X = \{\underline{\mathbf{x}} = (\mathbf{x}_i)_{i \in N} : \mathbf{x}_i \in \Re_+^l\}$$

Acting independently, any coalition can obtain any allocation that can be achieved by trading among itself so that

$$W(S) = \left\{\underline{\mathbf{x}} \in X : \sum_{i \in S} \mathbf{x}_i = \sum_{i \in S} \omega_i\right\}$$

To complete the description of the game, we extend the individual preference relations to the set of allocations X so that

$$\underline{\mathbf{x}} \succsim_i \underline{\mathbf{y}} \Leftrightarrow \mathbf{x}_i \succsim_i \mathbf{y}_i$$

The familiar *Edgeworth box* diagram illustrates the core of a two-person two-good exchange economy (figure 1.40).

Exercise 1.250
Every competitive equilibrium allocation $\underline{\mathbf{x}}^*$ belongs to the core of the corresponding market game.

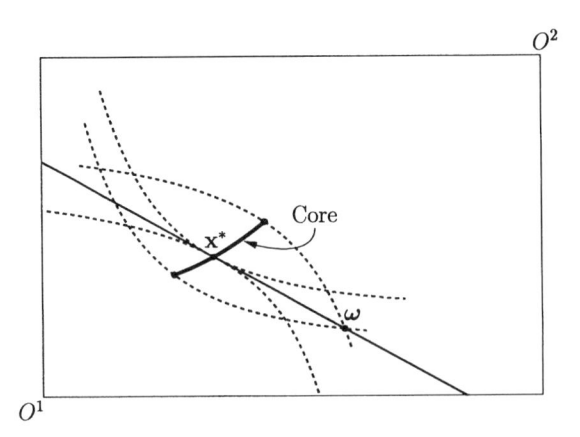

Figure 1.40
An Edgeworth box, illustrating an exchange economy with two traders and two goods

Exercise 1.251 (First theorem of welfare economics)
Every competitive equilibrium is Pareto efficient.

This theorem underlies the economist's faith in competitive markets. If an economic outcome is achieved through free trade in competitive markets, it is it is impossible to make any individual better off without harming another. There is no allocation which would make all the agents better off.

1.7 Conclusion

This chapter opened with a short introduction to the vocabulary of sets. We noted that the most familiar set, the real numbers $\Re$, exhibits three distinct properties: order, distance, and linearity. In succeeding sections we explored the consequences of generalizing each of these properties to more general sets. Ordered sets, posets, and lattices generalize the order properties of the numbers, the fact that numbers can be ranked by magnitude. Ranking is important for economics, since economists are continually comparing alternatives and searching for the best way of doing things. Metric spaces generalize the spatial properties of real numbers. Measurement of distance is also important to economists, since we want to know how far it is from one production plan to another, and to know whether or not we are getting closer to the desired point. Linear spaces generalize the algebraic properties of real numbers. Linearity is important, since averaging and scaling are two ways of generating new economic choices (production and consumption plans).

This individual exploration of the consequences of order, additivity, and distance is a powerful illustration of the utility of abstraction. As we noted in the preface, abstraction in mathematics serves the same function as model building in economics. Although most economic analysis takes place in the familiar set $\Re^n$, it is so commonplace that we tend to confuse order, algebraic, and geometric properties. Separating out these different aspects focuses attention on the essential aspects of a particular problem and sharpens our thinking.

While separation sharpens our focus, further insights can be obtained by combining the algebraic, geometric, and order structures. We have seen two good examples in this chapter. In section 1.5 we saw how the interplay of algebra and geometry contributed significantly to our understanding of the structure of finite-dimensional spaces. In the final section we explored the interaction of algebra, geometry, and order in preference relations, and showed how this led to new insights into consumer behavior.

1.8 Notes

To supplement this chapter, I particularly recommend Luenberger (1969) and Simmons (1963), which cover most of the material of this chapter (and much more besides) elegantly and lucidly. Klein (1973) covers similar material from the viewpoint of an economist. For a more concrete approach to mathematics for economists, Simon and Blume (1994) and Sydsaeter and Hammond (1995) are recommended. Debreu (1991) discusses the contribution of mathematics to the development of economic theory.

Halmos (1960) is a lucid introduction to set theory. The material on ordered sets is collated from many sources, which employ a variety of terminology. For the most part, we have used the terminology of Sen (1970a). Birkhoff (1973) is the standard reference for lattice theory. Our treatment is based largely on Topkis (1978) and Milgrom and Shannon (1994). The strong set order is called the induced set order by Topkis (1978).

Sen (1970a) provides a comprehensive and readable account of the problem of social choice, and Sen (1995) a recent review. Sen first noted the liberal paradox in Sen (1970b). Hammond (1976) investigates the relationship between the Rawlsian criterion of social justice and social choice.

For the most part, our encounters with game theory follow the approach in Osborne and Rubinstein (1994). Another standard reference is Fudenberg and Tirole (1991). Example 1.44 is adapted from Gately (1974). The deficit of a coalition (example 1.49) is usually called the "excess," although deficit seems more appropriate given the usual sign convention.

Binmore (1981) is a patient exposition of metric and topological ideas. The standard reference of topology is Kelley (1955). For normed linear spaces, see Luenberger (1969). Exercise 1.109 is adapted from Moulin (1986), who attributes it to Choquet.

1.8 Notes

The best reference for linear spaces is Halmos (1974). This classic text is consistent with the approach taken here and is very readable. The standard references on convexity are Rockafellar (1970) and Stoer and Witzgall (1970). The recent book by Panik (1993) is a useful compendium of results written with the economist in mind. Exercise 1.191 is adapted from Mas-Colell et al. (1995).

The Shapley-Folkman lemma is a good example of economics fertilizing mathematics. It was discovered by Lloyd Shapley and J. Folkman in answer to a problem posed by Ross Starr, arising from the latter's investigation of the implications of nonconvexity in economic models. It was first published in Starr (1969). An accessible account of its use in economics is given by Hildenbrand and Kirman (1976). (Unfortunately, this topic does not appear to have found its way into the second edition of this delightful book.)

The material on preference relations can be found in any advanced microeconomics text such as Kreps (1990), Mas-Colell et al. (1995), and Varian (1992). Proposition 1.5 on the existence of a maximal element in an ordered set is not widely cited. Border (1985) provides a useful overview of the relevant literature. Carter and Walker (1996) discuss the uniqueness of the nucleolus and outline an algorithm for its computation. Hildenbrand and Kirman (1976) is a concise and entertaining account of the relationship between Walrasian equilibria and the core in exchange economies.

2 Functions

While sets and spaces provide the basic characters of mathematical analysis, functions provide the plot. A function establishes a relationship or linkage between the elements in two or more sets. Of particular interest are functions that respect the structure of the sets that they associate. Functions that preserve the *order* of sets are called *monotone*, those that preserve the *geometry* are called *continuous*, and those that preserve the *algebraic* structure are called *linear*. In this chapter we explore monotone and continuous functions, while the next chapter is devoted to linear and related functions. In the course of this exploration, we encounter the major theorems founding mathematical economics: the maximum theorems (theorems 2.1, 2.3, 3.1), the separating hyperplane theorem (theorem 3.2), and Brouwer's fixed point theorem (theorem 2.6). The first section examines functions in general.

2.1 Functions as Mappings

2.1.1 The Vocabulary of Functions

A *function* $f: X \to Y$ is a rule that assigns to *every* element x of a set X (the domain) a *single* element of a set Y (the co-domain). Note that

- the definition comprises two sets (domain and co-domain) and a rule
- every element of X of x is assigned an element of Y
- only one element of Y is assigned to each $x \in X$

The mapping or assignment is usually denoted $y = f(x)$. The element $y \in Y$ that is assigned to a particular $x \in X$ is called the *image* of x under f. When f represents an economic model, the image y is frequently called the *dependent variable*, while x is called the *independent variable*. Synonyms for a function include *map, mapping*, and *transformation* (figure 2.1). A function $f: X \to X$ from a set X to itself is often called an *operator*.

The *range* of a function $f: X \to Y$ is the set of all elements in Y that are images of elements in X. Since it is the image of X, the range is denoted $f(X)$. Formally

$$f(X) = \{y \in Y : y = f(x) \text{ for some } x \in X\}$$

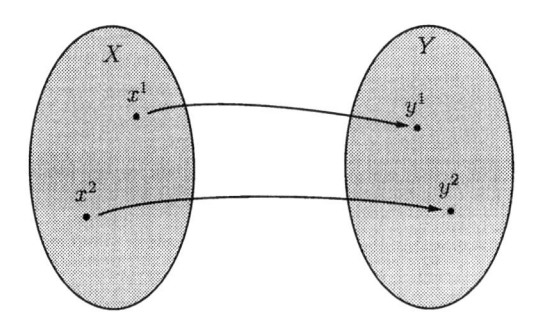

Figure 2.1
A function mapping X to Y

If every $y \in Y$ is the image of some $x \in X$, so that $f(X) = Y$, we say that f maps X *onto* Y. If every $x \in X$ maps to a distinct Y, so that $f(x) = f(x')$ implies that $x = x'$, we say that f is *one-to-one* or *univalent*.

The *graph* of a function $f: X \to Y$ is the set of all related pairs $(x, f(x))$ in $X \times Y$. Formally

$$\text{graph}(f) = \{(x, y) \in X \times Y : y = f(x), x \in X\}$$

This graphical representation of a function underscores the fact that a function $f: X \to Y$ is a special type of binary relation (section 1.2.1) on $X \times Y$ in which

- domain $f = X$
- for every $x \in X$, there is a unique $y \in Y$ such that $(x, y) \in f$

Example 2.1 Let $X = \{\text{members of a class}\}$ and $Y = \{\text{days of the year}\}$. The rule that assigns each member of the class to his or her birthday is a function. We note that

- everyone has a birthday
- nobody has two birthdays
- two people may have the same birthday
- not every day is someone's birthday

As this example illustrates, while every element of X must be assigned to some element of Y, not every element of Y need be assigned an element of X. In general, the range is a proper subset of the co-domain.

2.1 Functions as Mappings

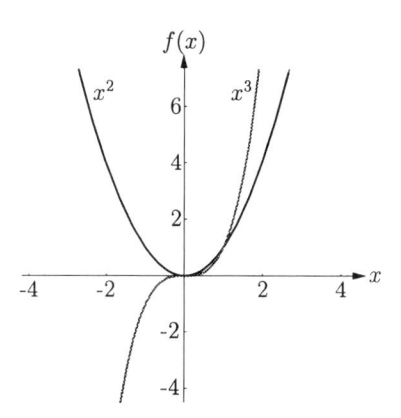

Figure 2.2
The functions $f(x) = x^2$ and $f(x) = x^3$

Exercise 2.1
Is the birthday mapping defined in the previous example *one-to-one* or *onto*?

Example 2.2 (Power function) Among the simplest functions encountered in economic analysis are the *power functions* $f_n: \Re \to \Re$ defined by

$$f_1(x) = x, \quad f_2(x) = x^2, \quad f_n(x) = xf_{n-1}(x) = x^n, \quad n = 3, 4 \ldots$$

which assign to every real number its nth power. Two power functions are illustrated in figure 2.2.

Example 2.3 (Rotation) The function $f: \Re^2 \to \Re^2$ defined by

$$f(x_1, x_2) = (x_1 \cos \theta - x_2 \sin \theta, \, x_1 \sin \theta + x_2 \cos \theta)$$

where θ is a number $0 \leq \theta < 2\pi$, "transforms" vectors in the plane $\Re^2$ by rotating them counterclockwise through the angle θ (figure 2.3). This function is in fact an operator, since it maps the plane $\Re^2$ into itself.

We sometimes depict a function by explicitly illustrating the map between the domain and co-domain, for example, as in figures 2.1 and 2.3. A function between finite sets may be expressed in a table. For example, the birthday mapping for a class of five is specified in the following table:

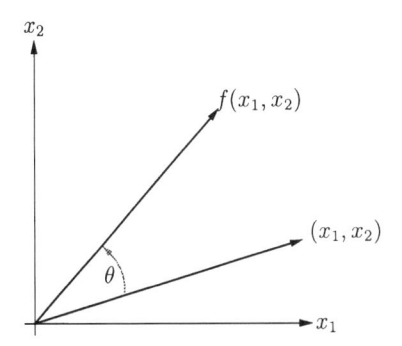

Figure 2.3
Rotation of a vector

John	3 February
Jason	22 June
Kathryn	16 March
Jenny	29 October
Chris	7 January

Prior to the prevalence of pocket calculators, many numerical functions were tabulated for use in calculations. Tables are normally used to represent finite strategic games (example 2.34). Numerical functions are usually represented by a mathematical formula or rule. Elementary numerical functions can be illustrated by drawing their graph (figure 2.2).

Example 2.4 (Demand function) In economics, it is common to deal with functions that cannot be specified by any table or rule. A familiar example from elementary economics is the *demand function* for a particular commodity, which specifies the quantity demanded for every price. Since prices and quantities are necessarily positive, the demand function f maps $\Re_+$ to $\Re_+$. The price p is the independent variable, while the quantity demanded $q = f(p)$ is the dependent variable. Rather than specifying a particular functional form (rule) for the function f, the economist is often content to specify certain properties for the function, such as being "downward sloping." The graph of a demand function is called the *demand curve* (figure 2.4).

In figure 2.2 we adopted the mathematician's convention of displaying the independent variable on the horizontal axis, and the dependent vari-

2.1 Functions as Mappings

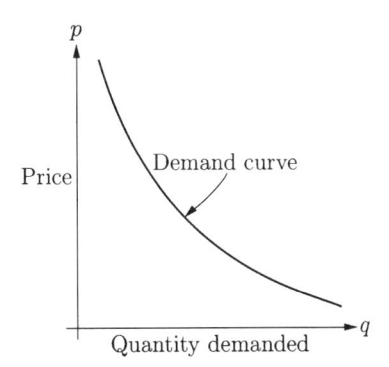

Figure 2.4
A downward sloping demand function

able on the vertical axis. Economists often employ the opposite convention (established by Alfred Marshall), putting the independent variable on the vertical axis. We followed the economist's convention in figure 2.4.

Example 2.5 (Constant and identity functions) The constant and identity functions are particularly simple functions. A *constant* function $f: X \to Y$ assigns all $x \in X$ to a single element $\bar{y}$ of Y, that is $f(X) = \{\bar{y}\}$. The *identity* function $I_X: X \to X$ assigns every element to itself, that is, $I_X(x) = x$ for every $x \in X$.

Given an operator $f: X \to X$, any $x \in X$ for which $f(x) = x$ is called a *fixed point* of f. For the identity function every point is a fixed point. However, an arbitrary operator may or may not have any fixed points. Since significant questions in economics (e.g., the existence of a Nash equilibrium) can be reduced to the existence of a fixed point of a suitable operator, we are interested in deducing conditions that guarantee that an operator has a fixed point. This question is addressed in section 2.4.

Exercise 2.2
Does the rotation operator (example 2.3) have any fixed points?

Any function $f: X \to Y$ induces a mapping between the subsets of X and Y. Thus for any $S \subset X$, the image $f(S)$ of S is

$$f(S) = \{y \in Y : y = f(x) \text{ for some } x \in X\}$$

Similarly for any $T \subset Y$, the *preimage* or *inverse image* $f^{-1}(T)$ of T is set of all $x \in X$ that are mapped into some $y \in T$, that is,

$$f^{-1}(T) = \{x \in X : f(x) \in T\}$$

When T comprises a single element $y \in Y$, it is customary to dispense with the brackets denoting the set $T = \{y\}$ so that the *preimage* of a single element $y \in Y$ is denoted $f^{-1}(y)$. The preimages of single points

$$f^{-1}(y) = \{x \in X : f(x) = y\}$$

are called *contours* of the function f.

Example 2.6 In the birthday mapping (example 2.1), f^{-1} (1 April) is the set of students in the class whose birthday is the 1st of April.

Exercise 2.3
The contours $\{f^{-1}(y) : y \in Y\}$ of a function $f: X \to Y$ partition the domain X.

For any particular $y \in Y$, its preimage $f^{-1}(y)$ may be

- empty
- consist of a single element
- consist of many elements

Where $f^{-1}(y)$ consists of one and only one element for every $y \in Y$, the preimage defines a function from $Y \to X$ which is called the *inverse function*. It is denoted f^{-1}.

Exercise 2.4
The function $f: X \to Y$ has an inverse function $f^{-1}: Y \to X$ if and only if f is one-to-one and onto.

Example 2.7 (Inverse demand function) In economic analysis it is often convenient to work with the inverse demand function $p = f^{-1}$ for a particular commodity, where $p(q)$ measures the price p at which the quantity q would be demanded.

If we have consecutive functions between matching sets, for example, $f: X \to Y$ and $g: Y \to Z$, the functions implicitly define a map between X and Z. This function is called the *composition* of f and g and is denoted by $g \circ f$. That is, $g \circ f: X \to Z$ is defined by

$$g \circ f(x) = g(f(x)) = g(y) \quad \text{where } y = f(x)$$

Exercise 2.5
If $f: X \to Y$ is one-to-one and onto with inverse f^{-1},
$$f^{-1} \circ f = I_X \quad \text{and} \quad f \circ f^{-1} = I_Y$$
where I_X and I_Y are the identity functions on X and Y respectively.

Function Spaces and Sequences

Sets of functions provide a fertile source of linear spaces.

Example 2.8 Let $F(X, Y)$ denote the set of *all* functions from X to Y. Suppose that Y is a linear space. For any $f, g \in F(X, Y)$, define $f + g$ and αf by

$$(f + g)(x) = f(x) + g(x)$$
$$(\alpha f)(x) = \alpha f(x)$$

Then $F(X, Y)$ is another linear space.

Example 2.9 (Polynomials) A polynomial of degree n is a function $f: \Re \to \Re$ defined by

$$f(x) = a_0 + a_1 x + a_2 x^2 + a_3 x^3 + \cdots + a_n x^n$$

It is a linear combination of power functions (example 2.2). We have previously shown that the set of all polynomials is a linear space (example 1.69).

If Y is a *normed* linear space, we can think about convergence of functions in $F(X, Y)$. Let (f^n) be a sequence of functions in $F(X, Y)$. If the sequence of images $(f^n(x))$ converges for every $x \in X$, we say that the sequence (f^n) *converges pointwise* to another function f defined by

$$f(x) = \lim_{n \to \infty} f^n(x) \quad \text{for every } x \in X$$

Convergence of functions is denoted $f^n \to f$. This implies that for every x and $\varepsilon > 0$ there exists N such that

$$\|f(x) - f^n(x)\| < \varepsilon \quad \text{for every } n \geq N$$

In general, N depends on x as well as ε. If there exists an N such that for every $x \in X$,

$$\|f(x) - f^n(x)\| < \varepsilon \quad \text{for every } n \geq N$$

then f^n *converges uniformly* f. Clearly, uniform convergence implies pointwise convergence, but not vice versa.

Example 2.10 (Exponential function) Consider the sequence of polynomials

$$f^n(x) = \sum_{k=0}^{n} \frac{x^k}{k!} = 1 + x + \frac{x^2}{2!} + \frac{x^3}{3!} + \cdots + \frac{x^n}{n!}$$

where $n!$ (called n *factorial*) is the product of the first n integers

$$n! = 1 \cdot 2 \cdot 3 \ldots (n-2)(n-1)n$$

For any $x \in \Re$,

$$|f^n(x) - f^m(x)| = \left| \sum_{k=m+1}^{n} \frac{x^k}{k!} \right| \leq \left| \frac{x^{m+1}}{(m+1)!} \sum_{k=0}^{n-m} \left(\frac{x}{m} \right)^k \right| \leq \frac{|x|^{m+1}}{(m+1)!} \sum_{k=0}^{n-m} \left(\frac{|x|}{m} \right)^k$$

For $n \geq m \geq 2|x|$, $|x|/m \leq 1/2$, and

$$|f^n(x) - f^m(x)| \leq \left(\tfrac{1}{2}\right)^{m+1} \left(1 + \tfrac{1}{2} + \tfrac{1}{4} + \cdots + \left(\tfrac{1}{2}\right)^{n-m}\right)$$

The sum inside the brackets is than 2 (exercise 1.206), and therefore

$$|f^n(x) - f^m(x)| \leq 2 \left(\frac{1}{2} \right)^{m+1} = \frac{1}{2^m} \to 0 \text{ as } m, n \to \infty$$

Clearly, $f^n(x)$ is a Cauchy sequence in $\Re$, and hence it converges to some $y \in \Re$. Since the sequence $f^n(x)$ converges for any $x \in \Re$, it defines a new function $f: \Re \to \Re$ where

$$f(x) = \lim_{n \to \infty} f^n(x) \tag{1}$$

Note that $f^n(0) = 1$ for every n, and therefore $f(0) = 1$. Known as the *exponential function*, the function f is often denoted e^x and written

$$e^x = 1 + \frac{x}{1} + \frac{x^2}{2!} + \frac{x^3}{3!} + \cdots = \sum_{n=0}^{\infty} \frac{x^n}{n!}$$

The exponential function is illustrated in figure 2.5. It has several useful properties, the most important of which is

2.1 Functions as Mappings

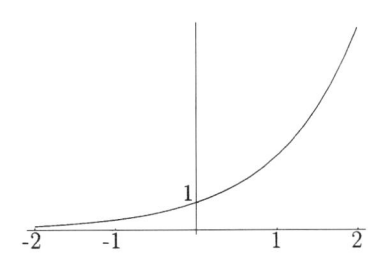

Figure 2.5
The exponential function

$$f(x_1 + x_2) = f(x_1)f(x_2) \quad \text{or} \quad e^{x_1+x_2} = e^{x_1}e^{x_2} \tag{2}$$

This will be proved in chapter 4 (exercise 4.40). Other properties are developed in exercises 2.6 and 2.7.

Exercise 2.6 (Properties of e^x)
Using (2), show that for every $x \in \Re$,

- $e^{-x} = 1/e^x$
- $e^x > 0$
- $e^x \to \infty$ as $x \to \infty$ and $e^x \to 0$ as $x \to -\infty$

This implies that the exponential function maps $\Re$ *onto* $\Re_+$.

Exercise 2.7
The exponential function is "bigger" than the power function, that is,

$$\lim_{x \to \infty} \frac{e^x}{x^n} = \infty \quad \text{for every } n = 1, 2, \ldots$$

[Hint: First show that $\lim_{n \to \infty}(e^x/x) = \infty$.]

Exercise 2.8
Show that the sequence of polynomials

$$f^n(x) = \sum_{k=0}^{n} \frac{x^k}{k!} = 1 + x + \frac{x^2}{2!} + \frac{x^3}{3!} \cdots + \frac{x^n}{n!}$$

converges uniformly on any compact subset $S \subset \Re$.

Functionals

In practice, the most common functions in economics are those which measure things, such as output, utility, profit. To distinguish this common case, functions whose values are real numbers have a special name. A real-valued function $f: X \to \Re$ is called a *functional*.

Remark 2.1 (Extended real-valued function) It is often analytically convenient (see example 2.28) to allow a function to take values in the extended real numbers $\Re^* = \Re \cup \{-\infty\} \cup \{+\infty\}$ (remark 1.6). Such a function $f: X \to \Re^*$ is called an *extended real-valued function*. For convenience we will allow the term functional to include extended real-valued functions.

More generally, the range of functionals may be real or complex numbers. In economics, complex functionals arise in dynamics models. However, they are beyond the scope of this book.

Since $\Re$ is naturally ordered, every functional $f: X \to \Re$ induces an ordering $\succsim_f$ on its domain, defined by

$$x_1 \succsim_f x_2 \Leftrightarrow f(x_1) \geq f(x_2)$$

with

$$x_1 \succ_f x_2 \Leftrightarrow f(x_1) > f(x_2)$$

and

$$x_1 \sim_f x_2 \Leftrightarrow f(x_1) = f(x_2)$$

Thus every functional $f: X \to \Re$ implicitly creates an ordered set $(X, \succsim_f)$. This ordering defines certain useful subsets of X, such as the *upper and lower contour sets* of f defined by

$$\succsim_f(a) = \{x \in X : f(x) \geq a\}$$
$$\precsim_f(a) = \{x \in X : f(x) \leq a\}$$

Similarly the *epigraph* of a functional $f: X \to \Re$ is the set of all points in $X \times \Re$ on or above the graph. Formally

$$\operatorname{epi} f = \{(x, y) \in X \times \Re : y \geq f(x), x \in X\}$$

The corresponding set of points on or below the graph is called the *hypograph*, which is defined as

$$\operatorname{hypo} f = \{(x, y) \in X \times \Re : y \leq f(x), x \in X\}$$

Using these concepts, the analysis of functionals can often be reduced to the analysis of properties of sets, utilizing the results developed in the previous chapter. For example, the contours of a linear functional are affine sets (hyperplanes) (section 1.4.3). Its upper and lower contour sets are called halfspaces. Similarly a function is concave if and only if its hypograph is a convex set.

Remark 2.2 (Functional analysis) So important are functionals that a whole branch of mathematics is devoted to their study. It is called *functional analysis*. We will encounter some of the principal results of this field in chapter 3.

Exercise 2.9
Let X be any set. Let $F(X)$ denote the set of all functionals on X. Show that $F(X)$ is a linear space.

Exercise 2.10
What is the zero element in the linear space $F(X)$?

A functional $f \in F(X)$ is *definite* if takes only positive or negative values. Specifically,

$$f \text{ is } \begin{cases} \text{strictly positive} \\ \text{nonnegative} \\ \text{nonpositive} \\ \text{strictly negative} \end{cases} \text{definite if } \begin{cases} f(x) > 0 \\ f(x) \geq 0 \\ f(x) \leq 0 \\ f(x) < 0 \end{cases} \text{for every } x \in X$$

A functional $f \in F(X)$ is *bounded* if there exists a number k such that $|f(x)| \leq k$ for every $x \in X$.

Example 2.11 (The space $B(X)$) For any set X, let $B(X)$ denote the set of all bounded functionals on X. Clearly, $B(X) \subseteq F(X)$. In fact $B(X)$ is a subspace of $F(X)$.

Exercise 2.11

1. $\|f\| = \sup_{x \in X} |f(x)|$ is a norm on $B(X)$.
2. $B(X)$ is a normed linear space.
3. $B(X)$ is a Banach space.

Example 2.12 A sequence in $B(X)$ converges uniformly if and only if $\|f^n - f\| \to 0$. That is, uniform converges corresponds to convergence of elements in the normed space $B(X)$.

2.1.2 Examples of Functions

In this section we introduce many other examples of functions encountered in economic analysis.

Examples from Mathematics

Example 2.13 (Indicator function) For any subset S of a set X, the *indicator function* $\chi_S \colon X \to \{0, 1\}$ of S is defined by

$$\chi_S(x) = \begin{cases} 1 & \text{if } x \in S \\ 0 & \text{if } x \notin S \end{cases}$$

Mathematicians sometimes call this the *characteristic function*. We reserve the latter term for a related concept in game theory (example 2.36).

We have already encountered some significant examples of functions in chapter 1, such as norms, metrics, and sequences.

Example 2.14 (Norm) Given a linear space X, a norm is a functional $\|\cdot\| \colon X \to \Re$ with the properties

1. $\|\mathbf{x}\| \geq 0$
2. $\|\mathbf{x}\| = 0$ if and only if $\mathbf{x} = 0$
3. $\|\alpha \mathbf{x}\| = |\alpha| \, \|\mathbf{x}\|$ for all $\alpha \in \Re$
4. $\|\mathbf{x} + \mathbf{y}\| \leq \|\mathbf{x}\| + \|\mathbf{y}\|$

The norm assigns to every element $\mathbf{x} \in X$ its size. Similarly, a metric ρ on a metric space X is a function from $X \times X$ to $\Re$ which assigns every pair of elements the distance between them.

Example 2.15 (Sequence) Given a set X, an infinite sequence (x^n) in X is a function from the set of integers $\Re$ to X, where $f(n) = x^n$. This clarifies the distinction between the elements of a sequence $x^1, x^2, \ldots$ and its range $f(\Re)$, comprising all those elements in X that are points in the sequence.

Example 2.16 (Countable set) A set X is called *countable* if it is the range of some sequence, that is, if there exists a function from the set of integers $\Re$ onto X. If there is not no such, the set is called *uncountable*. Clearly, the set $\Re$ is countable. It is a fundamental property of numbers that the set Q of all rational numbers is countable, while any interval of real numbers in uncountable.

Example 2.17 (Probability) Consider a random experiment with sample space S. A probability function P is a real-valued function (functional) on the set of events $\mathscr{S}$ with the properties

1. $P(E) \geq 0$ for every $E \in \mathscr{S}$
2. $P(S) = 1$
3. for any sequence $E^1, E^2, \ldots$ of mutually exclusive events ($E^m \cap E^n = \emptyset$)

$$P\left(\bigcup_{n=1}^{\infty} E^n\right) = \sum_{n=1}^{\infty} P(E^n)$$

$P(E)$ is called the probability of the event E.

When the sample space is finite, every subset of S is an event, and $\mathscr{S} = \mathscr{P}(S)$. Furthermore, using condition 3, we can define $P(E)$ by the probability of the elementary outcomes

$$P(E) = \sum_{s \in E} P(\{s\})$$

Where the sample space S is an infinite set (e.g., $\Re$), not all subsets of S can be events, and the probability function is defined only for a subcollection of $\mathscr{P}(S)$.

Example 2.18 The sample space of a single coin toss is $\{H, T\}$. The probability function for a fair coin is defined by

$$P(\{H\}) = P(\{T\}) = \tfrac{1}{2}$$

Exercise 2.12
The sample space for tossing a single die is $\{1, 2, 3, 4, 5, 6\}$. Assuming that the die is fair, so that all outcomes are equally likely, what is the probability of the event E that the result is even (See exercise 1.4)?

Example 2.19 (Random variable) Analysis of random processes is often simplified through the use of random variables. Any functional $f: S \to \Re$ whose domain is the sample space S of a random experiment is called a *random variable*. In probability theory it is conventional to denote a random variable by X.

Example 2.20 In many board games, progress at each turn is determined by the sum of two fair die. This is the random variable $X: \{1, 2, 3, 4, 5, 6\} \times$

$\{1,2,3,4,5,6\} \to \{2,3,\ldots,12\}$ defined by

$$X(m,n) = m + m$$

where m and n are randomly chosen from $\{1,2,3,4,5,6\}$.

Example 2.21 (Distribution function) Given a random variable $X\colon S \to \Re$, the *distribution function* of X is defined by the probability of the lower contour sets of X

$$F(a) = P(\{s \in S : X(s) \le a\})$$

Example 2.22 (Dynamical system) A *discrete dynamical system* is a set X together with an operator $f\colon X \to X$ which describes the evolution of the system. If the system is in state x^t at time t, the state at time $t+1$ is given by

$$x^{t+1} = f(x^t)$$

If the system begins at x^0, the subsequent evolution of the system is described by repeated application of the function f, that is,

$$x^1 = f(x^0)$$
$$x^2 = f(x^1) = f(f(x^0)) = f^2(x^0)$$
$$x^3 = f(x^2) = f(f(f(x^0))) = f^3(x^0)$$

and

$$x^{t+1} = f(x^t) = f^{t+1}(x^0)$$

The set X of possible states is called the *state space*. x^0 is called the initial position. Particular interest is attached to stationary points or *equilibria* of the dynamical system, where

$$x^{t+1} = f(x^t) = x^t$$

Equilibria are simply the fixed points of the function f.

Example 2.23 (Lag operator) The lag operator is commonly employed in econometrics and in the exposition of dynamic models. Suppose that $(x^0, x^1, x^2, \ldots) \in X^\infty$ is a sequence of observations or economic states. The *lag operator* L generates a new sequence $(y^0, y^1, y^2, \ldots)$, where each y^t is equal to value of x in the previous period. That is,

$$y^1 = Lx^1 = x^0$$
$$y^2 = Lx^2 = x^1$$
$$y^3 = Lx^3 = x^2$$

and

$$y^t = Lx^t = x^{t-1}$$

Given any sequence of observations, the lag operator generates a new sequence in which each observation is shifted one period. The lag operator is a function on the set X^∞ of all sequences, that is, $L: X^\infty \to X^\infty$.

Examples from Economics

Example 2.24 (Production function) In classical producer theory, where the firm produces a single output from n inputs, the technology can be represented by the input requirement set

$$V(y) = \{\mathbf{x} \in \Re_+^n : (y, -\mathbf{x}) \in Y\}$$

which measures the inputs necessary to produce y units of output (example 1.8). Equivalently the relationship inputs and outputs can be expressed by the production function, which specifies the maximum output that can be obtained from given inputs. Formally the *production function f* maps the set of feasible input vectors $\Re_+^n$ to the set of feasible outputs $\Re$ and is defined by

$$f(\mathbf{x}) = \sup\{y : \mathbf{x} \in V(y)\}$$

Example 2.25 (Distance function) The efficiency of any feasible production plan $(y, -\mathbf{x}) \in Y$ can be measured by the distance between $\mathbf{x}$ and the boundary of $V(y)$. Given any technology $V(y)$, the distance function is defined as

$$F(y, \mathbf{x}) = \sup\left\{\lambda > 0 : \frac{1}{\lambda}\mathbf{x} \in V(y)\right\}$$

Example 2.26 (Objective function) Most economic models pose one or more optimization problems. The decision maker has some control over a list (vector) of *choice* or *decision variables* $\mathbf{x}$. The outcome of any choice also depends on the values of one or more exogenous *parameters* θ. The combined effect of the decision variables and parameters is measured by a functional $f: X \times \Theta \to \Re$, which is called the *objective function*. X is the set of feasible values of the decision variables and Θ the set of parameters (the parameter space). Typically the decision maker seeks to maximize the value of the objective function for given parameters, so the optimization problem can be formulated as choosing $\mathbf{x} \in X$ to maximize $f(\mathbf{x}, \boldsymbol{\theta})$ given $\boldsymbol{\theta}$, or succinctly

$$\max_{\mathbf{x} \in X} f(\mathbf{x}, \boldsymbol{\theta}) \tag{3}$$

Example 2.27 (Competitive firm) A competitive firm buys and sells at fixed prices $\mathbf{p} = (p_1, p_2, \ldots, p_n)$. Its profit depends on both the prices $\mathbf{p}$ and the production plan $\mathbf{y}$ it chooses. Specifically, the profit (net revenue) of the production plan $\mathbf{y}$ is given by $f(\mathbf{y}, \mathbf{p}) = \sum_i p_i y_i$. To maximize profit, the firm will seek that feasible production plan $\mathbf{y} \in Y$ that maximizes $f(\mathbf{y}, \mathbf{p})$. Therefore the behavior of a profit-maximizing competitive firm can be represented by the maximization problem

$$\max_{\mathbf{y} \in Y} f(\mathbf{y}, \mathbf{p})$$

The function $f(\mathbf{y}, \mathbf{p}) = \sum_i p_i y_i$ is the firm's objective function, $\mathbf{y}$ are the decision variables and $\mathbf{p}$ the parameters.

Example 2.28 (Value function) The optimization problem (3) implicitly defines a functional on the set of parameters Θ that determines the best performance that can be attained for different values of the parameters. This functional $v: \Theta \to \Re^*$, which is defined by

$$v(\boldsymbol{\theta}) = \sup_{\mathbf{x} \in X} f(\mathbf{x}, \boldsymbol{\theta})$$

is called the *value function*.

The value function is properly an extended real-valued function (remark 2.1). Allowing v to take values in the extended real numbers $\Re^*$ ensures that the function v is well-defined. For any $\boldsymbol{\theta} \in \Theta$, the set $S_\theta = \{f(\mathbf{x}, \boldsymbol{\theta}) : \mathbf{x} \in X\}$ is a subset of $\Re$. S_θ always has an upper bound in $\Re^*$ (remark 1.6). $v(\boldsymbol{\theta}) = +\infty$ for every $\boldsymbol{\theta}$ for which S_θ is unbounded.

Remark 2.3 The value function uses a number of aliases. In the economics literature it is sometimes termed the *maximum value function*. This emphasizes its optimal nature but is inappropriate in minimization problems. Specific instances of the value function in economics have names appropriate to their circumstances, such as the profit function, the cost function, and the indirect utility function. In the mathematics and mathematical programming literature, a name like the *perturbation function* might be used.

Example 2.29 (Profit function) The value function for the problem of a competitive firm (example 2.27)

$$\Pi(\mathbf{p}) = \sup_{\mathbf{y} \in Y} f(\mathbf{p}, \mathbf{y}) = \sup_{\mathbf{y} \in Y} \sum_i p_i y_i$$

is known as the firm's *profit function*. As we will show in chapter 6, much of the behavior of a competitive firm can be deduced from the properties of its profit function.

Exercise 2.13
Where the firm produces just a single output, it is common to distinguish output from inputs. To do this, we reserve p for the price of the output, and let the vector or list $\mathbf{w} = (w_1, w_2, \ldots, w_n)$ denote the prices of the inputs. Using this convention, define the profit function for a profit-maximizing competitive firm producing a single output.

The following exercise shows the importance of allowing the value function to take infinite values.

Exercise 2.14 (Constant returns to scale)
Consider a competitive firm with a constant returns to scale technology $Y \subseteq \Re^n$ (example 1.101). Let $f(\mathbf{p}, \mathbf{y}) = \sum_i p_i y_i$ denote the net revenue (profit) of adopting production plan $\mathbf{y}$ with prices $\mathbf{p}$.

1. If production is profitable at prices $\mathbf{p}$, that is, there exists some $\mathbf{y} \in Y$ such that $f(\mathbf{y}, \mathbf{p}) > 0$, then $\Pi(\mathbf{p}) = +\infty$.

2. Show that the profit function takes only three values, that is, for every $\mathbf{p} \in \Re^n_+$,

$$\Pi(\mathbf{p}) = 0 \quad \text{or} \quad \Pi(\mathbf{p}) = +\infty \quad \text{or} \quad \Pi(\mathbf{p}) = -\infty$$

Example 2.30 (Constrained optimization) In most optimization problems, the choice of $\mathbf{x}$ is constrained to some subset $G(\theta) \subseteq X$ depending on the value of the parameters θ. The general constrained maximization problem can be formulated as choosing $\mathbf{x} \in G(\theta)$ so as to maximize the objective function $f(\mathbf{x}, \boldsymbol{\theta})$, which can be expressed succinctly as

$$\max_{\mathbf{x} \in G(\boldsymbol{\theta})} f(\mathbf{x}, \boldsymbol{\theta}) \tag{4}$$

The corresponding value function $v \colon \Theta \to \Re^*$ is defined as

$$v(\boldsymbol{\theta}) = \sup_{\mathbf{x} \in G(\boldsymbol{\theta})} f(\mathbf{x}, \boldsymbol{\theta})$$

Adopting the convention that $\sup \varnothing = -\infty$, then $v(\boldsymbol{\theta})$ is defined even where the *feasible set* $G(\boldsymbol{\theta}) = \varnothing$.

For given parameter values $\boldsymbol{\theta}$, the solution of the constrained maximization problem is a choice of the decision variables $\mathbf{x}^* \in G(\boldsymbol{\theta})$ such that

$$f(\mathbf{x}^*, \boldsymbol{\theta}) \geq f(\mathbf{x}, \boldsymbol{\theta}) \quad \text{for every } \mathbf{x} \in G(\boldsymbol{\theta})$$

in which case $\mathbf{x}^*$ satisfies the equation

$$v(\boldsymbol{\theta}) = f(\mathbf{x}^*, \boldsymbol{\theta}) \tag{5}$$

Chapter 5 is devoted to techniques for solving constrained optimization problems.

Sometimes an optimization problem is formulated in such a way that the decision maker wishes to minimize rather than maximize the objective function. An example is the firm's cost minimization problem (example 2.31). The general constrained minimization problem is

$$\min_{\mathbf{x} \in G(\boldsymbol{\theta})} f(\mathbf{x}, \boldsymbol{\theta})$$

and the corresponding value function $v \colon \Theta \to \Re^*$ is defined by

$$v(\boldsymbol{\theta}) = \inf_{\mathbf{x} \in G(\boldsymbol{\theta})} f(\mathbf{x}, \boldsymbol{\theta})$$

Since

$$\min_{\mathbf{x} \in G(\boldsymbol{\theta})} f(\mathbf{x}, \boldsymbol{\theta}) = \max_{\mathbf{x} \in G(\boldsymbol{\theta})} -f(\mathbf{x}, \boldsymbol{\theta})$$

minimization problems require no generalization in technique.

Exercise 2.15
For given $\boldsymbol{\theta} \in \Theta$, verify that $\mathbf{x}^* \in G(\boldsymbol{\theta})$ is a solution to (4) if and only if it satisfies (5).

Remark 2.4 (Max versus sup) You will often encounter the value function for a constrained optimization problem defined by

$$v(\boldsymbol{\theta}) = \max_{\mathbf{x} \in G(\boldsymbol{\theta})} f(\mathbf{x}, \boldsymbol{\theta}) \tag{6}$$

using *max* rather than *sup*. Strictly speaking, this is a different meaning of the abbreviation *max* than we have used in the expression

$$\max_{\mathbf{x} \in G(\boldsymbol{\theta})} f(\mathbf{x}, \boldsymbol{\theta}) \tag{7}$$

The *max* in (7) is a verb (maximize), whereas the *max* in (6) is a noun (maximum or maximal element). It is useful to keep this distinction in mind.

By virtue of exercise 2.15, the expression (6) is well defined *provided* that an optimal solution exists. However, we favor the more robust expression

$$v(\boldsymbol{\theta}) = \sup_{\mathbf{x} \in G(\boldsymbol{\theta})} f(\mathbf{x}, \boldsymbol{\theta})$$

since it ensures that the value function is well-defined without this proviso. It also helps to clearly distinguish the noun from the verb.

Example 2.31 (Cost function) In another useful model of the producer, also relevant to the analysis of monopolies and oligopolies, the firm purchases its inputs at fixed prices and seeks to minimize the cost of production. For simplicity, assume that the firm produces a single output. The total cost of the input bundle $\mathbf{x}$ is $\sum_{i=1}^{n} w_i x_i$. To produce any output level y, the firm seeks the input combination $\mathbf{x} \in V(y)$ that minimizes the cost of producing y. The decision variables are input choices $\mathbf{x} \in \Re_+^n$, while the parameters are input prices $\mathbf{w}$ and output y. We can model the cost-minimizing firm as a constrained minimization problem

$$\min_{\mathbf{x} \in V(y)} \sum_{i=1}^{n} w_i x_i$$

The value function for this problem, defined by

$$c(\mathbf{w}, y) = \inf_{\mathbf{x} \in V(y)} \sum_{i=1}^{n} w_i x_i$$

is called the *cost function*.

In the rest of this section, we specify in more detail a particular optimization model that is more sophisticated than standard economics models such as the model of a consumer (example 1.113) or a competitive firm (example 2.27). We introduce it now since it provides an ideal example to illustrate many of the concepts introduced in this chapter, including the fixed point theorem used to establish the existence of an optimal solution (section 2.4).

Example 2.32 (Dynamic programming) Dynamic programming is a special type of constrained optimization problem which takes the form

$$\max_{x_1, x_2, \ldots} \sum_{t=0}^{\infty} \beta^t f(x_t, x_{t+1})$$

subject to $x_{t+1} \in G(x_t)$, $t = 0, 1, 2, \ldots$

$x_0 \in X$ given

Starting from an initial point $x_0 \in X$, the problem is to make a sequence of choices $x_1, x_2, \ldots$ from a set X. The feasible choice x_{t+1} at period $t+1$ is constrained by the choice in the previous period, so that $x_{t+1} \in G(x_t) \subseteq X$. The functional $f(x_t, x_{t+1}): X \times X \to \Re$ measures the return in period t if x_{t+1} is chosen when the state is x_t. Future returns are discounted at the rate β with $0 \leq \beta < 1$. Any sequence $\mathbf{x} = (x_0, x_1, x_2, \ldots)$ in X is called a *plan*. Let X^∞ denote the set of all sequences (plans). The objective is to choose a plan $\mathbf{x}$ so as to maximize the present value (example 1.109) of the total return $\sum_{t=0}^{\infty} \beta^t f(x_t, x_{t+1})$.

Let

$$\Gamma(x_0) = \{\mathbf{x} \in X^\infty : x_{t+1} \in G(x_t), t = 0, 1, 2, \ldots\}$$

denote the set of plans that is *feasible*, starting from the initial point x_0. The feasible set depends on a single parameter x_0, the initial state. Let $U(\mathbf{x})$ denote the total return from feasible plan $\mathbf{x} \in \Gamma(x_0)$, that is,

$$U(\mathbf{x}) = \sum_{t=0}^{\infty} \beta^t f(x_t, x_{t+1})$$

Then the dynamic programming problem can be expressed as a standard constrained optimization problem (example 2.30)

$$\max_{\mathbf{x} \in \Gamma(x_0)} U(\mathbf{x})$$

2.1 Functions as Mappings

What distinguishes this constrained optimization problem from the preceding examples is the infinite planning horizon. If the planning horizon were finite, this problem would be no different in principle to the preceding examples and could be solved in a straightforward manner using the techniques in chapter 5. However, the infinite planning horizon is often an essential ingredient of the model, and necessitates the use of different solution techniques. One fruitful approach uses the value function.

The value function for the dynamic programming problem measures the best that can be achieved from any initial point x_0. It is defined by

$$v(x_0) = \sup_{\mathbf{x} \in \Gamma(x_0)} U(\mathbf{x})$$

Provided that u is bounded, $G(x)$ is nonempty for every $x \in X$ and $0 \leq \beta < 1$, the value function is a bounded functional on X that satisfies the equation

$$v(x) = \sup_{y \in G(x)} \{f(x,y) + \beta v(y)\} \quad \text{for every } x \in X \tag{8}$$

This is known as *Bellman's equation* (exercise 2.16).

A feasible plan $\mathbf{x}^* \in \Gamma(x_0)$ is *optimal* if

$$U(\mathbf{x}^*) \geq U(\mathbf{x}) \quad \text{for every } \mathbf{x} \in \Gamma(x_0)$$

in which case $v(x_0) = U(\mathbf{x}^*)$ (exercise 2.15). The right-hand side of equation (8) defines on operator T on the space $B(X)$ (exercise 2.18), namely

$$(Tv)(x) = \sup_{y \in G(x)} \{f(x,y) + \beta v(y)\}$$

The functional equation (8) can be written

$$v(x) = (Tv)(x)$$

That is, the value function v is a fixed point of the operator T.

Example 2.33 (Optimal economic growth) A particular application of dynamic programming is provided by the following model of optimal economic growth widely used in macroeconomics (Stokey and Lucas 1989). Time is divided into a sequence of periods (months or years). A single good is produced using a technology that requires two inputs, capital k and labor l. In each period the quantity y_t of output produced is given by

$$y_t = f(k_t, l_t)$$

where k_t is the stock of capital and l_t the quantity of labor available at the beginning of the period. This output y_t can be allocated between current consumption c_t and gross investment i_t so that

$$c_t + i_t = y_t$$

Capital depreciates at a constant rate $0 < \delta < 1$ so that the capital stock in the next period becomes

$$k_{t+1} = (1 - \delta)k_t + i_t$$

Labor supply is assumed to be constant. For convenience we assume that $l_t = 1$ in every period, and we define

$$F(k_t) = f(k_t, 1) + (1 - \delta)k_t$$

to be the total supply of goods available at the end of period k, comprising current output $y_t = f(k_t, 1)$ plus undepreciated capital $(1 - \delta)k_t$. The supply of goods must be allocated between current consumption c_t and investment in next period's capital k_{t+1}, so that for every t,

$$c_t + k_{t+1} = F(k_t)$$

or

$$c_t = F(k_t) - k_{t+1} \tag{9}$$

Investment increases future output at the expense of current consumption.

The benefits of consumption c_t in each period are measured by the instantaneous utility function $u(c_t)$ (example 2.58). Future utility is discounted β per period. The problem is to choose the optimal trade-off between consumption and investment in each period so as to maximize total discounted utility

$$\sum_{t=0}^{\infty} \beta^t u(c_t)$$

To cast the problem in the form of the previous example, substitute (9) into the objective function to obtain

$$\max \sum_{t=0}^{\infty} \beta^t u(F(k_t) - k_{t+1})$$

2.1 Functions as Mappings

In each period the choice between consumption c_t and future capital k_{t+1} is constrained by the available output $F(k_t)$:

$$0 \leq c_t, k_{t+1} \leq F(k_t)$$

The optimal growth policy is a sequence $((c_0, k_1), (c_1, k_2), \ldots)$ of consumption and investment pairs that maximizes total utility. It is analytically convenient to regard the future capital stock k_{t+1} as the decision variable, leaving the residual for current consumption c_t according to (9). Therefore optimal growth in this economy can be modeled as choosing a sequence $\mathbf{k} = (k_1, k_2, k_3, \ldots)$ of capital stocks to solve

$$\max \sum_{t=0}^{\infty} \beta^t u(F(k_t) - k_{t+1})$$

subject to $\quad 0 \leq k_{t+1} \leq F(k_t) \qquad (10)$

given k_0.

Let

$$\Gamma(k_0) = \{(k_1, k_2, k_3, \ldots) : 0 \leq k_{t+1} \leq F(k_t), t = 0, 1, 2, \ldots\}$$

denote the set of *feasible* investment plans, which depends on a single parameter k_0, the initial capital stock. The value function for the optimal growth problem is

$$v(k_0) = \sup_{\mathbf{k} \in \Gamma(k_0)} \sum_{t=0}^{\infty} \beta^t u(F(k_t) - k_{t+1}) \qquad (11)$$

The value function measures the total utility that can be derived from an initial capital stock of k_0, presuming that the allocation between consumption and investment is made optimally at each period. Similarly $v(k_1)$ is the total utility that could be derived by optimal investment, starting with a capital stock of k_1. Therefore in period 0 the best that can be done is to choose k_1 to solve

$$\max_{k_1 \in [0, F(k_0)]} u(F(k_0) - k_1) + \beta v(k_1) \qquad (12)$$

where $u(F(k_0) - k_1)$ is the utility derived from consumption in period 0 and $\beta v(k_1)$ is the total utility attainable from capital stock of k_1 in period 1, discounted to period 0.

Assume, for the moment, that the value function v is known. It is then straightforward to solve (12) to determine the optimal consumption c_0 and k_1 in the first period. In the second period the decision maker faces the analogous problem

$$\max_{k_2 \in [0, F(k_1)]} u(F(k_1) - k_2) + \beta v(k_2)$$

and so on, for subsequent periods. Knowledge of the value function enables the decision maker to decompose the multi-period problem (10) into a sequence of single-period optimization problems (12). The optimal growth problem can be solved by finding the value function v defined by (11).

Observe that the value function v defined by (11) is also the value function for the optimization problem (12); that is, the v must satisfy the equation

$$v(k_0) = \sup_{k_1 \in [0, F(k_0)]} u(F(k_0) - k_1) + \beta v(k_1)$$

Indeed, for an optimal investment sequence $(k_0, k_1, k_2, \ldots)$, this equation must hold in all periods, that is,

$$v(k_t) = \sup_{k_{t+1} \in [0, F(k_t)]} u(F(k_t) - k_{t+1}) + \beta v(k_{t+1}), \quad t = 0, 1, 2, \ldots$$

Consequently we can dispense with the superscripts, giving rise to Bellman's equation

$$v(k) = \sup_{0 \le y \le F(k)} u(F(k) - y) + \beta v(y) \quad \text{for every } k \ge 0$$

Since the unknown in this equation is a functional v rather than single point k, it is called a *functional equation*.

Exercise 2.16 (Bellman's equation)
Let

$$v(x_0) = \sup_{\mathbf{x} \in \Gamma(x_0)} U(\mathbf{x})$$

be the value function for the dynamic programming problem (example 2.32). Assume that

- f is bounded on $X \times X$
- $G(x)$ is nonempty for every $x \in X$

Show that v is a bounded functional on X (i.e., $v \in B(X)$) that satisfies the equation

$$v(x) = \sup_{y \in G(x)} \{f(x,y) + \beta v(y)\}$$

for every $x \in X$.

The previous exercise showed that the value function satisfies Bellman's equation. The next exercise shows that every optimal plan must satisfy Bellman's equation at each stage.

Exercise 2.17 (Principle of optimality)
Let

$$v(x_0) = \sup_{\mathbf{x} \in \Gamma(x_0)} U(\mathbf{x})$$

be the value function for the dynamic programming problem (example 2.32). Assume that

- f is bounded on $X \times X$
- $G(x)$ is nonempty for every $x \in X$

Show that the plan $\mathbf{x}^* = (x_0, x_1^*, x_2^*, \ldots) \in \Gamma(x_0)$ is optimal if and only if it satisfies Bellman's equation

$$v(x_t^*) = f(x_t^*, x_{t+1}^*) + \beta v(x_{t+1}^*), \qquad t = 0, 1, 2, \ldots \tag{13}$$

Exercise 2.18
In the dynamic programming problem (example 2.32), assume that

- f is bounded on $X \times X$
- $G(x)$ is nonempty for every $x \in X$

Show that the function T defined by

$$(Tv)(x) = \sup_{y \in G(x)} \{f(x,y) + \beta v(y)\}$$

is an operator on the space $B(X)$ (example 2.11).

Examples from Game Theory

Example 2.34 (Payoff function) It is customary to assign numerical values or *payoffs* to each of the outcomes in a strategic game (section

1.2.6). For example, in Rock–Scissors–Paper (exercise 1.5), if we assign the value 1 for a win, 0 for a draw, and −1 for a loss, the game can be represented in the familiar tabular or matrix form

		Chris		
		Rock	Scissors	Paper
Jenny	Rock	0, 0	1, −1	−1, 1
	Scissors	−1, 1	0, 0	1, −1
	Paper	1, −1	−1, 1	0, 0

The first entry in each cell represents the payoff to Jenny when she chooses that row and Chris chooses the column. The second entry is the corresponding payoff to Chris. For example, if Jenny chooses Rock and Chris chooses Scissors, Jenny wins. Her payoff is 1, while the payoff to Chris is −1.

For each player the mapping from strategies to payoffs is called the *payoff function* of each player. In this game Jenny's payoff function is the function $u_J: S_1 \times S_2 \to \Re$, whose values are given by the first entries in the above table. For example,

$$u_J(\text{Rock, Scissors}) = 1$$

Example 2.35 (Cournot oligopoly) In the standard Cournot model of oligopoly, n firms produce a homogeneous product. The strategic choice of each firm is to choose its output level y_i. Since the product is homogeneous, the market price p depends only on the total output $Y = y_1 + y_2 + \cdots + y_n$ according to the inverse demand function $p(Y)$. The revenue of an individual firm is determined by its own output y_i and the market price p, which depends on the output of all the other firms. Therefore the oligopoly is a game in which the payoff function of firm i is

$$u_i(y_i; \mathbf{y}_{-i}) = p(Y) - c_i(y_i)$$

where $c_i(y_i)$ is the cost function of firm i. $\mathbf{y}_{-i}$ denotes the output choices of the other firms. Provided that the cost functions satisfy appropriate conditions, this game can be shown to have a Nash equilibrium configuration of output choices $\mathbf{y}^* = (y_1^*, y_2^*, \ldots, y_n^*)$, which is known as the *Cournot equilibrium*.

Example 2.36 (Characteristic function) A coalitional game with transferable payoff (example 1.46) comprises

- a finite set N of players
- for every coalition $S \subseteq N$, a real number $w(S)$ that represents the worth of the coalition S (if it acts alone)

The function $w \colon \mathscr{P}(N) \to \mathfrak{R}$, which assigns to every coalition $S \subseteq N$ its worth $w(S)$, is called the *characteristic function* of the game. By convention $w(\emptyset) = 0$.

Exercise 2.19 (Three-person majority game)
Suppose that the allocation of $1 among three persons is to be decided by majority vote. Specify the characteristic function.

Example 2.37 (Value of a game) In section 1.6 we showed that the *nucleolus* (example 1.49) identifies precisely one outcome in every TP-coalitional game. In effect the nucleolus defines a function on the space of games $\mathscr{G}^N$ (example 1.70). It is an example of a value. A *value* for TP-coalitional games is a function φ defined on $\mathscr{G}^N$ that identifies a feasible allocation for every game. Formally any function $\varphi \colon \mathscr{G}^N \to \mathfrak{R}^n$ is a value if $\sum_{i \in N}(\varphi w)_i = w(N)$. Another prominent value for TP-coalitional games is the *Shapley value* (example 3.6). These values differ in their properties. In particular, the Shapley value is a linear function, whereas the nucleolus is nonlinear.

2.1.3 Decomposing Functions

The domain of most functions encountered in economics is a product space. That is, if f maps X to Y, the domain X can usually be decomposed into a product of simpler spaces

$$X = X_1 \times X_2 \times \cdots \times X_n$$

Similarly the co-domain Y can often be decomposed into

$$Y = Y_1 \times Y_2 \times \cdots \times Y_m$$

These decompositions can be helpful in exploring the structure of the function f.

For example, suppose that $f \colon X_1 \times X_2 \times \cdots \times X_n \to Y$, and choose some point $\mathbf{x}^0 = (x_1^0, x_2^0, \ldots, x_n^0) \in X$. The function $\hat{f}_i$ defined by

$$\hat{f}_i(t) = f(x_1^0, x_2^0, \ldots, x_{i-1}^0, t, x_{i+1}^0, \ldots, x_n^0)$$

maps $X_i \to Y$. The function $\hat{f}_i$ allows us to explore the implications of allowing one factor to vary, while holding all the others constant. It implements the economist's notion of ceteris paribus. Sometimes we will use the notation $\hat{f}(t; \mathbf{x}_{-i})$ to indicate such a decomposition, where the variables following the semicolon are regarded as constant.

Example 2.38 (Total cost function) A firm's cost function $c(\mathbf{w}, y): \Re_+^n \times \Re \to \Re$ measures the minimum cost of producing output level y when the input prices are $\mathbf{w} = (w_1, w_2, \ldots, w_n)$ (see example 2.31). Frequently we are interested in analyzing just the impact of output on costs. Holding input prices $\mathbf{w}$ constant, the function $\hat{c}: \Re \to \Re$,

$$\hat{c}(y) = c(\mathbf{w}, y)$$

is the familiar total cost function of elementary economics (figure 2.6).

Sometimes it is appropriate to allow a subset of the variables to vary, while holding the remainder constant. This is illustrated in the following example.

Example 2.39 (Short-run production function) A firm produces a single output using n inputs. Its technology is described by the production function $y = f(x_1, x_2, \ldots, x_n)$. Suppose that some inputs are fixed in the short-run. Specifically, suppose that we can decompose the list of inputs $\mathbf{x}$ into two sublists $(\mathbf{x}_f, \mathbf{x}_v)$, where $\mathbf{x}_f \in X_f$ are inputs that are fixed in the

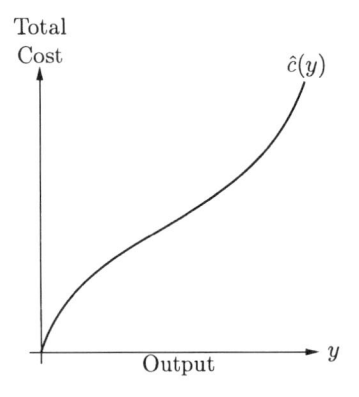

Figure 2.6
A total cost function

short-run and $\mathbf{x}_v \in X_v$ are variable inputs. The function $\hat{f} \colon X_v \to \Re$ defined by

$$\hat{f}(\mathbf{x}_v; \mathbf{x}_f) = f(\mathbf{x}_v, \mathbf{x}_f)$$

is known as the *short-run production function*. It describes the output obtainable from various levels of the variable factors $\mathbf{x}_v$, while holding the fixed inputs $\mathbf{x}_f$ constant.

When the co-domain Y is a product space $Y = Y_1 \times Y_2 \times \cdots \times Y_m$, it is useful to decompose a function $\mathbf{f} \colon X \to Y$ into m components $\mathbf{f} = (f_1, f_2, \ldots, f_m)$, where each component $f_i \colon X \to Y_i$. For any point $x \in X$, its image is

$$\mathbf{f}(x) = (f_1(x), f_2(x), \ldots, f_m(x))$$

In line with our notational convention, we use a bold font to designate a function whose co-domain is a product space. Almost invariably the co-domain Y is a subset of $\Re^m$, so each component $f_i \colon X \to \Re$ is a functional on X. It is often convenient to alternate between these two different perspectives, sometimes viewing $\mathbf{f}$ as a function from X to $\Re^m$ and other times regarding $\mathbf{f}$ as a list of functionals on X. These different perspectives are illustrated in the following example.

Example 2.40 (Constrained optimization) In the general constrained optimization problem (example 2.30)

$$\max_{\mathbf{x} \in G(\theta)} f(\mathbf{x}, \boldsymbol{\theta})$$

the constraint set $G(\theta)$ can often be represented a function $\mathbf{g} \colon X \times \Theta \to Y \subseteq \Re^m$, so the general constrained maximization problem becomes

$$\max_{\mathbf{x} \in X} f(\mathbf{x}, \boldsymbol{\theta})$$

subject to $\quad \mathbf{g}(\mathbf{x}, \boldsymbol{\theta}) \leq \mathbf{0}$

Sometimes it is convenient to think of the constraint as a single function $\mathbf{g} \colon X \times \Theta \to \Re^m$. At other times it is more convenient to decompose $\mathbf{g}$ into m separate constraints (functionals) $g_j \colon X \times \Theta \to \Re, j = 1, 2, \ldots, m$, with

$$g_1(\mathbf{x}, \boldsymbol{\theta}) \leq 0, \quad g_2(\mathbf{x}, \boldsymbol{\theta}) \leq 0, \quad \ldots, \quad g_m(\mathbf{x}, \boldsymbol{\theta}) \leq 0$$

We will take advantage of this decomposition in chapter 5.

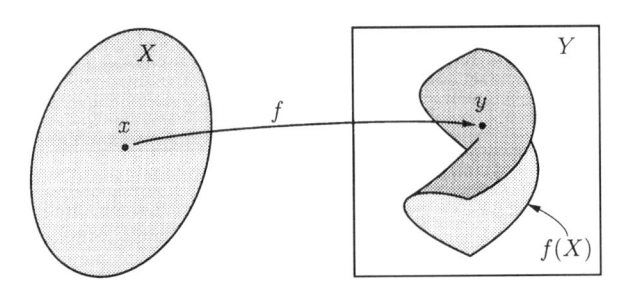

Figure 2.7
Illustrating a function from $\mathscr{R}^n \to \mathscr{R}^m$

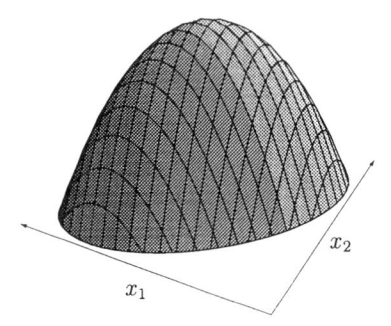

Figure 2.8
A function from $\mathscr{R}^2 \to \mathscr{R}$

2.1.4 Illustrating Functions

Some functions can be illustrated directly (figure 2.3) or by means of their graph (figures 2.2, 2.4, 2.5, and 2.8). The dimensionality of most economic models precludes such simple illustrations, and it is necessary to resort to schematic illustrations such as figures 2.1 and 2.7.

Many functions that we meet in economics, including objective functions, payoff functions, production and cost functions, are functionals. In other words, they are real-valued functions $f: X \to \Re$. Where the domain $X \subseteq \Re^2$, the graph of f,

$$\text{graph}(f) = \{((x_1, x_2), y) \in \Re^2 \times \Re = \Re^3) : y = f(x_1, x_2)\}$$

is a surface in $\Re^3$. With imagination this can be illustrated on a two-dimensional page (figure 2.8). Even where the economic model requires

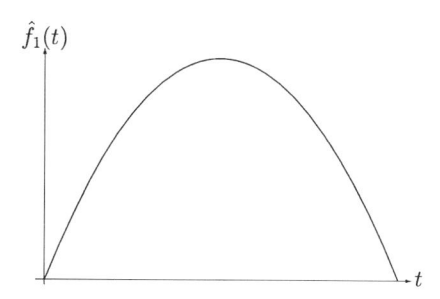

Figure 2.9
A vertical cross section of figure 2.8

more than two decision variables, we often use illustrations like figure 2.8 to help cement ideas. Alternatively, with higher dimensions and more general spaces, we sometimes depict the graph schematically, allowing a single horizontal axis to represent the domain (figure 5.11).

The decompositions discussed in the previous section can be very useful in illustrating higher-dimensional functions. For example, for any functional $f \in F(X)$, where $X \subseteq \Re^n$, the function

$$\hat{f}_i(t) = f(t; \mathbf{x}^0_{-i}) = f(x_1^0, x_2^0, \ldots, x_{i-1}^0, t, x_{i+1}^0, \ldots, x_n^0)$$

for any $\mathbf{x}^0 \in X$ maps $\Re$ to itself. The graph of $\hat{f}_i$ can be depicted as a curve in the plane (figure 2.9). It provides a *vertical cross section* of the graph of f, parallel to the ith axis. Figure 2.9 shows a vertical cross section of figure 2.8. The total cost curve (figure 2.6) provides another example.

Exploiting the linear structure of $\Re^n$, another useful cross section of a function $f \in F(\Re^n)$ is defined by

$$h(t) = f(t\mathbf{x}^0) = f(tx_1^0, tx_2^0, \ldots, tx_n^0)$$

where $\mathbf{x}^0$ is an arbitrary point in $\Re^n$. Again, h maps $\Re$ into $\Re$. Its graph is vertical cross section of the graph of f, this time along a ray through the origin. This cross section is particularly useful in describing technologies —it represents changing the scale of production while leaving input proportions fixed.

Recall that the contours $f^{-1}(y)$ of a function $f: X \to Y$ partition the domain X. Another useful way to explore the geometry of a function is to depict some contours (figure 2.10). These correspond to *horizontal* cross

x_2

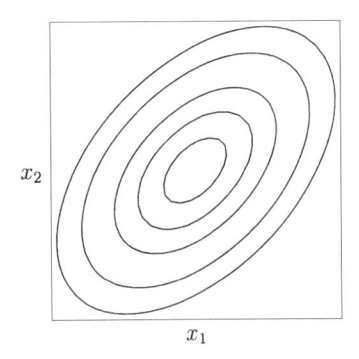

x_1

Figure 2.10
Horizontal cross sections (contours) of figure 2.8

sections of the graph. The use of contours to reduce dimensionality is familiar in topographical and weather maps. Contours are equally useful in economics.

Remark 2.5 (Tomography) Tomography is the technique of obtaining a planar image of a cross section of a human body or other object. In a CAT (computer-assisted tomography) scan, a sequence of parallel cross sections is obtained. These images enable the radiologist to construct a detailed picture of the interior of the body. Economists use an analogous technique, deducing the structure of a multidimensional function by mentally combining judicious cross sections.

Example 2.41 (Anatomy of the production surface) Economists frequently resort to horizontal and vertical cross sections to describe the properties of a technology as represented by a production function. Figure 2.11 illustrates a Cobb-Douglas production function

$$f(x_1, x_2) = x_1^{a_1} x_2^{a_2}, \qquad a_1 + a_2 = 1$$

together with three useful cross sections. Alongside the surface, the second quadrant depicts a sequence of horizontal cross sections or contours, known as *isoquants*. They represent the different combinations of x_1 and x_2 that can be used to produce a particular output level, illustrating the substitutability between the inputs. The third quadrant is a vertical cross section parallel to the x_1 axis. It shows the output produced by varying the amount of x_1 while holding x_2 constant, illustrating the diminishing

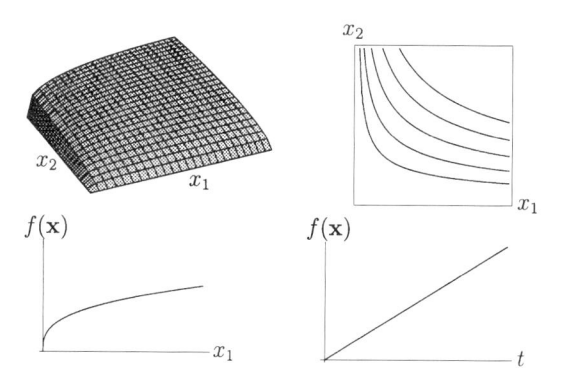

Figure 2.11
A Cobb-Douglas function and three useful cross sections

marginal product of factor 1. Alongside this, the fourth quadrant shows a vertical cross section along an expansion path (ray through the origin). It shows the output obtained by changing the scale of production, holding the input ratios fixed. In this particular case, the cross section is linear, illustrating the constant returns to scale of the Cobb-Douglas technology when $a_1 + a_2 = 1$.

2.1.5 Correspondences

The budget set $X(\mathbf{p}, m)$ of a consumer (example 1.113) is a subset of the consumption set X. The composition of the budget set, the bundles that are affordable, depends on the prices of all goods $\mathbf{p}$ and the consumer's income m. In fact affordability determines a function from set of feasible prices and incomes to the set of all subsets of the consumption set ($\mathcal{P}(X)$). This situation, where the co-domain of a function is the power set of another set, occurs frequently in economics. It justifies a slight generalization of the concept of a function that is known as a correspondence.

Given two sets X and Y, a *correspondence* φ is a rule that assigns to every element $x \in X$ a nonempty subset $\varphi(x)$ of Y. Every correspondence φ between X and Y can be viewed simply as a function from X to $\mathcal{P}(Y)$. Alternatively, it can be viewed as a multi-valued function, since any x can be associated with more than one $y \in Y$. Although a correspondence $\varphi: X \rightrightarrows Y$ is a proper function between X and $\mathcal{P}(Y)$, it is the relationship between X and Y that we wish to emphasize, which creates the need to introduce a new concept. We will denote a correspondences by

178   Chapter 2   Functions

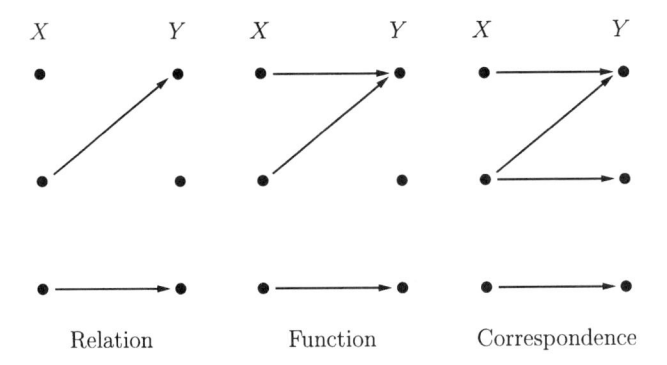

Figure 2.12
Comparing a relation, a correspondence, and a function

$\varphi: X \rightrightarrows Y$, using the double arrow to distinguish it from a function between X and Y. Correspondences, which arise so naturally in economic analysis, have an unjustified reputation for difficulty, which we would like to dispel. In fact we have already met several examples of correspondences. We discuss some of these after the following remark.

Remark 2.6 (Correspondences, functions, and relations) Recall that a relation between two sets X and Y is a subset of their product $X \times Y$, that is, a collection of pairs of elements from X and Y. A correspondence $\varphi \rightrightarrows X \to Y$ is a relation between X and Y in which every $x \in X$ is involved, that is, whose domain is the whole of X. A function is relation (in fact a correspondence) in which every $x \in X$ is involved only once; that is, every $x \in X$ has a unique relative in Y. For example, in figure 2.12, the left-hand panel is merely a relation, since not all points in X are related to Y. The right-hand panel is a correspondence, since every $x \in X$ is related to some $y \in Y$. It is not a function, since there is an x that is related to more than one y. The middle panel is a legitimate function, since every x is related to one and only one y.

By convention, we do not distinguish between a correspondence in which every image set contains a single element and a function.

Example 2.42 (Upper and lower contour sets) If $\succsim$ is an order relation on a set X, the upper contour sets

$$\succsim(a) = \{x \in X : x \succsim a\}$$

specify a correspondence on X, as do the sets $\succ(a)$. Similarly the lower contours sets $\precsim(a)$, $\prec(a)$ and the indifference sets $\sim(a)$ are all correspondences on X.

Example 2.43 (Preimage) Given any function $f\colon X \to Y$, the preimage

$$f^{-1}(y) = \{x \in X : f(x) = y\}$$

defines a correspondence between the range $f(X)$ of f and X.

Example 2.44 (Input requirement set) For a single-output technology Y, the input requirements sets

$$V(y) = \{\mathbf{x} \in \Re_+^n : (y, -\mathbf{x}) \in Y\}$$

define a correspondence between output levels $y \in \Re_+$ and the set of all nonnegative input levels $\Re_+^n$.

Example 2.45 (Coalitional game) In a coalitional game (section 1.2.6), the relation which specifies the subset of outcomes over which each coalition is decisive is a correspondence between the set of coalitions and the set of outcomes. Therefore a coalitional game comprises

- a finite set N of *players*
- a set X of *outcomes*
- for each player $i \in N$ a preference relation $\succsim_i$ on the set of outcomes X
- a correspondence $W\colon \mathscr{P}(N) \rightrightarrows X$ that specifies for each coalition S the set $W(S)$ of outcomes over which it is decisive

The correspondence W is typically called the *characteristic function* of the game (despite being a correspondence), although it is conventional to use a capital W to contrast with the little w used to denote the characteristic function of a game with transferable payoff.

A correspondence $\varphi\colon X \rightrightarrows Y$ is called *closed-valued* if $\varphi(x)$ is closed for every $x \in X$. Similarly φ is *compact-valued* if $\varphi(x)$ compact- and *convex-valued* if $\varphi(x)$ convex for every $\mathbf{x} \in X$. Alternatively, we say that φ has closed, compact, or convex sections.

Example 2.46 (Budget correspondence) For fixed prices $\mathbf{p}$ and income m, the consumer's budget set

$$X(\mathbf{p}, m) = \{\mathbf{x} \in X : p_1 x_1 + p_2 x_2 + \cdots + p_n x_n \leq m\}$$

is a subset of the consumption set X. The budget set depends on both prices $\mathbf{p}$ and income m. Let P denote the set of all price and income pairs for which the budget set is not empty, that is,

$$P = \{(\mathbf{p}, m) \in \Re^n \times \Re : X(\mathbf{p}, m) \neq \emptyset\}$$

Affordability determines a correspondence $X(\mathbf{p}, m) : P \rightrightarrows X$ between the parameter set P and the consumption set X, which is called the budget correspondence. The budget correspondence $X(\mathbf{p}, m)$ is convex-valued (exercise 1.232). It is also closed-valued and compact-valued provided all prices are positive $\mathbf{p} > \mathbf{0}$ (exercise 1.231).

Example 2.47 (Demand correspondence) In example 1.115 we showed the existence of an optimal choice $\mathbf{x}^*$ for a consumer with continuous preferences. For given prices $\mathbf{p}$ and income m in P, there may be more than one optimal choice. The way in which the set of optimal choices $\mathbf{x}^*(\mathbf{p}, m)$ varies with prices and income defines a correspondence from P to the consumption set X that is called the *demand correspondence* of the consumer.

Exercise 2.20
Assume that the consumer's preferences are continuous and strictly convex. Show that the demand correspondence is single valued. That is, the demand correspondence is a *function* mapping $P \to X$.

Example 2.48 (Best response correspondence) In a strategic game (section 1.2.6), the optimal choice of any player depends on the strategies of the other players. The set of strategies of player i that constitutes her best response to the strategies of the other players $\mathbf{s}_{-i}$ is called player i's *best response correspondence*

$$B_i(\mathbf{s}_{-i}) = \{s_i \in S_i : (s_i, \mathbf{s}_{-i}) \succsim_i (s_i', \mathbf{s}_{-i}) \text{ for every } s_i' \in S_i\}$$

Since player i may have more than one optimal response to any $\mathbf{s}_{-i}$, B_i is a correspondence between S_{-i} and S_i (rather than a function). The best response correspondence maps S_{-i} into the power set $\mathscr{P}(S_i)$ of S_i.

For some purposes (e.g., example 2.52) it is convenient to regard the domain of each player's best response correspondence as the whole strategy space S rather than just S_{-i}. This can be done by simply ignoring the

S_i dimension of the domain. The extended best response correspondence is then defined identically, namely

$$B_i(\mathbf{s}) = \{s_i \in S_i : (s_i, \mathbf{s}_{-i}) \succsim_i (s'_i, \mathbf{s}_{-i}) \text{ for every } s'_i \in S_i\}$$

In effect, the extended best response correspondence is constant on S_i. Game theorists often refer to B_i as the best response *function* rather than the more correct best response *correspondence*, even when they know that B_i is not strictly a function.

Exercise 2.21
$\mathbf{s}^* = (s_1^*, s_2^*, \ldots, s_n^*)$ is a Nash equilibrium if and only if

$$s_i^* \in B(\mathbf{s}^*) \qquad \text{for every } i \in N$$

Exercise 2.22 (Rationalizability)
In a strategic game, a strategy of player i is *justifiable* if it is a best response to some possible (mixed) strategy (example 1.98) of the other players, that is,

$$s_i \text{ is justifiable} \Leftrightarrow s_i \in B_i(\Sigma_{-i})$$

where Σ_{-i} is the set of mixed strategies of the opposing players (example 1.110). Let B_i^1 denote the set of justifiable strategies of player i. Then

$$B_i^1 = B_i(\Sigma_{-i})$$

A strategy of player i is rationalizable if it is justifiable using a belief that assigns positive probability only to strategies of $j \neq i$ that are justifiable, if these strategies are justified using beliefs that assign positive probability only to justifiable strategies of i, and so on. To formalize this definition, define the sequence of justifiable strategies

$$B_i^n = B_i(B_{-i}^{n-1})$$

The set of *rationalizable strategies* for player i is $R_i = \bigcap_{n=0}^{\infty} B_i^n$. That is, the set of rationalizable strategies is those that are left after iteratively discarding unjustified strategies. Show that when S is compact and $\succsim_i$ is continuous, there are rationalizable strategies for every game, that is, $R_i \neq \emptyset$. [Hint: Use the nested intersection theorem (exercise 1.117).]

Exercise 2.23
Every Nash equilibrium is rationalizable.

Example 2.49 (Solution correspondence) The general constrained optimization problem (example 2.30)

$$\max_{\mathbf{x} \in G(\boldsymbol{\theta})} f(\mathbf{x}, \boldsymbol{\theta})$$

defines a correspondence between Θ and X, known as the *solution correspondence*, which specifies the optimal choice of the decision variables $\mathbf{x}$ for varying values of the parameters $\boldsymbol{\theta}$. Formally let the solution correspondence is defined as

$$\varphi(\boldsymbol{\theta}) = \arg\max_{\mathbf{x} \in G(\boldsymbol{\theta})} f(\mathbf{x}, \boldsymbol{\theta})$$

where arg max denotes the set of elements of $G(\boldsymbol{\theta})$ that maximize $f(\mathbf{x}, \boldsymbol{\theta})$. Economists are very interested in the properties that φ inherits from f and $G(\boldsymbol{\theta})$. The demand correspondence (example 2.47) and best response correspondence (example 2.48) are particular examples of solution correspondences.

Exercise 2.24
Show that the value function (example 2.28) can be alternatively defined by

$$v(\boldsymbol{\theta}) = f(\mathbf{x}^*, \boldsymbol{\theta}) \quad \text{for } \mathbf{x}^* \in \varphi(\boldsymbol{\theta})$$

Most of the vocabulary of functions applies to correspondences with little change. The domain of a correspondence between X and Y is of course X. The range is

$$\varphi(X) = \{ y \in Y : y \in \varphi(x) \text{ for some } x \in X \}$$

The graph of a correspondence φ between X and Y is

$$\text{graph}(\varphi) = \{(x, y) \in X \times Y : y \in \varphi(x)\}$$

The graph of a hypothetical correspondence is illustrated by the shaded area in figure 2.13.

A correspondence φ is *closed* if its graph is closed in $X \times Y$, that is, for every pair of sequences $x^n \to x$ and $y^n \to y$ with $y^n \in \varphi(x^n)$, $y \in \varphi(x)$. Similarly a correspondence φ is *convex* if its graph is a convex subset of $X \times Y$; that is, for every $x^1, x^2 \in X$ and corresponding $y^1 \in \varphi(x^1)$ and $y^2 \in \varphi(x^2)$,

$$\alpha y^1 + (1 - \alpha) y^2 \in \varphi(\alpha x^1 + (1 - \alpha) x^2)$$

2.1 Functions as Mappings

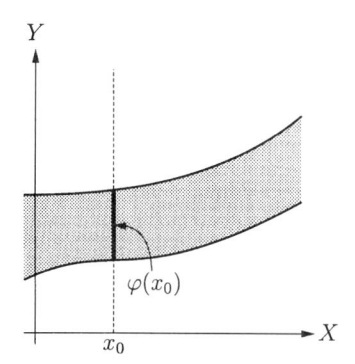

Figure 2.13
The graph of a correspondence

Every closed correspondence is closed-valued and every convex correspondence is convex-valued, but the converse is false. A closed-valued and convex-valued correspondence may be neither closed nor convex, as illustrated in the following example.

Example 2.50 Let $X = [0, 1]$. Define $\varphi : X \rightrightarrows X$ by

$$\varphi(x) = \begin{cases} \{x\} & 0 \leq x < 1 \\ \{0\} & x = 1 \end{cases}$$

φ is both closed- and convex-valued for every $x \in X$. However, φ is neither closed nor convex. The sequence (x^n, y^n) defined by $x^n = y^n = 1 - 1/n$ belongs to graph(φ). However the sequence converges to $(1, 1) \notin$ graph(φ). Therefore graph(φ) is not closed, nor is φ convex. The point $(0, 0)$ and $(1, 0)$ belong to graph(φ), but there convex combination $(\frac{1}{2}, 0)$ does not.

Example 2.51 (Input requirement sets) For a single-output technology, the input requirement sets $V(y)$,

$$V(y) = \{\mathbf{x} \in \Re_+^n : (y, -\mathbf{x}) \in Y\}$$

define a correspondence (Example 2.44) between desired output ($y \in \Re_+$) and required inputs ($\mathbf{x} \in \Re_+^n$). The graph of this correspondence is almost but not quite the production possibility set Y. The graph of the correspondence is

$$\text{graph}(V) = \{(y, \mathbf{x}) \in \Re_+ \times \Re_+^n : x \in V(y)\}$$

while the production possibility set Y is

$$Y = \{(y, -\mathbf{x}) \in \Re_+ \times \Re_+^n : x \in V(y)\}$$

the distinction being required by the convention that inputs are specified as negative quantities in production plans. For some purposes it is more convenient to use graph(V) rather than Y.

In producer theory it is conventional to assume that the input requirement set $V(y)$ is convex for every y (example 1.163). In other words, we usually assume that the the input requirements sets define a convex-valued correspondence. In general, V is not further assumed to be a convex correspondence. However, if the technology is restricted so that the production possibility set Y is convex, then V is a convex correspondence (exercise 2.25).

Exercise 2.25
Let Y be the production possibility set for a single-output technology and $V(y)$ denote the corresponding input requirements sets

$$V(y) = \{\mathbf{x} \in \Re_+^n : (y, -\mathbf{x}) \in Y\}$$

Then Y is convex if and only if $V(y)$ is a convex correspondence.

Exercise 2.26
Suppose that the constraint correspondence $G(\theta)$ in the constrained optimization problem (example 2.30)

$$\max_{\mathbf{x} \in G(\theta)} f(\mathbf{x}, \theta)$$

is defined by a set of inequalities (example 2.40)

$$g_1(\mathbf{x}, \theta) \leq 0, \ g_2(\mathbf{x}, \theta) \leq 0, \ \ldots, \ g_m(\mathbf{x}, \theta) \leq 0$$

If each functional $g_j(\mathbf{x}, \theta) \in F(X \times \Theta)$ is convex jointly in $\mathbf{x}$ and θ, then the correspondence

$$G(\theta) = \{\mathbf{x} \in X : g_j(\mathbf{x}, \theta) \leq 0, j = 1, 2, \ldots, m\}$$

is convex.

Individual correspondences can be combined in analogous ways to functions. If $\varphi: X \rightrightarrows Y$ and $\psi: Y \rightrightarrows Z$ are two correspondences, their *composition* $\psi \circ \varphi$ is a correspondence between X and Z, which is defined by

$$\psi \circ \varphi(x) = \bigcup_{y \in \varphi(x)} \psi(y)$$

If $\varphi_i \colon X \rightrightarrows Y_i$, $i = 1, 2, \ldots, n$, is a collection of correspondences with common domain X, their *product* is a correspondence between X and the Cartesian product $\prod Y_i$ defined by

$$\varphi(x) = \prod_i \varphi_i(x)$$

Where the co-domains Y_i belong to a linear space, their *sum* is a correspondence $\varphi \colon X \rightrightarrows \sum Y_i$ defined by

$$\varphi(x) = \sum_i \varphi_i(x)$$

Correspondences also invite some operations which are inapplicable to functions. If $\varphi \colon X \rightrightarrows Y$ is a correspondence between X and a convex set Y, its convex hull (conv φ) is another correspondence between X and Y defined by

$$(\text{conv } \varphi)(x) = \text{conv}(\varphi(x)) \quad \text{for every } x \in X$$

Similarly, where Y is a metric space, the closure of φ is a correspondence $\bar{\varphi} \colon X \to Y$ defined by

$$\bar{\varphi}(x) = \overline{\varphi(x)} \quad \text{for every } x \in X$$

Economic models often establish a correspondence between a set X and itself. A *fixed point* of a correspondence $\varphi \colon X \to X$ is an element that belongs to its own image set, that is an $x \in X$ such that $x \in \varphi(x)$.

Example 2.52 (Nash equilibrium as a fixed point.) Consider a strategic game of n players with strategy space $S = s_1 \times s_2 \times \cdots \times s_n$. In exercise 2.21, we showed that $\mathbf{s}^* = (s_1, s_2, \ldots, s_n) \in S$ is Nash equilibrium if and only if

$$s_i^* \in B_i(\mathbf{s}^*) \quad \text{for every } i \in N$$

The Cartesian product of the individual best response correspondences defines a correspondence φ on the whole strategy space S given by

$$\varphi(\mathbf{s}) = B_1(\mathbf{s}) \times B_2(\mathbf{s}) \times \cdots \times B_n(\mathbf{s})$$

s^* is Nash equilibrium if and only if $s^* \in \varphi(s^*)$, that is s^* is a fixed point of φ. Therefore the search for a Nash equilibrium can be reduced to the search for a fixed point of an appropriate mapping. An equilibrium will be ensured if the mapping can be guaranteed to have a fixed point. Section 2.4 discusses the necessary conditions for existence of a fixed point.

Selections

Given a correspondence φ between X and Y, we can always construct a function by choosing some $y \in \varphi(x)$ for every x, since $\varphi(x)$ is nonempty for every x. Any function constructed from a correspondence in this way is called a *selection*. We use the notation $f \in \varphi$ to denote that f is a selection from the correspondence φ. Unless the correspondence is in fact a function, there will be many selections from any correspondence.

2.1.6 Classes of Functions

As we remarked in opening this chapter, we are especially interested in functions that respect the structure of their domains. Of course, that structure can take various forms. Functions that respect the order structure of their domains are called monotone functions. Continuous functions preserve the geometry of the spaces that they link, while linear functions preserve the algebraic structure. In the next section we investigate monotone functions and correspondences, while in section 2.3 we deal with continuous functions and correspondences. Linear and related functions are explored in chapter 3. In the absence of further qualification, the domain and range are assumed to appropriately structured sets. In section 2.2 (monotone functions) all sets are assumed to be ordered sets. Similarly in section 2.3 all sets are assumed to be metric spaces.

2.2 Monotone Functions

A function between ordered sets X and Y is called monotone if it respects the order of X and Y. f is *increasing* if it preserves the ordering so that

$$x_2 \succsim_X x_1 \Rightarrow f(x_2) \succsim_Y f(x_1)$$

where $\succsim_X$ and $\succsim_Y$ are the orders on X and Y respectively. f is strictly increasing if *in addition*

$$x_2 \succ_X x_1 \Rightarrow f(x_2) \succ_Y f(x_1)$$

On the other hand, $f \colon X \to Y$ is *decreasing* if it reverses the ordering

$$x_2 \succsim_X x_1 \Rightarrow f(x_2) \precsim_Y f(x_1)$$

It is *strictly decreasing* if in addition

$$x_2 \succ_X x_1 \Rightarrow f(x_2) \prec_Y f(x_1)$$

f is *monotone* if it is either increasing or decreasing. Some authors use the term monotone increasing, although the first adjective is redundant.

Exercise 2.27 (Identity function)
Show that the identity function I_X (example 2.5) is strictly increasing.

Remark 2.7 For mappings between arbitrary ordered sets, the mathematical terms *isotone* and *antitone* for increasing and decreasing functions respectively are more appropriate. However, most monotone functions in economics are real-valued, and the terms increasing and decreasing are conventional.

Many authors use *nondecreasing* in place of increasing, reserving increasing for strictly increasing. Our terminology carries some risk of confusion. For example, a constant function is "increasing." On the other hand, our terminology is internally consistent (an increasing function preserves the weak order) and less cumbersome.

The following properties of monotone functions are used frequently.

Exercise 2.28
If $f \colon X \to Y$ and $g \colon Y \to Z$ are increasing functions, so is their composition $g \circ f \colon X \to Z$. Moreover, if f and g are both strictly increasing, then so is $g \circ f$.

Exercise 2.29
If X and Y are totally ordered (chains) and $f \colon X \to Y$ is strictly increasing, then f has a strictly increasing inverse $f^{-1} \colon f(X) \to X$.

Exercise 2.30
If $f \colon X \to \Re$ is increasing, $-f \colon X \to \Re$ is decreasing.

Exercise 2.31
If $f, g \in F(X)$ are increasing, then

- $f+g$ is increasing
- αf is increasing for every $\alpha \geq 0$

Therefore the set of all increasing functionals on a set X is a cone in $F(X)$. Moreover, if f is strictly increasing, then

- $f+g$ is strictly increasing
- αf is strictly increasing for every $\alpha > 0$

Exercise 2.32
If f and g are strictly positive definite and strictly increasing functionals on X, then so is their product fg defined by

$$(fg)(x) = f(x)g(x)$$

Example 2.53 (The power function) The power functions f_n defined by

$$f_n(x) = x^n, \quad n = 1, 2, 3, \ldots$$

are strictly increasing on $\Re_+$. First f_1 is the identity function and therefore strictly increasing (exercise 2.27). f_1 is also strictly positive definite on $\Re_{++}$. By the previous exercise,

$$f_2(x) = f_1(x)f_1(x)$$

is strictly increasing and strictly positive definite on $\Re_{++}$. Similarly

$$f_3(x) = f_1(x)f_2(x)$$

is strictly increasing and strictly positive definite on $\Re_{++}$. Continuing in this fashion using exercise 2.32, we can demonstrate that for every n,

$$f_n(x) = f_1(x)f_{n-1}(x) \tag{14}$$

is strictly increasing and strictly positive definite on $\Re_{++}$.

Note also that $f_1(0) = 0$ and therefore $f_n(0) = 0$ for every n by (14). Since f_n is strictly positive definite on $\Re_{++}$, $f_n(0) < f_n(x)$ for every $x \in \Re_{++}$. Furthermore $0 < x$ for every $x \in \Re_{++}$. We conclude that f_n is strictly increasing on $\Re_+$.

Remark 2.8 (Induction) Example 2.53 illustrates the common technique of *proof by induction*, applicable when seeking to demonstrate that a property belongs to every member of a sequence. We first prove that the

2.2 Monotone Functions

first member of the sequence has the property. Then we show that the nth member of the sequence has the property if member $n-1$ has the property. This is known as the *inductive step*, exemplified by applying exercise 2.32 to (14). Together these steps prove that every member of the sequence has the property.

Example 2.54 (Exponential function) We now let f^n denote the sequence of polynomials

$$f^n = \sum_{k=0}^{n} \frac{x^k}{k!}$$

Example 2.53 and exercise 2.31 shows that f^n is strictly increasing on $\mathfrak{R}_+$ for every $n = 1, 2, \ldots$ This implies that the exponential function (example 2.10)

$$e^x = \lim_{n \to \infty} f^n(x) = \sum_{n=0}^{\infty} \frac{x^n}{n!}$$

is also strictly increasing on $\mathfrak{R}_+$ (exercise 2.33). Since (exercise 2.6) $e^{-x} = 1/e^x$,

$$x_1 < x_2 < 0 \Rightarrow 0 < -x_2 < -x_1 \Rightarrow f(-x_2) < f(-x_1)$$

$$\Rightarrow \frac{1}{e^{x_2}} < \frac{1}{e^{x_1}} \Rightarrow e^{x_1} < e^{x_2}$$

Therefore the exponential function e^x is strictly increasing on $\mathfrak{R}$.

Exercise 2.33
Show that e^x is strictly increasing on $\mathfrak{R}_+$. [Hint: $e^x = 1 + x + \lim_{n \to \infty} g^n(x)$, $g^n(x) = \sum_{k=2}^{n} x^k/k!$, $n > 2$.]

Example 2.55 (Log function) In exercise 2.6 we showed that

$$e^x \to \infty \quad \text{as } x \to \infty \quad \text{and} \quad e^x \to 0 \quad \text{as } x \to -\infty$$

This implies that e^x maps $\mathfrak{R}$ onto $\mathfrak{R}_+$. Furthermore we have just shown that e^x is strictly increasing. Therefore (exercise 2.29), the exponential function has a strictly increasing inverse log: $\mathfrak{R}_+ \to \mathfrak{R}$ defined by

$$y = \log x \Leftrightarrow x = e^y$$

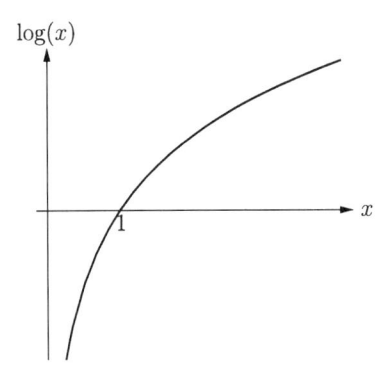

Figure 2.14
The log function

which is illustrated in figure 2.14. Log is an abbreviation of logarithm. Property (2) of the exponential function implies that

$$\log(x_1 x_2) = \log x_1 + \log x_2 \tag{15}$$

Prior to the development of pocket calculators and personal computers, (15) was used to facilitate numerical calculations. Students, engineers and others involved in nontrivial calculations were equipped with tables of logarithms, enabling them to convert multiplication problems to easier addition.

Example 2.56 (General power function) The log function enables us to extend the definition of the power function to noninteger exponents. For every $x \in \Re_+$, $e^{\log x}$ is the identity function, that is,

$$x = e^{\log x}$$

and we define the general power function $f: \Re_+ \to \Re$ by

$$f(x) = x^a = e^{a \log x}, \quad a \in \Re \tag{16}$$

Exercise 2.34
The general power function $f(x) = x^a$ is strictly increasing on $\Re_+$ for all $a > 0$ and strictly decreasing for $a < 0$.

Example 2.57 (Cobb-Douglas function) In economic analysis, the formula or rule that specifies a particular function is known as a *functional form*. One of the most popular functional forms in economics is the Cobb-

Douglas function $f: \Re_+^n \to \Re_+$ defined by

$$f(\mathbf{x}) = x_1^{a_1} x_2^{a_2} \ldots x_n^{a_n}, \qquad a_i > 0$$

The Cobb-Douglas function is the product of general power functions. Therefore (exercise 2.32) it is strictly increasing on $\Re_+^n$.

Exercise 2.35 (CES function)
Another popular functional form in economics is the CES function, $f: \Re_+^n \to \Re_+$ defined by

$$f(\mathbf{x}) = (\alpha_1 x_1^\rho + \alpha_2 x_2^\rho + \cdots \alpha_n x_n^\rho)^{1/\rho}, \qquad \alpha_i > 0, \rho \neq 0$$

Show that the CES function is strictly increasing on $\Re_+^n$.

Example 2.58 (Utility function) A strictly increasing functional u on a weakly ordered set $(X, \succsim)$ is called a *utility function*. A utility function is said to *represent* the preference relation $\succsim$, since (exercise 2.36)

$$x_2 \succsim x_1 \Leftrightarrow u(x_2) \geq u(x_1)$$

Exercise 2.36
Let $u: X \to \Re$ be a strictly increasing function on the weakly ordered set $(X, \succsim)$. Show that

$$x_2 \succsim x_1 \Leftrightarrow u(x_2) \geq u(x_1)$$

Example 2.59 (Monotonic preferences) A utility function is strictly increasing with respect to the preference order $\succsim$ on X. When $X \subseteq \Re^n$, the preference order $\succsim$ may not necessarily be consistent with the natural order on $\Re^n$. If the two orders are consistent, the preference order is monotonic (section 1.6). If the preference $\succsim$ is weakly monotonic on X,

$$\mathbf{x} \geq \mathbf{y} \Rightarrow \mathbf{x} \succsim \mathbf{y} \Leftrightarrow u(\mathbf{x}) \geq u(\mathbf{x})$$

and u is in increasing on X. If $\succsim$ is strongly monotonic

$$\mathbf{x} \gneq \mathbf{y} \Rightarrow \mathbf{x} \succ \mathbf{y} \Leftrightarrow u(\mathbf{x}) > u(\mathbf{y})$$

and u is strictly increasing on X.

Example 2.60 (Monotonic transformation) Given any functional f on X and a strictly increasing functional $g: \Re \to \Re$, their composition $g \circ f: X \to \Re$ is called a *monotonic transformation* of f. A monotonic transformation preserves the ordering $\succsim_f$ implied by f.

This terminology ("monotonic transformation") is at odds with our definition of monotone (weakly increasing or decreasing) but is well entrenched in the economics literature. Synonyms include "monotone transformation," "monotone increasing transformation," and "positive monotonic transformation". A typical application is given in the exercise 2.37.

Exercise 2.37 (Invariance to monotonic transformations)
Let $u: X \to \Re$ be a utility function representing the preference relation $\succsim$. Show that every monotonic transformation $g \circ u$ is a utility function representing the same preferences. We say that utility representation is invariant to monotonic transformations.

Remark 2.9 (Existence of a utility function) Continuity (section 1.6) is a necessary and sufficient condition for the existence of a utility function representing a given preference relation. However, a general proof of this fact is quite complicated, requiring both topological and order-theoretic ideas. A simple constructive proof can be given when $X = \Re^n_+$ and preferences are strongly monotonic (exercise 2.38).

Exercise 2.38
Let $\succsim$ be a continuous preference relation on $\Re^n_+$. Assume that $\succsim$ is strongly monotonic. Let Z denote the set of all bundles that have the same amount of all commodities (figure 2.15), that is, $Z = \{\mathbf{z} = z\mathbf{1} : z \in \Re_+\}$ where $\mathbf{1} = (1, 1, \ldots, 1)$.

1. For any $\mathbf{x} \in \Re^n_+$, show that

a. the sets $Z^+_\mathbf{x} = \succsim(\mathbf{x}) \cap Z$ and $Z^-_\mathbf{x} = \precsim(\mathbf{x}) \cap Z$ are nonempty and closed

b. $Z^+_\mathbf{x} \cap Z^-_\mathbf{x} \neq \emptyset$ [Hint: Z is connected.]

c. there exists $\mathbf{z}_\mathbf{x} \in Z$ which is indifferent to $\mathbf{x}$

d. $\mathbf{z}_\mathbf{x} = z_\mathbf{x}\mathbf{1}$ is unique

2. For every $\mathbf{x} \in \Re$, define $z_\mathbf{x}$ to be the scale of $\mathbf{z}_\mathbf{x} \sim \mathbf{x}$. That is, $\mathbf{z}_\mathbf{x} = z_\mathbf{x}\mathbf{1}$. The assignment $u(\mathbf{x}) = z_\mathbf{x}$ defines a function $u: \Re^n_+ \to \Re$ that represents the preference ordering $\succsim$.

Exercise 2.39
Remark 2.9 implies that the lexicographic preference relation (example 1.114) cannot be represented by a utility function, since the lexicographic

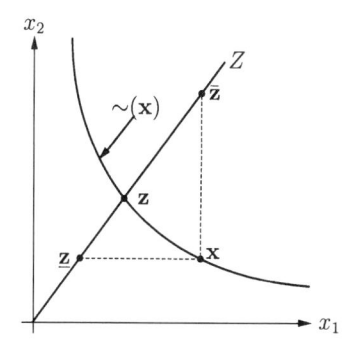

Figure 2.15
Constructive proof of the existence of a utility function

preference ordering is not continuous. To verify this, assume, to the contrary, that u represents the lexicographic ordering $\succsim_L$ on $\Re^2$.

1. For every $\mathbf{x}_1 \in \Re$ there exists a rational number $r(x_1)$ such that

$$u(x_1, 2) > r(x_1) > u(x_1, 1)$$

2. This defines an increasing function r from $\Re$ to the set $\mathfrak{Q}$ of rational numbers.

3. Obtain a contradiction.

Example 2.61 (Payoff function) A function $u_i \colon A \to \Re$ is a payoff function for player i in the strategic game $(N, A, (\succsim_1, \succsim_2, \ldots, \succsim_n))$ (example 2.34) if u_i represents the preferences of player i, that is,

$$\mathbf{a}_2 \succsim_i \mathbf{a}_1 \Leftrightarrow u_i(\mathbf{a}_2) \geq u_i(\mathbf{a}_1)$$

So a payoff function is simply a utility function over the set of action profiles A. The necessary conditions for existence of a payoff function for players in a game are those for the existence of a utility function, namely completeness and continuity of the preference relation $\succsim_i$.

Exercise 2.40 (Zero-sum game)
Suppose that $u_1 \colon A \in \Re$ represents the preferences of the player 1 in a two-person strictly competitive game (example 1.50). Then the function $u_2 = -u_1$ represents the preferences of the player 2 and

$$u_1(\mathbf{a}) + u_2(\mathbf{a}) = 0 \quad \text{for every } \mathbf{a} \in A$$

Consequently a strictly competitive game is typically called a *zero-sum game*.

Example 2.62 (Superadditive game) Monotonicity is a natural assumption for the characteristic function of a TP-coalitional game, reflecting the presumption that bigger coalition can achieve anything achievable by a smaller coalition. In fact a stronger presumption is customary. The characteristic function of a TP-coalitional game is *superadditive* if distinct coalitions cannot lose by acting jointly, that is, for all $S, T \subseteq N$, $S \cap T = \emptyset$,

$$w(S \cup T) \geq w(S) + w(T)$$

Exercise 2.41
Show that superadditivity implies monotonicity, that is, if $v \colon \mathscr{P}(N) \to \mathfrak{R}$ is superadditive, then v is monotonic.

Example 2.63 (Monotone operator) Given an arbitrary set X, the set $F(X)$ of all functionals on X is a linear space (exercise 2.9). There is a natural partial order on $F(X)$ that is defined as follows: For any $f, g \in F(X)$,

$$f \succsim g \Leftrightarrow f(x) \geq g(x) \quad \text{for every } x \in X$$

Let $A \subseteq F(X)$ be a set of functionals on a space X. A *monotone operator* is a function $T \colon A \to A$ that preserves the natural order of A, that is,

$$f \succsim g \Rightarrow Tf \succsim Tg$$

Use of the term *monotone operator* to describe an increasing function from a set A to itself is well-entrenched Stokey and Lucas (1989, p. 528), although the description *increasing operator* would be more consistent with our terminology. It is possible to conceive of a decreasing operator, although its behavior would be confusing.

Example 2.64 (Dynamic programming) In introducing the dynamic programming problem (example 2.32, exercise 2.18), we encountered the operator

$$(Tv)(x) = \sup_{y \in G(x)} f(x, y) + \beta v(y)$$

on the space $B(X)$ of bounded functionals on the set X. There is a natural partial order on $B(X)$ such that

$$w \succsim v \Leftrightarrow w(x) \geq v(x) \quad \text{for every } x \in X$$

The operator T preserves this order, that is,

$$w \succsim v \Rightarrow Tw \succsim Tv$$

That is, T is a monotone operator.

Exercise 2.42
Show that the operator $T: B(X) \to B(X)$ defined by

$$(Tv)(x) = \sup_{y \in G(x)} f(x,y) + \beta v(y)$$

is increasing.

2.2.1 Monotone Correspondences

Extending the concept of monotonicity to a correspondence $\varphi: X \rightrightarrows Y$ requires an order on the *subsets* of Y. One useful order is set inclusion. We say that a correspondence is *ascending* if

$$x_2 \succsim_X x_1 \Rightarrow \varphi(x_2) \supseteq \varphi(x_1)$$

It is *descending* if

$$x_2 \succsim_X x_1 \Rightarrow \varphi(x_2) \subseteq \varphi(x_1)$$

Example 2.65 (Budget set) The budget correspondence $X(\mathbf{p}, m)$ is ascending in income, since

$$m_2 \geq m_1 \Rightarrow X(\mathbf{p}, m_2) \supseteq X(\mathbf{p}, m_1)$$

It is descending in prices, since

$$\mathbf{p}_2 \geq \mathbf{p}_1 \Rightarrow X(\mathbf{p}_2, m) \subseteq X(\mathbf{p}_1, m)$$

Example 2.66 (Input requirement sets) The input requirement sets of a single output technology (example 2.44)

$$V(y) = \{\mathbf{x} \in \Re_+^n : (y, -\mathbf{x}) \in Y\}$$

are an ascending correspondence provided the technology exhibits free disposal (exercise 1.12).

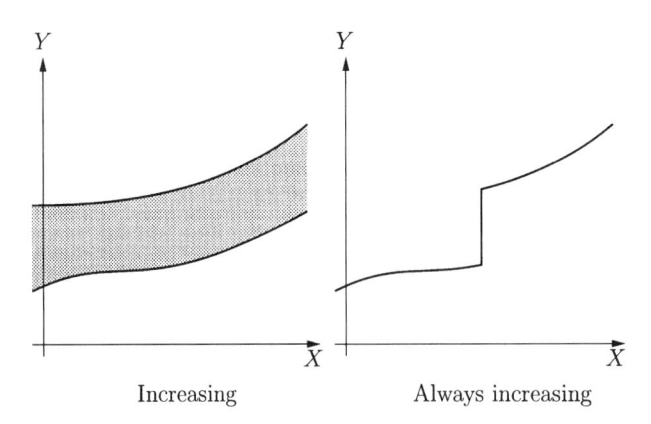

Figure 2.16
Weakly and strongly increasing correspondences

Alternatively, when Y is a lattice, we can use the strong set order (section 1.2.4) to order Y. We will say that correspondence $\varphi\colon X \rightrightarrows Y$ is *increasing* if it preserves the strong set order, that is,

$$x_2 \succsim_X x_1 \Rightarrow \varphi(x_2) \succsim_S \varphi(x_1)$$

where $\succsim_S$ is the strong set order induced on $\mathscr{P}(Y)$ by $\succsim_Y$. That is, $\varphi\colon X \rightrightarrows Y$ is increasing if for every $x_2 \succsim x_1$,

$$y_1 \wedge y_2 \in \varphi(x_1) \quad \text{and} \quad y_1 \vee y_2 \in \varphi(x_2)$$

for every $y_1 \in \varphi(x_1)$ and $y_2 \in \varphi(x_2)$. It is *decreasing* if it reverses the strong set order, that is,

$$x_2 \succsim_X x_1 \Rightarrow \varphi(x_2) \precsim_S \varphi(x_1)$$

A correspondence is *monotone* if it is either increasing or decreasing.

With a one-dimensional domain and range, it is straightforward to illustrate monotone correspondences (figure 2.16). However, in general, the concept of monotonicity is more subtle as the following example illustrates.

Example 2.67 (Budget set) The budget correspondence is *not* monotone. Consider figure 2.17, which shows the budget set for two commodities at two different income levels $m_2 > m_1$ (with prices constant). The com-

2.2 Monotone Functions

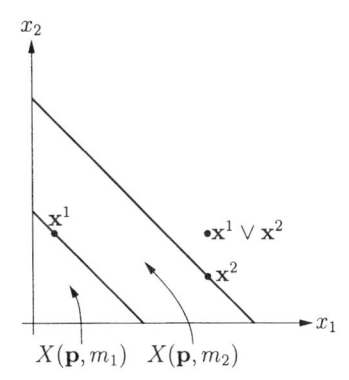

Figure 2.17
The budget correspondence is not monotone

modity bundle $\mathbf{x}_1$ is affordable at m_1 and $\mathbf{x}_2$ (with more of good 1 and less of good 2) is affordable at m_2. However, the commodity bundle $\mathbf{x}_1 \vee \mathbf{x}_2$ is not affordable at m_2. Hence the budget correspondence is not monotone.

Exercise 2.43
For $\Theta \in \Re$, if $g \in F(\Theta)$ is increasing, then the correspondence

$$G(\theta) = \{x : 0 \leq x \leq g(\theta)\}$$

is increasing.

The significance of this definition of monotonicity for correspondences is that every monotone correspondence has a monotone selection. This and other useful properties of monotone correspondences are detailed in the following exercises.

Exercise 2.44
Let φ be an increasing correspondence from X to Y, and let $x_1, x_2 \in X$ with $x_2 \succsim x_1$. Then

- for every $y_1 \in \varphi(x_1)$ there exists $y_2 \in \varphi(x_2)$ with $y_2 \succsim y_1$
- for every $y_2 \in \varphi(x_2)$ there exists $y_1 \in \varphi(x_1)$ with $y_2 \succsim y_1$

Exercise 2.45 (Increasing selection)
If $\varphi: X \rightrightarrows Y$ is increasing and every $\varphi(x)$ is a sublattice, there exists an increasing selection $f \in \varphi$.

Exercise 2.46
If $\varphi_i\colon X \rightrightarrows Y_i$, $i = 1, 2, \ldots, n$, is a collection of increasing correspondences with common domain X, their product $\varphi\colon X \rightrightarrows \prod Y_i$ defined by

$$\varphi(x) = \prod_i \varphi_i(x)$$

is also increasing.

Exercise 2.47
If $\varphi_i\colon X \rightrightarrows Y_i$, $i = 1, 2, \ldots, n$, is a collection of increasing correspondences with common domain X, and their intersection $\varphi\colon X \rightrightarrows \bigcap Y_i$ defined by $\varphi(x) = \bigcap_i \varphi_i(x)$ is nonempty for every $x \in X$, then φ is also increasing.

A stronger concept of monotonicity is also useful. A correspondence $\varphi\colon X \rightrightarrows Y$ is *always increasing* if

$$x_1 \succsim_X x_2 \Rightarrow y_1 \succsim_Y y_2 \quad \text{for every } y_1 \in \varphi(x_1) \text{ and } y_2 \in \varphi(x_2)$$

A correspondence is always increasing if and only if every selection is increasing. Note that this concept does not require that Y is a lattice.

Exercise 2.48
$\varphi\colon X \rightrightarrows Y$ is always increasing if and only if every selection $f \in \varphi$ is increasing.

2.2.2 Supermodular Functions

Monotonicity restricts the behavior of a function on comparable elements. It places no restriction on the action of the function with respect to noncomparable elements. For the special case of functionals on lattices, we can define a related property, called supermodularity, which restricts the behavior of the functional over its entire domain. A functional $f\colon X \to \Re$ on a lattice X is *supermodular* if every $x_1, x_2 \in X$,

$$f(x_1 \vee x_2) + f(x_1 \wedge x_2) \geq f(x_1) + f(x_2) \tag{17}$$

f is strictly supermodular if every noncomparable $x_1, x_2 \in X$,

$$f(x_1 \vee x_2) + f(x_1 \wedge x_2) > f(x_1) + f(x_2)$$

A functional f is (strictly) submodular if $-f$ is (strictly) supermodular.

As we will see, supermodularity formalizes the useful economic notion of complementarity (example 2.70). In strategic games it expresses the important idea of *strategic complementarity* (example 2.71). A TP-coalitional game in which the characteristic function is supermodular is called a *convex game*, which has special and very useful properties (example 2.69).

Remark 2.10 (Function or functional) Strictly speaking, this section should be entitled "supermodular functionals" because the concept of supermodularity relies on the linear structure of $\Re$ and is therefore only defined for real-valued functions. However, the terminology supermodular function has become established in the literature, and to insist on functional would seem unnecessarily pedantic. Similar usage is even more firmly established for convex and concave functions (section 3.7), which also implicitly refer to real-valued functions only. Exercise 2.57 presents a strictly ordinal property that can be used to generalize supermodularity to any function between ordered sets.

Exercise 2.49
Every functional on a chain is supermodular.

The following properties are analogous to those for monotone functions (exercises 2.31 and 2.32).

Exercise 2.50
If $f, g \in F(X)$ are supermodular, then

- $f + g$ is supermodular
- αf is supermodular for every $\alpha \geq 0$

Therefore the set of all supermodular functions on a set X is a cone in $F(X)$.

Exercise 2.51
If f and g are nonnegative definite, increasing, and supermodular functionals on X, then so is their product fg defined by

$$(fg)(x) = f(x)g(x)$$

Exercises 2.49 to 2.51 are useful in constructing supermodular functions.

Example 2.68 (Cobb-Douglas) Since $\Re_+$ is a chain, the power function $f(x) = x_i^{a_i}$ is supermodular on $\Re_+$ (exercise 2.49), and therefore (exercise 2.51) the Cobb-Douglas function

$$f(\mathbf{x}) = x_1^{a_1} x_2^{a_2} \ldots x_n^{a_n}, \qquad a_i \geq 0$$

is supermodular on $\Re_+^n$.

Exercise 2.52 (CES function)
Show that the CES function

$$f(\mathbf{x}) = (\alpha_1 x_1^\rho + \alpha_2 x_2^\rho + \cdots \alpha_n x_n^\rho)^{1/\rho}, \qquad \alpha_i > 0, \rho \neq 0$$

is supermodular on $\Re_+^n$.

Exercise 2.53 (Economies of scope)
If a firm produces many products, a straightforward generalization of the cost function $c(\mathbf{w}, \mathbf{y})$ measures the cost of producing the list or vector of outputs $\mathbf{y}$ when input prices are $\mathbf{w}$. The production technology displays economies of joint production or *economies of scope* at $\mathbf{y} = (y_1, y_2, \ldots, y_m)$ if the total cost of producing all the outputs separately is greater than the cost of producing the outputs jointly, that is,

$$\sum_{j=1}^m c(\mathbf{w}, y_j \mathbf{e}_j) > c(\mathbf{w}, \mathbf{y})$$

where $y_j \mathbf{e}_j = (0, 0, \ldots, y_j, 0 \ldots 0)$ is the output vector consisting of y_j units of good j. Show that the technology displays economies of scope if the cost function is strictly submodular in $\mathbf{y}$. Assume zero fixed costs.

Example 2.69 (Convex games) A TP-coalitional game is *convex* if its characteristic function is supermodular, that is,

$$w(S \cup T) + w(S \cap T) \geq w(S) + w(T) \qquad \text{for every } S, T \subseteq N \tag{18}$$

The set of convex games is a convex cone in the set of all TP-coalitional games $\mathscr{G}^N$ (exercise 2.50). Convexity in a game reflects increasing returns to cooperation (exercise 2.55). Convex games occur naturally in many applications, and they have special properties. In particular, the core (example 1.45) is nonempty, and contains the Shapley value (example 3.6), which coincides with the nucleolus (example 1.49).

Exercise 2.54
Every convex game is superadditive.

Exercise 2.55
Show that a TP-coalitional game (N, w) is convex if and only if

$$w(T \cup \{i\}) - w(T) \geq w(S \cup \{i\}) - w(S)$$

for every $i \in N$ and for every $S \subset T \subset N \setminus \{i\}$

The marginal contribution of every player increases with the size of the coalition to which the player is joined.

Exercise 2.56
Is the cost allocation game (exercise 1.66) convex?

Exercise 2.57 (Quasisupermodularity)
The definition of supermodularity utilizes the linear structure of $\Re$. Show that supermodularity implies the following strictly ordinal property

$$f(x_1) \geq f(x_1 \wedge x_2) \Rightarrow f(x_1 \vee x_2) \geq f(x_2)$$

and

$$f(x_1) > f(x_1 \wedge x_2) \Rightarrow f(x_1 \vee x_2) > f(x_2)$$

for every $x_1, x_2 \in X$.

Increasing Differences

Another property closely related to supermodularity is useful when dealing with functionals whose domain can be decomposed into two sets, as for example, the objective function of a constrained optimization problem (example 2.30) or the payoff function in a strategic game (example 2.34). Suppose that $f: X \times Y \to \Re$ is supermodular. For any $x_1, x_2 \in X$, and $y_1, y_2 \in Y$ with $x_2 \succsim_X x_1$ and $y_2 \succsim_Y y_1$,

$$(x_1, y_2) \wedge (x_2, y_1) = (x_1, y_1)$$

$$(x_1, y_2) \vee (x_2, y_1) = (x_2, y_2)$$

Evaluating (17) at (x_1, y_2) and (x_2, y_1), supermodularity implies that

$$f(x_2, y_2) + f(x_1, y_1) \geq f(x_1, y_2) + f(x_2, y_1)$$

Rearranging the inequality, we observe that

$$f(x_2, y_2) - f(x_1, y_2) \geq f(x_2, y_1) - f(x_1, y_1)$$

which motivates the following definition. Given two posets X and Y, a functional $f: X \times Y \to \Re$ displays *increasing differences* in (x, y) if, for all

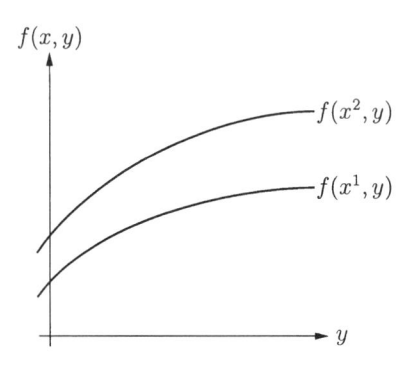

[Figure 2.18: A supermodular function displays increasing differences]

$x_2 \gtrsim x_1$, the difference $f(x_2, y) - f(x_1, y)$ is increasing in y. It has *strictly increasing differences* if $f(x_2, y) - f(x_1, y)$ is strictly increasing in y (see figure 2.18).

Exercise 2.58
Let $f: X \times Y \to \Re$. Show that f displays increasing differences if and only if

$$f(x_2, y_2) - f(x_2, y_1) \geq f(x_1, y_2) - f(x_1, y_1)$$

that is, the difference $f(x, y_2) - f(x, y_1)$ is increasing in x. Therefore the order of the comparison in the definition increasing differences is irrelevant. This is analogous to Young's theorem (theorem 4.2) for smooth functions.

The concepts of supermodularity and increasing differences are closely related. Both concepts formalize the notion of complementarity. The preceding discussion showed that any supermodular function on a product space displays increasing differences. Conversely, where the component sets are totally ordered (chains), increasing differences implies supermodularity, so the two properties coincide (exercise 2.59). This equivalence generalizes to finite products and hence applies to $\Re^n$, the domain of many economic models. The property of increasing differences is more readily applicable in economic models and easier to verify, while supermodularity is more tractable mathematically. Proposition 4.2 gives a useful characterization of smooth supermodular functions in terms of the second derivative.

Exercise 2.59
Let f be a functional on $X \times Y$ where X and Y are chains. Show that f has increasing differences in (x,y) if and only if f is supermodular on $X \times Y$.

Example 2.70 (Complementary inputs) The technology of a single output producer can be represented by a production function (example 2.24) $f: \Re_+^n \to \Re$. The production is supermodular if and only if it displays increasing differences. The (discrete) marginal product of input i is the additional product obtained by adding another unit of input

$$\text{MP}_i(\mathbf{x}) = f(\mathbf{x} + \mathbf{e}_i) - f(\mathbf{x})$$

where $\mathbf{e}_i$ is the ith unit vector in $\Re^n$. The production function is supermodular if and only if the marginal product $f(\mathbf{x} + \mathbf{e}_i) - f(\mathbf{x})$ of every input i is an increasing function of all the other inputs. This captures the economic idea of complementary inputs.

Example 2.71 (Supermodular games) A *supermodular game* is a strategic game in which

- every strategy set S_i is a lattice
- the payoff functions $u_i: S_i \times S_{-i} \to \Re$ are supermodular on S_i
- u_i display increasing differences in $(s_i, \mathbf{s}_{-i})$

Fortunately, many games meet these requirements, since supermodular games are particularly well behaved. They always have a pure strategy Nash equilibrium (example 2.92), and the set of Nash equilibria is a lattice (exercise 2.118).

Example 2.72 (Coordination failure in a macro model) Some recent work in macroeconomics attributes aggregate fluctuations to "coordination failures." A typical example is the following simple search model. Trade takes place by barter coordinated by a stochastic matching process. The payoff for any individual player depends on the probability of meeting a trading partner, which in turn is determined by search effort of all the players. Specifically, the probability of meeting a suitable trading partner is $s_i \sum_{i \neq j} s_j$, where s_i denotes the search effort of player i. If $\alpha > 0$ denotes the gain from successful trade and $c(s_i)$ the cost of search, player i's payoff function is

$$u_i(s_i, \mathbf{s}_{-i}) = \alpha s_i \sum_{i \neq nj} s_j - c(s_i)$$

u_i is supermodular in s_i. Furthermore, as player i increases her search activity from s_i^1 to s_i^2, her payoff increases, ceteris paribus, by

$$u_i(s_i^2, \mathbf{s}_{-i}) - u_i(s_i^1, \mathbf{s}_{-i}) = \alpha \sum_{i \neq j} s_j(s_i^2 - s_i^1)$$

which is clearly increasing in s_j. Therefore this is a supermodular game. In general, the game has multiple equilibria. Those equilibria with lower search activity have smaller aggregate output.

Exercise 2.60 (Bertrand oligopoly)
In the standard Bertrand model of oligopoly n firms each produce a differentiated product. The demand q_i for the product of the ith firm depends on its own price and the price charged by all the other firms, that is,

$$q_i = f(p_i, \mathbf{p}_{-i})$$

If each firm's production cost is measured by the cost function $\bar{c}_i$, firm i's payoff function is

$$u_i(p_i, \mathbf{p}_{-i}) = p_i f(p_i, \mathbf{p}_{-i}) - c_i(f(p_i, \mathbf{p}_{-i}))$$

In the simplest specification the demand functions are linear

$$f(p_i, \mathbf{p}_{-i}) = a_i - b_i p_i + \sum_{j \neq i} d_{ij} p_j$$

with $b_i > 0$ and the firm's produce at constant marginal cost $\bar{c}_i$, so the payoff functions are

$$u_i(p_i, \mathbf{p}_{-i}) = (p_i - \bar{c}_i) f(p_i, \mathbf{p}_{-i})$$

Show that if the goods are *gross substitutes* ($d_{ij} > 0$ for every i,j), the Bertrand oligopoly model with linear demand and constant marginal costs is a supermodular game.

Exercise 2.61 (Single-crossing condition)
Increasing differences implies the following ordinal condition, which is known as the *single-crossing condition*. For every $x_2 \gtrsim x_1$ and $y_2 \gtrsim y_1$,

$$f(x_2, y_1) \geq f(x_1, y_1) \Rightarrow f(x_2, y_2) \geq f(x_1, y_2)$$

and

$$f(x_2, y_1) > f(x_1, y_1) \Rightarrow f(x_2, y_2) > f(x_1, y_2)$$

2.2.3 The Monotone Maximum Theorem

In formulating economic models as optimization problems (example 2.30), economists are primarily interested in determining the way in which the optimal solution varies with the parameters. A powerful tool in this quest is provided by the following theorem.

Theorem 2.1 (Monotone maximum theorem) *Let $\Theta^* \subseteq \Theta$ denote the set of parameter values for which a solution to the problem*

$$\max_{\mathbf{x} \in G(\boldsymbol{\theta})} f(\mathbf{x}, \boldsymbol{\theta})$$

exists. If X is a lattice, Θ a poset and

- *the objective function $f \colon X \times \Theta \to \Re$ is supermodular in X*
- *f displays increasing differences in $(\mathbf{x}, \boldsymbol{\theta})$*
- *and the constraint correspondence $G \colon \Theta \rightrightarrows X$ is increasing in $\boldsymbol{\theta}$*

then the solution correspondence $\varphi \colon \Theta^* \rightrightarrows X$ defined by

$$\varphi(\boldsymbol{\theta}) = \arg\max_{\mathbf{x} \in G(\boldsymbol{\theta})} f(\mathbf{x}, \boldsymbol{\theta})$$

is increasing. Furthermore, if objective function is increasing in $\mathbf{x}$ and $\boldsymbol{\theta}$, the value function

$$v(\boldsymbol{\theta}) = \sup_{\mathbf{x} \in G(\boldsymbol{\theta})} f(\mathbf{x}, \boldsymbol{\theta})$$

is increasing.

Proof Let $\boldsymbol{\theta}_1, \boldsymbol{\theta}_2$ belong to Θ^* with $\boldsymbol{\theta}_2 \gtrsim \boldsymbol{\theta}_1$. Choose any optimal solutions $\mathbf{x}_1 \in \varphi(\boldsymbol{\theta}_1)$ and $\mathbf{x}_2 \in \varphi(\boldsymbol{\theta}_2)$. To show that φ is monotone, we have to show that $\mathbf{x}_1 \vee \mathbf{x}_2 \in \varphi(\boldsymbol{\theta}_2)$ and $\mathbf{x}_1 \wedge \mathbf{x}_2 \in \varphi(\boldsymbol{\theta}_1)$.

Since the constraint set $G(\boldsymbol{\theta})$ is monotone, $\mathbf{x}_1 \vee \mathbf{x}_2 \in G(\boldsymbol{\theta}_2)$ and $\mathbf{x}_1 \wedge \mathbf{x}_2 \in G(\boldsymbol{\theta}_1)$. That is, both $\mathbf{x}_1 \vee \mathbf{x}_2$ and $\mathbf{x}_1 \wedge \mathbf{x}_2$ are feasible. To show that they are optimal, consider the following sequence of inequalities.

Supermodularity implies that

$$f(\mathbf{x}_1 \vee \mathbf{x}_2, \theta_2) + f(\mathbf{x}_1 \wedge \mathbf{x}_2, \theta_2) \geq f(\mathbf{x}_1, \theta_2) + f(\mathbf{x}_2, \theta_2)$$

which can be rearranged to give

$$f(\mathbf{x}_1 \vee \mathbf{x}_2, \theta_2) - f(\mathbf{x}_2, \theta_2) \geq f(\mathbf{x}_1, \theta_2) - f(\mathbf{x}_1 \wedge \mathbf{x}_2, \theta_2)$$

Increasing differences applied to the right-hand side implies that

$$f(\mathbf{x}_1, \theta_2) - f(\mathbf{x}_1 \wedge \mathbf{x}_2, \theta_2) \geq f(\mathbf{x}_1, \theta_1) - f(\mathbf{x}_1 \wedge \mathbf{x}_2, \theta_1)$$

Combining these two inequalities we have

$$f(\mathbf{x}_1 \vee \mathbf{x}_2, \theta_2) - f(\mathbf{x}_2, \theta_2) \geq f(\mathbf{x}_1, \theta_1) - f(\mathbf{x}_1 \wedge \mathbf{x}_2, \theta_1) \tag{19}$$

However, $\mathbf{x}_1$ and $\mathbf{x}_2$ are optimal for their respective parameter values, that is,

$$f(\mathbf{x}_2, \theta_2) \geq f(\mathbf{x}_1 \vee \mathbf{x}_2, \theta_2) \Rightarrow f(\mathbf{x}_1 \vee \mathbf{x}_2, \theta_2) - f(\mathbf{x}_2, \theta_2) \leq 0$$
$$f(\mathbf{x}_1, \theta_1) \geq f(\mathbf{x}_1 \wedge \mathbf{x}_2, \theta_1) \Rightarrow f(\mathbf{x}_1, \theta_1) - f(\mathbf{x}_1 \wedge \mathbf{x}_2, \theta_1) \geq 0$$

Substituting in (19), we conclude that

$$0 \geq f(\mathbf{x}_1 \vee \mathbf{x}_2, \theta_2) - f(\mathbf{x}_2, \theta_2) \geq f(\mathbf{x}_1, \theta_1) - f(\mathbf{x}_1 \wedge \mathbf{x}_2, \theta_1) \geq 0$$

The inequality must be an equality with

$$f(\mathbf{x}_1 \vee \mathbf{x}_2, \theta_2) = f(\mathbf{x}_2, \theta_2), \quad f(\mathbf{x}_1 \wedge \mathbf{x}_2, \theta_1) = f(\mathbf{x}_1, \theta_1)$$

That is, $\mathbf{x}_1 \vee \mathbf{x}_2 \in \varphi(\theta_2)$ and $\mathbf{x}_1 \wedge \mathbf{x}_2 \in \varphi(\theta_1)$. Furthermore, if f is increasing,

$$v(\theta_2) = f(\mathbf{x}_1 \vee \mathbf{x}_2, \theta_2) \geq f(\mathbf{x}_1 \wedge \mathbf{x}_2, \theta_1) = v(\theta_1)$$

since $(\mathbf{x}_1 \vee \mathbf{x}_2, \theta_2) \succsim (\mathbf{x}_1 \wedge \mathbf{x}_2, \theta_1)$. The value function is increasing. □

Corollary 2.1.1 *If in addition to the hypotheses of the previous theorem, the feasible set is a lattice for every $\theta \in \Theta^*$, then the set of optimal solutions*

$$\varphi(\theta) = \arg \max_{\mathbf{x} \in G(\theta)} f(\mathbf{x}, \theta)$$

is a sublattice of X for every $\theta \in \Theta$ and φ has an increasing selection.

Exercise 2.62
Prove corollary 2.1.1.

2.2 Monotone Functions

Example 2.73 (Private value auction) In a first-price private value auction, each bidder has a value θ for an object which is known only to him. If he bids an amount x and is successful, he receives utility $u(\theta - x)$. Let $p(x)$ denote the probability of winning with a bid of x. Then his problem is to choose x given θ to maximize his expected utility, that is,

$$\max_{x \in [0, \theta]} u(\theta - x) p(x)$$

The constraint correspondence $G(\theta) = [0, \theta]$ is increasing (exercise 2.43) in θ, and the objective function is supermodular in x (exercise 2.49). If u is strictly concave, then u displays strictly increasing differences in (x, θ) (exercise 3.129). By theorem 2.1, the optimal bids belong to an increasing correspondence. Further (corollary 2.1.1), since $G(\theta)$ is a lattice for every θ, there exists an increasing selection (bidding function). Note that this conclusion is independent of the properties of p, which reflects the probability distribution of values among the bidders.

Corollary 2.1.2 *If, in addition to the hypotheses of theorem 2.1, the objective function displays strictly increasing differences in* $(\mathbf{x}, \boldsymbol{\theta})$, *the optimal correspondence*

$$\varphi(\boldsymbol{\theta}) = \arg \max_{\mathbf{x} \in G(\boldsymbol{\theta})} f(\mathbf{x}, \boldsymbol{\theta})$$

is always increasing. Every selection from $\varphi(\boldsymbol{\theta})$ *is an increasing function of* $\boldsymbol{\theta}$.

Exercise 2.63
Prove corollary 2.1.2. [Hint: Assume that $\mathbf{x}_2 \not\gtrsim \mathbf{x}_1$, and derive a contradiction.]

Example 2.74 (Supermodular games) In any strategic game, player i's best response correspondence (example 2.48) is the solution of a maximization problem, namely

$$B(\mathbf{s}_{-i}) = \arg \max_{s_i \in S_i} u_i(s_i, \mathbf{s}_{-i})$$

If the game is supermodular (example 2.71), the optimization problem meets the requirements of theorem 2.1, with $X = S_i$, $\Theta = S_{-i}$, $f = u_i$ and G equal to the identity correspondence. The theorem establishes that B is increasing in $\mathbf{s}_i$. In particular, this means that there exists an increasing selection $f \in B$, which can be used to establish the existence of an equilibrium.

Furthermore, if the payoff functions $u_i(s_i, \mathbf{s}_{-i})$ display strictly increasing differences in $(s_i, \mathbf{s}_{-i})$, then the best response correspondences are always increasing (corollary 2.1.2). Every selection is increasing so that for every $s_1 \in B(\mathbf{s}_{-1})$ and $s_2 \in B(\mathbf{s}_{-2})$, $\mathbf{s}^2_{-i} \succsim \mathbf{s}^1_{-i}$ implies that $s_i^2 \succsim s_i^1$.

The requirements of theorem 2.1 are severe, especially the requirement that the feasible set $G(\theta)$ be increasing. When the feasible set is independent of θ, this implicitly requires that the feasible set be a lattice, which precludes the application of theorem 2.1 to some common models in microeconomic theory such as example 2.31. In other cases, although the feasible set varies with the parameters, the relationship is not monotone. We provide some weaker results that can be applied in these cases.

Proposition 2.1 (Increasing maximum theorem) *If $f: X \times \Theta \to \Re$ is increasing in θ, the value function*

$$v(\theta) = \sup_{x \in G} f(\mathbf{x}, \theta)$$

is also increasing in θ.

Proof Assume $\theta_2 \succsim \theta_1 \in \Theta^*$, and let $\mathbf{x}_2$ and $\mathbf{x}_1$ be corresponding optimal solutions. Then

$$f(\mathbf{x}_2, \theta_2) \geq f(\mathbf{x}_1, \theta_2)$$

and

$$v(\theta_2) = f(\mathbf{x}_2, \theta_2) \geq f(\mathbf{x}_1, \theta_2) \geq (f(\mathbf{x}_1, \theta_1) = v(\theta_1) \qquad \square$$

Example 2.75 (Cost function) The cost function (example 2.31) of a firm producing output y purchasing inputs at fixed prices $\mathbf{w} = (w_1, w_2, \ldots, w_n)$ is

$$c(\mathbf{w}, y) = \inf_{\mathbf{x} \in V(y)} \sum_{i=1}^{n} w_i x_i = - \sup_{\mathbf{x} \in V(y)} \sum_{i=1}^{n} (-w_i x_i)$$

and the objective function $\sum -w_i x_i$ is increasing in $-\mathbf{w}$. However, for fixed output y, the input requirement set is not a lattice. Therefore, we cannot apply theorem 2.1. We can apply proposition 2.1, which implies that $-\sup_{\mathbf{x} \in V(y)} \sum_{i=1}^n (-w_i x_i)$ is increasing in $-\mathbf{w}$, and therefore the cost function

$$c(\mathbf{w}, y) = \inf_{\mathbf{x} \in V(y)} \sum_{i=1}^n w_i x_i$$

is increasing in $\mathbf{w}$.

Proposition 2.2 (Ascending maximum theorem) *If f is independent of $\boldsymbol{\theta}$ and $G(\boldsymbol{\theta})$ is ascending, the value function*

$$v(\boldsymbol{\theta}) = \sup_{\mathbf{x} \in G(\boldsymbol{\theta})} f(\mathbf{x})$$

is increasing in $\boldsymbol{\theta}$.

Proof Assume that $\boldsymbol{\theta}_2 \succsim \boldsymbol{\theta}_1 \in \Theta$. Since $G(\boldsymbol{\theta})$ is ascending, $G(\boldsymbol{\theta}_1) \subseteq G(\boldsymbol{\theta}_2)$ and therefore

$$v(\boldsymbol{\theta}_2) = \sup_{\mathbf{x} \in G(\boldsymbol{\theta}_2)} f(\mathbf{x}) \geq \sup_{\mathbf{x} \in G(\boldsymbol{\theta}_1)} f(\mathbf{x}) = v(\boldsymbol{\theta}_1) \qquad \square$$

Example 2.76 (Indirect utility function) For fixed $\mathbf{p}$, the budget correspondence $X(\mathbf{p}, m)$ is ascending in m. Therefore

$$v(\mathbf{p}, m_2) = \sup_{\mathbf{x} \in X(\mathbf{p}, m_2)} u(\mathbf{x}) \geq \sup_{\mathbf{x} \in X(\mathbf{p}, m_1)} u(\mathbf{x}) = v(\mathbf{p}, m_1)$$

The indirect utility function is increasing in m.

Exercise 2.64
Show that the indirect utility function

$$v(\mathbf{p}, m) = \sup_{\mathbf{x} \in X(\mathbf{p}, m)} u(\mathbf{x})$$

is decreasing in $\mathbf{p}$.

Example 2.77 (Cost function) Assuming free disposal, the input requirement sets are ascending (exercise 1.12). For fixed input prices $\mathbf{w}$, the cost function

$$c(\mathbf{w}, y) = \inf_{\mathbf{x} \in V(y)} \sum_{i=1}^n w_i x_i = -\sup_{\mathbf{x} \in V(y)} \sum_{i=1}^n (-w_i x_i)$$

is increasing in y.

The following exercise refines theorem 2.1, to show that quasisupermodularity and the strict crossing condition are both necessary and sufficient for monotone comparative statics.

Exercise 2.65
Consider the general constrained maximization problem where X is a lattice, Θ a poset and the feasible set G is independent of θ. The optimal solution correspondence

$$\varphi(\theta, G) = \arg \max_{\mathbf{x} \in G} f(\mathbf{x}, \theta)$$

is increasing in (θ, G) if and only if

- f is quasisupermodular in X
- and f satisfies the single crossing condition

2.3 Continuous Functions

Roughly speaking, a function is continuous if small changes in input (the independent variable) produce only small changes in output (the dependent variable). Continuity of the physical world makes life bearable. When you make a small adjustment in the volume control of your stereo system, you do not expect to be deafened by a vast change in loudness. In riding a bicycle, a small change in posture does not produce a dramatic change in altitude. By and large, physical systems are continuous. Continuity is equally important for economic analysis. Throughout this section the domain and co-domain will be metric spaces.

A function between metric spaces is continuous if the images of neighboring points are neighbors. Formally a function $f: X \to Y$ is *continuous* at x_0 in X if for every neighborhood T of $f(x_0)$, there exists a corresponding neighborhood S of x_0 such that $f(S) \subseteq T$. f is *continuous* if it is continuous at all x_0 in X. Continuous functions are important because they respect the geometric structure of the domain and co-domain.

Remark 2.11 An equivalent definition of continuity is: A function $f: X \to Y$ is continuous at x_0 if for every $\varepsilon > 0$ there exist a $\delta > 0$ such that for every $x \in X$,

$$\rho(x, x_0) \leq \delta \Rightarrow \rho(f(x), f(x_0)) < \varepsilon$$

The next three exercises provide equivalent characterizations that are often useful in practice.

Exercise 2.66
$f: X \to Y$ is continuous if and only if the inverse image of any open subset of Y is an open subset of X.

Exercise 2.67
$f: X \to Y$ is continuous if and only if the inverse image of any closed subset of Y is a closed subset of X.

Exercise 2.68
$f: X \to Y$ is continuous if and only if $f(x) = \lim_{n \to \infty} f(x^n)$ for every sequence $x^n \to x$.

Care must be taken to distinguish between continuous and open mappings. A function $f: X \to Y$ is continuous if $f^{-1}(T)$ is open in X whenever T is open in Y. It is called an *open mapping* if $f(S)$ is open in Y whenever S is open in X. An open mapping preserves open sets. If an open mapping has inverse, then the inverse is continuous (exercise 2.69). In general, continuous functions are not open mappings (example 2.78). However, every continuous function on a compact domain is an open mapping (exercise 2.76), as is every bounded linear function (proposition 3.2).

Example 2.78 (A continuous function that is not an open mapping) The function $f: \Re \to \Re$ defined by $f(x) = x^2$ is continuous. However, its range $f(\Re) = \Re_+$ is closed (not open) in $\Re$. Therefore it is not an open mapping.

Exercise 2.69
Let $f: X \to Y$ be one-to-one and onto. Suppose that f is an open mapping. Then f has a continuous inverse $f^{-1}: Y \to X$.

Remark 2.12 (Homeomorphism) A one-to-one continuous open function f of X onto Y is called a *homeomorphism*. Since it is one-to-one and onto, f has an inverse. Since f is open, the inverse f^{-1} is a continuous mapping from Y onto X. Homeomorphic spaces are indistinguishable geometrically, and differ only in the nature of their elements.

Exercise 2.70 (Closed graph)
If f is a continuous function from X to Y, the graph of f,

$$\text{graph}(f) = \{(x,y) : y = f(x), x \in X\}$$

is a closed subset of $X \times Y$.

The converse of this result is not true in general. The following exercise details a partial converse in the special case in which the range Y is compact. Later we show that converse also holds for linear functions (exercise 3.37), a fundamental result which is known as the closed graph theorem.

Exercise 2.71
Suppose that Y is compact. $f: X \to Y$ is continuous if and only if

$$\text{graph}(f) = \{(x,y) : y = f(x), x \in X\}$$

is a closed subset of $X \times Y$.

Exercise 2.72
If $f: X \to Y$ and $g: Y \to Z$ are continuous function, so is their composition $g \circ f: X \to Z$.

Most of the functions that we encounter in practice are continuous. Trivially, constant and identity functions are continuous. Typical functional forms, such as the Cobb-Douglas function, are continuous (example 2.81). The norm on a normed linear space is continuous (exercise 2.73). One of the most important theorems in this book (theorem 2.3) shows that the solution of a constrained optimization problem is continuous provided the structure of the problem is continuous.

Exercise 2.73
Let X be a normed linear space. The norm $\| \cdot \|$ is a continuous function on X.

Exercise 2.74 (Utility functions)
Let $\succsim$ be a continuous preference relation on $\Re_+^n$. Assume that $\succsim$ is strongly monotonic. There exists a continuous function $u: \Re_+^n \to \Re$ which represents the preferences.
[Hint: Show that the function u defined in exercise 2.38 is continuous.]

Example 2.79 (Path) Given a set X, any continuous function $f: \Re \to X$ is called a *path*. In a sense, a path is the opposite of a continuous functional, mapping $\Re \to X$ rather than $X \to \Re$. Paths arise in dynamic models where the dependent variable is often time. The terminology comes from the physical world where the motion of any object traces a path in $\Re^3$.

Example 2.80 (Nucleolus) The nucleolus (example 1.49) is a value (example 2.37), a function Nu: $\mathscr{G}^N \to \Re^n$ such that $\text{Nu}(N, w) \in X = \{\mathbf{x} \in \Re^n : \sum_{i \in N} x_i = w(N)\}$. The nucleolus is in fact a continuous function. That is, if (N, w^n) is a sequence of games converging to a game (N, w), and $\mathbf{x}^n$ is the nucleolus of the each game (N, w^n), then $\mathbf{x} = \lim \mathbf{x}^n$ is the nucleolus of the game (N, w) (Schmeidler 1969). The significance of continuity is that the nucleolus is relatively insensitive to small changes in the characteristic function. This is important in practice since the specification of a game is seldom known with precision. We can be confident that small errors in the measurement of the worth of specific coalitions will not result in drastic changes in the suggested outcome.

Continuous functions preserve two of the most significant topological properties.

Proposition 2.3 *Let $f: X \to Y$ be continuous.*

- *$f(X)$ is compact if X is compact*
- *$f(X)$ is connected if X is connected*

Exercise 2.75
Prove proposition 2.3.

Exercise 2.76
Suppose that X is compact and f is a continuous one-to-one function from X onto Y. Then f is an open mapping, which implies that f^{-1} is continuous and f is a homeomorphism.

2.3.1 Continuous Functionals

Some properties of continuity can be sharpened when applied to functionals, which are the most frequently encountered functions. First, we have a convenient characterization of continuity in terms of the upper and lower contour sets.

Exercise 2.77
A functional $f: X \to \Re$ is continuous if and only if its upper

$$\succsim_f(a) = \{x: f(x) \geq a\}$$

and lower contour sets

$$\precsim_f(a) = \{x: f(x) \leq a\}$$

are both closed.

Remark 2.13 We noted earlier (section 2.1.1) that every functional induces an ordering on the its domain X. An immediate implication of the previous result is that a continuous functional induces a continuous ordering. This shows that continuity is a necessary as well as a sufficient condition for the existence of a continuous utility function (exercise 2.74).

Next, we show that standard algebraic operations on functionals preserve continuity. These results can be used to show some familiar functional forms in economics are continuous.

Exercise 2.78
If f, g are continuous functionals on a metric space X, then

- $f + g$ is continuous
- αf is continuous for every $\alpha \in \Re$

Therefore the set of all continuous functionals on X is a linear space.

Exercise 2.79 If f, g are continuous functionals on a metric space X, then their product fg defined by $(fg)(x) = f(x)g(x)$ is continuous.

Remark 2.14 We could follow a similar agenda to that in section 2.2 to demonstrate that common functional forms are continuous. The identity function is clearly continuous. Repeated application of exercise 2.79 shows that the power functions are continuous. Exercise 2.78 shows that every polynomial of power functions is continuous. From there we can deduce that the exponential function (example 2.10) is continuous, which in turn implies that the log function (example 2.55) is continuous. Exercise 2.72 then shows that the general power function (example 2.56) is continuous. Instead, we will take this for granted for now. In chapter 5 we will show that these functions are differentiable, which implies that they are continuous.

2.3 Continuous Functions

Example 2.81 The Cobb-Douglas function

$$f(\mathbf{x}) = x_1^{a_1} x_2^{a_2} \ldots x_n^{a_n}, \qquad a_i > 0$$

is continuous on $\Re_+^n$, since it is the product of general power functions.

Exercise 2.80 (CES function)
Show that the CES function

$$f(\mathbf{x}) = (\alpha_1 x_1^\rho + \alpha_2 x_2^\rho + \cdots \alpha_n x_n^\rho)^{1/\rho}, \qquad \alpha_i > 0 \text{ and } \rho \neq 0$$

is continuous on $\Re_+^n$.

Exercise 2.81
Given two functionals f and g on X, define

$$(f \vee g)(x) = \max\{f(x), g(x)\}$$
$$(f \wedge g)(x) = \min\{f(x), g(x)\}$$

If f and g are continuous, then so are $f \vee g$ and $f \wedge g$.

Applied to functionals, proposition 2.3 yields three important corollaries. The first, a counterpart of proposition 1.5 known as the *Weierstrass theorem*, gives sufficient conditions for a constrained optimization problem to have a solution. The second corollary, known as the intermediate value theorem, should be well known from elementary calculus. The third corollary (exercise 2.84) shows that every continuous functional on a compact set is bounded.

A functional $f: X \to \Re$ achieves a maximum at a point $x^* \in X$ if $f(x^*) \geq f(x)$ for every $x \in X$. Similarly it achieves a minimum at x_* if $f(x_*) \geq f(x)$ for every $x \in X$.

Theorem 2.2 (Weierstrass theorem) *A continuous functional on a compact set achieves a maximum and a minimum.*

Proof Let $M = \sup_{x \in X} f(x)$. There exists a sequence x^n in X with $f(x^n) \to M$. Since X is compact, there exists a convergent subsequence $x^m \to x^*$ and $f(x^m) \to M$. However, since f is continuous, $f(x^m) \to f(x^*)$. We conclude that $f(x^*) = M$. □

Exercise 2.82
Use proposition 2.3 to provide an alternative proof of theorem 2.2. [Hint: See the proof of theorem 1.5.]

Exercise 2.83 (Intermediate value theorem)
Let f be a continuous functional on a connected space X. For every $x_1, x_2 \in X$ and $c \in \Re$ such that $f(x_1) < c < f(x_2)$, there exists $x \in X$ such that $f(x) = c$.

Exercise 2.84
Every continuous functional on a compact metric space X is bounded.

More generally, when X is not compact, the set of continuous functionals form a closed subset of the set of bounded functionals.

Exercise 2.85 (The space $C(X)$)
Given a metric space X, the $C(X)$ denote the set of all bounded, continuous functionals on X. Show that

- $C(X)$ is a linear subspace of $B(X)$
- $C(X)$ is closed (in $B(X)$)
- $C(X)$ is a Banach space with the sup norm

$$\|f\| = \sup_{x \in X} |f(x)|$$

For certain applications somewhat weaker or stronger forms of continuity are appropriate or necessary. These generalization are dealt with in the next two sections. Then we extend the notion of continuity to correspondences, where we find that some of the standard equivalences (exercise 2.70) diverge.

2.3.2 Semicontinuity

Continuous functionals are characterized by the property that both upper $\{x : f(x) \geq \alpha\}$ and lower $\{x : f(x) \leq \alpha\}$ contour sets are closed. A functional $f: X \to \Re$ is said to be *upper semicontinuous* if its upper contour sets $\{x : f(x) \geq \alpha\}$ are closed. Similarly f is *lower semicontinuous* if its lower contour sets $\{x : f(x) \leq \alpha\}$ are closed. An upper semicontinuous function is illustrated in figure 2.19. An upper (or lower) semicontinuous function can have jumps, but the jumps must all be in one direction.

Exercise 2.86
f is upper semicontinuous $\Leftrightarrow -f$ is lower semicontinuous.

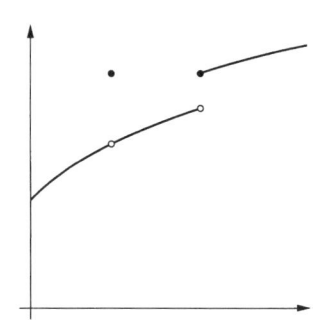

Figure 2.19
An upper semicontinuous function

Exercise 2.87
A function f is continuous if and only if it is both upper and lower semicontinuous.

The following exercise, which should be compared to exercise 2.70, provides equivalent characterizations of semicontinuity which are useful in practice.

Exercise 2.88
For any $f: X \to \Re$, the following conditions are equivalent:

1. f is upper semicontinuous.
2. $f(x) \geq \lim_{n \to \infty} f(x^n)$ for every sequence $x^n \to x$.
3. The hypograph of f is closed in $X \times \Re$.

Semicontinuity, as opposed to the more restrictive continuity, is often assumed in economic analysis, since it is sufficient to guarantee the existence of a maximum in a constrained optimization model. This is a consequence of the following result, which shows that semicontinuous functions obey a form of the Weierstrass theorem.

Exercise 2.89
An upper semicontinuous functional on a compact set achieves a maximum.

2.3.3 Uniform Continuity

Completeness was noticeably absent from the list of properties preserved by continuous mappings (proposition 2.3). This is because mere continuity is insufficient to preserve Cauchy sequences (example 2.82). For this reason, a slight strengthening of continuity is of particular significance in analysis. A function $f: X \to Y$ is *uniformly continuous* if for every $\varepsilon > 0$ there exist a $\delta > 0$ such that for every $x, x_0 \in X$,

$$\rho(x, x_0) \leq \delta \Rightarrow \rho(f(x), f(x_0)) < \varepsilon \tag{20}$$

Remark 2.15 (Uniform continuity versus continuity) The distinction between the definitions of continuity (see remark 2.11) and uniform continuity is subtle but significant. For mere continuity the choice of δ necessary to satisfy (20) may depend on x_0 as well as ε. Uniform continuity imposes the additional restriction that for every ε there exists a δ that satisfies (20) uniformly over the entire space X. Note, however, that the concepts are equivalent on compact domains (exercise 2.91).

Example 2.82 Let $f: [0, 1) \to \Re$ be the defined by $f(x) = x/(1-x)$. f is continuous but not uniformly continuous. The sequence $x^n = 1 - 1/n$ is a Cauchy sequence, its image $f(x^n) = n - 1$ is not.

Exercise 2.90
Let $f: X \to Y$ be uniformly continuous. If (x^n) is a Cauchy sequence in X, $(f(x^n))$ is a Cauchy sequence in Y.

Exercise 2.91
A continuous function on a compact domain is uniformly continuous.

In economic analysis, uniform continuity typically takes a slightly stronger form. A function $f: X \to Y$ is *Lipschitz (continuous)* if there is a constant β such that for every $x, x_0 \in X$,

$$\rho(f(x), f(x_0)) \leq \beta \rho(x, x_0)$$

β is called the *Lipschitz constant* or *modulus*.

Exercise 2.92
A Lipschitz function is uniformly continuous.

We frequently encounter a particularly strong form of Lipschitz continuity where the function maps a metric space into itself with modulus less than one. Such a function, which maps points closer together, is called a contraction. Specifically, an operator $f: X \to X$ is called a *contraction mapping* if (or simply a *contraction*) if there exists a constant β, $0 \leq \beta < 1$, such that

$$\rho(f(x), f(x_0)) \leq \beta \rho(x, x_0)$$

for every $x, x_0 \in X$. Contraction mappings are valuable in economic analysis since they can easily be shown to have a unique fixed point (theorem 2.5).

Example 2.83 (Dynamic programming) The dynamic programming problem (example 2.32)

$$\max_{x_1, x_2, \ldots} \sum_{t=0}^{\infty} \beta^t f(x_t, x_{t+1})$$

subject to $\quad x_{t+1} \in G(x_t), t = 0, 1, 2, \ldots, x_0 \in X$

gives rise to an operator

$$(Tv)(x) = \sup_{y \in G(x)} \{f(x, y) + \beta v(y)\}$$

on the space $B(X)$ of bounded functionals (exercise 2.18). Provided the discount rate $\beta < 1$, T is a contraction mapping with modulus β. To see this, assume that $v, w \in B(X)$. Since $B(X)$ is a normed linear space (exercise 2.11), for every $y \in X$,

$$v(y) - w(y) = (v - w)(y) \leq \|v - w\|$$

or

$$v(y) \leq w(y) + \|v - w\|$$

Consequently for any $\beta \geq 0$,

$$\beta v(y) \leq \beta w(y) + \beta \|v - w\|$$

and

$$\begin{aligned}(Tv)(x) &= \sup_{y \in G(x)} f(x, y) + \beta v(y) \\ &\leq \sup_{y \in G(x)} f(x, y) + \beta w(y) + \beta \|v - w\| \\ &= Tw(x) + \beta \|v - w\|\end{aligned}$$

or

$$(Tv - Tw)(x) = Tv(x) - Tw(x) \leq \beta \|v - w\|$$

Since this is true for every $x \in X$,

$$\|Tv - Tw\| = \sup_x (Tv - Tw)(x) \leq \beta \|v - w\|$$

T is a contraction with modulus β.

The only specific features of the operator T in the preceding example that are required to demonstrate that it is a contraction are the facts that T is increasing (exercise 2.42) and future returns are discounted. The following exercise, which captures these properties, is useful in identifying contraction mappings in economic models.

Exercise 2.93 (Sufficient conditions for a contraction)
Let $B(X)$ be the space of bounded functionals on a metric space X (example 2.11). Let $T: B(X) \to B(X)$ be an increasing function with property that for every constant $c \in \Re$,

$$T(f + c) = T(f) + \beta c \quad \text{for every } f \in B(X) \tag{21}$$

for some $0 \leq \beta < 1$. Show that T is a contraction with modulus β.

Exercise 2.94
Show that operator T in example 2.83 satisfies the conditions of the previous exercise.

Remark 2.16 (Isometry) Another special case of a Lipschitz function f is one that preserves distance so that

$$\rho(f(x_1), f(x_2)) = \rho(x_1, x_2)$$

Such a function is called an isometry. Isometric spaces are essentially equivalent as metric spaces, differing only in the nature of their points.

Equicontinuity

Uniform continuity applies to a single function. An even stronger notion is useful in characterizing sets of functions. A set F of continuous functions defined on a compact metric space X is *equicontinuous* if for every $\varepsilon > 0$ there exists a $\delta > 0$ such that for every $x, x_0 \in X$, and $f \in F$,

$$\rho(x, x_0) \leq \delta \Rightarrow \rho(f(x), f(x_0)) < \varepsilon$$

That is, a family F of continuous functions is equicontinuous if each function f is uniformly continuous and the continuity is uniform for all functions in F.

The most important application of equicontinuity is in characterizing compact subsets of $C(X)$. Recall that a closed subspace of a complete metric space is compact if and only if it is totally bounded (exercise 1.113). Also we have previously shown that $C(X)$ is complete. Therefore a subset of $C(X)$ will be compact if and only if it is totally bounded, which is the case provided it is bounded and equicontinuous.

Exercise 2.95 (Ascoli's theorem)
Let X be a compact metric space. A closed subspace of $C(X)$ is compact if and only if it is bounded and equicontinuous. [Hint: Adapt exercise 1.113).]

Exercise 2.96
If $F \subseteq C(X)$ is equicontinuous, then so is $\bar{F}$.

2.3.4 Continuity of Correspondences

A function is continuous where small changes in input produce small changes in output. We formalized this by requiring that neighboring images arise from neighboring points. Defining continuity for correspondences is a little more complicated, since there is a possible ambiguity regarding the identity of the neighbors. Specifically, there are two reasonable definitions of the inverse image of any set. The following example illustrates the issue.

Example 2.84 Consider the strategic game

		Player 2			
		t_1	t_2	t_3	t_4
	s_1	1, 1	1, 1	0, 0	0, 0
Player 1	s_2	0, 0	2, 2	2, 2	0, 0
	s_3	1, 0	1, 0	0, 0	3, 3

Player 2's best response correspondence φ_2 is

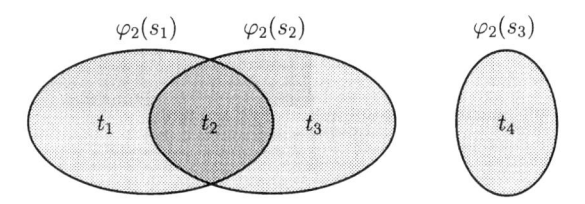

Figure 2.20
The best response correspondence of player 2

$\varphi_2(s_1) = \{t_1, t_2\}$

$\varphi_2(s_2) = \{t_2, t_3\}$

$\varphi_2(s_3) = \{t_4\}$

which is illustrated in figure 2.20.

Clearly, the inverse image of t_4 is s_3. Player 2's optimal response is t_4 if and only if 1 plays s_3. However, what should we regard as the inverse image of $\{t_2, t_3\}$? $\{s_2\}$ is the set of strategies of player 1 which ensure a response in $\{t_2, t_3\}$. We see that a best response in $\{t_2, t_3\}$ is possible when 1 chooses either s_1 or s_2. Our definition of continuity will vary depending on whether we regard s_1 as an element of the inverse image of $\{t_2, t_3\}$.

Given a correspondence $\varphi: X \rightrightarrows Y$, the *upper* (or *strong*) *inverse* of $T \subseteq Y$ is

$$\varphi^+(T) = \{x \in X : \varphi(x) \subseteq T\}$$

The *lower* (or *weak*) *inverse* is

$$\varphi^-(T) = \{x \in X : \varphi(x) \cap T \neq \emptyset\}$$

The upper inverse includes only assured precursors of $\varphi(x)$, while the lower inverse includes all possible precursors.

Example 2.85 In the previous example

$\varphi_2^+(\{t_2, t_3\}) = \{s_2\}$

$\varphi_2^-(\{t_2, t_3\}) = \{s_1, s_2\}$

Exercise 2.97
Let $\varphi\colon X \rightrightarrows Y$. For every $T \subseteq Y$,
$$\varphi^+(T) = [\varphi^-(T^c)]^c$$

Exercise 2.98
Regarding a correspondence $\varphi\colon X \rightrightarrows Y$ as a function from X to $\mathscr{P}(Y)$, the natural inverse is
$$\varphi^{-1}(T) = \{x \in X : \varphi(x) = T\}$$
Show that for every $T \in \varphi(X)$,
$$\varphi^{-1}(T) \subseteq \varphi^+(T) \subseteq \varphi^-(T)$$

Unfortunately, the natural inverse φ^{-1} is not very useful as its composition is erratic (see the following exercise).

Exercise 2.99
For the game in example 2.84, calculate $\varphi_2^{-1}, \varphi_2^+, \varphi_2^-$ for the sets $\{t_1\}$, $\{t_2\}$, $\{t_1, t_2\}$, $\{t_2, t_3\}$, and $\{t_1, t_2, t_3\}$.

The two definitions of inverse image give rise to two definitions of continuity for correspondences. A correspondence is said to be upper hemicontinuous if, whenever x_0 is in the upper inverse of an open set, so is a neighborhood of x_0. Similarly a correspondence is lower hemicontinuous if, whenever x_0 is in the lower inverse of an open set, so is a neighborhood of x_0.

Formally a correspondence $\varphi\colon X \rightrightarrows Y$ is *upper hemicontinuous* (uhc) *at* x_0 if for every open set T containing $\varphi(x_0)$, there exists a neighborhood S of x_0 such that $\varphi(x) \subset T$ for every $x \in S$. φ is upper hemicontinuous if it is uhc at every $x_0 \in X$. A uhc correspondence cannot suddenly become much larger or "explode" for a small change in x. The correspondence illustrated in figure 2.21 is not uhc at x_0, since there are neighboring points of x_0 for which $\varphi(x)$ lies outside a small open set T containing $\varphi(x_0)$.

A correspondence φ is *lower hemicontinuous* (lhc) *at* x_0 if for every open set T meeting $\varphi(x_0)$, there exists a neighborhood S of x_0 such that $\varphi(x) \cap T \neq \emptyset$ for every $x \in S$. A lhc correspondence cannot suddenly contract or "implode." The correspondence illustrated in figure 2.22 is not lhc at x_0, since there are neighboring points of x_0 for which $\varphi(x)$ does not meet the open set T.

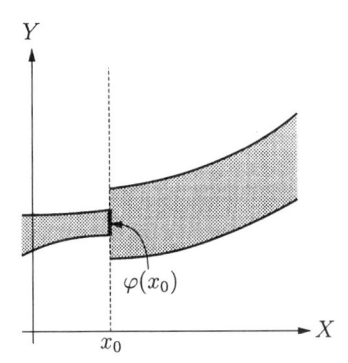

Figure 2.21
φ is not uhc at x_0

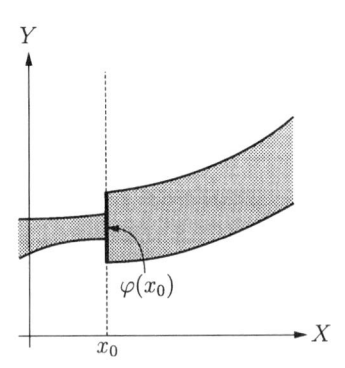

Figure 2.22
φ is not lhc at x_0

Finally a correspondence φ is *continuous* at x_0 if it is both upper hemicontinuous and lower hemicontinuous at x_0.

Remark 2.17 (Hemicontinuity or semicontinuity) Many authors use the term semicontinuity to describe the continuity of correspondences, which risks confusion with the distinct concept of semicontinuity of functionals (section 2.3.2).

Example 2.86 Let $X = [0, 2]$. The correspondence $\varphi \colon X \rightrightarrows X$ defined by

$$\varphi(x) = \begin{cases} \{1\} & 0 \le x < 1 \\ X & 1 \le x \le 2 \end{cases}$$

is uhc but not lhc at $x = 1$. If T is an open set containing $\varphi(1) = X$, then T contains $\varphi(x)$ for every $x \in X$. Therefore φ is uhc at 1. To see that φ is not lhc at $x = 1$, consider the open interval $T = (3/2, 2)$. Clearly, $\varphi(1) \cap T \neq \emptyset$ but $\varphi(x) \cap T = \emptyset$ for every $x < 1$. Therefore φ is not lhc at $x = 1$. Note that φ is continuous for every $x \neq 1$ (exercise 2.101).

Exercise 2.100
Let $X = [0, 2]$. Show that the correspondence $\varphi \colon X \rightrightarrows X$ defined by

$$\varphi(x) = \begin{cases} \{1\} & 0 \leq x \leq 1 \\ X & 1 < x \leq 2 \end{cases}$$

is lhc but not uhc at $x = 1$.

Exercise 2.101 (Constant correspondence)
Let K be any subset of Y. The constant correspondence $\varphi \colon X \to Y$ defined by

$$\varphi(x) = K \quad \text{for every } \mathbf{x} \in X$$

is continuous.

Example 2.87 (Matching Pennies) Consider the following strategic game

		Player 2	
		H	T
Player 1	H	1, −1	−1, 1
	T	−1, 1	1, −1

which is usually known as Matching Pennies. The game has no pure strategy equilibrium.

Let σ_1 denote the probability with which player 1 plays H. If player 1 is more likely to choose H ($\sigma_1 > 1/2$), player 2 should respond with T. Conversely, if player 1 is more likely to choose T ($\sigma_1 < 1/2$), player 2 should respond with H. However, if 1 is equally likely to choose H or T, any response is equally useful. Therefore player 2's best response correspondence $\varphi_2 \colon [0, 1] \rightrightarrows [0, 1]$ is given by

$$\varphi_2(\sigma_1) = \begin{cases} 1 & \text{if } \sigma_1 < \frac{1}{2} \\ [0, 1] & \text{if } \sigma_1 = \frac{1}{2} \\ 0 & \text{if } \sigma_1 > \frac{1}{2} \end{cases}$$

φ_2 is uhc, but it is not lhc at $s_1 = \frac{1}{2}$. For example, if $T = (\frac{1}{4}, \frac{3}{4})$

$$\varphi_2^+(T) = \emptyset \quad \text{and} \quad \varphi_2^-(T) = \{\tfrac{1}{2}\}$$

As in the case of continuity of functions, we have useful characterizations in terms of open sets and in terms of sequences. A correspondence is upper hemicontinuous if the upper inverse images of open sets are open. It is lower hemicontinuous if the lower inverse images of open sets are open. Both conditions arise in applications, and neither condition implies the other.

Exercise 2.102
A correspondence $\varphi \colon X \rightrightarrows Y$ is

- uhc $\Leftrightarrow \varphi^+(T)$ is open for every open set T
- lhc $\Leftrightarrow \varphi^-(T)$ is open for every open set T

Exercise 2.103
A correspondence $\varphi \colon X \rightrightarrows Y$ is

- uhc $\Leftrightarrow \varphi^-(T)$ is closed for every closed set T
- lhc $\Leftrightarrow \varphi^+(T)$ is closed for every closed set T

Exercise 2.104
A compact-valued correspondence $\varphi \colon X \rightrightarrows Y$ is uhc if and only if for every sequence $x^n \to x$ in X and every sequence $(y^n) \in Y$ with $y^n \in \varphi(x^n)$, there exists a subsequence of y^n that converges to $y \in \varphi(x)$.

Exercise 2.105
A correspondence $\varphi \colon X \rightrightarrows Y$ is lhc if and only if for every sequence $x^n \to x$ in X and for every $y \in \varphi(x)$, there exists a sequence $y^n \to y$ with $y^n \in \varphi(x^n)$.

Upper hemicontinuity of a correspondence is often confused with the property of having a closed graph. The two properties are distinct (example 2.88), although they are equivalent for closed-valued correspondences into a compact space (exercise 2.107).

Example 2.88 (Closed graph versus upper hemicontinuity) The correspondence $\varphi \colon \Re_+ \rightrightarrows \Re$ defined by

$$\varphi(x) = \begin{cases} \left\{\dfrac{1}{x}\right\} & \text{if } x > 0 \\ \{0\} & \text{if } x = 0 \end{cases}$$

is closed but is not uhc at 0.

To see this, note that the set $\{(x, 1/x) : x > 0\}$ is closed in $\Re_+ \times \Re$, and hence so also is graph$(\varphi) = \{(x, 1/x) : x > 0\} \cup (0, 0)$. Note also that for every sequence $x^n \to 0$, $y^n \in \varphi(x^n)$ does not converge.

The constant correspondence $\varphi \colon \Re \rightrightarrows \Re$ defined by

$$\varphi(x) = (0, 1)$$

is uhc but not closed. It is uhc, since for every x and every $T \supseteq \varphi(x)$, $\varphi^+(T) = \Re$. Exercise 2.107 does not apply, since φ is not closed-valued.

The next two exercises should be compared with the corresponding results for functions (exercises 2.70 and 2.71).

Exercise 2.106
Let $\varphi \colon X \rightrightarrows Y$.

1. If φ is closed, then φ is closed-valued.
2. If φ is closed-valued and uhc, then φ is closed.
3. If Y is compact and φ closed, then φ is uhc.

Exercise 2.107 (Closed equals upper hemicontinuous)
Suppose that Y is compact. The correspondence $\varphi \colon X \rightrightarrows Y$ is closed if and only if it is closed-valued and uhc.

The following exercise is a useful generalization of the previous result. It will be used to prove the continuous maximum theorem (theorem 2.3).

Exercise 2.108
If $\varphi_1 \colon X \rightrightarrows Y$ is closed and $\varphi_2 \colon X \rightrightarrows Y$ is uhc and compact-valued, $\varphi = \varphi_1 \cap \varphi_2$ is uhc and compact-valued.

Example 2.89 (Budget correspondence is uhc) Let P denote the domain of the budget correspondence, that is, the set of all prices and incomes pairs for which some consumption is feasible

$$P = \left\{ (\mathbf{p}, m) \in \Re^n \times \Re : \min_{\mathbf{x} \in X} \sum_{i=1}^{m} p_i x_i \leq m \right\}$$

The graph of budget correspondence $X(\mathbf{p}, m)$ (example 2.46),

$$\text{graph}(X) = \left\{ (\mathbf{p}, m, \mathbf{x}) \in P \times X : \sum_{i=1}^{m} p_i x_i \leq m \right\}$$

is closed in $P \times X$ (see exercise 1.231). Consequently, if the consumption set X is compact, the budget correspondence $X(\mathbf{p}, m)$ is uhc (exercise 2.107).

Exercise 2.109 (Budget correspondence is continuous)
Assume that the consumption set X is nonempty, compact, and convex. Let

$$X(\mathbf{p}, m) = \left\{ \mathbf{x} \in X : \sum_{i=1}^{m} p_i x_i \leq m \right\}$$

denote the budget correspondence. Choose any $(\mathbf{p}, m) \in P$ such that $m > \min_{\mathbf{x} \in X} \sum_{i=1}^{m} p_i x_i$, and let T be an open set such that $X(\mathbf{p}, m) \cap T \neq \emptyset$. For $n = 1, 2, \ldots$, let

$$B_n(\mathbf{p}, m) = \{ (\mathbf{p}', m') \in P : \|\mathbf{p} - \mathbf{p}'\| + |m - m'| < 1/n \}$$

denote the sequence of open balls about $(\mathbf{p}, m)$ of radius $1/n$.

1. Show that there exists $\tilde{\mathbf{x}} \in T$ such that $\sum_{i=1}^{n} p_i \tilde{x}_i < m$.
2. Suppose that $X(\mathbf{p}, m)$ is not lhc. Show that this implies that
   a. there exists a sequence $((\mathbf{p}^n, m^n))$ in P such that that

   $(\mathbf{p}^n, m^n) \in B_n(\mathbf{p}, m) \quad \text{and} \quad X(\mathbf{p}^n, m^n) \cap T = \emptyset$

   b. there exists N such that $\tilde{\mathbf{x}} \in X(\mathbf{p}^N, m^N)$
   c. $\tilde{\mathbf{x}} \notin T$
3. Conclude that $X(\mathbf{p}, m)$ is lhc at $(\mathbf{p}, m)$.
4. The budget correspondence is continuous for every $\mathbf{p} \neq \mathbf{0}$ such that $m > \inf_{\mathbf{x} \in X} \sum_{i=1}^{m} p_i x_i$.

Remark 2.18 The assumption in exercise 2.109 that the consumption set is compact is unrealistic and stronger than necessary. It suffices to assume that the X is closed and bounded from below (Debreu 1959).

Exercise 2.110 is fundamental, while exercise 2.111 is given for its own interest.

Exercise 2.110
Let $\varphi\colon X \rightrightarrows Y$ be uhc and compact-valued. Then $\varphi(K)$ is compact if K is compact.

Exercise 2.111
If X is a compact space and $\varphi\colon X \rightrightarrows X$ uhc and compact-valued such that $\varphi(x)$ is nonempty for every x, then there exists a compact nonempty subset K of X such that $\varphi(K) = K$.

Exercise 2.112 (Product of correspondences)
Let φ_i, $i = 1, 2, \ldots, n$, be a collection of compact-valued and uhc correspondences $\varphi_i\colon X \rightrightarrows Y_i$. The product correspondence $\varphi\colon S \rightrightarrows Y$, $Y = Y_1 \times Y_2 \times \cdots \times Y_n$ defined by

$$\varphi(x) = \varphi_1(x) \times \varphi_2(x) \times \cdots \times \varphi_n(x)$$

is compact-valued and uhc.

Continuous Selections

As we stated before, given a correspondence $\varphi\colon X \rightrightarrows Y$, we can always construct a selection, that is, a function $f\colon X \to Y$, such that $f(x) \in \varphi(x)$ for every $x \in X$. If the correspondence φ is continuous, can we make a *continuous* selection? The answer is yes, provided that X is compact and φ has closed convex values. In fact lower hemicontinuity suffices and upper hemicontinuity is not required. Straightforward proofs of this result, known as the *Michael selection theorem*, can be found in Border (1985, p. 70) and Hildenbrand and Kirman (1976, p. 203).

2.3.5 The Continuous Maximum Theorem

The continuous maximum theorem is usually known simply as *the* maximum theorem. It is one of the most frequently used theorems in mathematical economics. It gives sufficient conditions to impose a constrained optimization model to ensure that an optimal solution exists and varies continuously with the parameters.

Theorem 2.3 (Continuous maximum theorem) *Consider the general constrained maximization problem*

$$\max_{\mathbf{x} \in G(\boldsymbol{\theta})} f(\mathbf{x}, \boldsymbol{\theta})$$

If the objective function $f: X \times \Theta \to \Re$ *is continuous and the constraint correspondence* $G: \Theta \rightrightarrows X$ *continuous and compact-valued, then the value function* $v: \Theta \to \Re$,

$$v(\theta) = \sup_{\mathbf{x} \in G(\theta)} f(\mathbf{x}, \theta)$$

is continuous and the optimal correspondence

$$\varphi(\theta) = \arg\max_{\mathbf{x} \in G(\theta)} f(\mathbf{x}, \theta)$$

is nonempty, compact-valued, and upper hemicontinuous.

Proof

$\varphi(\theta)$ **is nonempty for every** θ Since $G(\theta)$ is compact for every θ and f is continuous, $\varphi(\theta)$ is nonempty (theorem 2.2).

φ **is closed-valued** For any $\theta \in \Theta$, let $(\mathbf{x}^n)$ be sequence in $\varphi(\theta)$ which converges to $\mathbf{x}$. Since $\mathbf{x}^n \in \varphi(\theta)$, $f(\mathbf{x}^n) = v(\theta)$ for every n. Moreover

- $G(\theta)$ compact implies that $\mathbf{x} \in G(\theta)$
- f continuous implies that $f(\mathbf{x}, \theta) = \lim_{n \to \infty} f(\mathbf{x}^n, \theta) = v(\theta)$

We conclude that $\mathbf{x} \in \varphi(\theta)$ and that therefore $\varphi(\theta)$ is closed (exercise 1.107).

φ **is compact-valued** $\varphi(\theta)$ is a closed subset of a compact set $G(\theta)$. Therefore $\varphi(\theta)$ is compact for every θ (exercise 1.111).

φ **is closed** Let $\theta^n \to \theta$ be a sequence of parameters and $\mathbf{x}^n \in \varphi(\theta^n)$ a corresponding sequence of maximizers with $\mathbf{x}^n \to \mathbf{x}$. We have to show that $\mathbf{x} \in \varphi(\theta)$.

We first note that $\mathbf{x}$ is feasible, that is $\mathbf{x} \in G(\theta)$, since $\mathbf{x}^n \in G(\theta)$ and G is closed (exercise 2.106). Suppose that $\mathbf{x}$ is not maximal, that is $\mathbf{x} \notin \varphi(\theta)$. Then there exists some $\mathbf{z} \in G(\theta)$ with $f(\mathbf{z}, \theta) > f(\mathbf{x}, \theta)$. By lower hemicontinuity of G, there exists a sequence $\mathbf{z}^n \to \mathbf{z}$ with $\mathbf{z}^n \in G(\theta^n)$. Since $f(\mathbf{z}, \theta) > f(\mathbf{x}, \theta)$, there must exist some n such that $f(\mathbf{z}^n, \theta^n) > f(\mathbf{x}^n, \theta^n)$, contradicting the hypothesis that $\mathbf{x}^n \in \varphi(\theta^n)$. This contradiction establishes that $\mathbf{x}$ is maximal, that is, $\mathbf{x} \in \varphi(\theta)$.

φ **is uhc** Since $\varphi(\theta) \subseteq G(\theta)$, $\varphi = \varphi \cap G$. We have just shown that φ is closed, and we assumed that G is uhc and compact-valued. Therefore φ is uhc (exercise 2.108).

$v(\boldsymbol{\theta})$ **is continuous** Continuity of f implies that $v(\boldsymbol{\theta}^n) = f(\mathbf{x}^n, \boldsymbol{\theta}^n) \to f(\mathbf{x}, \boldsymbol{\theta}) = v(\boldsymbol{\theta})$. □

Example 2.90 (Consumer theory) The consumer's problem (example 1.113) is to choose an affordable consumption bundle $\mathbf{x} \in X$ to maximize satisfaction. Provided that the consumer's preferences are continuous, they can be represented by a continuous utility function $u: X \to \Re$ (exercise 2.74), and the consumer's problem can be expressed by the following constrained optimization problem

$$\max_{\mathbf{x} \in X(\mathbf{p},m)} u(\mathbf{x})$$

where $X(\mathbf{p}, m) = \{\mathbf{x} \in X : \sum_{i=1}^{m} p_i x_i \leq m\}$ is the consumer's budget constraint.

Assume that prices $\mathbf{p} > \mathbf{0}$ and $m > \inf_{\mathbf{x} \in X} \sum_{i=1}^{m} p_i x_i$. Then the budget correspondence is compact-valued and continuous (exercise 2.109). With these assumptions, the consumer's problem satisfies the requirements of the continuous maximum theorem (theorem 2.3), ensuring that the indirect utility function

$$v(\mathbf{p}, m) = \sup_{\mathbf{x} \in X(\mathbf{p},m)} u(\mathbf{x})$$

is continuous and the demand correspondence (example 2.47)

$$\mathbf{x}^*(\mathbf{p}, m) = \arg \max_{\mathbf{x} \in X(\mathbf{p},m)} u(\mathbf{x})$$

is nonempty, compact-valued and upper hemicontinuous.

Furthermore, if the consumer's preference relation is strictly convex (example 1.116), the consumer's demand correspondence $\mathbf{x}(\mathbf{p}, m)$ is a continuous *function* (see example 3.62).

Exercise 2.113 (Dynamic programming)
The dynamic programming problem (example 2.32)

$$\max_{x_1, x_2, \ldots} \sum_{t=0}^{\infty} \beta^t f(x_t, x_{t+1})$$

subject to $\quad x_{t+1} \in G(x_t), t = 0, 1, 2, \ldots, x_0 \in X$

gives rise to an operator

$$(Tv)(x) = \sup_{y \in G(x)} \{f(x,y) + \beta v(y)\}$$

on the space $B(X)$ of bounded functionals (exercise 2.16). Assuming that

- f is bounded and continuous on $X \times X$
- $G(x)$ is nonempty, compact-valued, and continuous for every $x \in X$

show that T is an operator on the space $C(X)$ of bounded continuous functionals on X (exercise 2.85), that is $Tv \in C(X)$ for every $v \in C(X)$.

2.4 Fixed Point Theorems

Fixed point theorems are powerful tools for the economic theorist. They are used to demonstrate the existence of a solution to an economic model, which establishes the consistency of the model and highlights the requirements minimal requirements to ensure a solution. The classic applications of fixed point theorems in economics involve the existence of market equilibria in an economy and the existence of Nash equilibria in strategic games. They are also applied in dynamic models, a fundamental tool in macroeconomic analysis.

Fixed point theorems are essentially existence theorems. They guarantee that a particular model (which fulfills the conditions of the theorem) has a solution, but they tell us nothing about the identity and properties of the solution. However, the theory underlying fixed point theorems can be used to provide practical guidance on the actual computation of solutions.

2.4.1 Intuition

Recall that a fixed point of a mapping from a set X to itself is an element $x \in X$, which is its own image. That is, x is a fixed point of $f: X \to X$ if and only if $f(x) = x$. A fixed point theorem specifies the minimal properties on X and f that are required to ensure that there exists at least one fixed point for every qualifying function.

The fundamental intuition of a fixed point theorem is illustrated in figure 2.23, which depicts a function from the interval $[0,1]$ to itself. The graph of the function must connect the left-hand side of the box to the right-hand side. A fixed point occurs whenever the curve crosses the

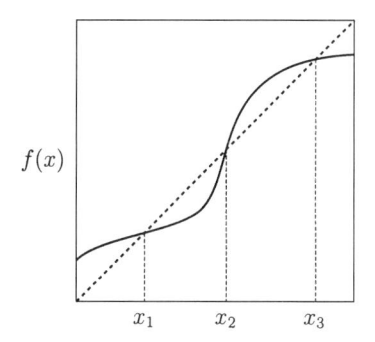

Figure 2.23
A function with three fixed points

45 degree line. The function illustrated in figure 2.23 has three fixed points.

There are three fundamental classes of fixed point theorems, which differ in the structure that is required of the underlying spaces. The Tarski fixed point theorem and its corollaries (section 2.4.2) rely solely on the order structure of X and the monotonicity of f. The Banach fixed point theorem (section 2.4.3) utilizes metric space structure, requiring completeness of the metric space X and a strong form of continuity for f. The most powerful theorem, the Brouwer theorem (section 2.4.4) combines linear and metric structure in a potent cocktail. We deal with each class of theorems in turn.

2.4.2 Tarski Fixed Point Theorem

Our first fixed point theorem has minimal assumptions—an increasing function on a complete lattice. This is sufficient to establish the existence of a pure strategy Nash equilibrium in a supermodular game.

Theorem 2.4 (Tarski's fixed point theorem) *Every increasing function $f: X \to X$ on a complete lattice $(X, \succsim)$ has a greatest and a least fixed point.*

Proof Let

$$M = \{x \in X : f(x) \precsim x\}$$

Note that M contains all the fixed points of f. M is not empty, since $\sup X \in M$. Let $\tilde{x} = \inf M$. We claim that $\tilde{x}$ is a fixed point of f.

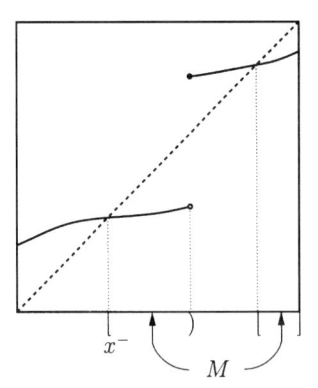

Figure 2.24
Illustrating the proof of the Tarksi theorem

First, we show that $\tilde{x} \in M$. Since $\tilde{x}$ is the greatest lower bound of M and f is increasing,

$$\tilde{x} \precsim x \quad \text{and} \quad f(\tilde{x}) \precsim f(x) \precsim x \quad \text{for every } x \in M$$

Therefore $f(\tilde{x})$ is also a lower bound for M. Since $\tilde{x}$ is the greatest lower bound of M, we must have

$$f(\tilde{x}) \precsim \tilde{x} \tag{22}$$

and so $\tilde{x} \in M$.

Since f is increasing, (22) implies that

$$f(f(\tilde{x})) \precsim f(\tilde{x})$$

and therefore $f(\tilde{x}) \in M$ and (since $\tilde{x} = \inf M$)

$$f(\tilde{x}) \succsim \tilde{x} \tag{23}$$

Together, (22) and (23) (and the fact that $\succsim$ is antisymmetric) imply that

$$\tilde{x} = f(\tilde{x})$$

That is, $\tilde{x}$ is a fixed point of f. Furthermore every fixed point of f belongs to M. So $\tilde{x} = \inf M$ is the least fixed point of f. Similarly we can show that $\sup\{x \in X : f(x) \succsim x\}$ is the greatest fixed point of f. □

Corollary 2.4.1 *Let f be an increasing function on a complete lattice. The set of fixed points of f is a complete lattice.*

Corollary 2.4.2 (Zhou's theorem) *Let $\varphi\colon X \rightrightarrows X$ be an increasing correspondence on a complete lattice X. If $\varphi(x)$ is a complete sublattice of X for every $x \in X$, then the set of fixed points of φ is a nonempty complete lattice.*

It is important to note that while the set of fixed points of an increasing function or correspondence on X forms a complete lattice, it is not necessarily a sublattice of the X. The distinction is illustrated in the following example.

Example 2.91 Let X be the lattice $\{1,2,3\} \times \{1,2,3\}$ and f be a function that maps the points $(2,2), (3,2), (2,3)$, to $(3,3)$ and maps all other points to themselves. The set E of fixed points of f is a complete lattice where, for example, $\sup_E\{(2,1),(1,2)\} = (3,3)$. Note, however, that E is not a sublattice of X. For example, $\sup_X\{(1,2),(2,1)\} = (2,2) \notin E$.

Exercise 2.114
To prove corollary 2.4.1, let $f\colon X \to X$ be an increasing function on a complete lattice $(X, \succsim)$, and let E be the set of fixed points of f. For any $S \subseteq E$ define

$S^* = \{x \in X : x \succsim s \text{ for every } s \in S\}$

S^* is the set of all upper bounds of S in X. Show that

1. S^* is a complete sublattice.
2. $f(S^*) \subset S^*$.
3. Let g be the restriction of f to the sublattice S^*. g has a least fixed point $\tilde{x}$.
4. $\tilde{x}$ is the least upper bound of S in E.
5. E is a complete lattice.

Exercise 2.115
Let $\varphi\colon X \rightrightarrows X$ be an increasing correspondence on a complete lattice X. Assume that $\varphi(x)$ is a complete sublattice of X for every $x \in X$. Let E be the set of fixed points of φ, and define

$M = \{x \in X : \text{there exists } y \in \varphi(x) \text{ such that } y \precsim x\}$

Note that $E \subseteq M$ and $M \neq \emptyset$, since $\sup X \in M$. Let $\tilde{x} = \inf M$. Show that

1. For every $x \in M$, there exists some $z_x \in \varphi(\tilde{x})$ such that $z_x \precsim x$.
2. Let $\tilde{z} = \inf\{z_x\}$. Then
a. $\tilde{z} \precsim \tilde{x}$
b. $\tilde{z} \in \varphi(\tilde{x})$
3. $\tilde{x} \in M$.
4. There exists some $y \in \varphi(\tilde{z})$ such that $y \precsim \tilde{z} \in \varphi(\tilde{x})$. Hence $\tilde{z} \in M$.
5. $\tilde{x} \in E \neq \emptyset$.
6. $\tilde{x}$ is the least fixed point of φ.

Exercise 2.116
To prove corollary 2.4.2, let $S \subseteq E$ and $s^* = \sup S$.

1. For every $x \in S$ there exists some $z_x \in \varphi(s^*)$ such that $z_x \succsim x$.
2. Let $z^* = \sup z_x$. Then
a. $z^* \precsim s^*$
b. $z^* \in \varphi(s^*)$
3. Define $S^* = \{x \in X : x \succsim s \text{ for every } s \in S\}$. S^* is the set of all upper bounds of S in X. S^* is a complete lattice.
4. Define $\mu \colon S^* \rightrightarrows S^*$ by $\mu(x) = \varphi(x) \cap \psi(x)$ where $\psi \colon S^* \rightrightarrows S^*$ is the constant correspondence $\psi(x) = S^*$ for every $x \in S^*$. Show that
a. $\mu(x) \neq \emptyset$ for every $x \in S^*$.
b. $\mu(x)$ is a complete sublattice for every $x \in S^*$.
c. μ is increasing on S^*.
5. μ has a least fixed point $\tilde{x}$.
6. $\tilde{x}$ is the least upper bound of S in E.
7. E is a nonempty complete lattice.

We can use the Tarski fixed point theorem to provide a simple proof of the existence of a pure strategy Nash equilibrium in a supermodular game.

Example 2.92 (Supermodular games) Recall that a strategic game is supermodular if

- every strategy set A_i is a lattice
- the payoff functions $u_i(a_i, \mathbf{a}_{-i})$ are supermodular in a_i
- and display increasing differences in $a_i, \mathbf{a}_{-i}$

Assume further that either

- the strategy spaces A_i are finite or
- the strategy spaces are compact and the payoff functions u_i are upper semicontinuous in $\mathbf{a}$

These assumptions imply that each player i's best response correspondence

$$B_i(\mathbf{a}_{-i}) = \arg \max_{a_i \in A_i} u_i(a_i, \mathbf{a}_{-i})$$

is nonempty (exercise 2.89) and increasing in $\mathbf{a}_{-i}$ (theorem 2.1). This implies that for every player i there exists an increasing selection $f_i \in B(\mathbf{s}_{-i})$, a best response *function* that is increasing the opponents's actions. Define $f: A \to A$ by

$$f(\mathbf{a}) = f_1(\mathbf{a}_{-1}) \times f_2(\mathbf{a}_{-2}) \times \cdots \times f_n(\mathbf{a}_{-n})$$

Then f is increasing (exercise 2.46) on the complete lattice A and therefore has a fixed point $\mathbf{a}^*$ such that $\mathbf{a}^* = f(\mathbf{a}^*)$ or

$$\mathbf{a}^* \in B(\mathbf{a}^*)$$

That is, $\mathbf{a}^*$ is a Nash equilibrium of the game.

Not only is the set of Nash equilibria nonempty, it contains a largest and a smallest equilibrium (in the product order on A). For every $\mathbf{a}_{-i}$, player i's best response set $B(\mathbf{a}_{-i})$ is a (nonempty) sublattice of A_i (corollary 2.1.1). Therefore it has a greatest element $\bar{a}_i$, that is,

$$\bar{a}_i \succsim_i a_i \quad \text{for every } a_i \in B(\mathbf{a}_{-i})$$

Let $\bar{f}_i: A_{-i} \to A$ be the selection of $B_i(\mathbf{a}_i)$ consisting of the greatest elements, that is,

$$\bar{f}_i(\mathbf{a}_{-i}) = \bar{a}_i$$

$\bar{f}_i$ is increasing for every i (exercise 2.117). Applying theorem 2.4, we see that the product mapping $\bar{f}: A \to A$ defined by

$$\bar{f}(\mathbf{a}) = \bar{f}_1(\mathbf{a}_{-1}) \times \bar{f}_2(\mathbf{a}_{-2}) \times \cdots \times \bar{f}_n(\mathbf{a}_{-n})$$

has a greatest fixed point $\bar{\mathbf{a}}^*$ where

$$\bar{\mathbf{a}}^* = \sup\{\mathbf{a} \in A : \mathbf{a} \precsim \bar{f}(\mathbf{a})\}$$

Let $\mathbf{a}^* \in E$ be any Nash equilibrium. Then $\mathbf{a}_i^* \in B(\mathbf{a}_{-i}^*)$, and therefore $\bar{f}_i(\mathbf{a}^*) \succsim_i \mathbf{a}_i^*$ for every i. So we have

$$\bar{\mathbf{a}}^* = \sup\{\mathbf{a} \in A : \bar{f}(\mathbf{a}) \succsim \mathbf{a}\} \succsim \mathbf{a}^*$$

Therefore $\bar{\mathbf{a}}^*$ is the greatest Nash equilibrium. Similarly there exists a least Nash equilibrium $\underline{\mathbf{a}}^*$.

Exercise 2.117
Show that $\bar{f}_i$ is increasing for every i.

Exercise 2.118
Show that the best response correspondence

$$B(\mathbf{a}) = B_1(\mathbf{a}_{-1}) \times B_2(\mathbf{a}_{-2}) \times \cdots \times B_n(\mathbf{a}_{-n})$$

of a supermodular game satisfies the conditions of Zhou's theorem (corollary 2.4.2). Therefore the set of Nash equilibria of a supermodular game is a complete lattice.

2.4.3 Banach Fixed Point Theorem

Our second fixed point theorem applies to a contraction mapping (section 2.3.3) on a complete metric space. The Banach fixed point theorem is a simple and powerful theorem with a wide range of application, including iterative methods for solving linear, nonlinear, differential, and integral equations.

Theorem 2.5 (Banach fixed point theorem) *Every contraction mapping $f \colon X \to X$ on a complete metric space has a unique fixed point.*

Proof Let $\beta < 1$ denote the Lipschitz constant of f. Select an arbitrary $x^0 \in X$. Define the sequence (x^n) by setting

$$x^{n+1} = f(x^n), \quad n = 0, 1, 2, \ldots$$

(x^n) **is a Cauchy sequence**

2.4 Fixed Point Theorems

$$\rho(x^{n+1}, x^n) = \rho(f(x^n), f(x^{n-1}))$$
$$\leq \beta \rho(x^n, x^{n-1})$$
$$\leq \beta^2 \rho(x^{n-1}, x^{n-2})$$
$$\ldots$$
$$\leq \beta^n \rho(x^1, x^0)$$

Using the triangle inequality and the formula for the sum of a geometric series (exercise 1.205)

$$\rho(x^n, x^{n+m}) \leq \rho(x^n, x^{n+1}) + \rho(x^{n+1}, x^{n+2}) + \cdots + \rho(x^{n+m-1}, x^{n+m})$$
$$\leq (\beta^n + \beta^{n+1} + \cdots + \beta^{n+m-1})\rho(x^0, x^1)$$
$$\leq \frac{\beta^n}{1-\beta}\rho(x^0, x^1) \to 0 \text{ as } n \to \infty$$

Therefore (x^n) is a Cauchy sequence.

x^n converges to x in X Since X is complete, there exists some $x \in X$ such that $x^n \to x$.

x is a fixed point Since f is a contraction, it is uniformly continuous, and therefore

$$f(x) = \lim_{n \to \infty} f(x^n) = \lim_{n \to \infty} x^{n+1} = x$$

x is the only fixed point Suppose that $x = f(x)$ and $z = f(z)$. Then

$$\rho(x, z) = \rho(f(x), f(z)) \leq \beta \rho(x, z)$$

which implies that $x = z$. □

The Banach theorem does more than ensure the existence of a unique fixed point. It provides a straightforward algorithm for computing the fixed point by repeated application of f to an arbitrary starting point x^0, computing the sequence

$$x^{n+1} = f(x^n) = f^n(x^0)$$

Whereas many iterative algorithms are sensitive to the initial value, with a contraction mapping, convergence is ensured from any starting point

$x^0 \in X$. Furthermore the following corollary gives useful error bounds for this procedure.

Corollary 2.5.1 *Let $f: X \to X$ be a contraction mapping on the complete metric space X. Let (x^n) be the sequence constructed from an arbitrary starting point x^0, and let $x = \lim x^n$ be the unique fixed point. Then*

$$\rho(x^n, x) \leq \frac{\beta^n}{1-\beta} \rho(x^0, x^1)$$

$$\rho(x^n, x) \leq \frac{\beta}{1-\beta} \rho(x^{n-1}, x^n)$$

Exercise 2.119
Prove corollary 2.5.1.

Exercise 2.120
Example 1.64 outlined the following algorithm for computing the square root of 2:

$$x^0 = 2, \quad x^{n+1} = \frac{1}{2}\left(x^n + \frac{2}{x^n}\right)$$

Verify that

- the function $f(x) = \frac{1}{2}\left(x + \frac{2}{x}\right)$ is a contraction mapping on the set $X = \{x \in \Re : x \geq 1\}$
- the fixed point of f is $\sqrt{2}$

Estimate how many iterations are required to ensure that the approximation error is less than 0.001.

The following result is often useful in establishing the properties of the fixed point of a particular model.

Corollary 2.5.2 *Let $f: X \to X$ be a contraction mapping on the complete metric space X with fixed point x. If S is a closed subset of X and $f(S) \subseteq S$, then $x \in S$.*

Exercise 2.121
Prove corollary 2.5.2.

Corollary 2.5.3 (N-stage contraction) Let $f: X \to X$ be an operator on a complete metric space X. Suppose that for some integer N, the function $f^N: X \to X$ is a contraction. Then f has a unique fixed point.

Exercise 2.122
Prove corollary 2.5.3.

Exercise 2.123 (Continuous dependence on a parameter)
Let X and Θ be metric spaces, and let $f: X \times \Theta \to X$ where

- X is complete
- for every $\theta \in \Theta$, the function $f_\theta(x) = f(x, \theta)$ is contraction mapping on X with modulus β
- f is continuous in θ, that is for every $\theta_0 \in \Theta$, $\lim_{\theta \to \theta_0} f_\theta(x) = f_{\theta_0}(x)$ for every $x \in X$

Then f_θ has a unique fixed point x_θ for every $\theta \in \Theta$ and $\lim_{\theta \to \theta_0} x_\theta = x_{\theta_0}$.

Although there are many direct methods for solving systems of linear equations, iterative methods are sometimes used in practice. The following exercise outlines one such method and devises a sufficient condition for convergence.

Exercise 2.124
Suppose that the linear model (section 3.6.1)

$A\mathbf{x} = \mathbf{c}$

has been scaled so that $a_{ii} = 1$ for every i. Show the following:

1. Any solution is a fixed point of the mapping $f(\mathbf{x}) = (I - A)\mathbf{x} + c$.
2. f is a contraction provided A has strict diagonal dominance, that is, $|a_{ii}| > \sum_{j \neq i} |a_{ij}|$.

[Hint: Use the sup norm.]

Dynamic Programming

We now show how the Banach fixed point theorem can be applied to the dynamic programming problem (example 2.32)

$$\max_{x_1, x_2, \ldots} \sum_{t=0}^{\infty} \beta^t f(x_t, x_{t+1})$$

subject to $\quad x_{t+1} \in G(x_t), t = 0, 1, 2, \ldots, x_0 \in X$ (24)

Let

$$\Gamma(x_0) = \{\mathbf{x} \in X^\infty : x_{t+1} \in G(x_t), t = 0, 1, 2, \ldots\}$$

denote the set of plans which are *feasible* starting from the initial point x_0. Assuming that

- f is bounded on $X \times X$
- $G(x)$ is nonempty for every $x \in X$

we have previously shown (exercise 2.16) that the value function v defined by

$$v(x_0) = \sup_{\mathbf{x} \in \Gamma(x_0)} U(\mathbf{x})$$

satisfies Bellman's equation

$$v(x) = \sup_{y \in G(x)} \{f(x_0, y) + \beta v(y)\} \quad \text{for every } x \in X \quad (25)$$

Consequently v is a fixed point of the operator

$$(Tv)(x) = \sup_{y \in G(x)} \{f(x, y) + \beta v(y)\}$$

Furthermore T is a contraction mapping (example 2.83) on the complete metric space $B(X)$ (exercise 2.11). Therefore it has a *unique* fixed point (theorem 2.5). In other words, the Banach fixed point theorem establishes that the value function is the unique solution of Bellman's equation (25).

To prove the existence of an optimal solution to the dynamic programming problem, we need to establish the continuity of the value function

$$v(x_0) = \sup_{\mathbf{x} \in \Gamma(x_0)} U(\mathbf{x})$$

where

$$U(\mathbf{x}) = \sum_{t=0}^{\infty} \beta^t f(x_t, x_{t+1})$$

denotes the total return from plan $\mathbf{x} \in \Gamma(x_0)$. We cannot appeal directly to the continuous maximum theorem (theorem 2.3), since the set of feasible plans $\Gamma(x_0)$ is not compact. However, we can apply corollary 2.5.2. To do this, we strengthen the assumptions on (24) to include

- f is bounded *and continuous* on $X \times X$
- $G(x)$ is nonempty, *compact-valued, and continuous* for every $x \in X$
- $0 \leq \beta < 1$

Then

- the operator

$$(Tv)(x) = \sup_{y \in G(x)} \{f(x,y) + \beta v(y)\}$$

is a contraction on $B(X)$ (example 2.83)
- $C(X)$ is a closed subset of $B(X)$ (exercise 2.85)
- $T(C(X)) \subseteq C(X)$ (exercise 2.113)

By corollary 2.5.2, the unique fixed point v of T belongs to $C(X)$. The value function of the dynamic programming problem defined by

$$v(x_0) = \sup_{\mathbf{x} \in \Gamma(x_0)} U(\mathbf{x})$$

is continuous. In the next exercise we use the continuity of the value function to demonstrate the existence of optimal plans in the dynamic programming problem.

Exercise 2.125 (Existence of an optimal plan)
Let v be the value function for the dynamic programming problem (example 2.32)

$$\max_{x_1, x_2, \ldots} \sum_{t=0}^{\infty} \beta^t f(x_t, x_{t+1})$$

subject to $\quad x_{t+1} \in G(x_t), t = 0, 1, 2, \ldots, x_0 \in X$

Assume that

- f is bounded and continuous on $X \times X$
- $G(x)$ is nonempty, compact-valued, and continuous for every $x \in X$
- $0 \leq \beta < 1$

Define the correspondence $\varphi: X \rightrightarrows X$ by

$$\varphi(x) = \{y \in G(x) : v(x) = f(x,y) + \beta v(y)\}$$

φ describes the set of solutions to Bellman's equation (exercise 2.17) at any $x \in X$. Show that

1. $\varphi(x) = \arg\max_{y \in G(x)} \{f(x,y) + \beta v(y)\}$.
2. $\varphi(x)$ is nonempty, compact-valued and uhc.
3. There exists a sequence $\mathbf{x}^* = (x_0, x_1^*, x_2^*, \ldots)$ such that $x_{t+1}^* \in \varphi(x_t^*)$.
4. $\mathbf{x}^*$ is an optimal plan.

By imposing additional structure on the problem, we can show that optimal plan is monotone. In exercise 3.158 we give sufficient conditions for the optimal plan to be unique.

Exercise 2.126 (*Monotonicity of optimal plans*)
Consider a dynamic programming problem that satisfies all the assumptions of the previous exercise. In addition assume that the state space X is a lattice on which

- $f(x,y)$ is supermodular in y
- $f(x,y)$ displays strictly increasing differences in (x,y)
- $G(x)$ is increasing

Show that

1. $\varphi(x)$ is always increasing.
2. Consequently every optimal plan $(x_0, x_1^*, x_2^*, \ldots)$ is a monotone sequence.

Example 2.93 (Optimal economic growth)
As it stands, the optimal economic growth model (example 2.33) does not fulfill the requirements of exercise 2.125, since the utility function u may be unbounded on its domain $\Re_+$. Rather than artificially impose boundedness, it is more common to adopt a restriction on the technology that is akin to diminishing marginal productivity. We assume that there exists an upper bound to investment $\bar{k}$ above which productivity is negative. Specifically, we assume that

- there exists $\bar{k} > 0$ such that $F(k) \leq k$ for every $k \geq \bar{k}$

In addition we assume that

- u is continuous on $\Re_+$
- F is continuous and increasing on $\Re_+$ with $F(0) = 0$
- $0 \leq \beta < 1$

Let $X = [0, \bar{k}]$. Assume that $k_0 \in X$. Then $F(k) \in X$ for every $k \in X$. Without loss of generality, we may restrict analysis to X. Then

- u is bounded on X (exercise 2.84)
- $u(F(k_t) - k_{t+1})$ is bounded and continuous on $X \times X$
- $G(k) = [0, F(k)]$ is nonempty, compact and continuous for every $k \in X$

Exercise 2.125 establishes that there exists an optimal growth policy $(k_0, k_1^*, k_2^*, \ldots)$ for every starting point k_0.

2.4.4 Brouwer Fixed Point Theorem

The most useful fixed point theorem in mathematical economics is the Brouwer fixed point theorem and its derivatives. The Brouwer theorem asserts that every continuous function on a compact convex set in a normed linear space has a fixed point. In this section we present and prove the Brouwer theorem, derive some important extensions, and outline the most important applications—the existence of competitive equilibrium and the existence of a Nash equilibrium in a noncooperative game.

The Brouwer theorem is intuitively obvious and easy to prove in $\Re$. Consider the continuous function $f: [0, 1] \to [0, 1]$ illustrated in figure 2.25. Its graph is a curve joining the left-hand side of the box to the right-hand side. If the function is continuous, its graph has no gaps and thus must cross the diagonal at some point. Every such intersection is a fixed point. Exercise 2.127 formalizes this proof.

Exercise 2.127
Let $f: [0, 1] \to [0, 1]$ be continuous. Show that f has a fixed point. [Hint: Apply the intermediate value theorem (exercise 2.83) to $g(x) = f(x) - x$.]

In higher dimensions the Brouwer theorem is much less intuitive and correspondingly harder to prove. To appreciate its profundity, take a cup of coffee and gently swirl it around to mix thoroughly, being careful not to introduce any turbulence. (Unfortunately, you cannot stir the coffee with a spoon, since the transformation would no longer be continuous.)

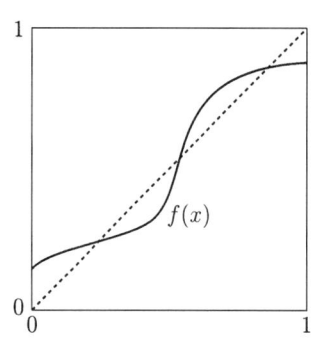

Figure 2.25
Brouwer's theorem in $\mathscr{R}$

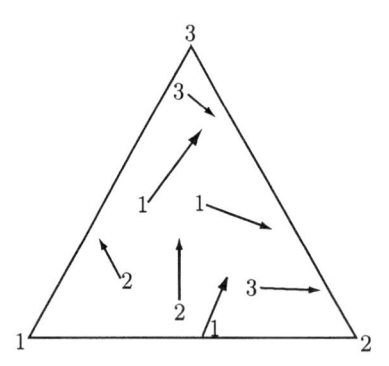

Figure 2.26
Illustrating an operator on the two-dimensional simplex

No matter how long you swirl, at least one "molecule" must end up exactly where it started.

Our approach to proving Brouwer's theorem utilizes Sperner's lemma on admissibly labeled simplicial partitions (proposition 1.3). We first illustrate the approach on the two-dimensional simplex. A function on the two-dimensional simplex can be illustrated by using arrows to connect selected points and their images. Label each point with the label of the vertex from which it points away (figure 2.26). Where the arrow points away from two vertices (e.g., on the boundary), choose one of them. We can label each vertex of a simplicial partition in this way. By construction, such a labeling constitutes an admissible labeling. For any simplicial partition, Sperner's lemma ensures that there is a always exists a completely labeled subsimplex, that is, a subsimplex that has vertices at which the function points in each of the three directions. If we take a sequence of increasingly fine partitions, we will find a point at which it appears the function is pointing in all three directions at once. This is only possible if in fact it is a fixed point. We now make this argument rigorous.

We first show that an operator on a simplex conveniently defines a admissible labeling of the points of the simplex. Let S be the n-dimensional simplex with vertices $\{\mathbf{x}_0, \mathbf{x}_1, \ldots, \mathbf{x}_n\}$. Recall that every point $\mathbf{x} \in S$ has a unique representation as a convex combination of the vertices

$$\mathbf{x} = \alpha_0 \mathbf{x}_0 + \alpha \mathbf{x}_1 + \cdots + \alpha_n \mathbf{x}_n$$

with $\alpha_i \geq 0$ and $\alpha_0 + \alpha_1 + \cdots + \alpha_n = 1$ (exercise 1.159). The coefficients $\alpha_0, \alpha_1, \ldots, \alpha_n$ are called the barycentric coordinates of $\mathbf{x}$. Similarly the image $f(\mathbf{x})$ of $\mathbf{x}$ under f has a unique representation

$$f(\mathbf{x}) = \beta_0 \mathbf{x}_0 + \beta \mathbf{x}_1 + \cdots + \beta_n \mathbf{x}_n$$

with $\beta_i \geq 0$ and $\beta_0 + \beta_1 + \cdots + \beta_n = 1$. Given any function $f \colon S \to S$, a label in the set $\{0, 1, \ldots, n\}$ can be assigned to every point in the simplex S using the rule

$$\mathbf{x} \mapsto \min\{i : \beta_i \leq \alpha_i \neq 0\}$$

where α_i and β_i are the barycentric coordinates of $\mathbf{x}$ and $f(\mathbf{x})$ respectively. This assignment satisfies the requirements of an admissible labeling for the application of Sperner's lemma (exercise 2.128).

Exercise 2.128
Let $f \colon S \to S$ be an operator on an n simplex with vertices $\{\mathbf{x}_0, \mathbf{x}_1, \ldots, \mathbf{x}_n\}$. Suppose that the elements of S are labeled using the rule

$$\mathbf{x} \mapsto \min\{i : \beta_i \leq \alpha_i \neq 0\}$$

where α_i and β_i are the barycentric coordinates of $\mathbf{x}$ and $f(\mathbf{x})$ respectively. Show that

1. The rule assigns a label in $\{0, 1, \ldots, n\}$ to every $\mathbf{x} \in S$.

2. Each vertex of S retains its own label.

3. Each vertex on a face of S receives a label corresponding to one of the vertices of that face.

Hence the rule generates an admissible labeling of the simplex.

Theorem 2.6 (Brouwer's theorem) *Let S be a nonempty, compact, convex subset of a finite dimensional normed linear space. Every continuous function $f: S \to S$ has a fixed point.*

Proof We assume for simplicity that S is a simplex. The extension to an arbitrary compact convex set is given in exercise 2.129. We proceed by constructing a sequence of increasingly fine simplicial partitions of S which eventually "trap" the fixed point.

Let Λ^k, $k = 1, 2, \ldots$, be a sequence of simplicial partitions of S in which the maximum diameter of the subsimplices tend to zero as $k \to \infty$. For each vertex $\mathbf{x}^k$ of Λ_k, assign a label $i \in \{0, 1, \ldots, n\}$ using the labeling rule

$$\mathbf{x}^k \mapsto \min\{i : \beta_i^k \leq \alpha_i^k \neq 0\}$$

where α_i^k and β_i^k are the barycentric coordinates of $\mathbf{x}^k$ and $f(\mathbf{x}^k)$ respectively. Every partition is admissibly labeled (exercise 2.128).

By Sperner's lemma (proposition 1.3), each partition Λ_k has a completely labeled subsimplex. That is, there is a simplex with vertices $\mathbf{x}_0^k, \mathbf{x}_1^k, \ldots, \mathbf{x}_n^k$ such that

$$\beta_i^k \leq \alpha_i^k \tag{26}$$

for the vertex $\mathbf{x}_i^k$. In other words, every vertex of the completely labeled subsimplex satisfies (26) in its corresponding coordinate.

Since S is compact, each sequence $\mathbf{x}_i^k$ has a convergent subsequence $\mathbf{x}_i^{k'}$. Moreover, since the diameters of the subsimplices converge to zero, these subsequences must converge to the same point, say $\mathbf{x}^*$. That is,

$$\lim_{k' \to \infty} \mathbf{x}_i^{k'} = \mathbf{x}^*, \quad i = 0, 1, \ldots, n$$

Since f is continuous, their images also converge:

$$\lim_{k' \to \infty} f(\mathbf{x}_i^{k'}) = f(\mathbf{x}^*), \quad i = 0, 1, \ldots, n$$

This implies that the corresponding barycentric coordinates also converge:

$$\alpha_i^k \to \alpha_i^* \quad \text{and} \quad \beta_i^k \to \beta_i^*, \quad i = 0, 1, \ldots, n$$

where α_i^* and β_i^* are the barycentric coordinates of $\mathbf{x}^*$ and $f(\mathbf{x}^*)$ respectively. Since for every $i = 0, 1, \ldots, n$, there exist coordinates such that $\beta_i^{k'} \leq \alpha_i^{k'}$ for every k, we have $\beta_i^* \leq \alpha_i^*$ for every coordinate $i = 0, 1, \ldots, n$.

Since $\sum \beta_i^* = \sum \alpha_i^* = 1$, this implies that

$$\beta_i^* = \alpha_i^*, \quad i = 0, 1, \ldots, n$$

In other words,

$$f(\mathbf{x}^*) = \mathbf{x}^*$$

$\mathbf{x}^*$ is a fixed point of f. □

Example 2.94 (Markov chains) Let T be the $n \times n$ transition matrix of a finite Markov process (section 3.6.4). The set of state distributions

$$S = \left\{ \mathbf{p} \in \Re^n : \sum_i p_i = 1 \right\}$$

is precisely the $(n-1)$-dimensional standard simplex (example 1.95), which is nonempty, convex, and compact. T is a linear operator on the finite-dimensional space S and is therefore continuous (exercise 3.31). Applying the Brouwer theorem, T has a fixed point $\mathbf{p}$

$$T\mathbf{p} = \mathbf{p}$$

which is a stationary distribution of the Markov process. Consequently every Markov chain has a stationary distribution.

For any $S \subset T$ in a metric space, a continuous function $r: T \to S$ is called a *retraction* of T onto S if $r(\mathbf{x}) = \mathbf{x}$ for every $\mathbf{x} \in S$. In chapter 3 we will show (exercise 3.74) that every set in a finite-dimensional normed linear space can be retracted onto its closed convex subsets.

Exercise 2.129
Generalize the proof of the Brouwer theorem to an arbitrary compact convex set as follows. Let $f: S \to S$ be a continuous operator on a nonempty, compact, convex subset of a finite-dimensional normed linear space.

1. Show that there exists a simplex T containing S.

2. By exercise 3.74, there exists a continuous retraction $r: T \to S$. Show that $f \circ r: T \to T$ and has a fixed point in $\mathbf{x}^* \in T$.

3. Show that $\mathbf{x}^* \in S$ and therefore $f(\mathbf{x}^*) = \mathbf{x}^*$.

Consequently f has a fixed point $\mathbf{x}^*$.

Exercise 2.130
Where is the convexity of S required in the previous exercise?

Exercise 2.131
To show that each of the hypotheses of Brouwer's theorem is necessary, find examples of functions $f \colon S \to S$ with $S \subseteq \Re$ that do not have fixed points, where

1. f is continuous and S is convex but not compact
2. f is continuous and S is compact but not convex
3. S is compact and convex but f is not continuous

The following proposition, which is equivalent to Brouwer's theorem, asserts that it is impossible to map the unit ball continuously on to its boundary.

Exercise 2.132 (No-retraction theorem)
Let B denote the closed unit ball in a finite-dimensional normed linear space

$$B = \{ \mathbf{x} \in X : \|\mathbf{x}\| \leq 1 \}$$

and let S denote its boundary, that is,

$$S = \{ \mathbf{x} \in X : \|\mathbf{x}\| = 1 \}$$

There is no continuous function $r \colon B \to S$ such that $r(\mathbf{x}) = \mathbf{x}$ for every $\mathbf{x} \in S$.

Exercise 2.133
Let $f \colon B \to B$ be a continuous operator on the closed unit ball B in a finite-dimensional normed linear space. Show that the no-retraction theorem implies that f has a fixed point.

Exercise 2.134
Prove that the no-retraction theorem is equivalent to Brouwer's theorem.

The following proposition, due to Knaster, Kuratowki, and Mazurkiewicz (K-K-M), is equivalent to the Brouwer theorem. It is often used as a step on the way to the Brouwer theorem. It is more useful than the Brouwer theorem in some applications.

Proposition 2.4 (K-K-M theorem) Let $A_0, A_1, \ldots, A_n$ be closed subsets of an n-dimensional simplex S with vertices $\mathbf{x}_0, \mathbf{x}_1, \ldots, \mathbf{x}_n$. If for every $I \subseteq \{0, 1, \ldots, n\}$ the face $\mathrm{conv}\{\mathbf{x}_i : i \in I\}$ is contained in the corresponding union $\bigcup_{i \in I} A_i$, then the intersection $\bigcap_{i=0}^n A_i$ is nonempty.

Exercise 2.135
Prove the K-K-M theorem directly, using Sperner's lemma.

Exercise 2.136
Prove that the K-K-M theorem is equivalent to Brouwer's theorem, that is,

K-K-M theorem ⇔ Brouwer's theorem

The classic application of the Brouwer theorem in economics is to prove the existence of competitive equilibrium. We extract the mathematical essence in the following corollary, and then show how it applies to competitive equilibrium in example 2.95.

Corollary 2.6.1 (Excess demand theorem) Let $\mathbf{z}: \Delta^{n-1} \to \Re^n$ be a continuous function satisfying $\mathbf{p}^T \mathbf{z}(\mathbf{p}) = 0$ for every $\mathbf{p} \in \Delta^{n-1}$. Then there exists $\mathbf{p}^* \in \Delta^{n-1}$ such that $\mathbf{z}(\mathbf{p}^*) \leq 0$.

Proof Define the function $g: \Delta^{l-1} \to \Delta^{l-1}$ by

$$g_i(\mathbf{p}) = \frac{p_i + \max(0, z_i(\mathbf{p}))}{1 + \sum_{j=1}^l \max(0, z_j(\mathbf{p}))}$$

g is continuous (exercises 2.78, 2.79, 2.81). By Brouwer's theorem, there exists a fixed point $\mathbf{p}^*$ such that $g(\mathbf{p}^*) = \mathbf{p}^*$. Given $\mathbf{p}^T \mathbf{z}(\mathbf{p}) = 0$ for every $\mathbf{p} \in \Delta^{n-1}$, it is easy to show (exercise 2.137) that

$$g(\mathbf{p}^*) = \mathbf{p}^* \Rightarrow \mathbf{z}(\mathbf{p}^*) \leq 0 \qquad \square$$

Example 2.95 (Existence of competitive equilibrium) A competitive equilibrium $(\mathbf{p}^*, \underline{\mathbf{x}}^*)$ in an exchange economy (example 1.117) is a set of prices $\mathbf{p}^*$ and an allocation $\underline{\mathbf{x}}^* = (\mathbf{x}_1^*, \mathbf{x}_2^*, \ldots \mathbf{x}_n^*)$ such that

- every consumer i chooses the optimal bundle in his budget set

$$\mathbf{x}_i^* \succsim \mathbf{x}_i \quad \text{for every } \mathbf{x}_i \in X(\mathbf{p}, m_i)$$

where $m_i = \sum_{j=1}^l p_j \omega_{ij}$

- aggregate demand is less than or equal to available supply

$$\sum_{i \in N} \mathbf{x}_i \le \sum_{i \in N} \omega_i$$

Assume that the consumers' preferences $\succsim_i$ are continuous and *strictly convex*. Then every consumer has a continuous demand *function* $\mathbf{x}(\mathbf{p}, m)$ indicating their optimal choice at given prices $\mathbf{p}$ (examples 2.90 and 3.62). Let $\mathbf{z}_i \colon R_+^l \to R^l$ denote consumer i's *excess demand function*

$$\mathbf{z}_i(\mathbf{p}) = \mathbf{x}_i(\mathbf{p}, m) - \omega_i$$

which measures his desired net trade in each commodity at the prices $\mathbf{p}$. Let $\mathbf{z}(\mathbf{p})$ denote the aggregate excess demand function

$$\mathbf{z}(\mathbf{p}) = \sum_{i=1}^{n} \mathbf{z}_i(\mathbf{p})$$

The aggregate excess demand function is continuous and homogeneous of degree zero (exercise 2.138), so only relative prices matter. We can normalize so that prices are restricted to the unit simplex Δ^{l-1}. Furthermore, provided that consumers' preferences are *nonsatiated*, the excess demand function satisfies the following identity known as *Walras's law* (exercise 2.139):

$$\mathbf{p}^T \mathbf{z}(\mathbf{p}) \equiv 0 \quad \text{for every } \mathbf{p} \quad \text{(Walras's law)}$$

The excess demand functions $\mathbf{z}(\mathbf{p})$ satisfy the conditions of corollary 2.6.1. Therefore there exists a price $\mathbf{p}^*$ such that

$$\mathbf{z}(\mathbf{p}^*) \le \mathbf{0} \tag{27}$$

$\mathbf{p}^*$ is a competitive equilibrium price (exercise 2.140), and $(\mathbf{p}^*, x(\mathbf{p}^*))$ is a competitive equilibrium.

The function

$$g_i(\mathbf{p}) = \frac{p_i + \max(0, z_i(\mathbf{p}))}{1 + \sum_{j=1}^{l} \max(0, z_i(\mathbf{p}))}$$

used in the proof of corollary 2.6.1 has a nice interpretation in this application—it increases the price of commodities in excess demand and lowers the price of those commodities in excess supply.

At first sight, (27) is a system of l inequalities in l unknowns. However, since $\mathbf{z}(\mathbf{p})$ satisfies Walras's law, there are only $l-1$ independent inequalities. On the other hand, $\mathbf{z}(\mathbf{p})$ is homogeneous of degree zero, so only relative prices matter. There are only $l-1$ relative prices. Therefore (27) is a system of $l-1$ independent inequalities in $l-1$ unknowns. If the excess demand function $\mathbf{z}$ were linear, we could apply the theory of section 3.6 to deduce a solution. It is precisely because the system (27) is nonlinear that we have to resort a more powerful fixed point argument.

Exercise 2.137
Let $\mathbf{z}: \Delta^{n-1} \to \Re^n$ be a continuous function satisfying $\mathbf{pz}(\mathbf{p}) = 0$ for every $\mathbf{p} \in \Delta^{n-1}$ and

$$g_i(\mathbf{p}) = \frac{p_i + \max(0, z_i(\mathbf{p}))}{1 + \sum_{j=1}^{l} \max(0, z_i(\mathbf{p}))}$$

Show that

$$g(\mathbf{p}^*) = \mathbf{p}^* \Rightarrow \mathbf{z}(\mathbf{p}^*) \leq \mathbf{0}$$

Exercise 2.138 (Properties of the excess demand function)
Show that the aggregate excess demand function $\mathbf{z}(\mathbf{p})$ is continuous and homogeneous of degree zero.

Exercise 2.139 (Walras's law)
Assuming that the consumers' preference relations $\succsim_i$ are nonsatiated and strictly convex, show that the aggregate excess demand function $\mathbf{z}(\mathbf{p})$ satisfies Walras's law

$$\mathbf{p}^T \mathbf{z}(\mathbf{p}) \equiv 0 \quad \text{for every } \mathbf{p}$$

[Hint: Use exercise 1.248.]

Remark 2.19 (Strong and weak forms of Walras's law) The previous result is known as the strong form of Walras's law. Homogeneity alone implies the analogous weak form of Walras's law

$$\mathbf{p}^T \mathbf{z}(\mathbf{p}) \leq 0 \quad \text{for every } \mathbf{p}$$

but this alone is inadequate to support our proof of existence of equilibrium. In addition to homogeneity the strong form of Walras's law requires that consumers spend all their income, which is implied by local nonsatiation.

Exercise 2.140
$\mathbf{p}^*$ is a competitive equilibrium price if $\mathbf{z}(\mathbf{p}^*) \leq \mathbf{0}$.

Remark 2.20 (Uzawa equivalence theorem) Corollary 2.6.1 abstracts the mathematical essence of the existence of equilibrium in a competitive exchange economy. We showed that this is implied by Brouwer's theorem. Uzawa (1962) proved the converse, namely that corollary 2.6.1 implies Brouwer's theorem, establishing their equivalence. This underlines the profundity of Brouwer's theorem, and it means that a fixed point argument is essential to proving existence of economic equilibrium. This cannot be done with simpler means.

Two generalizations of Brouwer's theorem are important in economics. The first extends the theorem to correspondences (Kakutani's theorem), while the second extends to infinite-dimensional spaces (Schauder's theorem). We consider these in turn.

Kakutani's Theorem

To use Brouwer's theorem to prove the existence of a competitive equilibrium in example 2.95 required that the consumers' optimal choices be unique (demand functions), which necessitated the unreasonable assumption that consumer preferences are strictly convex. To relax this assumption, and also to incorporate production into the economic system, requires an extension of the Brouwer theorem to correspondences. This extension was provided by Kakutani for precisely this purpose. It also allows us to prove a general existence theorem for games.

Theorem 2.7 (Kakutani's theorem) Let S be a nonempty, compact, convex subset of a finite dimensional normed linear space. Every closed, convex-valued correspondence $\varphi: S \rightrightarrows S$ has a fixed point.

Remark 2.21 Recall that a correspondence $\varphi: S \rightrightarrows S$ is closed if it graph is closed in $S \times S$. Since S is compact, this is equivalent to φ being closed-valued and uhc (exercise 2.107).

Proof We assume for simplicity that S is a simplex. The extension to an arbitrary compact convex set is given in exercise 2.142. We proceed by constructing a sequence of continuous functions that approximate a selection from the correspondence. By Brouwer's theorem, each of these functions has a fixed point, and these fixed points converge to a fixed point of the correspondence.

Let Λ^k, $k = 1, 2, \ldots$ be a sequence of simplicial partitions of S in which the maximum diameter of the subsimplices tend to zero as $k \to \infty$. Construct a sequence of continuous functions $f^k \colon S \to S$ that approximate φ, by assigning to each vertex $\mathbf{x}$ of the partition Λ^k a point in the set $\varphi(\mathbf{x})$ and then extending f^k linearly to the subsimplices. Specifically, if V^k denotes the set of all vertices of the subsimplices in Λ^k,

- For every vertex $\mathbf{x} \in V^k$, choose some $\mathbf{y} \in \varphi(\mathbf{x})$ and set $f(\mathbf{x}) = \mathbf{y}$.
- For every nonvertex $\mathbf{x} \in S \setminus V^k$, let $S^k \in \Lambda^k$ denote the subsimplex that contains $\mathbf{x}$. Let $\alpha_0^k, \alpha_1^k, \ldots, \alpha_n^k$ denote the barycentric coordinates (exercise 1.159) of $\mathbf{x}$ with respect to the vertices $\mathbf{x}_0^k, \mathbf{x}_1^k, \ldots, \mathbf{x}_n^k \in V^k$ of S^k. That is,

$$\mathbf{x} = \alpha_0^k \mathbf{x}_0^k + \alpha_1^k \mathbf{x}_1^k + \cdots + \alpha_n^k \mathbf{x}_n^k$$

and we define

$$f^k(\mathbf{x}) = \alpha_0^k f(\mathbf{x}_0^k) + \alpha_1^k f(\mathbf{x}_1^k) + \cdots + \alpha_n^k f(\mathbf{x}_n^k)$$

By Brouwer's theorem, each function f^k has a fixed point $\mathbf{x}^k$. Since S is compact, the sequence of fixed points $\mathbf{x}^k$ has a convergent subsequence $\mathbf{x}^{k'}$ that converges to a point $\mathbf{x}^* \in S$. Since each function f^k matches the correspondence at the vertices of the subsimplices the diameters of which converge to zero, it follows (exercise 2.141) that $\mathbf{x}^* \in \varphi(\mathbf{x}^*)$. That is, $\mathbf{x}^*$ is the required fixed point of the correspondence. □

Exercise 2.141
Verify that $\mathbf{x}^* = \lim_{k' \to \infty} \mathbf{x}^{k'}$ as defined in the preceding proof is a fixed point of the correspondence, that is $\mathbf{x}^* \in \varphi(\mathbf{x}^*)$.

Exercise 2.142
Generalize the proof of the Kakutani theorem to an arbitrary convex, compact set S. [Hint: See exercise 2.129.]

Example 2.96 (Existence of Nash equilibrium) A strategic game (section 1.2.6) comprises

- a finite set N of *players*
- for each player $i \in N$ a nonempty set S_i of strategies

- for each player $i \in N$ a *preference relation* $\succsim_i$ on the strategy space $S = S_1 \times S_2 \times \cdots \times S_n$

Assume that the strategy space of S_i of every player is nonempty, compact, and convex. Then the product $S = S_1 \times S_2 \times \cdots \times S_n$ is likewise nonempty, compact, and convex. Assume further that for each player i there exists a continuous, quasi-concave function $u_i \colon S \to \Re$ that represents the player's preferences in the sense that

$$(s, \mathbf{s}_{-i}) \succsim_i (s', \mathbf{s}_{-i}) \Leftrightarrow u_i(\mathbf{s}) \geq u_i(\mathbf{s}')$$

where $\mathbf{s}' = (s', \mathbf{s}_{-i})$. The best response correspondence of player i is

$$B_i(\mathbf{s}) = \{s \in S_i : (s, \mathbf{s}_{-i}) \succsim_i (s', \mathbf{s}_{-i}) \text{ for every } s' \in S_i\}$$

This can be alternatively defined as the solution correspondence of the maximization problem

$$B_i(\mathbf{s}) = \arg \max_{s_i \in S_i} u_i(\mathbf{s})$$

By the maximum theorems, each best response correspondence B_i is compact-valued and upper hemicontinuous (theorem 2.3) and convex-valued (theorem 3.1).

Let B denote the product of the individual player's best response correspondences. That is, for every $\mathbf{s} \in S$,

$$B(\mathbf{s}) = B_1(\mathbf{s}) \times B_2(\mathbf{s}) \times \cdots \times B_n(\mathbf{s})$$

Then B is a closed, convex-valued correspondence $B \colon S \rightrightarrows S$ (exercise 2.143). By Kakutani's theorem, B has a fixed point $\mathbf{s} \in S$ such that $\mathbf{s} \in B(\mathbf{s})$. That is,

$$s_i \in B_i(\mathbf{s}) \quad \text{for every } i \in N$$

$\mathbf{s} = (s_1, s_2, \ldots, s_n)$ is a Nash equilibrium of the game.

Exercise 2.143
Show that the best response correspondence $B \colon S \rightrightarrows S$ is closed and convex valued.

Remark 2.22 The existence theorem in the previous example applies to two important special cases:

finite games Assume that each player has a finite set of actions A_i, and let S_i denote the set of mixed strategies, that is, the set of all probability distributions over A_i (example 1.98). Let u_i denote the expected payoff from strategy **s**, that is,

$$u_i(\mathbf{s}) = \sum_j p_j u(a_i)$$

Since u is linear, it is continuous and quasiconcave.

Cournot oligopoly The payoff function u_i is the profit function

$$u_i(y_i, \mathbf{y}_{-i}) = p(Y)y_i - c(y_i)$$

where Y is total output and $p(Y)$ is the inverse demand curve (example 2.35). Provided that the demand and cost functions satisfy suitable conditions, the profit function u_i will be continuous and quasiconcave. If $\overline{Y}$ is an upper bound on feasible output, the strategy spaces can be taken to be $[0, \overline{Y}]$.

Exercise 2.144 (Uniqueness of Nash equilibrium)
Suppose, in addition to the hypotheses of example 2.96, that

- the players' payoff functions $u_i: S \to \Re$ are *strictly* quasiconcave
- the best response mapping $B: S \to S$ is a contraction

Then there exists a unique Nash equilibrium of the game.

Schauder's Theorem

We generalized Brouwer's theorem to an arbitrary convex, compact set S by mapping S to an enclosing simplex. To generalize to infinite-dimensional spaces, we adopt a similar technique. The following lemma shows that every compact set can be mapped continuously to a finite-dimensional convex set.

Exercise 2.145
Let K be a compact subset of a normed linear space X. For every $\varepsilon > 0$, there exists a finite-dimensional convex set $S \subseteq X$ and a continuous function $h: K \to S$ such that $S \subseteq \operatorname{conv} K$ and

$$\|h(\mathbf{x}) - \mathbf{x}\| < \varepsilon \quad \text{for every } \mathbf{x} \in K$$

[Hint: Exercise 1.112.]

Theorem 2.8 (Schauder's theorem) Let S be a nonempty, compact, convex subset of a normed linear space. Every continuous function $f \colon S \to S$ has a fixed point.

Proof $f(S)$ is compact (proposition 2.3). Applying the preceding lemma, we can approximate $f(S)$ by a sequence of finite-dimensional convex sets. Specifically, for $k = 1, 2, \ldots$ there exists a finite-dimensional convex set S^k and continuous function $h^k \colon f(S) \to S^k$ such that

$$\|h^k(\mathbf{x}) - \mathbf{x}\| < \frac{1}{k} \quad \text{for every } \mathbf{x} \in f(S)$$

Since S convex,

$$S^k \subseteq \operatorname{conv} f(S) \subseteq S$$

The function $g^k = h^k \circ f$ approximates f on S^k. That is (exercise 2.146),

- $g^k \colon S^k \to S^k$
- $\|g^k(\mathbf{x}) - f(\mathbf{x})\| \leq 1/k$ for every $\mathbf{x} \in S^k$

Furthermore g^k is continuous (exercise 2.72) and S^k is compact, convex, and finite dimensional. Applying Brouwer's theorem (theorem 2.6), we see that every function g^k has a fixed point $\mathbf{x}^k = g^k(\mathbf{x}^k)$. Every fixed point $\mathbf{x}^k \in S$. Since S is compact, there exists a convergent subsequence $\mathbf{x}^{k'} \to \mathbf{x}^* \in S$. Furthermore $f(\mathbf{x}^*) = \mathbf{x}^*$; that is, $\mathbf{x}^*$ is a fixed point of f (exercise 2.147). □

Exercise 2.146
Let $g^k = h^k \circ f$ as defined in the preceding proof. Show that

1. $g^k \colon S^k \to S^k$
2. $\|g^k(\mathbf{x}) - f(\mathbf{x})\| \leq 1/k$ for every $\mathbf{x} \in S^k$

Exercise 2.147
Verify that $\mathbf{x}^* = \lim_{k \to \infty} \mathbf{x}^k$ as defined in the preceding proof is a fixed point of f, that is, $f(\mathbf{x}^*) = \mathbf{x}^*$.

Schauder's theorem is frequently applied in cases where the underlying space is not compact. The following alternative version relaxes this condition to require that the image lie in a compact set. A function $f \colon X \to Y$ is called *compact* if $f(X)$ is contained in a compact set of Y.

Corollary 2.8.1 (Schauder's theorem—Alternative version) *Let S be a nonempty, closed, and bounded convex subset of a normed linear space. Every compact continuous operator $f: S \to S$ has a fixed point.*

Proof Let $A = \overline{\text{conv}(f(S))}$. Then A is a subset of S which is compact and convex. Furthermore $f(A) \subset A$. Therefore the restriction of f to A is a continuous operator on a compact, convex set. By Schauder's theorem, f has a fixed point which is automatically a fixed point of f on S. □

The alternative version implies the following result which is used in dynamic economic models.

Exercise 2.148
Let F be a nonempty, closed and bounded, convex subset of $C(X)$, the space of continuous functionals on a compact metric space X. Let $T: F \to F$ be a continuous operator on F. If the family $T(F)$ is equicontinuous, then T has a fixed point.

2.4.5 Concluding Remarks

We have presented a suite of fixed point theorems, the heavy artillery of the analyst's arsenal. The most powerful is Brouwer's theorem and its generalizations, whose essential requirements are continuity of the mapping together with compactness and convexity of underlying space. Banach's theorem shows that compactness can dispensed with by strengthening the continuity requirement, while Tarksi's theorem shows that even continuity is dispensable if we have monotonicity.

2.5 Notes

The general references cited in chapter 1 are also relevant for sections 2.1 and 2.3, to which should be added Berge (1963). Our presentation of dynamic programming is based on Stokey and Lucas (1989), who give numerous applications. Maor (1994) discusses the history of the exponential and log functions. The standard Cournot (example 2.35) and Bertrand (exercise 2.60) oligopoly models are explored in Shapiro (1989). The definition of rationalizability in exercise 2.22 differs from the standard definition, in that it allows for the actions of opponents to be correlated. See Osborne and Rubinstein (1994, pp. 57–58) for a discussion of this point.

In representing continuous preferences, the difficult part is not the existence of a utility function but its continuity. This has tripped some distinguished economists and generated a sizable literature; see Beardon and Mehta (1994) for references. The fundamental result is due to Debreu (1954, 1964). A concise account is given by Barten and Böhm (1982). The simple constructive proof for monotone preferences (exercises 2.38, 2.73) originated with Wold. Our treatment is adapted from Mas-Colell et al. (1995, p. 47).

The properties of supermodular functions were studied by Topkis (1978). Further references are given in chapter 6. The study of convex games (example 2.69) originated with Shapley (1971–1972). In the light of subsequent developments, the choice of adjective convex rather than supermodular to describe these games is unfortunate, since convexity and supermodularity are quite distinct properties. The single-crossing condition of exercise 2.61 is closely related to the "sorting" or "Spence-Mirrlees" condition which is often invoked in the literature on signaling and mechanism design. Example 2.74 adapted from Fudenberg and Tirole (1991, p. 492) and modeled on Diamond (1982). Exercise 2.64 is the principal result of Milgrom and Shannon (1994).

The monotone maximum theorem (theorem 2.1) is due to Topkis (1978) and the continuous maximum theorem (theorem 2.3) to Berge (1963). The latter is usually called simply the maximum theorem. It should be distinguished from the "maximum principle," which is a counterpart of the principle of optimality (exercise 2.17) for dynamic programming in continuous time.

Good treatments of the continuity of correspondences can be found in Border (1985), Ellickson (1993), and Sundaram (1996), from which we adapted some examples. Border (1985) is an excellent source on fixed point theorems for economists. Zeidler (1986) is also recommended for its clarity and thoroughness. The extension of Tarski's theorem to correspondences (corollary 2.4.2), due to Zhou (1994), parallels the generalization of Brouwer's theorem to Kakutani's theorem. Our proof of Kakutani's theorem follows Kakutani (1941). An alternative approach is to apply Brouwer's theorem to a continuous selection (see Border 1985, pp. 71–72 and Hildenbrand and Kirman 1976, pp. 201–204). Our derivation of Schauder's theorem is based on Zeidler (1986). Some economic applications of Schauder's theorem can be found in Stokey and Lucas (1989).

The primary role of Kakutani's theorem in economics is to establish the existence of competitive equilibrium in economies in which demand and supply correspondences are not single-valued. Proof of the existence of competitive equilibrium is one of the major accomplishments of mathematical economics. Our proof is a standard textbook account omitting much of the fine detail. Lucid introductory accounts are provided by Ellickson (1993), Mas-Colell et al. (1995), and Starr (1997), while Debreu (1959) and Arrow and Hahn (1971) are classics in the field. The survey by Debreu (1982) outlines the various approaches which have been used. Debreu (1982, pp. 719–720) and Starr (1997, pp. 136–138) discuss the Uzawa equivalence theorem (remark 2.20).

3 Linear Functions

A function $f \colon X \to Y$ between two linear spaces X and Y is *linear* if it preserves the linearity of the sets X and Y, that is, for all $\mathbf{x}_1, \mathbf{x}_2 \in X$, and $\alpha \in \Re$,

additivity $f(\mathbf{x}_1 + \mathbf{x}_2) = f(\mathbf{x}_1) + f(\mathbf{x}_2)$
homogeneity $f(\alpha \mathbf{x}_1) = \alpha f(\mathbf{x}_1)$

A linear function is often called a linear transformation and a linear function from a set X to itself is often called a linear operator. Throughout this chapter the domain and co-domain are assumed to be subsets of linear spaces.

Exercise 3.1
A function $f \colon X \to Y$ is linear if and only if
$$f(\alpha_1 \mathbf{x}_1 + \alpha_2 \mathbf{x}_2) = \alpha_1 f(\mathbf{x}_1) + \alpha_2 f(\mathbf{x}_2)$$
for all $\mathbf{x}_1, \mathbf{x}_2 \in X$, and $\alpha_1, \alpha_2 \in \Re$.

Exercise 3.2
Show that the set $L(X, Y)$ of all linear functions $X \to Y$ is a linear space.

Example 3.1 The function $f \colon \Re^2 \to \Re^2$ defined by
$$f(x_1, x_2) = (x_1 \cos \theta - x_2 \sin \theta, x_1 \sin \theta + x_2 \cos \theta), \qquad 0 \leq \theta < 2\pi$$
rotates any vector in the plane counterclockwise through the angle θ (figure 2.3). It is easily verified that f is linear. Linearity implies that rotating the sum of two vectors yields the same result as summing the rotated vectors.

Exercise 3.3
Show that f in example 3.1 is linear.

Exercise 3.4
Show that the function $f \colon \Re^3 \to \Re^2$ defined by
$$f(x_1, x_2, x_3) = (x_1, x_2, 0)$$
is a linear function. Describe this mapping geometrically.

Example 3.2 (The high-fidelity amplifier) Pure musical tones can be thought of as elements of a linear space. Pure tones can be combined (added) to produce complex tones and they can be scaled in amplitude to different volumes. An amplifier can be thought of as a function, transforming the inputs (electrical signals) into music (sound signals). An ideal amplifier would be a linear function, combining different pure tones faithfully and scaling their volumes proportionately. Real amplifiers suffer from various degrees of nonlinearity known as distortion. Generally, more expensive amplifiers produce better sound reproduction because they are more nearly linear.

Example 3.3 A *matrix* is a collection of similar elements (numbers, functions) arranged in a table. For example,

$$A = \begin{pmatrix} 1 & 5 & 10 \\ 2 & 15 & 25 \end{pmatrix}$$

is a 2×3 matrix of numbers, while

$$H = \begin{pmatrix} f_{11}(x) & f_{12}(x) \\ f_{21}(x) & f_{22}(x) \end{pmatrix}$$

is a 2×2 matrix of functions $f_{ij} \colon X \to Y$, $i, j = 1, 2$.

Any $m \times n$ matrix $A = (a_{ij})$ of numbers defines a linear mapping from $\Re^n \to \Re^m$ defined by

$$f(\mathbf{x}) = \begin{pmatrix} \sum_{j=1}^{n} a_{1j} x_j \\ \sum_{j=1}^{n} a_{2j} x_j \\ \vdots \\ \sum_{j=1}^{n} a_{mj} x_j \end{pmatrix}$$

This is usually compactly written as

$$f(\mathbf{x}) = A\mathbf{x}$$

Exercise 3.5
Describe the action of the mapping $f \colon \Re^2 \to \Re^2$ defined by

$$f(x_1, x_2) = \begin{pmatrix} 0 & 1 \\ 1 & 0 \end{pmatrix} \begin{pmatrix} x_1 \\ x_2 \end{pmatrix}$$

Example 3.4 (Portfolio investment) Example 1.82 introduced a simple linear model of financial assets. Suppose there exist a finite number A of financial assets or securities in which to invest. Each asset a is fully described by its return vector $\mathbf{r}_a = (r_{1a}, r_{2a}, \ldots, r_{Sa})$, which details the prospective return of asset a in each of the S possible states of the world.

Arranging the return vectors of the A financial assets into a table or matrix, we can form an $S \times A$ matrix of prospective returns

$$R = \begin{pmatrix} r_{11} & r_{12} & \cdots & r_{1A} \\ r_{21} & r_{22} & \cdots & r_{2A} \\ \vdots & \vdots & & \vdots \\ r_{S1} & r_{S2} & \cdots & r_{SA} \end{pmatrix}$$

where r_{sa} denote the return of asset a in state s. The matrix R is called the *return matrix*. The sth row of the matrix specifies the return to the various assets if state of the world s prevails. Similarly the ath column of the matrix specifies the return to asset a in the various states.

A *portfolio* $\mathbf{x} = (x_1, x_2, \ldots, x_A)$ is a list of amounts invested in the different assets. The function

$$f(\mathbf{x}) = R\mathbf{x}$$
$$= \begin{pmatrix} \sum_{a=1}^{A} r_{1a} x_a \\ \sum_{a=1}^{A} r_{2a} x_a \\ \vdots \\ \sum_{a=1}^{A} r_{Sa} x_a \end{pmatrix}$$

specifies the total return to the portfolio $\mathbf{x}$ in the various states. f is linear, so the combined return of two portfolios $\mathbf{x}^1$ and $\mathbf{x}^2$ is equal to the return of a combined portfolio $\mathbf{x}^1 + \mathbf{x}^2$. Similarly scaling the portfolio $\alpha \mathbf{x}$ changes the aggregate return proportionately. Linearity requires that potential returns are independent of the portfolio choice, a reasonable assumption for a small investor.

Example 3.5 (Transpose) The matrix obtained by interchanging rows and columns in an matrix A is known as the transpose of A, and denoted A^T. That is,

if $\quad A = \begin{pmatrix} a_{11} & a_{12} & \cdots & a_{1n} \\ a_{21} & a_{22} & \cdots & a_{2n} \\ \vdots & \vdots & \ddots & \vdots \\ a_{m1} & a_{m2} & \cdots & a_{mn} \end{pmatrix}$

then $A^T = \begin{pmatrix} a_{11} & a_{21} & \cdots & a_{m1} \\ a_{12} & a_{22} & \cdots & a_{m2} \\ \vdots & \vdots & \ddots & \vdots \\ a_{1n} & a_{2n} & \cdots & a_{mn} \end{pmatrix}$

If A represents a linear function from X to Y, A^T represents a linear function from Y to X.

Example 3.6 (Shapley value) Since the set of all TP-coalitional games is a linear space (example 1.70), it is natural to consider values (example 2.37) that respect this linearity. A *linear value* on the space of games G^N is a linear function $\varphi \colon G^N \to \Re^n$ such that $\sum_{i \in N}(\varphi w)_i = w(N)$. Linearity requires that for any two games $w, w' \in G^N$,

$$\varphi(w + w') = \varphi w + \varphi w'$$

$$\varphi(\alpha w) = \alpha \varphi w$$

Both aspects of linearity have natural interpretations in the context of coalitional games. Homogeneity requires that the solution be invariant to the units of measurement, while additivity requires the solution to be invariant to the degree of aggregation. These are natural requirements in many applications of coalitional games (e.g., the cost allocation game of exercise 1.66).

The Shapley value is a particular linear function on the space of TP-games $\mathscr{G}^N$. It is defined by

$$\varphi_i(w) = \sum_{S \subseteq N} \gamma_S(w(S) - w(S \setminus \{i\}))$$

where

$$\gamma_S = \frac{(s-1)!(n-s)!}{n!}$$

$s = |S|$ = number of players in coaliton S

$\varphi_i(w)$ = the allocation to player i at the outcome $\varphi(w)$

Since only those coalitions in which i is a member carry any weight in the preceding sum ($w(S) = w(S \setminus \{i\})$ if $i \notin S$), the formula for Shapley value is often more usefully written as

$$\varphi_i(w) = \sum_{S \ni i} \gamma_S(w(S) - w(S\setminus\{i\})) \tag{1}$$

Example 3.7 (Three-way market) A farmer, f, owns a block of land that is worth $1 million as a farm. There are two potential buyers

- a manufacturer m to whom it is worth $2 million as a plant site
- a subdivider s to whom it is worth $3 million

This situation can be modeled as a TP-coalitional game with $N = \{f, m, s\}$ and the characteristic function

$$w(\{f\}) = 1 \quad w(\{m\}) = 0 \quad w(\{s\}) = 0$$
$$w(\{f,m\}) = 2 \quad w(\{f,s\}) = 3 \quad w(\{m,s\}) = 0$$
$$w(N) = 3$$

The following table details the computation of the Shapley value for player f:

S	γ_S	$w(S)$	$w(S\setminus\{i\})$	$\gamma_S(w(S) - w(S\setminus\{i\}))$
$\{f\}$	$\frac{1}{3}$	1	0	$\frac{1}{3}$
$\{f,m\}$	$\frac{1}{6}$	2	0	$\frac{1}{3}$
$\{f,s\}$	$\frac{1}{6}$	3	0	$\frac{1}{2}$
$\{f,m,s\}$	$\frac{1}{3}$	3	0	1
$\varphi_f(w)$				$2\frac{1}{6}$

The Shapley value assigns a payoff of $2\frac{1}{6}$ to the farmer. Similar calculations reveal that the Shapley values of the manufacturer and the subdivider are $\frac{1}{6}$ and $\frac{2}{3}$ respectively. The Shapley value of this game is $\varphi(w) = (2\frac{1}{6}, \frac{1}{6}, \frac{2}{3})$.

Exercise 3.6
Show that the Shapley value φ defined by (1) is linear.

Exercise 3.7
Compute the Shapley value for the cost allocation game (exercise 1.66).

Exercise 3.8
Verify that the Shapley value is a feasible allocation, that is,

$$\sum_{i \in N} \varphi_i w = w(N)$$

This condition is sometimes called *Pareto optimality* in the literature of game theory.

Exercise 3.9
Two players i and j are *substitutes* in a game (N, w) if their contributions to all coalitions are identical, that is, if

$$w(S \cup \{i\}) = w(S \cup \{j\}) \quad \text{for every } S \subseteq N \setminus \{i,j\}$$

Verify that the Shapley value treats substitutes symmetrically, that is

i, j substitutes $\Rightarrow \varphi_i w = \varphi_j w$

Exercise 3.10
A player i is called a *null player* in a game (N, w) if he contributes nothing to any coalition, that is, if

$$w(S \cup \{i\}) = w(S) \quad \text{for every } S \subseteq N$$

Verify that the Shapley value of a null player is zero, that is,

i null $\Rightarrow \varphi_i w = 0$

Exercise 3.11
Recall that, for any coalition $T \subset N$, the T-unanimity game (example 1.48) $u_T \in G^N$ is

$$u_T(S) = \begin{cases} 1 & \text{if } T \subset S \\ 0 & \text{otherwise} \end{cases}$$

Compute the Shapley value of a T-unanimity game.

Exercise 3.12 (Potential function)
For any TP-coalitional game (N, w) the *potential function* is defined to be

$$P(N, w) = \sum_{T \subseteq N} \frac{1}{t} \alpha_T$$

where $t = |T|$ and α_T are the coefficients in the basic expansion of w (exercise 1.75). Show that

$$\varphi_i w = P(N, w) - P(N\setminus\{i\}, w)$$

$$P(N, w) = \frac{1}{n}\left(w(N) - \sum_{i \in N} P(N\setminus\{i\}, w)\right)$$

Consequently the potential function provides a straightforward recursive method for computing the Shapley value of game. [Hint: Use the linearity of φ, example 1.75 and exercises 3.8 and 3.11.]

3.1 Properties of Linear Functions

The requirements of linearity impose a great deal of structure on the behavior of linear functions. The elaboration of this structure is one of the most elegant and satisfying fields of mathematics.

Exercise 3.13
Every linear function $f: X \to Y$ maps the zero vector in X into the zero vector in Y. That is, $f(\mathbf{0}_X) = \mathbf{0}_Y$.

Exercise 3.14
If $f: X \to Y$ and $g: Y \to Z$ are linear functions, then so is their composition $g \circ f: X \to Z$.

Exercise 3.15
Show that a linear function maps subspaces to subspaces, and vice versa. That is, if S is a subspace of X, then $f(S)$ is a subspace of Y; if T is a subspace of Y, then $f^{-1}(T)$ is a subspace of X.

Associated with any linear function are two subspaces that are particularly important in analyzing the behavior of the function. The range $f(X)$ of a linear function is called the *image* of f. The inverse image of the zero element is called the *kernel*, that is,

$$\text{kernel } f = f^{-1}(\mathbf{0}) = \{\mathbf{x} \in X : f(\mathbf{x}) = \mathbf{0}\}$$

The dimension of the image is called the *rank* of f. The dimension of the kernel is called the *nullity* of f. If X is finite dimensional, then (exercise 3.24)

$$\text{rank } f + \text{nullity } f = \dim X \tag{2}$$

A linear function $f: X \to Y$ has *full rank* if

$$\text{rank } f(X) = \min\{\text{rank } X, \text{rank } Y\}$$

The rank of a matrix A is the rank of the linear transformation $f(\mathbf{x}) = A\mathbf{x}$ that it represents (example 3.3). An $m \times n$ matrix has full if rank if rank $A = \min\{m, n\}$.

Exercise 3.16
Suppose that $f: X \to Y$ is a linear function with rank $f =$ rank $Y \le$ rank X. Then f maps X *onto* Y.

Exercise 3.17
Show that the kernel of a linear function $f: X \to Y$ is a subspace of X.

The behavior of a linear function is essentially determined by way in which it maps the kernel.

Exercise 3.18
Suppose that $f: X \to Y$ is a linear function with kernel $f = \{\mathbf{0}\}$. Then f is *one-to-one*, that is,

$$f(\mathbf{x}_1) = f(\mathbf{x}_2) \Rightarrow \mathbf{x}_1 = \mathbf{x}_2$$

A linear function $f: X \to Y$ that has an inverse $f^{-1}: Y \to X$ is said to be *nonsingular*. A function that does not have an inverse is called *singular*.

Exercise 3.19
A linear function $f: X \to Y$ is nonsingular if and only if kernel $f = \{\mathbf{0}\}$ and $f(X) = Y$.

Exercise 3.20
The inverse of a (nonsingular) linear function is linear.

Exercise 3.21
If f, g are nonsingular linear functions, then so is their composition $g \circ f$ with

$$(g \circ f)^{-1} = f^{-1} \circ g^{-1}$$

The following converse of 3.14 is the linear version of the important implicit function theorem (theorem 4.5).

Exercise 3.22 (Quotient theorem)
If $f: X \to Y$ and $h: X \to Z$ are linear functions with kernel $f \subseteq$ kernel h, then there exists a linear function $g: f(X) \to Z$ such that $h = g \circ f$.

Exercise 3.23
Suppose that $f: X \to Y$ is a linear function and $B \subset X$ is a basis for X. Then $f(B)$ spans $f(X)$.

Exercise 3.23 implies that *any linear mapping is completely determined by its action on a basis for the domain*. This has an several useful consequences for finite-dimensional mappings. It implies that any linear mapping between finite-dimensional spaces can be represented by a matrix (proposition 3.1). It establishes the link between the rank and nullity of a linear mapping (2). A striking application in game theory is given by the next example.

Example 3.8 (Shapley value is unique) The Shapley value (example 3.6) is uniquely defined for T-unanimity games by (exercise 3.11)

$$\varphi_i(u_T) = \begin{cases} 1/t & i \in T \\ 0 & i \notin T \end{cases}$$

where $t = |T|$. Since these form a basis for $\mathscr{G}^N$ (exercise 1.146), $\{\varphi(u_T) : T \subseteq N\}$ spans $\varphi(G^N)$. φ is uniquely defined for all $w \in G^N$.

We previously demonstrated that the Shapley value defined by

$$\varphi_i(w) = \sum_{S \subseteq N} \gamma_S(w(S) - w(S \setminus \{i\})), \quad \gamma_S = \frac{(s-1)!(n-s)!}{n!}, s = |S| \quad (3)$$

is feasible, treats substitutes symmetrically and disregards null players (exercises 3.8–3.10). We conclude that (3) is the only linear function with these properties. The Shapley value on the space of TP-coalitional games is unique.

Exercise 3.24 (Rank theorem)
Suppose that $f: X \to Y$ is a linear function. If X is finite-dimensional, then

rank f + nullity f = dim X

Exercise 3.25
Suppose that $f: X \to Y$ is a linear function with rank f = rank $X \leq$ rank Y and dim $X < \infty$. Then f is one-to-one.

Recall example 3.3 showing that any matrix defines a linear mapping. Exercise 3.23 implies the converse: *any linear mapping* between finite-dimensional spaces can be represented by a $m \times n$ matrix (of numbers).

Proposition 3.1 (Matrix representation) Let $f: X \to Y$ be a linear mapping between an n-dimensional space X and m-dimensional space Y. Then, for every choice of bases for X and Y, there exists an $m \times n$ matrix of numbers $A = (a_{ij})$ that represents f in the sense that

$$f(\mathbf{x}) = A\mathbf{x} \quad \text{for every } \mathbf{x} \in X$$

where $A\mathbf{x}$ is as defined in example 3.3.

Exercise 3.26
Assuming that $X = \Re^n$ and $Y = \Re^m$, prove proposition 3.1 for the standard basis (example 1.79).

A matrix provides a means of describing completely, concisely and uniquely any finite-dimensional linear function. Note that the matrix representation depends on a choice of basis for X and Y. Unless a particular basis is specified, the usual basis is implied in the matrix representation of a linear function.

Example 3.9 Consider example 2.3. Given the usual basis for $\Re^2$, the matrix representing this function is

$$A = \begin{pmatrix} \cos\theta & -\sin\theta \\ \sin\theta & \cos\theta \end{pmatrix}$$

For a rotation of 90 degrees ($\theta = \pi/2$), the matrix is

$$A = \begin{pmatrix} 0 & -1 \\ 1 & 0 \end{pmatrix}$$

Exercise 3.27
Give a matrix representation with respect to the usual bases for the linear function in exercise 3.4.

Exercise 3.28
Describe the matrix representing the Shapley value. Specify the matrix for three-player games ($|N| = 3$).

If $f: X \to Y$ is a nonsingular linear function with matrix representation A, then the representation of the inverse function f^{-1} with respect to the same bases is called the *matrix inverse* of A and is denoted A^{-1}.

3.1.1 Continuity of Linear Functions

The continuity of linear functions between normed linear spaces illustrates again a subtle interplay of linearity and geometry. The fundamental insight is the uniformity of linear spaces, as illustrated in the following result. Throughout this section, X and Y are assumed to be normed linear spaces.

Exercise 3.29
A linear function $f: X \to Y$ is continuous if and only if it is continuous at **0**.

A linear function $f: X \to Y$ is *bounded* if there exists a constant M such that

$$\|f(\mathbf{x})\| \leq M \|\mathbf{x}\| \qquad \text{for every } \mathbf{x} \in X \tag{4}$$

Note that boundedness does not imply that the range $f(X)$ is bounded but rather that $f(S)$ is bounded for every bounded set S. As the following exercise demonstrates, boundedness is equivalent to continuity for linear functions. Consequently these two terms are used interchangeably in practice.

Exercise 3.30
A linear function $f: X \to Y$ is continuous if and only if it is bounded.

Fortunately every linear function on a finite-dimensional space is bounded and therefore continuous.

Exercise 3.31
A linear function $f: X \to Y$ is bounded if X has finite dimension. [Hint: Use lemma 1.1.]

Rewriting (4), we have that a linear function f is bounded if there exists a constant $M < \infty$ such that

$$\frac{\|f(\mathbf{x})\|}{\|\mathbf{x}\|} \leq M \qquad \text{for every } x \in X \tag{5}$$

The smallest constant M satisfying (5) is called the norm of f. It is given by

$$\|f\| = \sup_{\mathbf{x} \neq \mathbf{0}} \frac{\|f(\mathbf{x})\|}{\|\mathbf{x}\|}$$

Clearly,

$$\|f(\mathbf{x})\| \leq \|f\| \|\mathbf{x}\|$$

Exercise 3.32
If f is a bounded linear function, an equivalent definition of the least upper bound is

$$\|f\| = \sup_{\|\mathbf{x}\|=1} \|f(\mathbf{x})\|$$

Exercise 3.33
The space $BL(X, Y)$ of all bounded linear functions from X to Y is a normed linear space, with norm

$$\|f\| = \sup\{\|f(\mathbf{x})\| : \|\mathbf{x}\| = 1\}$$

It is a Banach space (complete normed linear space) if Y is complete.

The following proposition is an important result regarding bounded linear functions.

Proposition 3.2 (Open mapping theorem) *Assume that X and Y are complete (i.e., Banach spaces). Every bounded linear function from X onto Y is an open map. Consequently, if f is nonsingular, the inverse function f^{-1} is continuous.*

A proof of the general theorem is beyond the scope of this text. It is within our resources to prove the theorem in the important case in which X is finite-dimensional.

Exercise 3.34
Prove proposition 3.2 assuming that X is finite-dimensional as follows: Let B be the unit ball in X, that is,

$$B = \{\mathbf{x} : \|\mathbf{x}\| < 1\}$$

The boundary of B is the unit sphere $S = \{\mathbf{x} : \|\mathbf{x}\| = 1\}$. Show that

1. $f(S)$ is a compact subset of Y which does not contain $\mathbf{0}_Y$.
2. There exists an open ball $T \subseteq (f(S))^c$ containing $\mathbf{0}_Y$.
3. $T \subseteq f(B)$.
4. f is open.
5. If f is nonsingular, f^{-1} is continuous.

We now give three applications of proposition 3.2. The first formalizes a claim made in chapter 1, namely that the geometry of all finite-dimensional spaces is the same. There is essentially only *one* finite-dimensional normed linear space, and $\Re^n$ is a suitable manifestation of this space. The second (exercise 3.36) shows that a linear homeomorphism is bounded from below as well as above. The third application (exercise 3.37) shows that for linear maps, continuity is equivalent to having a closed graph (exercise 2.70). We will use proposition 3.2 again in section 3.9 to prove the separating hyperplane theorem.

Exercise 3.35
Let X be a finite-dimensional normed linear space, and $\{\mathbf{x}_1, \mathbf{x}_2, \ldots, \mathbf{x}_n\}$ any basis for X. The function $f \colon \Re^n \to X$ defined by

$$f(\alpha_1, \alpha_2, \ldots, \alpha_n) = \sum_{i=1}^{n} \alpha_i \mathbf{x}_i$$

is a linear homeomorphism (remark 2.12). That is,

- f is linear
- f is one-to-one and onto
- f and f^{-1} are continuous

[Hint: Use the norm $\|\alpha\|_1 = \sum_{i=1}^{m} |\alpha_i|$.]

Exercise 3.36
Let $f \colon X \to Y$ be a linear homeomorphism (remark 2.12). Then there exists constants m and M such that for all $\mathbf{x}_1, \mathbf{x}_2 \in X$,

$$m\|\mathbf{x}_1 - \mathbf{x}_2\| \leq \|f(\mathbf{x}_1) - f(\mathbf{x}_2)\| \leq M\|\mathbf{x}_1 - \mathbf{x}_2\|$$

Exercise 3.37 (Closed graph theorem)
Let X and Y be Banach spaces. Any linear function $f \colon X \to Y$ is continuous if and only if its graph

$$\text{graph}(f) = \{(\mathbf{x}, \mathbf{y}) : \mathbf{y} = f(\mathbf{x}), \mathbf{x} \in X\}$$

is a closed subset of $X \times Y$.

3.2 Affine Functions

Affine functions relate to linear functions in the same way as subspaces relate to affine sets. A function $f: X \to Y$ is *affine* if

$$f(\alpha \mathbf{x}_1 + (1 - \alpha)\mathbf{x}_2) = \alpha f(\mathbf{x}_1) + (1 - \alpha)f(\mathbf{x}_2)$$

for all $\mathbf{x}_1, \mathbf{x}_2 \in X$, and $\alpha \in \Re$. (Compare with exercise 3.1.) Affine functions preserve affine sets (lines, planes). Their graphs are translations of the graph of a linear function, and do not pass through the origin (unless the function is linear). The following example illustrates the distinction.

Example 3.10 The function $f: \Re \to \Re$ defined by

$$f(x) = 2x + 3$$

is an affine function. Its graph is a straight line in the Euclidean plane, with a vertical intercept of 3 (figure 3.1). Such functions are often incorrectly called linear. It is not linear because $f(0) = 3 \neq 0$.

Exercise 3.38
Show that $f(x) = 2x + 3$ violates both the additivity and the homogeneity requirements of linearity.

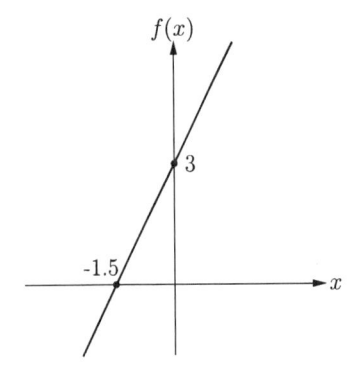

Figure 3.1
The graph of the affine function $f(x) = 2x + 3$

Exercise 3.39
A function $f\colon X \to Y$ is affine if and only if

$$f(\mathbf{x}) = g(\mathbf{x}) + \mathbf{y}$$

where $g\colon X \to Y$ is linear and $\mathbf{y} \in Y$

Exercise 3.40
Show that an affine function maps affine sets to affine sets, and vice versa. That is, if S is an affine subset of X, then $f(S)$ is an affine subset of Y; if T is an affine subset of Y, then $f^{-1}(T)$ is an affine subset of X.

Exercise 3.41
An affine function preserves convexity; that is, $S \subseteq X$ convex implies that $f(S)$ is convex.

3.3 Linear Functionals

Recall that a real-valued function $f\colon X \to \Re$ is called a functional. Linear functionals are the simplest and most prevalent linear functions. They assign a real number to every element of a linear space. For the economist, these assignments will often be interpreted as valuations and the linear functional as a *valuation function*. Linearity embodies the natural property that the value of two objects is equal to the sum of their individual values.

Example 3.11 Let $X \subseteq \Re^n$. For any $\mathbf{p} = (p_1, p_2, \ldots, p_n) \in \Re^n$, define the functional $f\colon X \to \Re$ by

$$f_{\mathbf{p}}(\mathbf{x}) = p_1 x_1 + p_2 x_2 + \cdots + p_n x_n$$

where $\mathbf{x} = (x_1, x_2, \ldots, x_n)$. Then $f_{\mathbf{p}}(\mathbf{x})$ is a linear function, that is for every $\mathbf{x}_1, \mathbf{x}_2 \in X$ $f_{\mathbf{p}}(\mathbf{x}_1 + \mathbf{x}_2) = f_{\mathbf{p}}(\mathbf{x}_1) + f_{\mathbf{p}}(\mathbf{x}_2)$ and $f_{\mathbf{p}}(\alpha \mathbf{x}_1) = \alpha f_{\mathbf{p}}(\mathbf{x}_1)$ for every $\alpha \in \Re$. Note how the linear functional depends on $\mathbf{p}$. Each $\mathbf{p} \in \Re^n$ defines a different linear functional (valuation). The linear functional has a natural interpretation as a valuation of X using the system of prices $\mathbf{p}$. The next two examples make this interpretation more explicit.

Example 3.12 If X is the consumption set, the function $c_p\colon X \to \Re$ defined by

$$c_\mathbf{p}(\mathbf{x}) = p_1 x_1 + p_2 x_2 + \cdots + p_n x_n$$

measures the cost of the commodity bundle **x** at prices **p**. Linearity implies that the joint cost of two different bundles is equal to the sum of the costs of the bundles separately and that the cost of bigger or smaller bundles is proportional to the cost of the original bundle.

Example 3.13 (Competitive firm) A producer is competitive if it takes the prices $\mathbf{p} = (p_1, p_2, \ldots, p_n)$ of all net outputs as given. If the producer adopts the production plan $\mathbf{y} = (y_1, y_2, \ldots, y_n)$, its net revenue or profit is

$$\Pi_\mathbf{p}(\mathbf{y}) = \sum_{i=1}^{n} p_i y_i$$

(Remember the convention that net inputs are negative.) The linear functional $\Pi_\mathbf{p}\colon Y \to \Re$ evaluates net revenue or profit of any production plan **y** at prices **p**. Each price vector **p** generates a different evaluation functional $\Pi_\mathbf{p}$. A profit-maximizing firm seeks to find that production plan $\mathbf{y}^* \in Y$ that maximizes net revenue (example 2.27). This necessarily requires it to produce efficiently (exercise 3.42). We use the term net revenue function to distinguish it from the related maximized profit function (example 2.29).

Exercise 3.42
If the production plan $\mathbf{y} \in Y$ maximizes profits at prices $\mathbf{p} > \mathbf{0}$, then **y** is efficient (example 1.61).

Example 3.14 (Expectation) Let $\mathfrak{X}$ be the set of all random variables (example 2.19) defined on a sample space S, that is, $\mathfrak{X} = F(S, \Re)$. *Expectation* E is a linear functional on $\mathfrak{X}$ with the properties

$$E(X) \geq 0 \quad \text{for every } X \geq 0 \text{ and } E(\mathbf{1}) = 1$$

where X is an arbitrary positive random variable in $\mathfrak{X}$ and **1** is the degenerate random variable that takes the value 1 for every outcome. (A convergence condition is also required; Whittle 1992, p. 15.) The value $E(X)$ of a particular random variable X is called the *expected value* of X. Linearity of expectation is commonly exploited in probability theory, for example, implying that

$$E(aX + b) = aE(X) + b$$

Exercise 3.43
Assume that the sample space S is finite. Then the expectation functional E takes the form

$$E(X) = \sum_{s \in S} p_s X(s)$$

with $p_s \geq 0$ and $\sum_{s \in S} p_s = 1$. $p_s = P(\{s\})$ is the probability of state s.

Example 3.15 (Shapley value) For any individual player i in a set N, her Shapley value φ_i is a linear functional on the space of games G^N, whose value $\varphi_i(w)$ can be interpreted as the expected value to i of playing the game w.

Example 3.16 (TP-coalitional games) Each coalition S in a TP-coalitional game implicitly defines a linear functional g_S on the space of outcomes X defined by

$$g_S(\mathbf{x}) = \sum_{i \in S} x_i$$

representing the total share of coalition S at the outcome $\mathbf{x}$. Note that this linear functional is defined on a different space to the preceding example.

Exercise 3.44
Let $X = C[0, 1]$ be the space of all continuous functions $x(t)$ on the interval $[0, 1]$. Show that the functional defined by

$$f(x) = x(\tfrac{1}{2})$$

is a linear functional on $C[0, 1]$.

Example 3.17 Another linear functional on the space $X = C[0, 1]$ of all continuous functions $x(t)$ on the interval $[0,1]$ is given by the integral, that is,

$$f(x) = \int_0^1 x(t)\,dt$$

The following result is fundamental for the economics of decentralization.

Exercise 3.45
Let $\{S_1, S_2, \ldots, S_n\}$ be a collection of subsets of a linear space X with $S = S_1 + S_2 + \cdots + S_n$. Let f be a linear functional on X. Then $\mathbf{x}^* = \mathbf{x}_1^* + \mathbf{x}_2^* + \cdots + \mathbf{x}_n^*$ maximizes f over S if and only if $\mathbf{x}_i^*$ maximizes f over S_i for every i. That is,

$$f(\mathbf{x}^*) \geq f(\mathbf{x}) \text{ for every } \mathbf{x} \in S \Leftrightarrow f(\mathbf{x}_i^*) \geq f(\mathbf{x}_i) \text{ for every } \mathbf{x}_i \in S_i \text{ for every } i$$

Example 3.18 Suppose that an economy consists of n producers each with a production possibility set $Y_i \subset \Re^m$. Assume that they produce without interaction, so that the aggregate production possibility set is $Y = Y_1 + Y_2 + \cdots + Y_n$. Then, applying the previous exercise, the aggregate production plan $\mathbf{y}^* = \mathbf{y}_1^* + \mathbf{y}_2^* + \cdots + \mathbf{y}_n^*$, $\mathbf{y}_j \in Y$ maximizes gross national product

$$\text{GNP} = \sum_{j=1}^{n} \sum_{i=1}^{m} p_i y_{ij}$$

at prices $\mathbf{p}$ if and only each producer maximizes her own profit $\sum_{i=1}^{m} p_i y_{ij}$ at $\mathbf{y}_j^*$.

3.3.1 The Dual Space

Example 3.17 and exercise 3.44 illustrate two distinct linear functionals on the same space. The set of all linear functionals on a linear space X is another linear space (exercise 3.2), which is called the *algebraic dual* of X; we will denote this by X'. The original space X is called the *primal space*. The set of all *continuous* linear functionals on a linear space X is called the *topological dual* or *conjugate* space of X and is denoted $X^* \subseteq X'$. Since this is of more practical importance, the adjective topological is usually omitted and the unqualified term dual space implies the topological dual. For finite-dimensional spaces (e.g., $\Re^n$), the distinction is vacuous, since all linear functionals on a finite-dimensional space are continuous (exercise 3.31). The following proposition is a special case of exercise 3.33.

Proposition 3.3 X^* *is a Banach space.*

Example 3.11 shows how to construct a host of linear functionals on $\Re^n$. It is a remarkable fact that all linear functionals on $\Re^n$ are constructed in this way. That is, *every* linear functional on $\Re^n$ is a valuation for some price system $\mathbf{p}$.

Proposition 3.4 (The dual of $\Re^n$) *For every linear functional $f \colon \Re^n \to \Re$, there exists an element $\mathbf{p} \in \Re^n$ such that*

$$f(x) = p_1 x_1 + p_2 x_2 + \cdots + p_n x_n$$

Proof Although this is a special case of proposition 3.3, it is insightful to prove the theorem directly. Let f be a linear functional on $X = \Re^n$, and let $\{\mathbf{e}_1, \mathbf{e}_2, \ldots, \mathbf{e}_n\}$ be the standard basis for $\Re^n$. Define $p_i = f(\mathbf{e}_i)$ for each $i = 1, 2, \ldots, n$. Any $\mathbf{x} \in X$ has the standard representation

$$\mathbf{x} = \sum_{i=1}^{n} x_i \mathbf{e}_i$$

and hence by linearity

$$f(x) = f\left(\sum_{i=1}^{n} x_i \mathbf{e}_i\right) = \sum_{i=1}^{n} x_i f(\mathbf{e}_i)$$

$$= \sum_{i=1}^{n} x_i p_i = p_1 x_1 + p_2 x_2 + \cdots + p_n x_n \qquad \square$$

This representation theorem is another application of the principle that the action of any linear mapping is summarized precisely by its action on a basis (exercise 3.23). It can be given an insightful economic interpretation. If we think of X as a commodity space, then the elements of the standard basis $\{\mathbf{e}_1, \mathbf{e}_2, \ldots, \mathbf{e}_n\}$ are unit quantities of each of the commodities. The p_i's are the values of each commodity, that is, their *prices*, and the linear functional prescribes the value of any commodity bundle for a given set of prices $\mathbf{p}$. Different price vectors give rise to different valuations (linear functionals), and every linear functional corresponds to a valuation function for a certain set of prices.

Remark 3.1 (Primal versus dual) Strictly speaking, the vector $\mathbf{p}$ in proposition 3.4 is an element of the dual space X^*, and we should carefully distinguish it from elements the primal space X. Indeed, some authors do this by distinguishing between column vectors (primal space) and row vectors (dual space). However, finite-dimensional spaces are self-dual, and there is an obvious identification between elements of X and elements of X^* with the same coordinates. This correspondence should be used with caution. It is peculiar to finite-dimensional linear spaces and is dependent on the choice of basis in each space. In general, the primal and dual spaces are mathematically distinct.

The distinction between the primal and dual spaces is clear in consumption space. Commodity bundles belong to the primal space X, whereas price lists belong to a different linear space X^*. While we can make a formal identification between commodity bundles and price lists as n dimensional vectors, they remain distinct types of objects. To put it bluntly, you cannot eat price lists. We are quite adept at manipulating prices and quantities mathematically but distinguish between them where necessary. We need to apply the same skill with finite-dimensional dual spaces.

Example 3.19 (Characteristic vector of a coalition) The set X of feasible outcomes in a TP-coalitional game (N, w) is a subset of $\Re^n$. The linear functional

$$g_S(\mathbf{x}) = \sum_{i \in S} x_i$$

measures the share of coalition S at the outcome $\mathbf{x} \in X$ (example 3.16). Corresponding to each coalition S, there exists a vector $\mathbf{e}_S \in \Re^n$ that represents this functional such that

$$g_S(\mathbf{x}) = \mathbf{e}_S^T \mathbf{x}$$

Here $\mathbf{e}_S$, which is called the *characteristic vector* of the coalition S, is defined by

$$(\mathbf{e}_S)_i = \begin{cases} 1 & \text{if } i \in S \\ 0 & \text{otherwise} \end{cases}$$

It identifies the members of the coalition S. Each characteristic vector corresponds to a vertex of the unit cube in $\Re^n$.

Things are more complicated in infinite-dimensional spaces, and not all dual spaces can be given a simple representation.

Example 3.20 (Dual of l_∞) In the dynamic programming problem (example 2.32), the choice set X is the set of infinite bounded sequences l_∞ (example 1.107). Those sequences $(x_1, x_2, x_3, \ldots)$ for which

$$\sum_{t=1}^{\infty} |x_t| < \infty$$

comprise a proper subspace of l_∞ which is denoted l_1. Every sequence $\mathbf{p} = (p_1, p_2, p_3, \ldots)$ in l_1 specifies a continuous linear functional $f_\mathbf{p}$ on l_∞ defined by

$$f_\mathbf{p}(\mathbf{x}) = \sum_{i=1}^{\infty} p_t x_t \tag{6}$$

Therefore l_1 is a subset of the dual space l_∞^*. We can think of the sequence $\mathbf{p} = (p_1, p_2, p_3, \ldots)$ as being a path of prices through time.

Unfortunately, l_1 is a proper subset of l_∞^*. There are linear functionals on l_∞ that cannot be given a simple representation of the form (6). This poses a problem for the use of l_∞ as the choice set for such models (see Stokey and Lucas 1989, pp. 460–461).

Exercise 3.46
Let c_0 denote the subspace of l_∞ consisting of all infinite sequences converging to zero, that is $c_0 = \{(x_t) \in l_\infty : x_t \to 0\}$. Show that

1. $l_1 \subset c_0 \subset l_\infty$
2. l_1 is the dual of c_0
3. l_∞ is the dual of l_1

The next two results will be used in subsequent applications. Exercise 3.48 implies the fundamental Lagrange multiplier rule of classical programming (chapter 5).

Exercise 3.47
Let X be a linear space and φ be a linear functional on the product space $X \times \Re$. Then φ has the representation

$$\varphi(x, t) = g(x) + \alpha t$$

where $g \in X'$ and $\alpha \in \Re$. [Hint: Show that $\varphi(x,t) = \varphi(x,0) + \varphi(0,1)t$.]

Exercise 3.48 (Fredholm alternative)
Let $f, g_1, g_2, \ldots, g_m$ be linear functionals on a linear space X. f is linearly dependent on $g_1, g_2, \ldots, g_m$, that is, $f \in \text{lin } g_1, g_2, \ldots, g_m$ if and only if

$$\bigcap_{j=1}^{m} \text{kernel } g_j \subseteq \text{kernel } f$$

[Hint: Define the function $G: X \to \Re^n$ by $G(\mathbf{x}) = (g_1(\mathbf{x}), g_2(\mathbf{x}), \ldots, g_m(\mathbf{x}))$ and apply exercise 3.22.]

3.3.2 Hyperplanes

In section 1.4.3 we defined *hyperplanes* as the largest proper affine subsets of a linear space X. We now develop an alternative characterization of hyperplanes as the contours of linear functionals. This intimate and useful correspondence between sets in the primal space X and elements in the dual space X' provides the foundation of the theory of duality.

Exercise 3.49
H is a hyperplane in a linear space X if and only if there exists a nonzero linear functional $f \in X'$ such that

$$H = \{\mathbf{x} \in X : f(\mathbf{x}) = c\}$$

for some $c \in \Re$.

We use $H_f(c)$ to denote the specific hyperplane corresponding to the c-level contour of the linear functional f.

Example 3.21 (Hyperplanes in $\Re^n$) Since every linear functional on $\Re^n$ corresponds to a valuation function for some price list $\mathbf{p}$ (proposition 3.4), hyperplanes in $\Re^n$ are sets of constant value. That is, a set H in $\Re^n$ is a hyperplane if and only if there exists some price list $\mathbf{p} \in \Re^n$ and constant c such that

$$H = \{\mathbf{x} : p_1 x_1 + p_2 x_2 + \cdots + p_n x_n = c\}$$

The zero hyperplane $c = 0$ is the subspace of all elements in X that are *orthogonal* to $\mathbf{p}$, that is,

$$H_\mathbf{p}(0) = \left\{\mathbf{x} \in \Re^n : \sum p_i x_i = 0\right\}$$

Other hyperplanes with the same price vector $\mathbf{p}$ consist of parallel translations of this subspace, with the distance from the origin increasing with $|c|$ (figure 3.2). The price vector $\mathbf{p}$ is called the *normal* to the hyperplane $H_\mathbf{p}(c)$. It has a geometric representation as a vector at right angles to the hyperplane.

Example 3.22 (Isoprofit lines) For a competitive firm the net revenue function

3.3 Linear Functionals

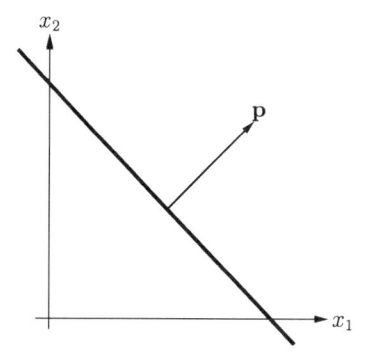

Figure 3.2
A hyperplane in $\Re^2$

$$\Pi_{\mathbf{p}}(\mathbf{y}) = \sum_{i=1}^{n} p_i y_i$$

is a linear functional on the production possibility set $Y \subset \Re^n$. The contours of the net revenue function

$$H_{\mathbf{p}}(c) = \{\mathbf{y} \in Y : \Pi_{\mathbf{p}}(\mathbf{y}) = c\}$$

are hyperplanes containing those production plans which yield a constant profit c. They are sometimes known as *isoprofit lines* (figure 3.3).

Excluding the special case in which the hyperplane is a subspace, the correspondence between hyperplanes in the primal space and linear functional in the dual space is unique.

Exercise 3.50
Let H be a hyperplane in a linear space that is not a subspace. Then there is a unique linear functional $f \in X'$ such that

$$H = \{\mathbf{x} \in X : f(\mathbf{x}) = 1\}$$

On the other hand, where H is a subspace, we have the following primitive form of the Hahn-Banach theorem (section 3.9.1).

Exercise 3.51
Let H be a maximal proper subspace of a linear space X and $\mathbf{x}_0 \notin H$. (H is a hyperplane containing $\mathbf{0}$). There exists a unique linear functional $f \in X'$

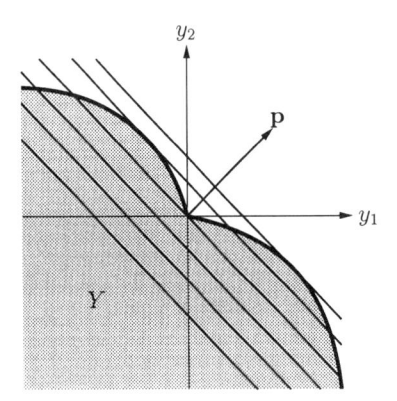

Figure 3.3
Isoprofit lines

such that

$$H = \{\mathbf{x} \in X : f(\mathbf{x}) = 0\} \quad \text{and} \quad f(\mathbf{x}_0) = 1$$

All linear functionals that share the same kernel differ only in their scale. If f is a linear functional with kernel V and $f(\mathbf{x}_0) = 1$, then for any $\lambda \neq 0$ the linear functional $g = \lambda f$ also has kernel V but $g(\mathbf{x}_0) = \lambda$. Conversely, if two linear functionals share the same kernel, they must be scalar multiples of one another (exercise 3.52). In this sense the linear functional corresponding to a particular hyperplane is only uniquely defined up to a scalar multiple. Selecting a particular linear functional from the class with a common kernel is known as *normalization*.

Remark 3.2 (Normalization) Since the hyperplane

$$H_{\lambda \mathbf{p}}(\lambda c) = \left\{ \mathbf{x} \in \Re^n : \sum \lambda p_i x_i = \lambda c \right\}$$

is identical to the hyperplane

$$H_{\mathbf{p}}(c) = \left\{ \mathbf{x} \in \Re^n : \sum p_i x_i = c \right\}$$

it is often useful to standardize the representation of a given hyperplane. This standardization is called *normalization*. Common normalizations include choosing $c = 1$, $\|\mathbf{p}\| = 1$, or $p_i = 1$ for some i. It is important to appreciate that normalizing involves nothing more than selecting one of the multitude of equivalent representations of a given hyperplane and implies no loss of generality. In an economic context in which hyperplanes correspond to valuations at given prices, normalization corresponds to selecting the scale of the general price level. Selecting good i as numéraire corresponds to the normalization $p_i = 1$.

Exercise 3.52 is a simple version of the Lagrange multiplier theorem (theorem 5.2) for constrained optimization. λ is the Lagrange multiplier. It is also a special case of exercise 3.48.

Exercise 3.52
For any $f, g \in X'$

kernel f = kernel $g \Leftrightarrow f = \lambda g$

for some $\lambda \in \Re \setminus \{0\}$.

Finally, we note that closed hyperplanes in X correspond precisely to continuous linear functionals in X'. That is, there is a one-to-one relationship between closed hyperplanes in X and elements of X^*.

Exercise 3.53
Let f be a nonzero linear functional on a normed linear space X. The hyperplane

$H = \{\mathbf{x} \in X : f(\mathbf{x}) = c\}$

is closed if and only if f is continuous.

3.4 Bilinear Functions

A function $f: X \times Y \to Z$ between linear spaces X, Y and Z is *bilinear* if it linear in each factor separately, that is, for all $\mathbf{x}, \mathbf{x}_1, \mathbf{x}_2 \in X$ and $\mathbf{y}, \mathbf{y}_1, \mathbf{y}_2 \in Y$,

$f(\mathbf{x}_1 + \mathbf{x}_2, \mathbf{y}) = f(\mathbf{x}_1, \mathbf{y}) + f(\mathbf{x}_2, \mathbf{y})$

(additivity)

$f(\mathbf{x}, \mathbf{y}_1 + \mathbf{y}_2) = f(\mathbf{x}, \mathbf{y}_1) + f(\mathbf{x}, \mathbf{y}_2)$

$f(\alpha \mathbf{x}, \mathbf{y}) = \alpha f(\mathbf{x}, \mathbf{y}) = f(\mathbf{x}, \alpha \mathbf{y})$ for every $\alpha \in \Re$

(homogeneity)

In other words, the partial functions $f_x: Y \to Z$ and $f_y: X \to Z$ are linear for every $\mathbf{x} \in X$ and $\mathbf{y} \in Y$ respectively.

Bilinear functions are one of the most common types of nonlinear functions. They are often used to represent the objective function in economic models. Bilinear functions are also encountered in the second-order conditions for optimization, since the second derivative of any smooth function is bilinear (section 4.4.1). Most of the bilinear functions that we will encounter are real-valued ($Z = \Re$), in which case we speak of *bilinear functionals*. Two important classes of bilinear functional that we will encounter in this book are the inner product and quadratic forms. These are introduced in separate sections below.

Example 3.23 The familiar product function $f: \Re^2 \to \Re$ defined by $f(x, y) = xy$ is bilinear, since

$$f(x_1 + x_2, y) = (x_1 + x_2)y = x_1 y + x_2 y = f(x_1, y) + f(x_2, y)$$

and

$$f(\alpha x, y) = (\alpha x)y = \alpha xy = \alpha f(x, y) \qquad \text{for every } \alpha \in \Re$$

Example 3.24 Any $m \times n$ matrix $A = (a_{ij})$ of numbers defines a bilinear functional on $\Re^m \times \Re^n$ by

$$f(\mathbf{x}, \mathbf{y}) = \sum_{i=1}^{m} \sum_{j=1}^{n} a_{ij} x_i y_j$$

Exercise 3.54
Show that the function defined in the previous example is bilinear.

There is an intimate relationship between bilinear functionals and matrices, paralleling the relationship between linear functions and matrices (theorem 3.1). The previous example shows that every matrix defines a bilinear functional. Conversely, every bilinear functional on finite dimensional spaces can be represented by a matrix.

Exercise 3.55 (Matrix representation of bilinear functionals)
Let $f: X \times Y \to \Re$ be a bilinear functional on finite-dimensional linear spaces X and Y. Let $m = \dim X$ and $n = \dim Y$. For every choice of bases for X and Y, there exists an $m \times n$ matrix of numbers $A = (a_{ij})$ that represents f in the sense that

$$f(\mathbf{x}, \mathbf{y}) = \sum_{i=1}^{m} \sum_{j=1}^{n} a_{ij} x_i y_j \qquad \text{for every } \mathbf{x} \in X \text{ and } \mathbf{y} \in Y$$

Example 3.25 Let X be any linear space and $Y = X'$ be the dual space. Then

$$f(\mathbf{x}, \mathbf{y}) = \mathbf{y}(\mathbf{x})$$

is a bilinear functional on $X \times X'$.

Exercise 3.56
Show that the function f defined in the preceding example is bilinear.

Exercise 3.57
Let $BiL(X \times Y, Z)$ denote the set of all continuous bilinear functions from $X \times Y$ to Z. Show that $BiL(X \times Y, Z)$ is a linear space.

The following result may seem rather esoteric but is really a straightforward application of earlier definitions and results. It will be used in the next chapter.

Exercise 3.58
Let X, Y, Z be linear spaces. The set $BL(Y, Z)$ of all bounded linear functions from Y to Z is a linear space (exercise 3.33). Let $BL(X, BL(Y, Z))$ denote the set of bounded linear functions from X to the set $BL(Y, Z)$. Show that

1. $BL(X, BL(Y, Z))$ is a linear space.
2. Let $\varphi \in BL(X, BL(Y, Z))$. For every $x \in X$, $\varphi_\mathbf{x}$ is a linear map from Y to Z. Define the function $f : X \times Y \to Z$ by

$$f(\mathbf{x}, \mathbf{y}) = \varphi_\mathbf{x}(\mathbf{y})$$

Show that f is bilinear, that is, $f \in BiL(X \times Y, Z)$.

3. For every $f \in BiL(X \times Y, Z)$, let $f_\mathbf{x}$ denote the partial function $f_\mathbf{x} : Y \to Z$ defined by

$$f_\mathbf{x}(\mathbf{y}) = f(\mathbf{x}, \mathbf{y})$$

Define

$$\varphi_f(\mathbf{x}) = f_\mathbf{x}$$

Show that $\varphi_f \in BL(X, BL(Y, Z))$.

This establishes a one-to-one relationship between the spaces $BiL(X \times Y, Z)$ and $BL(X, BL(Y, Z))$.

3.4.1 Inner Products

A bilinear functional f on the space $X \times X$ is called

symmetric if $f(\mathbf{x}, \mathbf{y}) = f(\mathbf{y}, \mathbf{x})$ for every $\mathbf{x}, \mathbf{y} \in X$

nonnegative definite if $f(\mathbf{x}, \mathbf{x}) \geq 0$ for every $\mathbf{x} \in X$

positive definite if $f(\mathbf{x}, \mathbf{x}) > 0$ for every $\mathbf{x} \in X$, $\mathbf{x} \neq \mathbf{0}$

Exercise 3.59 (Cauchy-Schwartz inequality)
Every symmetric, nonnegative definite bilinear functional f satisfies the inequality

$$(f(\mathbf{x}, \mathbf{y}))^2 \leq f(\mathbf{x}, \mathbf{x}) f(\mathbf{y}, \mathbf{y})$$

for every $\mathbf{x}, \mathbf{y} \in X$.

A symmetric, positive definite bilinear functional on a linear space X is called an *inner product*. It is customary to use a special notation to denote the inner product. We will use $\mathbf{x}^T \mathbf{y}$ to denote $f(\mathbf{x}, \mathbf{y})$ when f is an inner product. By definition, an inner product satisfies the following properties for every $\mathbf{x}, \mathbf{x}_1, \mathbf{x}_2, \mathbf{y} \in X$:

symmetry $\mathbf{x}^T \mathbf{y} = \mathbf{y}^T \mathbf{x}$

additivity $(\mathbf{x}_1 + \mathbf{x}_2)^T \mathbf{y} = \mathbf{x}_1^T \mathbf{y} + \mathbf{x}_2^T \mathbf{y}$

homogeneity $\alpha \mathbf{x}^T \mathbf{y} = \alpha \mathbf{x}^T \mathbf{y}$

positive definiteness $\mathbf{x}^T \mathbf{x} \geq 0$ and $\mathbf{x}^T \mathbf{x} = 0$ if and only if $\mathbf{x} = \mathbf{0}$

Remark 3.3 (Notation) A variety of notation is used for the inner product. The common choices $\mathbf{x} \cdot \mathbf{y}$ and $\langle \mathbf{x}, \mathbf{y} \rangle$ emphasize the symmetry of the function. However, our choice $\mathbf{x}^T \mathbf{y}$ will be advantageous in defining quadratic forms (section 3.5.3) and representing the derivative (chapter 4). We will find it convenient to use $\mathbf{x} \cdot \mathbf{y}$ in section 6.2.1.

A linear space equipped with an inner product is called an *inner product space*. Every inner product defines a norm (exercise 3.63) given by

$$\|\mathbf{x}\| = \sqrt{\mathbf{x}^T \mathbf{x}}$$

Consequently every inner product space is a normed linear space. A finite-dimensional inner product space is called a *Euclidean space* and a complete inner product space is called a *Hilbert space*.

Example 3.26 $\Re^n$ is a Euclidean space, with inner product $\mathbf{x}^T\mathbf{y} = \sum_{i=1}^{n} x_i y_i$.

Exercise 3.60
Every Euclidean space is complete, that is, a Hilbert space.

Exercise 3.61 (Cauchy-Schwartz inequality)
For every $\mathbf{x}, \mathbf{y}$ in an inner product space,

$$|\mathbf{x}^T\mathbf{y}| \le \|\mathbf{x}\| \|\mathbf{y}\|$$

Exercise 3.62
The inner product is a continuous bilinear functional.

Exercise 3.63
The functional $\|\mathbf{x}\| = \sqrt{\mathbf{x}^T\mathbf{x}}$ is a norm on X.

Exercise 3.64
Every element $\mathbf{y}$ in an inner product space X defines a continuous linear functional on X by $f_\mathbf{y}(\mathbf{x}) = \mathbf{x}^T\mathbf{y}$.

Exercise 3.65 (Existence of extreme points)
A nonempty compact convex set in an inner product space has at least one extreme point.

Exercise 3.66 (Parallelogram law)
In an inner product space

$$\|\mathbf{x} + \mathbf{y}\|^2 + \|\mathbf{x} - \mathbf{y}\|^2 = 2\|\mathbf{x}\|^2 + 2\|\mathbf{y}\|^2$$

Remark 3.4 An inner product space mimics the geometry of ordinary Euclidean space. It is the most structured of linear spaces. Not all normed linear spaces are inner product spaces (e.g., l^∞ in example 1.107 and $C(X)$ in exercise 3.67). In fact, a normed linear space is an inner product space if and only if its norm satisfies the parallelogram law (exercise 3.66), in which case the inner product can be recovered from the norm by the following *polarization identity*:

$$\mathbf{x}^T\mathbf{y} = \tfrac{1}{4}(\|\mathbf{x} + \mathbf{y}\|^2 - \|\mathbf{x} - \mathbf{y}\|^2)$$

Exercise 3.67
Show that $C(X)$ (exercise 2.85) is not an inner product space. [Hint: Let $X = [0, 1]$, and consider the functionals $x(t) = 1$ and $y(t) = t$.]

Two vectors **x** and **y** in an inner product space X are *orthogonal* if $\mathbf{x}^T\mathbf{y} = 0$. We symbolize this by $\mathbf{x} \perp \mathbf{y}$. The *orthogonal complement* $S^\perp$ of a subset $S \subset X$ as the set of all vectors that are orthogonal to every vector in S, that is,

$$S^\perp = \{\mathbf{x} \in X : \mathbf{x}^T\mathbf{y} = 0 \text{ for every } \mathbf{y} \in S\}$$

A set of vectors $\{\mathbf{x}_1, \mathbf{x}_2, \ldots, \mathbf{x}_n\}$ is called pairwise orthogonal if $\mathbf{x}_i \perp \mathbf{x}_j$ for every $i \neq j$. A set of vectors $\{\mathbf{x}_1, \mathbf{x}_2, \ldots, \mathbf{x}_n\}$ is called *orthonormal* if it is pairwise orthogonal and each vector has unit length so that

$$\mathbf{x}_i^T\mathbf{x}_j = \begin{cases} 1 & \text{if } i = j \\ 0 & \text{otherwise} \end{cases}$$

Example 3.27 (Orthonormal basis) Every orthonormal set is linearly independent (exercise 3.68). If there are sufficient vectors in the orthonormal set to span the space, the orthonormal set is called an *orthonormal basis*. The standard basis $\{\mathbf{e}_1, \mathbf{e}_2, \ldots, \mathbf{e}_n\}$ for $\Re^n$ (example 1.79) is an orthonormal basis, since

$$\mathbf{e}_i^T\mathbf{e}_j = \begin{cases} 1 & \text{if } i = j \\ 0 & \text{otherwise} \end{cases}$$

Exercise 3.68
Any pairwise orthogonal set of nonzero vectors is linearly independent.

Exercise 3.69
Let the matrix $A = (a_{ij})$ represent a linear operator with respect to an orthonormal basis $\mathbf{x}_1, \mathbf{x}_2, \ldots, \mathbf{x}_n$ for an inner product space X. Then

$$a_{ij} = \mathbf{x}_i^T f(\mathbf{x}_j) \quad \text{for every } i, j$$

A link between the inner product and the familiar geometry of $\Re^3$ is established in the following exercise, which shows that the inner product is a measure of the angle between two vectors.

Exercise 3.70
For any two nonzero elements **x** and **y** in an inner product space X, define the angle θ between **x** and **y** by

3.4 Bilinear Functions

$$\cos \theta = \frac{\mathbf{x}^T \mathbf{y}}{\|\mathbf{x}\| \|\mathbf{y}\|} \quad (7)$$

for $0 \le \theta \le \pi$. Show that

1. $-1 \le \cos \theta \le 1$
2. $\mathbf{x} \perp \mathbf{y}$ if and only if $\theta = 90$ degrees

The angle between two vectors is defined by (7) corresponds to the familiar notion of angle in $\Re^2$ and $\Re^3$.

Exercise 3.71 (Pythagoras)
If $\mathbf{x} \perp \mathbf{y}$, then

$$\|\mathbf{x} + \mathbf{y}\|^2 = \|\mathbf{x}\|^2 + \|\mathbf{y}\|^2$$

The next result provides the crucial step in establishing the separating hyperplane theorem (section 3.9).

Exercise 3.72 (Minimum distance to a convex set)
Let S be a nonempty, closed, convex set in a Euclidean space X and $\mathbf{y}$ a point outside S (figure 3.4). Show that

1. There exists a point $\mathbf{x}_0 \in S$ which is closest to $\mathbf{y}$, that is,

$$\|\mathbf{x}_0 - \mathbf{y}\| \le \|\mathbf{x} - \mathbf{y}\| \quad \text{for every } \mathbf{x} \in S$$

[Hint: Minimize $g(\mathbf{x}) = \|\mathbf{x} - \mathbf{y}\|$ over a suitable compact set.]

2. $\mathbf{x}_0$ is unique
3. $(\mathbf{x}_0 - \mathbf{y})^T (\mathbf{x} - \mathbf{x}_0) \ge 0$ for every $\mathbf{x} \in S$

Finite dimensionality is not essential to the preceding result, although completeness is required.

Exercise 3.73
Generalize the preceding exercise to any Hilbert space. Specifically, let S be a nonempty, closed, convex set in Hilbert space X and $\mathbf{y} \notin S$. Let

$$d = \inf_{x \in S} \|\mathbf{x} - \mathbf{y}\|$$

Then there exists a sequence $(\mathbf{x}^n)$ in S such that $\|\mathbf{x}^n - \mathbf{y}\| \to d$. Show that

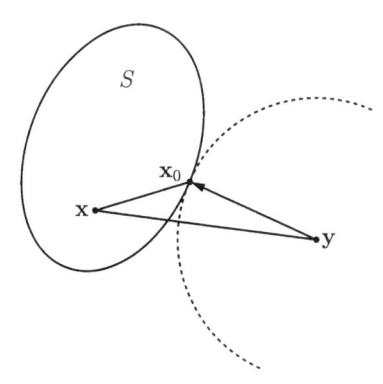

Figure 3.4
Minimum distance to a closed convex set

1. $(\mathbf{x}^n)$ is a Cauchy sequence.
2. There exists a *unique* point $\mathbf{x}_0 \in S$ which is closest to $\mathbf{y}$, that is,

$$\|\mathbf{x}_0 - \mathbf{y}\| \leq \|\mathbf{x} - \mathbf{y}\| \quad \text{for every } \mathbf{x} \in S$$

To complete this section, we give two important applications of exercise 3.72. Exercise 3.74 was used in chapter 2 to prove Brouwer's fixed point theorem (theorem 2.6).

Exercise 3.74 (Existence of a retraction)
Let S be a closed convex subset of a Euclidean space X and T be another set containing S. There exists a continuous function $g: T \to S$ that retracts T onto S, that is, for which $g(\mathbf{x}) = \mathbf{x}$ for every $\mathbf{x} \in S$.

Earlier (exercise 3.64) we showed that every element in an inner product space defines a distinct continuous linear functional on the space. We now show that for a complete linear space, every continuous linear functional takes this form.

Exercise 3.75 (Riesz representation theorem)
Let $f \in X^*$ be a continuous linear functional on a Hilbert space X. There exists a unique element $\mathbf{y} \in X$ such that

$$f(\mathbf{x}) = \mathbf{x}^T \mathbf{y} \quad \text{for every } \mathbf{x} \in X$$

[Hint: Show that there exists some $\mathbf{z} \perp S = \text{kernel } f$ and consider $\hat{S} = \{f(\mathbf{x})\mathbf{z} - f(\mathbf{z})\mathbf{x} : \mathbf{x} \in X\}$.]

Remark 3.5 (Reflexive normed linear space) Exercise 3.205 shows that dual X^* of a normed linear space X contains nonzero elements. Since X^* is a normed linear space in its own right (proposition 3.3), it too has a dual space denoted X^{**} which is called the second dual space of X. Every $\mathbf{x} \in X$ defines a linear functional F on X^* by

$$F(f) = f(\mathbf{x}) \quad \text{for every } f \in X^*$$

In general, X^{**} is bigger than X, that is there are linear functionals on X^* which cannot be identified with elements in X. A normed linear space is called *reflexive* if $X = X^{**}$, that is for every $F \in X^{**}$, there exists an $\mathbf{x} \in X$ such that

$$f(\mathbf{x}) = F(f) \quad \text{for every } f \in X^*$$

Every finite-dimensional space and every Hilbert space is reflexive.

Exercise 3.76
If X is a Hilbert space, then so is X^*.

Exercise 3.77
Every Hilbert space is reflexive.

Exercise 3.78 (Adjoint transformation)
Let $f \in L(X, Y)$ be a linear function between Hilbert spaces X and Y.

1. For every $\mathbf{y} \in Y$, define $f_\mathbf{y}(\mathbf{x}) = f(\mathbf{x})^T \mathbf{y}$. Then $f_\mathbf{y} \in X^*$.
2. There exists a unique $\mathbf{x}^* \in X$ such that $f_\mathbf{y}(\mathbf{x}) = \mathbf{x}^T \mathbf{x}^*$.
3. Define $f^* \colon Y \to X$ by $f^*(\mathbf{y}) = \mathbf{x}^*$. Then f^* satisfies

$$f(\mathbf{x})^T \mathbf{y} = \mathbf{x}^T f^*(\mathbf{y})$$

4. f^* is a linear function, known as the adjoint of f.

3.5 Linear Operators

Some important tools and results are available for linear operators, that is, linear functions from a set to itself. Since every linear operator on a finite-dimensional space can be represented (proposition 3.1) by a square matrix, the following can be seen alternatively as the theory of square matrices.

Example 3.28 (Identity operator) The *identity operator* $I: X \to X$ maps every point in X to itself, that is,

$$I(\mathbf{x}) = \mathbf{x} \quad \text{for every } \mathbf{x} \in X$$

If $\dim X = n < \infty$, the identity operator is represented (relative to any basis) by the *identity matrix* of order n,

$$I_n = \begin{pmatrix} 1 & 0 & \cdots & 0 \\ 0 & 1 & \cdots & 0 \\ \vdots & \vdots & \ddots & \vdots \\ 0 & 0 & \cdots & 1 \end{pmatrix}_{n \times n}$$

Exercise 3.79
Every linear operator $f: X \to X$ has at least one fixed point.

3.5.1 The Determinant

The set of all linear operators on a given space X is denoted $L(X, X)$. If X is finite-dimensional, there is a unique functional det on $L(X, X)$ with the following properties

- $\det(f \circ g) = \det(f) \det(g)$
- $\det(I) = 1$
- $\det(f) = 0$ if and only if f is nonsingular

for every $f, g \in L(X, X)$. This functional is known as the *determinant*. The last property is especially important, the determinant provides a simple means of distinguishing nonsingular operators. Note that the determinant is *not* a linear functional. In general, $\det(f + g) \neq \det(f) + \det(g)$.

Example 3.29 Let $\dim X = 1$. Every linear operator on X takes the form

$$f(\mathbf{x}) = a\mathbf{x}$$

for some $a \in \Re$. The functional $\varphi(f) = a$ satisfies the properties of the determinant. Therefore $\det(f) = \varphi(f) = a$.

The determinant of a square matrix is defined to be the determinant of the linear operator that it represents. The determinant of a matrix A can be computed recursively by the formula

$$\det(A) = \sum_{j=1}^{n}(-1)^{i+j} a_{ij} \det(A_{ij}) \qquad (8)$$

where A_{ij} is the $(n-1) \times (n-1)$ matrix obtained from A by deleting the ith row and jth column. This is known as "expansion along the ith row." Alternatively, the determinant can be calculated by "expansion down the jth column" using the formula

$$\det(A) = \sum_{i=1}^{n}(-1)^{i+j} a_{ij} \det(A_{ij}) \qquad (9)$$

It is a remarkable implication of the structure of linear operators that it does not matter which basis we use to represent the operator by a matrix, nor does it matter which row or column we use in the recursion. The determinant of a linear operator f is uniquely defined by (8) or (9) for any matrix representation A. To prove this would require a substantial diversion from our main path, and so the discussion is omitted. Suitable references are given at the end of the chapter.

Example 3.30 The determinant of the matrix

$$A = \begin{pmatrix} a_{11} & a_{12} \\ a_{21} & a_{22} \end{pmatrix}$$

is sometimes denoted by

$$|A| = \begin{vmatrix} a_{11} & a_{12} \\ a_{21} & a_{22} \end{vmatrix}$$

Expanding along the first row and using example 3.29, we have

$$\det A = \begin{vmatrix} a_{11} & a_{12} \\ a_{21} & a_{22} \end{vmatrix} = a_{11} \det(a_{22}) - a_{12} \det(a_{21})$$

$$= a_{11}a_{22} - a_{12}a_{21}$$

Example 3.31 (Triangular matrix) A matrix $A = (a_{ij})$ is called upper-triangular if $a_{ij} = 0$ for every $i > j$ and lower-triangular if $a_{ij} = 0$ for every $i < j$. Thus a triangular matrix has only zero elements on one side of the diagonal. The matrix

$$A = \begin{pmatrix} 1 & 0 & 3 \\ 0 & 2 & 1 \\ 0 & 0 & 3 \end{pmatrix}$$

is upper-triangular. Expanding down the first column, its determinant is

$$\det(A) = 1 \begin{vmatrix} 2 & 1 \\ 0 & 3 \end{vmatrix} + 0 \begin{vmatrix} 0 & 3 \\ 0 & 3 \end{vmatrix} + 0 \begin{vmatrix} 0 & 3 \\ 2 & 1 \end{vmatrix}$$

$$= 1(2 \times 3 - 0 \times 1) = 6$$

which is the product of the diagonal elements.

We record several useful properties of the determinant in the following proposition.

Proposition 3.5 (Properties of the determinant) *For any matrices A and B,*

1. $\det(AB) = \det(A)\det(B)$.
2. $\det(I) = 1$.
3. $\det(A) = 0$ *if and only if A is invertible.*
4. $\det A^{-1} = 1/\det(A)$.
5. *If A has a row of zeros, then* $\det(A) = 0$.
6. $\det(A^T) = \det(A)$.
7. *If A has two rows that are equal,* $\det A = 0$.
8. *If B is obtained from A by multiplying a row of A by a number* α, *then* $\det(B) = \alpha \det(A)$.
9. *If B is obtained from A by interchanging two rows,* $\det(B) = -\det(A)$.
10. *If B is obtained from A by adding a multiple of one row to a different row, then* $\det(B) = \det(A)$.
11. *If A is triangular,* $\det(A)$ *is the product of the diagonal entries.*

Proof The first three properties are simply translations of the properties of the determinant of an operator. Property 4 is proved in exercise 3.80. The remaining properties flow from the expansions (8) and (9) (Simon and Blume 1994, pp. 726–735). □

Exercise 3.80
Suppose that the matrix A has an inverse A^{-1}. Then $\det(A^{-1}) = 1/\det(A)$.

The determinant of a matrix is "linear in the rows" in the sense established in the following exercise. An analogous result holds for columns.

Exercise 3.81
Let A, B, and C be matrices that differ only in their ith row, with the ith row of C being a linear combination of the rows of A and B. That is,

$$A = \begin{pmatrix} \mathbf{a}_1 \\ \vdots \\ \mathbf{a}_i \\ \vdots \\ \mathbf{a}_n \end{pmatrix}, \quad B = \begin{pmatrix} \mathbf{a}_1 \\ \vdots \\ \mathbf{b}_i \\ \vdots \\ \mathbf{a}_n \end{pmatrix}, \quad C = \begin{pmatrix} \mathbf{a}_1 \\ \vdots \\ \alpha \mathbf{a}_i + \beta \mathbf{b}_i \\ \vdots \\ \mathbf{a}_n \end{pmatrix}$$

Then

$$\det(C) = \alpha \det(A) + \beta \det(B)$$

3.5.2 Eigenvalues and Eigenvectors

For linear operators, a generalization of the notion of a fixed point proves useful. Given a linear operator $f \colon X \to X$, a nonzero element $\mathbf{x} \in X$ is called an *eigenvector* if

$$f(\mathbf{x}) = \lambda \mathbf{x}$$

for some number $\lambda \in \Re$. The constant λ is called an *eigenvalue* of f. The synonyms characteristic vector and characteristic value are also used. The operator acts very simply on its eigenvectors, scaling them by a constant. If the eigenvalue λ corresponding to an eigenvector is one, then the eigenvector is a fixed point. The eigenvectors corresponding to a particular eigenvalue, together with the zero vector $\mathbf{0}_X$, form a subspace of X called an eigenspace (exercise 3.82).

Exercise 3.82
Show that the eigenvectors corresponding to a particular eigenvalue, together with the zero vector $\mathbf{0}_X$, form a subspace of X

Exercise 3.83 (Zero eigenvalues)
A linear operator is singular if and only if it has a zero eigenvalue.

Exercise 3.84

If **x** is an eigenvector of a linear operator f on an inner product space X with eigenvalue λ, then the eigenvalue λ can be expressed as

$$\lambda = \frac{f(\mathbf{x})^T \mathbf{x}}{\|\mathbf{x}\|}$$

An operator f on an inner product space X is called *symmetric* if

$$f(\mathbf{x})^T \mathbf{y} = \mathbf{x}^T f(\mathbf{y}) \quad \text{for every } \mathbf{x}, \mathbf{y} \in X$$

If the inner product space is finite-dimensional (Euclidean), the operator can be represented by a matrix (proposition 3.1), in the sense that

$$f(\mathbf{x}) = A\mathbf{x} \quad \text{for every } \mathbf{x} \in X$$

Provided that we use an orthonormal basis (example 3.27) for the representation, the operator is symmetric if and only if its associated matrix is a symmetric matrix, that is, $A = A^T$. Since many of the linear operators encountered in practice are represented by symmetric matrices, the properties of symmetric operators are important.

Exercise 3.85

Let f be a linear operator on a Euclidean space, and let the matrix $A = (a_{ij})$ represent f with respect to an orthonormal basis. Then f is a symmetric operator if and only if A is a symmetric matrix, that is, $A = A^T$.

Remark 3.6 (Self-adjoint operator) A symmetric operator on a Hilbert space is often called *self-adjoint*. The adjoint f^* of a linear operator f is defined by (exercise 3.78) $f(\mathbf{x})^T \mathbf{y} = \mathbf{x}^T f^*(\mathbf{y})$. If the operator f is symmetric, then

$$f(\mathbf{x})^T \mathbf{y} = \mathbf{x}^T f(\mathbf{y}) \quad \text{for every } \mathbf{x}, \mathbf{y} \in X$$

which implies that $f^* = f$.

Exercise 3.86

For a symmetric operator, the eigenvectors corresponding to distinct eigenvalues are orthogonal.

Remark 3.7 (Existence of eigenvalues) Not every linear operator has an eigenvalue (example 2.3). However, every symmetric operator on a

Euclidean space has eigenvalues and the corresponding eigenvectors form a basis for the space (proposition 3.6). To be able to analyze nonsymmetric operators, many texts on linear algebra resort to complex linear spaces, in which the scalars are complex numbers (Halmos 1974, p. 150). Then it can be shown that every finite-dimensional linear operator has an eigenvalue that may, however, be complex (Janich 1994, p. 156)

Exercise 3.87
Let f be a symmetric operator on a Euclidean space X. Let S be the unit sphere in X, that is $S = \{\mathbf{x} \in X : \|\mathbf{x}\| = 1\}$, and define $g: X \times X \to \Re$ by

$$g(\mathbf{x}, \mathbf{y}) = (\lambda \mathbf{x} - f(\mathbf{x}))^T \mathbf{y} \qquad \text{where } \lambda = \max_{\mathbf{x} \in S} f(\mathbf{x}^T)\mathbf{x} \qquad (10)$$

Show that

1. The maximum in (10) is attained at some $\mathbf{x}_0 \in S$. Therefore g is well-defined.

2. g is nonnegative definite.

3. g is symmetric.

4. $\mathbf{x}_0$ is an eigenvector of f.

Hence every symmetric operator on a Euclidean space has an eigenvector of norm 1. [Hint: Use exercise 3.59.]

The following key result is an existence theorem for eigenvalues. It shows that the eigenvalues of a symmetric linear operator on a Euclidean space X are real (as opposed to complex). Furthermore, although the eigenvalues may not be distinct, there are sufficient linearly independent eigenvectors to span the space.

Proposition 3.6 (Spectral theorem) *If f is a symmetric linear operator on a Euclidean space X, then X has an orthogonal basis comprising eigenvectors of f. The matrix representing f with respect to this basis is a diagonal matrix whose diagonal elements are the eigenvalues of f.*

Proof By exercise 3.87, there exists an eigenvector $\mathbf{x}_0$ of norm 1. Let $n = \dim X$. If $n = 1$, then $\mathbf{x}_1$ is a basis for X. Otherwise ($n > 1$), assume that the proposition is true for all spaces of dimension $n - 1$. Let $S = \{\mathbf{x}_0\}^\perp$. We claim (exercise 3.88) that

- S is a subspace of dimension $n - 1$
- $f(S) \subseteq S$.

Therefore f is a symmetric linear operator on S on a Euclidean space of dimension $n - 1$. By assumption, S has an orthonormal basis $\{\mathbf{x}_2, \mathbf{x}_3, \ldots, \mathbf{x}_n\}$ of eigenvectors. $\mathbf{x}_1$ is orthogonal to $\mathbf{x}_i$, $i = 2, 3, \ldots, n$, and therefore $\{\mathbf{x}_1, \mathbf{x}_2, \ldots, \mathbf{x}_n\}$ is a basis for X (exercise 3.68). Let the matrix A represent f with respect to the orthonormal basis $\{\mathbf{x}_1, \mathbf{x}_2, \ldots, \mathbf{x}_n\}$. By exercise 3.69,

$$a_{ij} = \mathbf{x}_i^T f(\mathbf{x}_j) = \lambda_j \mathbf{x}_i^T \mathbf{x}_j = \begin{cases} \lambda_i & i = j \\ 0 & i \neq j \end{cases} \qquad \square$$

Exercise 3.88
Let S be defined as in the preceding proof. Show that

1. S is a subspace of dimension $n - 1$
2. $f(S) \subseteq S$

Exercise 3.89 (Determinant of symmetric operator)
The determinant of symmetric operator is equal to the product of its eigenvalues.

2.5.3 Quadratic Forms

Let X be a Euclidean space. A functional $Q: X \to \Re$ is called a *quadratic form* if there exists a symmetric linear operator $f: X \to X$ such that

$$Q(\mathbf{x}) = \mathbf{x}^T f(\mathbf{x}) \qquad \text{for every } \mathbf{x} \in X$$

Quadratic forms are amongst the simplest nonlinear functionals we encounter. They play an important role in optimization (chapter 5).

Example 3.32 The function

$$Q(x_1, x_2) = x_1^2 + 4x_1 x_2 + x_2^2$$

is a quadratic form on $\Re^2$. The matrix

$$A = \begin{pmatrix} 1 & 2 \\ 2 & 1 \end{pmatrix}$$

defines a symmetric linear operator on $\Re^2$,

$$f(x_1, x_2) = \begin{pmatrix} 1 & 2 \\ 2 & 1 \end{pmatrix} \begin{pmatrix} x_1 \\ x_2 \end{pmatrix} = \begin{pmatrix} x_1 + 2x_2 \\ 2x_1 + x_2 \end{pmatrix}$$

and

$$\mathbf{x}^T f(\mathbf{x}) = (x_1, x_2)^T \begin{pmatrix} x_1 + 2x_2 \\ 2x_1 + x_2 \end{pmatrix} = x_1^2 + 4x_1 x_2 + x_2^2 = Q(x_1, x_2)$$

As the previous example suggests, any $n \times n$ symmetric matrix $A = (a_{ij})$ of numbers defines a quadratic form by

$$Q(\mathbf{x}) = \sum_{i=1}^{n} \sum_{j=1}^{n} a_{ij} x_i x_j$$

which is usually compactly written as $Q(\mathbf{x}) = \mathbf{x}^T A \mathbf{x}$. Conversely, every quadratic form can be represented by a symmetric matrix. As with linear functions (proposition 3.1), the specific matrix which represents a given quadratic form depends upon the choice of basis for X. For a fixed basis there is a one-to-one relationship between quadratic forms Q and their representing matrices A specified by $Q(\mathbf{x}) = \mathbf{x}^T A \mathbf{x}$. Accordingly we usually do not distinguish between a quadratic form and its matrix representation.

Exercise 3.90
Let the matrix $A = (a_{ij})$ represent a linear operator f with respect to the orthonormal basis $\mathbf{x}_1, \mathbf{x}_2, \ldots, \mathbf{x}_n$. Then the sum

$$Q(\mathbf{x}) = \sum_{i=1}^{n} \sum_{j=1}^{n} a_{ij} x_i x_j$$

defines a quadratic form on X, where $x_1, x_2, \ldots, x_n$ are the coordinates of $\mathbf{x}$ relative to the basis.

Example 3.33 (Quadratic forms on $\Re^2$) The general two-dimensional quadratic form $Q: \Re^2 \to \Re$,

$$Q(x_1, x_2) = a_{11} x_1^2 + 2a_{12} x_1 x_2 + a_{22} x_2^2$$

is represented by the matrix

$$A = \begin{pmatrix} a_{11} & a_{12} \\ a_{21} & a_{22} \end{pmatrix} \quad \text{where } a_{12} = a_{21}$$

Exercise 3.91 (Principal axis theorem)
For any quadratic form $Q(\mathbf{x}) = \mathbf{x}^T A \mathbf{x}$, there exists a basis $\mathbf{x}^1, \mathbf{x}^2, \ldots, \mathbf{x}^n$ and numbers such that

$$Q(\mathbf{x}) = \lambda_1 x_1^2 + \lambda_2 x_2^2 + \cdots + \lambda_n x_n^2$$

where $\lambda_1, \lambda_2, \ldots, \lambda_n$ are the eigenvalues of A and $x_1, x_2, \ldots, x_n$ are the coordinates of $\mathbf{x}$ relative to the basis $\mathbf{x}_1, \mathbf{x}_2, \ldots, \mathbf{x}_n$.

Recall that a functional is definite if it takes only positive or negative values (section 2.1.1). Definite quadratic forms are important in practice. However, no quadratic form Q can be strictly definite (exercise 3.93). Consequently we say that quadratic form $Q: X \to \Re$ is

$$\begin{Bmatrix} \text{positive} \\ \text{negative} \end{Bmatrix} \text{ definite if } \begin{Bmatrix} Q(\mathbf{x}) > 0 \\ Q(\mathbf{x}) < 0 \end{Bmatrix} \text{ for every } \mathbf{x} \neq \mathbf{0} \text{ in } X$$

Similarly it is

$$\begin{Bmatrix} \text{nonnegative} \\ \text{nonpositive} \end{Bmatrix} \text{ definite if } \begin{Bmatrix} Q(\mathbf{x}) \geq 0 \\ Q(\mathbf{x}) \leq 0 \end{Bmatrix} \text{ for every } \mathbf{x} \text{ in } X$$

Otherwise, the quadratic form is called *indefinite*. Similarly a symmetric matrix is called positive (negative) definite if it represents positive (negative) definite quadratic form. That is, a symmetric matrix A is

$$\begin{Bmatrix} \text{positive} \\ \text{nonnegative} \\ \text{negative} \\ \text{nonpositive} \end{Bmatrix} \text{ definite if } \begin{Bmatrix} \mathbf{x}^T A \mathbf{x} > 0 \\ \mathbf{x}^T A \mathbf{x} \geq 0 \\ \mathbf{x}^T A \mathbf{x} < 0 \\ \mathbf{x}^T A \mathbf{x} \leq 0 \end{Bmatrix} \text{ for every } \mathbf{x} \neq \mathbf{0} \text{ in } X$$

Remark 3.8 (Semidefinite quadratic forms) It is common in economics to describe a nonnegative definite quadratic form as *positive semidefinite*. Similarly a nonpositive definite quadratic form is called *negative semidefinite*. That is, a quadratic form Q is

$$\begin{Bmatrix} \text{positive} \\ \text{negative} \end{Bmatrix} \text{ semidefinite if } \begin{Bmatrix} Q(\mathbf{x}) \geq 0 \\ Q(\mathbf{x}) \leq 0 \end{Bmatrix} \text{ for every } \mathbf{x} \text{ in } X$$

We use the former terminology as it is more descriptive.

Example 3.34 The two-dimensional quadratic form

$$Q(x_1, x_2) = a_{11} x_1^2 + 2a_{12} x_1 x_2 + a_{22} x_2^2 \tag{11}$$

is

$$\left\{\begin{matrix}\text{positive}\\ \text{negative}\end{matrix}\right\} \text{ definite if and only if } \left\{\begin{matrix}a_{11} > 0\\ a_{11} < 0\end{matrix}\right\} \text{ and } a_{11}a_{22} > a_{12}^2 \qquad (12)$$

It is

$$\left\{\begin{matrix}\text{nonnegative}\\ \text{nonpositive}\end{matrix}\right\} \text{ definite if and only if } \left\{\begin{matrix}a_{11}, a_{22} \geq 0\\ a_{11}, a_{22} \leq 0\end{matrix}\right\} \text{ and } a_{11}a_{22} \geq a_{12}^2$$
$$(13)$$

Exercise 3.92

1. Show that the quadratic form (11) can be rewritten as

$$Q(x_1, x_2) = a_{11}\left(x_1 + \frac{a_{12}}{a_{11}}x_2\right)^2 + \left(\frac{a_{11}a_{22} - a_{12}^2}{a_{11}}\right)x_2^2$$

assuming that $a_{11} \neq 0$. This procedure is known as "completing the square."

2. Deduce (12).
3. Deduce (13).

This is an example of the principal axis theorem (exercise 3.91).

Exercise 3.93
Show that $Q(\mathbf{0}) = 0$ for every quadratic form Q.

Since every quadratic form passes through the origin (exercise 3.93), a positive definite quadratic form has a unique minimum (at **0**). Similarly a negative definite quadratic form has a unique maximum at **0**. This hints at their practical importance in optimization. Consequently we need criteria to identify definite quadratic forms and matrices. Example 3.34 provides a complete characterization for 2×2 matrices. Conditions for definiteness in higher-dimensional spaces are analogous but more complicated (e.g., see Simon and Blume 1994, pp. 375–386; Sundaram 1996, pp. 50–55; Takayama 1985, pp. 121–123; Varian 1992, pp. 475–477.) Some partial criteria are given in the following exercises.

Exercise 3.94
A positive (negative) definite matrix is nonsingular.

Exercise 3.95
A positive definite matrix $A = (a_{ij})$ has a positive diagonal, that is,

A positive definite $\Rightarrow a_{ii} > 0$ for every i

One of the important uses of eigenvalues is to characterize definite matrices, as shown in the following exercise.

Exercise 3.96
A symmetric matrix is

$$\begin{Bmatrix} \text{positive} \\ \text{nonnegative} \\ \text{negative} \\ \text{nonpositive} \end{Bmatrix} \text{ definite if and only if all eigenvalues are } \begin{Bmatrix} \text{positive} \\ \text{nonnegative} \\ \text{negative} \\ \text{nonpositive} \end{Bmatrix}$$

Exercise 3.97
A nonnegative definite matrix A is positive definite if and only if it is nonsingular.

3.6 Systems of Linear Equations and Inequalities

Many economic models are linear, comprising a system a linear equations or inequalities

$$a_{11}x_1 + a_{12}x_2 + \cdots + a_{1n}x_n = c_1$$
$$a_{21}x_1 + a_{22}x_2 + \cdots + a_{2n}x_n = c_2$$
$$\vdots$$
$$a_{m1}x_1 + a_{m2}x_2 + \cdots + a_{mn}x_n = c_m$$

or

$$a_{11}x_1 + a_{12}x_2 + \cdots + a_{1n}x_n \leq c_1$$
$$a_{21}x_1 + a_{22}x_2 + \cdots + a_{2n}x_n \leq c_2$$
$$\vdots$$
$$a_{m1}x_1 + a_{m2}x_2 + \cdots + a_{mn}x_n \leq c_m$$

3.6 Systems of Linear Equations and Inequalities

Solving the model requires finding values for the variables $x_1, x_2, \ldots, x_n$ that satisfy the m equations or inequalities simultaneously. A linear model is called *consistent* if has such a solution. Otherwise, the model is called *inconsistent*.

Matrices can be used to represent these linear systems more compactly, as in

$$A\mathbf{x} = \mathbf{c} \quad \text{or} \quad A\mathbf{x} \leq \mathbf{c} \tag{14}$$

where

$$A = \begin{pmatrix} a_{11} & a_{12} & \cdots & a_{1n} \\ a_{21} & a_{22} & \cdots & a_{2n} \\ \vdots & \vdots & \ddots & \vdots \\ a_{m1} & a_{m2} & \cdots & a_{mn} \end{pmatrix}$$

is a matrix of *coefficients*, $\mathbf{x} = (x_1, x_2, \ldots, x_n) \in \Re^n$ is a list of the *variables*, and $\mathbf{c} = (c_1, c_2, \ldots, c_n) \in \Re^m$ is a list of *constants*.

Example 3.35 (Leontief input-output model) Consider the linear production model (example 1.103) with n commodities in which

- each activity produces only one output
- each commodity is an output of only one activity.

Let $\mathbf{a}_i = (a_{i1}, a_{i2}, \ldots, a_{in})$ denote the production plan for producing one unit of commodity i. Then a_{ij}, $i \neq j$ represents the quantity of commodity j required to produce one unit of commodity i. By definition, $a_{ii} = 1$ for every i. ($a_{ii} = 1$ is the net output of i in activity i, after allowing for any use of good i in producing itself.) Since each activity produces only one input, $a_{ij} \leq 0$ for every $i \neq j$.

Let x_i denote the scale or intensity of activity i. Then x_i denotes the *gross output* of commodity i and $a_{ij}x_i$ is the amount of good j required to produce x_i units of i. However, each commodity is used in the production of other goods. If each of the n activities is operated at scale x_j, the *net output* of good i is

$$y_i = \sum_{j=1}^{n} a_{ij} x_j$$

and the total net output of the economy is

$$\mathbf{y} = A\mathbf{x} \qquad (15)$$

where A is the $n \times n$ matrix whose rows $\mathbf{a}_i$ comprise the basic activities. It is called the *technology matrix*. A particular net output $\mathbf{y}$ will be feasible provided there exists a nonnegative solution to the linear system (15). That is, feasibility requires an intensity vector $\mathbf{x}$ that satisfies the system of equations and inequalities

$$A\mathbf{x} = \mathbf{y}, \qquad \mathbf{x} \geq 0$$

We note that most presentations of the Leontief input–output model start with a nonnegative matrix $\hat{A}$ listing the input requirements to produce one unit of each output (e.g., Simon and Blume 1994, pp. 110–13; Gale 1960). Then the technology matrix A is given by $I - \hat{A}$, where I is the $n \times n$ identity matrix.

It is often fruitful to view a linear model as a linear function $f(\mathbf{x}) = A\mathbf{x}$ from $\mathfrak{R}^n$ to $\mathfrak{R}^m$ (example 3.3). In this section we catalog some of the implications of the theory of linear functions for linear models such as (14), dealing in turn with equation and inequalities.

3.6.1 Equations

A vector $\mathbf{x}$ will be a solution of the system of equations

$$A\mathbf{x} = \mathbf{c} \qquad (16)$$

if and only if f maps $\mathbf{x}$ into $\mathbf{c}$. Consequently the linear model (16) will have a solution if and only if $\mathbf{c} \in f(X)$, the image of X. For any $c \in f(X)$ the set of all solutions to (16) is simply the inverse image of $\mathbf{c}$, $f^{-1}(\mathbf{c})$.

A special case of a linear system occurs when $\mathbf{c} = \mathbf{0}$. Such a system is called *homogeneous*. The set of solutions to the homogeneous linear system

$$A\mathbf{x} = \mathbf{0} \qquad (17)$$

is the kernel of the linear mapping $f(\mathbf{x}) = A\mathbf{x}$. We know that the kernel of any linear mapping is a subspace (exercise 3.17), which implies that

- $\mathbf{0}$ is always a solution of (17). It is called the *trivial solution*.
- If $\mathbf{x}_1, \mathbf{x}_2$ are solutions of (17), then their sum $\mathbf{x}_1 + \mathbf{x}_2$ is also a solution.
- The homogeneous system has a nontrivial solution $\mathbf{x} \neq \mathbf{0}$ if and only rank $f =$ rank $A < n$.

Exercise 3.98
Verify these assertions directly.

The general linear equation system (16) with $\mathbf{c} \neq \mathbf{0}$ is called a *nonhomogeneous* system of equations. The set of solutions to a nonhomogeneous system form an affine subset of X. This implies that

- If $\mathbf{x}_1, \mathbf{x}_2$ are solutions of (16), then their difference $\mathbf{x}_1 - \mathbf{x}_2$ is a solution of the corresponding homogeneous system (17).
- The set of all solutions to the nonhomogeneous system 16 takes the form $\mathbf{x}_p + K$, where $\mathbf{x}_p$ is any *particular solution* to the nonhomogeneous system 16 and K is the set of all solutions to the corresponding homogeneous system (17) (the kernel of f).
- if $\mathbf{0}$ is the only solution of homogeneous system (17), then the nonhomogeneous system (16) has a unique solution.

Exercise 3.99
Verify these assertions directly.

Exercise 3.100
The set of solutions to a nonhomogeneous system of linear equations $A\mathbf{x} = \mathbf{c}$ is an affine set.

The converse is also true.

Exercise 3.101
Every affine set in $\Re^n$ is the solution set of a system of a linear equations.

We conclude that there are three possible cases for the number of solutions to a linear equation system. A system of linear equations (16) may have

No solution $\mathbf{c} \notin f(\Re^n)$

A unique solution $\mathbf{c} \in f(\Re^n)$ and kernel $f = \{\mathbf{0}\}$ (or rank $A = n$)

An infinity of solutions $\mathbf{c} \in f(\Re^n)$ and kernel $f \neq \{\mathbf{0}\}$ (rank $A < n$)

In the first case the system is inconsistent, and there is not much more to said. The second and third cases are consistent systems.

$\mathbf{x} = (x_1, x_2, \ldots, x_n)$ is a solution of the linear system $A\mathbf{x} = \mathbf{c}$ if and only if

$$\begin{pmatrix} c_1 \\ c_2 \\ \vdots \\ c_m \end{pmatrix} = x_1 \begin{pmatrix} a_{11} \\ a_{21} \\ \vdots \\ a_{m1} \end{pmatrix} + x_2 \begin{pmatrix} a_{12} \\ a_{22} \\ \vdots \\ a_{m2} \end{pmatrix} + \cdots + x_n \begin{pmatrix} a_{1n} \\ a_{2n} \\ \vdots \\ a_{mn} \end{pmatrix}$$

that is,

$$\mathbf{c} = x_1 A_1 + x_2 A_2 + \cdots + x_n A_n \tag{18}$$

where A_i are the columns of A. Thus the system (16) is consistent if and only if $\mathbf{c} \in \lin\{A_1, A_2, \ldots, A_n\}$, which is called the *column space* of A. Furthermore the equation system $A\mathbf{x} = \mathbf{c}$ has a solution for every $\mathbf{c} \in \Re^m$ provided that the columns of A span $\Re^m$. This requires that rank $A = m$, the number of equations.

In the third case (multiple solutions) there are fewer equations than variables, and the system is said to be *underdetermined*. Practitioners are most interested in the solutions with the fewest number of nonzero components. If rank $A = m < n$, any $\mathbf{c} \in \Re^m$ can be expressed as a linear combination of at most m columns which form a basis for $\Re^m$. That is, there exist solutions (18) with $x_j \neq 0$ for at most m columns of A. A solution with at most m nonzero components is called a *basic solution* of (16), since the nonzero components corresponds to the elements of a basis for $\Re^m$.

Exercise 3.102
Prove that the linear equation system

$$x_1 + 3x_2 = c_1$$

$$x_1 - x_2 = c_2$$

has a unique solution for every choice of c_1, c_2.

Exercise 3.103 (Cramer's rule)
Let $A\mathbf{x} = \mathbf{c}$ be a linear equation system with A a nonsingular square $n \times n$ matrix (rank $A = n$). For every $\mathbf{c} \in \Re^n$ there exists a unique solution $\mathbf{x} = (x_1, x_2, \ldots, x_n)$ given by

$$x_j = \frac{\det(B_j)}{\det(A)}, \quad j = 1, 2, \ldots, n$$

where

$$B_j = \begin{pmatrix} a_{11} & \dots & c_1 & \dots & a_{1n} \\ \vdots & & \vdots & & \vdots \\ a_{n1} & \dots & c_n & \dots & a_{nn} \end{pmatrix}$$

is the matrix obtained by replacing the jth column of A with $\mathbf{c}$. [Hint: Subtract $\mathbf{c}$ from the jth column of A and apply exercise 3.81.]

Cramer's rule (exercise 3.103) is not a practical method of solving large systems of equations. However, it can be used to analyze how the solution $\mathbf{x}$ varies with changes in $\mathbf{c}$. It is an important tool comparative statics (example 6.14).

Exercise 3.104
Show that

$$\begin{pmatrix} a & b \\ c & d \end{pmatrix}^{-1} = \frac{1}{\Delta} \begin{pmatrix} d & -b \\ -c & a \end{pmatrix}$$

where $\Delta = \det(A) = ad - bc$.

Example 3.36 (Portfolio investment) Example 3.4 introduced a simple linear model of portfolio investment comprising A risky assets or securities and S states of the world. If r_{sa} denotes the return of asset a in state s and $\mathbf{x} = (x_1, x_2, \dots, x_A)$ is a list of amounts invested in different assets, the total return $f(\mathbf{x})$ of a portfolio $\mathbf{x}$ is given by

$$f(\mathbf{x}) = R\mathbf{x} = \begin{pmatrix} \sum_{a=1}^{A} r_{1a} x_a \\ \sum_{a=1}^{A} r_{2a} x_a \\ \vdots \\ \sum_{a=1}^{A} r_{Sa} x_a \end{pmatrix}$$

where

$$R = \begin{pmatrix} r_{11} & r_{12} & \dots & r_{1A} \\ r_{21} & r_{22} & \dots & r_{2A} \\ \multicolumn{5}{c}{\dotfill} \\ r_{S1} & r_{S2} & \dots & r_{SA} \end{pmatrix}$$

is the matrix of prospective returns. The sth component of $f(\mathbf{x})$, $\sum_{a=1}^{A} r_{sa} x_a$, is the total return of portfolio $\mathbf{x}$ in state s.

Note that it is allowable for x_a to be negative for some assets. If $x_a > 0$, then the investor holds a *long* position in asset a, entitling her to receive $r_{sa}x_a$ if state s pertains. On the other hand, a negative x_a indicates a *short position*, in which the investor effectively borrows x_a units of assets a and promises to pay back $r_{sa}x_a$ in state s.

A portfolio **x** is called *riskless* if it provides the same rate of return in every state, that is,

$$\sum_{a=1}^{A} r_{1a}x_a = \sum_{a=1}^{A} r_{2a}x_a = \cdots = \sum_{a=1}^{A} r_{Sa}x_a = \bar{r}$$

In other words, **x** is a riskless portfolio if it satisfies the equation

$$R\mathbf{x} = \bar{r}\mathbf{1}$$

for some $\bar{r} \in \Re$ where $\mathbf{1} = (1, 1, \ldots, 1)$. A sufficient condition for the existence of a riskless portfolio is that rank $R = S$, that is,

- There are at least as many assets as states ($A \geq S$).
- The prospective returns of at least S assets are linearly independent.

In other words, the existence of a riskless portfolio is guaranteed provided that there are a sufficient number of assets whose returns are independent across states.

Exercise 3.105
A portfolio is called *duplicable* if there is a different portfolio $\mathbf{y} \neq \mathbf{x}$ which provides exactly the same returns in every state, that is, $R\mathbf{x} = R\mathbf{y}$ or

$$\sum_{a=1}^{n} r_{sa}x_a = \sum_{a=1}^{n} r_{sa}y_a \quad \text{for every } s \in S$$

Show that every portfolio is duplicable if rank $R < A$.

Exercise 3.106
A state $\bar{s}$ is called *insurable* if there exists a portfolio **x** which has a positive return if state $\bar{s}$ occurs and zero return in any other state, that is,

$$\sum_{a=1}^{A} r_{\bar{s}a}x_a > 0 \quad \text{and} \quad \sum_{a=1}^{A} r_{sa}x_a = 0, \quad s \neq \bar{s}$$

Show that every state is insurable if and only if rank $R = S$.

Example 3.37 (Arrow-Debreu securities) Recall that *Arrow-Debreu* securities are hypothetical financial assets that pay $1 if and only if a particular state of the world occurs (example 1.82). Therefore the payoff profile of the s Arrow-Debreu security is $\mathbf{e}_s = (0,\ldots,1,\ldots,0)$, where the 1 occurs in the location s. Suppose that there is a full set of Arrow-Debreu securities, that is there exists an Arrow-Debreu security $\mathbf{e}_s$ for every state s, $s = 1, 2, \ldots S$. Then $A \geq S$ and rank $R = S$. From the preceding exercises, we conclude that

- there exists a riskless portfolio
- every portfolio is duplicable
- every state is insurable

Indeed, any pattern of payoffs (across different states) can be constructed by an appropriate portfolio of Arrow-Debreu securities. (The Arrow-Debreu securities span the payoff space $f(\Re^A) = \Re^S$.)

Assuming that investors only care about the final distribution of wealth, any two portfolios that provide the same pattern of returns must have the same value. Therefore, in equilibrium, the price of any financial asset a must be equal to the value of the corresponding portfolio of Arrow-Debreu securities that yield the same distribution of payoffs. Consequently any security can be valued if we know the price of each Arrow-Debreu security. That is, if p_a is the price of security a with payoff vector $(r_{1a}, r_{2a}, \ldots, r_{Sa})$ and π_s is the price of the s Arrow-Debreu security, then in equilibrium

$$p_a = \sum_{s=1}^{S} r_{sa} \pi_s$$

A single linear equation

$$a_{i1}x_1 + a_{i2}x_2 + \cdots + a_{in}x_n = c_i$$

defines a hyperplane in $\Re^n$ (example 3.21), and it is often convenient to think of a system of linear equations (16) as a finite collection of hyperplanes. The solution to the system corresponds to the intersection of these hyperplanes. Figure 3.5 illustrates the possibilities for three equations in three unknowns. Each equation defines a plane in $\Re^3$. These planes may intersect in a single point (unique solution), a line, a plane, or not intersect at all (no solution).

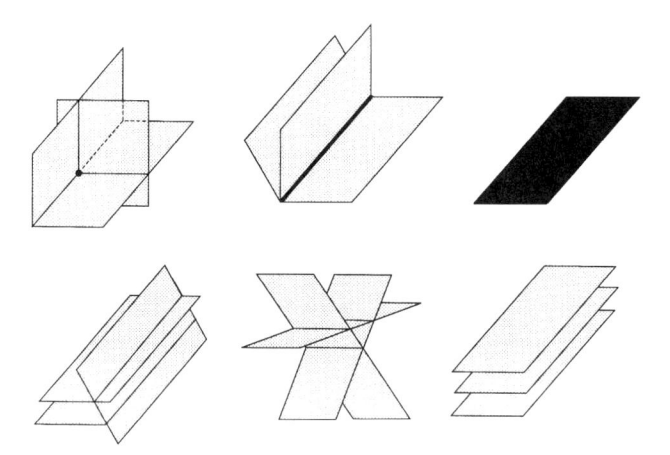

Figure 3.5
The solutions of three equations in three unknowns

Exercise 3.107
Draw analogous diagrams illustrating the possible cases for a system of three equations in two unknowns.

Exercise 3.108
Every affine subset of $\Re^n$ is the intersection of a finite collection of hyperplanes.

3.6.2 Inequalities

A solution to system of linear inequalities

$$A\mathbf{x} \leq \mathbf{c} \tag{19}$$

is a vector $\mathbf{x} = (x_1, x_2, \ldots, x_n)$ that satisfies the inequalities simultaneously. While the solution set of a system of equations is an affine set (subspace when $\mathbf{c} = \mathbf{0}$), the set of solutions to a system of linear inequalities (19) is a convex set. When the system is homogeneous, the set of solutions is a convex cone.

Exercise 3.109
The set of solutions to a system of linear inequalities $A\mathbf{x} \leq \mathbf{c}$ is a convex set.

Exercise 3.110
The set of solutions to a homogeneous system of linear inequalities $A\mathbf{x} \leq \mathbf{0}$ is a convex cone.

Each inequality $\mathbf{a}_i^T \mathbf{x} \leq c_i$ defines a halfspace in $\Re^n$ (section 3.9). Therefore the set of solutions $S = \{x : Ax \leq c\}$, which satisfy a system of linear inequalities is the intersection of the m halfspaces. We will show later (section 3.9.2) that this implies that the solution set S is a particularly simple convex set, a polytope which is the convex hull of a finite number of points.

Example 3.38 Consider the following system of linear inequalities:

$3x_1 + 8x_2 \leq 12$

$x_1 + x_2 \leq 2$

$2x_1 \leq 3$

Each of the inequalities defines a halfspace in $\Re^2$. For example, the set of all points satisfying the inequality $3x_1 + 8x_2 \leq 12$ is the region below and to the left of the line (figure 3.6a) $3x_1 + 8x_2 = 12$. The set of points satisfying all three inequalities simultaneously is the set that lies below and to the left of the three lines

$3x_1 + 8x_2 = 12$

$x_1 + x_2 = 2$

$2x_1 = 3$

This is the shaded region in figure 3.6b. Frequently we are only concerned with nonnegative solutions to a system of inequalities. This is the set bounded by the axes and the lines (hyperplanes) associated with each of the inequalities. It is the shaded set in figure 3.6c

Any system of inequalities can be transformed into a equivalent system of equations and nonnegativity conditions, by adding another variable to each equations. For example, the inequality system

$a_{11}x_1 + a_{12}x_2 + \cdots + a_{1n}x_n \leq c_1$

$a_{21}x_1 + a_{22}x_2 + \cdots + a_{2n}x_n \leq c_2$

$$\vdots$$

$a_{m1}x_1 + a_{m2}x_2 + \cdots + a_{mn}x_n \leq c_m$

is equivalent to the system of equations

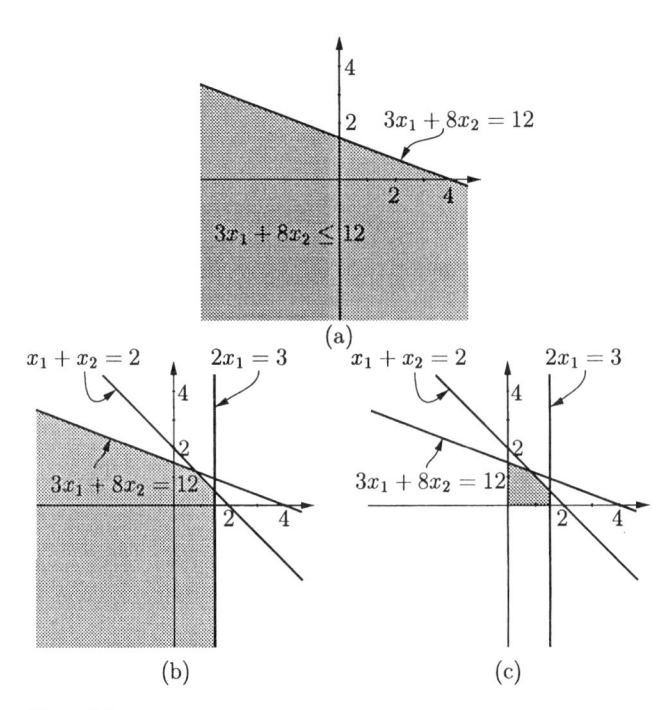

Figure 3.6
Systems of inequalities

$$a_{11}x_1 + a_{12}x_2 + \cdots + a_{1n}x_n + x_{n+1} = c_1$$
$$a_{21}x_1 + a_{22}x_2 + \cdots + a_{2n}x_n + x_{n+2} = c_2$$
$$\vdots$$
$$a_{m1}x_1 + a_{m2}x_2 + \cdots + a_{mn}x_n + x_{n+m} = c_m$$

(20)

together with the requirement

$$x_{n+1} \geq 0, \quad x_{n+2} \geq 0, \ldots, x_{n+m} \geq 0$$

The additional variables x_{n+i} are called *slack variables*, since they measure the degree of slack in the corresponding ith inequality. This transformation of inequalities to equations is especially common in optimization techniques, such as linear programming (section 5.4.4). The transformed systems has two important characteristics:

- There are fewer equations than variables.
- Some of the variables are restricted to be nonnegative.

Commonly the original variables $x_1, x_2, \ldots, x_n$ are also required to be nonnegative, and linear systems of the form

$$A\mathbf{x} = \mathbf{c} \tag{21}$$

$$\mathbf{x} \geq 0 \tag{22}$$

are especially prevalent in practice. Since the system (21) necessarily has more variables $m + n$ than equations m, it will usually be underdetermined. If there exists any feasible solution, there will be multiple solutions. The simplest of these solutions will be basic feasible solutions.

Earlier we showed that it is always possible to reduce a solution

$$\mathbf{c} = x_1 A_1 + x_2 A_2 + \cdots + x_n A_n$$

to a linear equation system (21) by eliminating redundant columns, reducing the number of nonzero components of $\mathbf{x}$ to m. However, it is not clear that this reduction can be done without violating the nonnegativity constraint (22). In the following exercise we show that any feasible solution to (21) and (22) can be reduced to a basic feasible solution. This result, which has important practical consequences, is often called the *fundamental theorem of linear programming*.

Exercise 3.111 (*Fundamental theorem of linear programming*)
Let $\mathbf{x}$ be a feasible solution to the linear system

$$A\mathbf{x} = \mathbf{c}, \quad \mathbf{x} \geq 0 \tag{23}$$

where A is an $m \times n$ matrix and $\mathbf{c} \in \Re^m$. Then

$$\mathbf{c} = x_1 A_1 + x_2 A_2 + \cdots + x_n A_n$$

where $A_i \in \Re^m$ are the columns of A and and $x_i \geq 0$ for every i. Without loss of generality, assume that the first k components are positive and the rest are zero, that is,

$$\mathbf{c} = x_1 A_1 + x_2 A_2 + \cdots + x_k A_k$$

with $k \leq n$ and $x_i > 0$ for every $a = 1, 2, \ldots k$.

1. The columns $\{A_1, A_2, \ldots, A_k\}$ are vectors in $\Re^m$. If the columns $\{A_1, A_2, \ldots, A_k\}$ are linearly independent, then $k \leq m$ and there exists a basic feasible solution.

2. If the vectors $\{A_1, A_2, \ldots, A_k\}$ are linearly dependent,

a. There exists a nonzero solution to the homogeneous system

$$y_1 A_1 + y_2 A_2 + \cdots + y_k A_k = \mathbf{0}$$

b. For $t \in \Re$ define $\hat{\mathbf{x}} = \mathbf{x} - t\mathbf{y}$. $\hat{\mathbf{x}}$ is a solution to the nonhomogeneous system

$$A\mathbf{x} = \mathbf{c}$$

c. Let

$$t = \min_j \left\{ \frac{x_j}{y_j} : y_j > 0 \right\}$$

Then $\hat{\mathbf{x}} = \mathbf{x} - t\mathbf{x}$ is a feasible solution, that is, $\hat{\mathbf{x}} \geq 0$.

d. There exists h such that

$$\mathbf{c} = \sum_{\substack{j=1 \\ j \neq h}}^{k} \hat{x}_j A_J$$

$\hat{\mathbf{x}}$ is a feasible solution with one less positive component.

3. If there exists a feasible solution, there exists a basic feasible solution.

Remark 3.9 Later we will show that the basic feasible solutions of nonnegative system like (23) correspond to the extreme points of the convex solution set, and that any optimal solution of a linear program will occur at an extreme point. Therefore the search for optimal solutions to a linear program can be confined to extreme points. The simplex algorithm for linear programming is an efficient method for moving from one basic feasible solution to another.

The fundamental theorem can be used to give an elegant and straightforward proof of the Shapley-Folkman theorem (remark 1.20), as outlined in the following exercise.

Exercise 3.112 (Shapley-Folkman theorem)
Let $\{S_1, S_2, \ldots, S_n\}$ be a collection of nonempty (possibly nonconvex) subsets of an m-dimensional linear space, and let $\mathbf{x} \in \text{conv} \sum_{i=1}^{n} S_i$. Then

1. $\mathbf{x} = \sum_{i=1}^n \mathbf{x}_i$, where $\mathbf{x}_i \in \text{conv } S_i$.
2. $\mathbf{x} = \sum_{i=1}^n \sum_{j=1}^{l_i} a_{ij} x_{ij}$, where $x_{ij} \in S_i$, $a_{ij} > 0$, $\sum_{j=1}^{l_i} a_{ij} = 1$.
3. $\mathbf{z} = \sum_{i=1}^n \sum_{j=1}^{l_i} a_{ij} z_{ij}$, where

$$\mathbf{z} = \begin{pmatrix} \mathbf{x} \\ 1 \end{pmatrix}, \quad z_{ij} = \begin{pmatrix} x_{ij} \\ \mathbf{e}_i \end{pmatrix}, \quad \mathbf{1}, \mathbf{e}_i \in \Re^n$$

4. $\mathbf{z} = \sum_{i=1}^n \sum_{j=1}^{l_i} b_{ij} z_{ij}$ with $b_{ij} \geq 0$ and $b_{ij} > 0$ for at most $m+n$ components.
5. Define $\hat{\mathbf{x}}_i = \sum_{j=1}^{l_i} b_{ij} x_{ij}$. Then $\hat{\mathbf{x}}_i = \text{conv } S_i$ and $\mathbf{x} = \sum_{i=1}^n \hat{\mathbf{x}}_i$. Show that all but at most $\hat{\mathbf{x}}_i$ actually belong to S_i.

3.6.3 Input–Output Models

In the input-output model (example 3.35), a necessary condition for a given net output $\mathbf{y}$ to be feasible is that technology matrix A is nonsingular (rank $A = n$). However, nonsingularity is not sufficient, since it does not guarantee that the corresponding intensity vector

$$\mathbf{x} = A^{-1} \mathbf{y}$$

is nonnegative. An input-output system A is said to be *productive* if it is capable of producing a positive amount of all commodities, that is, if the inequality system

$$A\mathbf{x} > 0$$

has any nonnegative solution. In the following exercises, we show that

- A necessary and sufficient condition for input-output system A to have a nonnegative solution for any output $\mathbf{y}$ is that A is productive.
- The system A is productive if and only if A is nonsingular and A^{-1} is nonnegative.

Remark 3.10 The first conclusion states that if there is *any* feasible way of producing positive quantities of all commodities, then it is possible to produce any output vector. In other words, it is possible to produce arbitrarily large quantities of any of the goods in any proportions. While this is somewhat surprising, we should recall that there are no resources constraints and the system is entirely self-contained. Real economies have resource constraints that limit the quantity of feasible outputs (example 3.39).

Exercise 3.113
Assume that A is productive. Show that

1. $A\mathbf{z} \geq 0$ implies $\mathbf{z} \geq 0$
2. A is nonsingular
3. for every $\mathbf{y} \geq 0$, the system $A\mathbf{x} = \mathbf{y}$ has a unique nonnegative solution

[Hint: Consider the matrix $B = I - A \geq \mathbf{0}$.]

Exercise 3.114
The system A is productive if and only if A^{-1} exists and is nonnegative.

The essential characteristic of the technology matrix is that its off-diagonal elements are nonpositive. Any $n \times n$ matrix A is called a *Leontief matrix* if $a_{ij} \leq 0$, $i \neq j$. In the following exercise we extend the properties of input-output system to arbitrary Leontief matrices.

Exercise 3.115 (Leontief matrices)
Let A be an $n \times n$ matrix with $a_{ij} \leq 0$, $i \neq j$. Then the following conditions are mutually equivalent:

1. There exists some $\mathbf{x} \in \Re_+^n$ such that $A\mathbf{x} > 0$.
2. For any $\mathbf{c} \in \Re_+^n$, there exists an $\mathbf{x} \in \Re_+^n$ such that $A\mathbf{x} = \mathbf{c}$.
3. A is nonsingular and $A^{-1} \geq 0$.

Example 3.39 (Primary inputs) An input that is not produced by any activity is called a *primary input*. The standard Leontief model has a single primary input, which is required by all activities. It is conventionally called "labor." If there are primary inputs, then the economy cannot produce arbitrarily large quanities of output. However, it can still produce in arbitrary proportions if the technology is productive.

Exercise 3.116
Augment the input-output model to include a primary commodity ("labor"). Let a_{0j} denote the labor required to produce one unit of commodity j. Show that there exists a price system $\mathbf{p} = (p_1, p_2, \ldots, p_n)$ such that the profit of each activity (industry) is zero.

3.6.4 Markov Chains

A *stochastic process* is a dynamical system (example 2.22) in which the transitions from state to state are random. A *Markov chain* is a discrete stochastic process in which

- the state space is finite
- the probabilities of transitions from one state to another are fixed and independent of time

Let $S = \{s_1, s_2, \ldots, s_n\}$ denote the finite set of states. Let t_{ij} denote the probability that if the system is in state j at some period, it will move to state i in the next period. t_{ij} is called the *transition probability* from state j to i. Since t_{ij} is a probability, we have

$$0 \le t_{ij} \le 1, \quad i,j = 1, 2, \ldots, n$$

Furthermore, since the system must be in some state $s_i \in S$ at every period,

$$t_{1j} + t_{2j} + \cdots + t_{nj} = 1$$

The vector $t_j = (t_{1j}, t_{2j}, \ldots, t_{nj})$ is the probability distribution of the state of the system at time $t+1$ given that it is in state j in period t.

The important assumption of the model is that the transition probabilities t_{ij} are constant through time, so the state of the system at time $t+1$ depends only on the state at time t (and not on the state at any earlier time). This is called the *Markov* assumption. A stochastic model with this assumption is called a *Markov process*. A Markov process with a finite number of states is called a *Markov chain*.

Let $T = (t_{ij})$ be the matrix of transition probabilities. T is called the *transition matrix*. By construction, the transition matrix is nonnegative. Furthermore the entries in each column sum to 1. Any matrix with these properties is called a *Markov matrix* or *stochastic matrix*.

At any point of time, we can describe the state of the stochastic system by the probability that it is any given state. Let $(p_1^t, p_2^t, \ldots, p_n^t)$, $\sum p_j^t = 1$, denote the probability distribution of states at time t. p_j^t is the probability that the system is in state s_j at time t. Given the distribution p^t of states at time t, the expected distribution at time $t+1$ is

$$p^{t+1} = Tp^t$$

Furthermore the distribution at time $t+k$ is

$$p^{t+k} = T^k p^t$$

where $T^k = \overbrace{TT\ldots T}^{k \text{ times}}$. The stochastic behavior of the system is entirely determined by the transition matrix T. The fact that T is a stochastic matrix circumscribes the possible behavior of the dynamical system.

While it is convenient to analyze this as a standard linear dynamical system, it should be emphasized that **p** in not really the state of Markov process. At any point in time the process is in one of the n distinct states $s_1, s_2, \ldots, s_n$. The vector **p** lists the probabilities that the process is in the various states. A distribution **p** is called a *stationary distribution* of the Markov chain with transition matrix T if

$$\mathbf{p} = T\mathbf{p}$$

that is, **p** is a fixed point of the linear mapping defined by T. In chapter 2 we used the Brouwer fixed point theorem to show that every Markov chain has a stationary distribution (example 2.94). Example 3.87 gives an alternative proof based on the separating hyperplane theorem.

Example 3.40 (Labor market turnover) Hall (1972) modeled turnover in the US labor force as a Markov process. Using survey data in 1966, he estimated that a 30-year-old married white male employee living in New York had a 0.22 percent chance of becoming unemployed in any given week. A similar unemployed male had a 13.6 percent chance of obtaining another job in the same period. This implied the transition probabilities listed in table 3.1. Similar estimates were obtained for a range of different categories based on age, gender, race, and location. Hall used these estimated probabilities in a simple Markov model to explain the differences in unemployment rates of different groups. For example, the higher un-

Table 3.1
Transition probabilities in the U.S. labor force

| | Currently | |
Remaining/becoming	Employed	Unemployed
Black males		
Employed	0.9962	0.1025
Unemployed	0.0038	0.8975
White males		
Employed	0.9978	0.1359
Unemployed	0.0022	0.8641

employment rate experienced by black men can be attributed to their higher probability of becoming unemployed in any period as well as a lower probability of becoming employed again. Note that the Markov assumption is very strong in this example, since the transition probabilities presumably vary through time with employment experience and the state of the labor market.

Exercise 3.117
What steady state unemployment rates are implied by the transition probabilities in table 3.1?

Exercise 3.118
A magazine maintains a mailing list containing both current subscribers and potential subscribers. Experience has shown that sending a letter to all the individuals on the list will induce 60 percent of current subscribers to renew their subscriptions. In addition the letter will sell subscriptions to 25 percent of the potential subscribers who are not actual subscribers.

1. Write out the transition matrix for this stochastic process.

2. Suppose that 40 percent of the mailing list comprise actual subscribers. How many subscriptions or renewals can be expected from another mailing.

3.7 Convex Functions

Recall that a linear functional f on a linear space X satisfies the twin conditions of additivity and homogeneity:

$f(\mathbf{x}_1 + \mathbf{x}_2) = f(\mathbf{x}_1) + f(\mathbf{x}_2)$

$f(\alpha \mathbf{x}_1) = \alpha f(\mathbf{x}_1)$ for every $\alpha \in \Re$

For many purposes in economics, linearity is too restrictive. For example, linear production functions imply constant returns to scale, and linear utility functions imply that the consumer is never satiated no matter how much she consumes of any good. Convex and homogeneous functions generalize some of the properties of linear functions, providing more suitable functional forms (figure 3.7).

A real-valued function f defined on a convex set S of a linear space X is convex if the value of the function along a line joining any two points $\mathbf{x}_1$

Linear functions
$$f(\alpha \mathbf{x}_1 + (1-\alpha)\mathbf{x}_2) = \alpha f(\mathbf{x}_1) + (1-\alpha)f(\mathbf{x}_2)$$

implies ↙   ↘ implies

Additivity
$$f(\mathbf{x}_1 + \mathbf{x}_2) = f(\mathbf{x}_1) + f(\mathbf{x}_2)$$

Homogeneity
$$f(\alpha \mathbf{x}) = \alpha f(\mathbf{x})$$

generalizes to

Concave functions
$$f(\alpha \mathbf{x}_1 + (1-\alpha)\mathbf{x}_2) \leq \alpha f(\mathbf{x}_1) + (1-\alpha)f(\mathbf{x}_2)$$

Homogeneous functions
$$f(\alpha \mathbf{x}) = \alpha^k f(\mathbf{x}), \alpha > 0$$

generalizes to

Quasiconcave functions
$$f(\alpha \mathbf{x}_1 + (1-\alpha)\mathbf{x}_2) \leq \min\{f(\mathbf{x}_1), f(\mathbf{x}_2)\}$$

Homothetic functions
$$f(\mathbf{x}_1) = f(\mathbf{x}_2) \implies f(\alpha \mathbf{x}_1) = f(\alpha \mathbf{x}_2), \alpha > 0$$

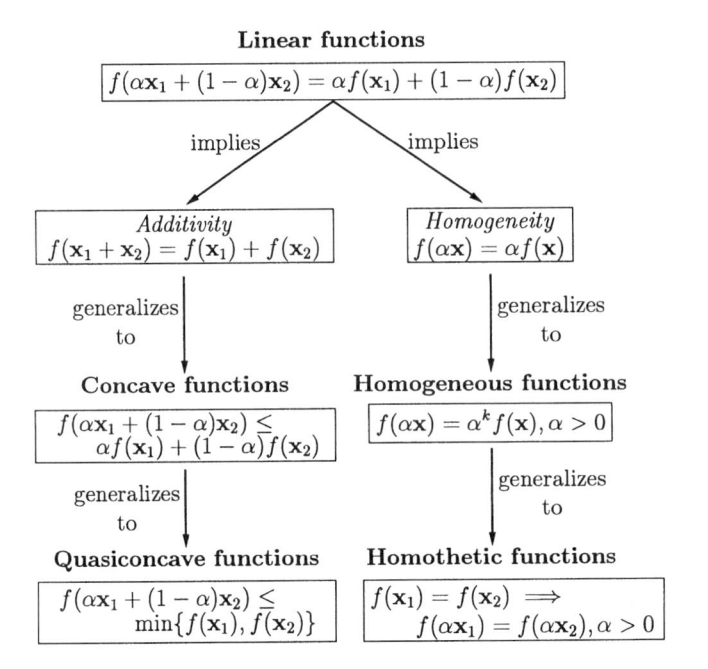

Figure 3.7
Generalizing linear functions

and $\mathbf{x}_2$ is never greater than a weighted average of its value at the two endpoints. Formally the function $f: S \to \Re$ is *convex* if for every $\mathbf{x}_1, \mathbf{x}_2$ in S

$$f(\alpha \mathbf{x}_1 + (1-\alpha)\mathbf{x}_2) \leq \alpha f(\mathbf{x}_1) + (1-\alpha)f(\mathbf{x}_2) \quad \text{for every } 0 \leq \alpha \leq 1 \quad (24)$$

This condition relaxes the additivity requirement of a linear function, and dispenses with homogeneity. A function is *strictly convex* if the inequality is strict; that is, for every $\mathbf{x}_1, \mathbf{x}_2$ in S with $\mathbf{x}_1 \neq \mathbf{x}_2$,

$$f(\alpha \mathbf{x}_1 + (1-\alpha)\mathbf{x}_2) < \alpha f(\mathbf{x}_1) + (1-\alpha)f(\mathbf{x}_2) \quad \text{for every } 0 < \alpha < 1$$

Remark 3.11 Strictly speaking, we should refer to convex *functionals*, but "convex function" is more usual (remark 2.10).

Example 3.41 Two familiar convex functions x^2 and e^x are illustrated in figure 3.8. Note how a line joining any two points on the curve lies everywhere above the curve.

3.7 Convex Functions

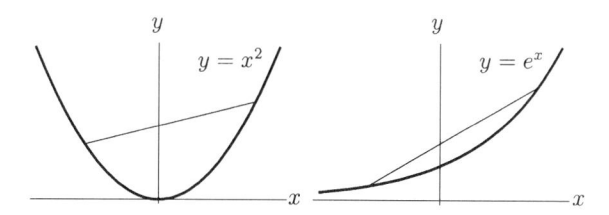

Figure 3.8
Two examples of convex functions

Exercise 3.119
Show that x^2 is convex on $\Re$.

Exercise 3.120 (Power function)
Show that the power functions $f(x) = x^n$, $n = 1, 2, \ldots$ are convex on $\Re_+$.

Example 3.42 (Profit function)   The profit function of a competitive firm (example 2.29)

$$\Pi(\mathbf{p}) = \sup_{\mathbf{y} \in Y} \sum_i p_i y_i$$

measures the maximum profit which the firm can earn given prices $\mathbf{p}$ and technology Y. To show that it is a convex function of $\mathbf{p}$, suppose that $\mathbf{y}_1$ maximizes profit at prices $\mathbf{p}_1$ and $\mathbf{y}_2$ maximizes profit at $\mathbf{p}_2$. For some $\alpha \in [0, 1]$, let $\bar{\mathbf{p}}$ be the weighted average price, that is,

$$\bar{\mathbf{p}} = \alpha \mathbf{p}_1 + (1 - \alpha) \mathbf{p}_2$$

Now suppose that $\bar{\mathbf{y}}$ maximizes profits at $\bar{\mathbf{p}}$. Then

$$\Pi(\bar{\mathbf{p}}) = \bar{\mathbf{p}}^T \bar{\mathbf{y}} = (\alpha \mathbf{p}_1 + (1 - \alpha) \mathbf{p}_2)^T \bar{\mathbf{y}} = \alpha \mathbf{p}_1^T \bar{\mathbf{y}} + (1 - \alpha) \mathbf{p}_2^T \bar{\mathbf{y}}$$

But since $\mathbf{y}_1$ and $\mathbf{y}_2$ maximize profit at $\mathbf{p}_1$ and $\mathbf{p}_2$ respectively,

$$\alpha \mathbf{p}_1^T \bar{\mathbf{y}} \leq \alpha \mathbf{p}_1^T \mathbf{y}_1 = \alpha \Pi(\mathbf{p}_1)$$
$$(1 - \alpha) \mathbf{p}_2^T \bar{\mathbf{y}} \leq (1 - \alpha) \mathbf{p}_2^T \mathbf{y}_2 = (1 - \alpha) \Pi(\mathbf{p}_2)$$

so

$$\Pi(\bar{\mathbf{p}}) = \Pi(\alpha \mathbf{p}_1 + (1 - \alpha) \mathbf{p}_2) = \alpha \mathbf{p}_1^T \bar{\mathbf{y}} + (1 - \alpha) \mathbf{p}_2^T \bar{\mathbf{y}} \leq \alpha \Pi(\mathbf{p}_1) + (1 - \alpha) \Pi(\mathbf{p}_2)$$

This establishes that the profit function Π is convex in $\mathbf{p}$.

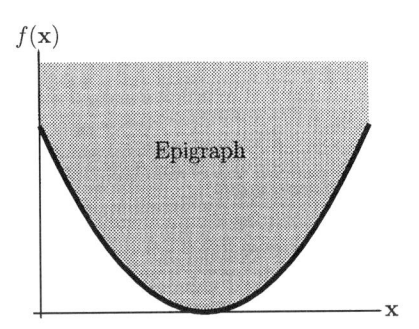

Figure 3.9
The epigraph of a convex function is a convex set

Geometrically the graph of a convex function lies below the line joining any two points of the graph. This provides an intimate and fruitful connection between convex functions and convex sets. Recall that the *epigraph* of a functional $f: X \to \Re$ is the set of all points in $X \times \Re$ on or above the graph, that is,

$$\text{epi } f = \{(\mathbf{x}, y) \in X \times \Re : y \geq f(\mathbf{x}), \mathbf{x} \in X\}$$

Convex functions are precisely those functions with convex epigraphs (figure 3.9).

Proposition 3.7 *A function* $f: S \to \Re$ *is convex if and only if* epi f *is convex.*

Proof Assume f is convex, and let $(\mathbf{x}_1, y_1), (\mathbf{x}_2, y_2) \in \text{epi } f$ so that

$$f(\mathbf{x}_1) \leq y_1 \quad \text{and} \quad f(\mathbf{x}_2) \leq y_2$$

For any $\alpha \in [0, 1]$ define

$$\bar{\mathbf{x}} = \alpha \mathbf{x}_1 + (1 - \alpha) \mathbf{x}_2, \quad \bar{y} = \alpha y_1 + (1 - \alpha) y_2$$

Since f is convex,

$$f(\bar{\mathbf{x}}) = f(\alpha \mathbf{x}_1 + (1 - \alpha) \mathbf{x}_2) \leq \alpha f(\mathbf{x}_1) + (1 - \alpha) f(\mathbf{x}_2) \leq \alpha y_1 + (1 - \alpha) y_2 = \bar{y}$$

Therefore $(\bar{\mathbf{x}}, y) = \alpha(\mathbf{x}_1, y_1) + (1 - \alpha)(\mathbf{x}_2, y_2) \in \text{epi } f$; that is, epi f is convex.

Conversely, assume that epi f is convex. Let $\mathbf{x}_1, \mathbf{x}_2 \in S$, and define

$$y_1 = f(\mathbf{x}_1) \quad \text{and} \quad y_2 = f(\mathbf{x}_2)$$

Then $(\mathbf{x}_1, y_1), (\mathbf{x}_2, y_2) \in \text{epi } f$. For any $\alpha \in [0, 1]$ define

$$\bar{\mathbf{x}} = \alpha \mathbf{x}_1 + (1 - \alpha)\mathbf{x}_2, \quad \bar{y} = \alpha y_1 + (1 - \alpha)y_2$$

Since epi f is convex,

$$(\bar{\mathbf{x}}, y) = \alpha(\mathbf{x}_1, y_1) + (1 - \alpha)(\mathbf{x}_2, y_2) \in \text{epi } f$$

and therefore $f(\bar{\mathbf{x}}) \leq \bar{y}$, that is,

$$f(\alpha \mathbf{x}_1 + (1 - \alpha)\mathbf{x}_2) \leq \alpha y_1 + (1 - \alpha)y_2 = \alpha f(\mathbf{x}_1) + (1 - \alpha)f(\mathbf{x}_2)$$

f is convex. □

Proposition 3.7 implies another useful characterization of convex functions. A function is convex if and only if every vertical cross section (section 2.1.3) is convex.

Corollary 3.7.1 (Convex cross sections) *For any $f \in F(S)$ and $\mathbf{x}_1, \mathbf{x}_2 \in S$, let $h \in F[0, 1]$ be defined by*

$$h(t) = f((1 - t)\mathbf{x}_1 + t\mathbf{x}_2)$$

Then f is convex if and only if h is convex for every $\mathbf{x}_1, \mathbf{x}_2 \in S$.

Proof

$$\text{epi } h = \{(t, y) \in [0, 1] \times \Re : h(t) \leq y\}$$

and we observe that

$$(t, y) \in \text{epi } h \Leftrightarrow (\mathbf{x}, y) \in \text{epi } f$$

where $\mathbf{x} = (1 - t)\mathbf{x}_1 + t\mathbf{x}_2$, and therefore

$$\text{epi } h \text{ is convex} \Leftrightarrow \text{epi } f \text{ is convex}$$

Therefore h is convex if and only if f is convex. □

Exercise 3.121
Prove corollary 3.7.1 directly from the definition (24) without using proposition 3.9.

Exercise 3.122 (Jensen's inequality)
A function $f: S \to \Re$ is convex if and only if

$$f(\alpha_1 \mathbf{x}_1 + \alpha_2 \mathbf{x}_2 + \cdots + \alpha_n \mathbf{x}_n) \leq \alpha_1 f(\mathbf{x}_1) + \alpha_2 f(\mathbf{x}_2) + \cdots + \alpha_n f(\mathbf{x}_n) \quad (25)$$

for all $\alpha_1, \alpha_2, \ldots, \alpha_n \geq 0$, $\sum_{i=1}^{n} \alpha_i = 1$ [Hint: Use proposition 3.7.]

Exercise 3.123
Show that

$$x_1^{\alpha_1} x_2^{\alpha_2} \ldots x_n^{\alpha_n} \leq \alpha_1 x_1 + \alpha_2 x_2 + \cdots + \alpha_n x_n$$

for every $x_1, x_2, \ldots, x_n \in \Re_+$. Deduce that the arithmetic mean of a set of positive numbers is always greater than or equal to the geometric mean, that is,

$$\bar{x} = \frac{1}{n} \sum_{i=1}^{n} x_i \geq (x_1 x_2 \ldots x_n)^{1/n}$$

[Hint: Use that fact that e^x is convex (example 3.41).]

Example 3.43 (Price stabilization) The fact that the profit function of a competitive firm is convex has some surprising ramifications. For example, it implies that price stabilization will reduce average profits. Suppose that prices are random, taking the values $(\mathbf{p}_1, \mathbf{p}_2, \ldots, \mathbf{p}_n)$ with probabilities $(\alpha_1, \alpha_2, \ldots, \alpha_n)$. On average, the competitive firm will earn the expected profit

$$\bar{\Pi} = \sum_{i=1}^{n} \alpha_i \Pi(\mathbf{p}_i)$$

Now suppose that the prices are stabilized at the average price

$$\bar{\mathbf{p}} = \sum_{i=1}^{n} \alpha_i \mathbf{p}_i$$

Since the profit function is convex, Jensen's inequality implies that

$$\Pi(\bar{\mathbf{p}}) \leq \bar{\Pi} = \sum_{i=1}^{n} \alpha_i \Pi(\mathbf{p}_i)$$

Price stabilization reduces expected profit. The intuition is straightforward. When the price is allowed to vary, the firm can tailor its production to the prevailing prices in each period. When the price is stabilized, the firm is not encouraged to respond optimally to price variations.

Even more common in economics are concave functions, which are characterized by reversing the inequality in (24). A function $f: S \to \Re$ is *concave* if for every $\mathbf{x}_1, \mathbf{x}_2$ in S,

3.7 Convex Functions

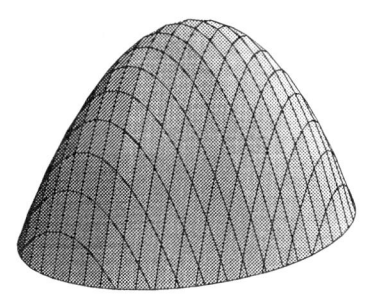

Figure 3.10
A concave function

$$f(\alpha \mathbf{x}_1 + (1-\alpha)\mathbf{x}_2) \geq \alpha f(\mathbf{x}_1) + (1-\alpha)f(\mathbf{x}_2) \quad \text{for every } 0 \leq \alpha \leq 1 \quad (26)$$

A function is *strictly concave* if the inequality is strict; that is, for every $\mathbf{x}_1, \mathbf{x}_2$ in S with $\mathbf{x}_1 \neq \mathbf{x}_2$,

$$f(\alpha \mathbf{x}_1 + (1-\alpha)\mathbf{x}_2) > \alpha f(\mathbf{x}_1) + (1-\alpha)f(\mathbf{x}_2) \quad \text{for every } 0 < \alpha < 1$$

Reversing the inequality corresponds to turning the graph of the function upside down. Therefore a function is concave if and only if its hypograph is convex. The graph of a concave function on $\Re^2$ looks like an upturned bowl (figure 3.10).

Example 3.44 (Power function) The general power function $f\colon \Re_+ \to \Re$ is (example 2.56)

$$f(x) = x^a, \quad a \in \Re$$

Figure 3.11 illustrates the graph of x^a for various values of a. Consistent with these illustrations, we will verify in chapter 4 (example 4.38) that the power function is strictly concave if $0 < a < 1$ and strictly convex if $a < 0$ or $a > 1$. It is both concave and convex when $a = 0$ and $a = 1$.

Exercise 3.124
f is concave if and only if $-f$ is convex.

There is an analogous relation between concave functions and convex sets. A function is concave if and only if its *hypograph*—the set of all points on or below the graph—is convex.

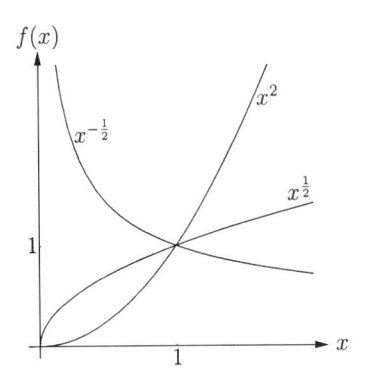

Figure 3.11
The power function x^a for different values of a

Exercise 3.125
A function $f: S \to \Re$ is concave if and only if hypo f is convex.

Example 3.45 (Inverse functions) Let $f: \Re \to \Re$ be invertible with inverse $g = f^{-1}$. Then

$$\text{hypo } f = \{(x,y) \in \Re^2 : y \leq f(x)\} = \{(x,y) : g(y) \leq x\}$$

while

$$\text{epi } g = \{(y,x) \in \Re^2 : g(y) \leq x\}$$

We observe that

hypo f convex $\Leftrightarrow$ epi g convex

Therefore f is concave if and only if g is convex.

Example 3.46 (Production function) The technology of a firm producing a single output from n inputs can be represented by its production function f (example 2.24) where $y = f(\mathbf{x})$ is the maximum output attainable from inputs $\mathbf{x}$. If the production function is concave, the technology exhibits nonincreasing returns to scale. It exhibits decreasing returns to scale if the technology is strictly concave.

Example 3.47 (Production possibility set) The relationship between a production function f and the underlying *production possibility set* Y

(example 1.7) is complicated by the convention that inputs have negative sign in Y. Given a production function f define the function

$$g(\mathbf{x}) = f(-\mathbf{x}) \quad \text{for every } \mathbf{x} \in \Re^n_-$$

the production possibility set Y is the hypograph of the function g. The production function f is concave if and only if the production possibility set Y is convex.

Exercise 3.126 (Cost function)
Show that the cost function $c(\mathbf{w}, y)$ of a competitive firm (example 2.31) is concave in input prices $\mathbf{w}$.

Exercise 3.127 (Lifetime consumption)
Suppose that a consumer retires with wealth w and wishes to choose remaining lifetime consumption stream $c_1, c_2, \ldots, c_T$ to maximize total utility

$$U = \sum_{t=1}^{T} u(c_t) \quad \text{with} \quad \sum_{t=1}^{T} c_t \leq w$$

Assuming that the consumer's utility function u is concave, show that it is optimal to consume a constant fraction $\bar{c} = w/T$ of wealth in each period. [Hint: Use Jensen's inequality (exercise 3.122).]

The following result is useful. For instance, it provides a simple proof of exercise 3.129, a result we used in example 2.74 and will use again in exercise 3.159.

Exercise 3.128
If f is convex on $\Re$

$$f(x_1 - x_2 + x_3) \leq f(x_1) - f(x_2) + f(x_3)$$

for every $x_1 \leq x_2 \leq x_3 \in \Re$. The inequality is strict if f is strictly convex and reversed if f is concave.

Exercise 3.129
If $f \in F(\Re)$ is strictly concave, $f(x - y)$ displays strictly increasing differences in (x, y).

If a function is both convex and concave, it must be affine.

Exercise 3.130
A functional is affine if and only if it is simultaneously convex and concave.

Finally, on a normed linear space, it is useful to define convexity locally. A functional f on a normed linear space X is *locally convex* at $\mathbf{x}_0$ if there exists a convex neighborhood S of $\mathbf{x}_0$ such that for every $\mathbf{x}_1, \mathbf{x}_2 \in S$,

$$f(\alpha \mathbf{x}_1 + (1-\alpha)\mathbf{x}_2) \leq \alpha f(\mathbf{x}_1) + (1-\alpha)f(\mathbf{x}_2) \quad \text{for every } 0 \leq \alpha \leq 1$$

In other words, a function is locally convex at $\mathbf{x}_0$ if its restriction to a neighborhood of $\mathbf{x}_0$ is convex. Analogously, it is strictly locally convex at $\mathbf{x}_0$ if the inequality is strict, and locally concave at $\mathbf{x}_0$ if the inequality is reversed.

Example 3.48 The power function $f(x) = x^3$ is neither convex nor concave on $\Re$ (figure 2.2). It is locally convex on $\Re_+$ and locally concave on $\Re_-$.

3.7.1 Properties of Convex Functions

We first note here some useful rules for combining convex functions. Analogous rules apply for concave functions. In particular, we note that the minimum of concave functions is concave (see exercise 3.132).

Exercise 3.131
If $f, g \in F(X)$ are convex, then

- $f + g$ is convex
- αf is convex for every $\alpha \geq 0$

Therefore the set of convex functions on a set X is a cone in $F(X)$. Moreover, if f is strictly convex, then

- $f + g$ is strictly convex
- αf is strictly convex for every $\alpha > 0$

Example 3.49 (Exponential function) Let

$$f^n(x) = 1 + \frac{x}{1} + \frac{x^2}{2} + \frac{x^3}{6} + \cdots + \frac{x^n}{n!}$$

f^n is convex for every $n = 1, 2, \ldots$ by exercises 3.120 and 3.131. That is,

$$f^n(\alpha x_1 + (1-\alpha)x_2) \leq \alpha f^n(x_1) + (1-\alpha)f^n(x_2), \qquad n = 1, 2, \ldots$$

Therefore for any $x_1, x_2 \in \Re$,

$$\exp(\alpha x_1 + (1-\alpha)x_2) = \lim_{n \to \infty} f^n(\alpha x_1 + (1-\alpha)x_2)$$
$$\leq \lim_{n \to \infty} (\alpha f^n(x_1) + (1-\alpha)f^n(x_2))$$
$$= \alpha e^{x_1} + (1-\alpha)e^{x_2}$$

We conclude that e^x is convex on $\Re_+$. In fact e^x is strictly convex on $\Re$, which we will show in the next chapter.

Example 3.50 (Log function) The log function $\log(x)$ is the inverse (example 2.55) of the exponential function. Since the exponential function is convex, the log function $\log(x)$ is concave (example 3.45).

Exercise 3.132

If f and g are convex functions defined on a convex set S, the function $f \vee g$ defined by

$$(f \vee g)(\mathbf{x}) = \max\{f_1(\mathbf{x}), f_2(\mathbf{x})\} \qquad \text{for every } \mathbf{x} \in S$$

is also convex on S.

Exercise 3.133 (Composition)

If $f \in F(X)$ and $g \in F(\Re)$ with g increasing, then

f and g convex $\Rightarrow g \circ f$ convex

f and g concave $\Rightarrow g \circ f$ concave

Example 3.51 (Log transformation) Logarithmic transformations are often used in analysis. It is nice to know that they preserve concavity, since log is both concave (example 3.50) and increasing (example 2.55). Therefore (exercise 3.133), assuming that f is nonnegative definite,

f concave $\Rightarrow \log f$ concave

Example 3.51 has a useful converse.

Exercise 3.134

If f is nonnegative definite

$\log f$ convex $\Rightarrow f$ convex

Exercise 3.135
If f is a strictly positive definite concave function, then $1/f$ is convex. If f is a strictly negative definite convex function, then $1/f$ is concave.

Exercise 3.133 has a counterpart for supermodular functions (section 2.2.2).

Exercise 3.136 (Composition)
Suppose that $f \in F(X)$ is monotone and $g \in F(\Re)$ is increasing. Then

f supermodular and g convex $\Rightarrow g \circ f$ supermodular

f submodular and g concave $\Rightarrow g \circ f$ submodular

Continuity of Convex Functions

We noted earlier the close relationship between continuity and boundedness for linear functions. Linear functionals are continuous if and only if they are bounded. An analogous requirement applies to convex and concave functions. Clearly, a function that is continuous at any point must be bounded in a neighborhood of that point. This necessary condition turns out to be sufficient.

Proposition 3.8 (Continuity of convex functions) Let f be a convex function defined on an open convex set S in a normed linear space. If f is bounded from above in a neighborhood of a single point $\mathbf{x}_0 \in S$, then f is continuous on S.

Proof Exercise 3.140. □

The following important corollary implies that any convex function on a finite-dimensional space is continuous on the interior of its domain.

Corollary 3.8.1 Let f be a convex function on an open convex set S in a finite-dimensional normed linear space. Then f is continuous.

Remark 3.12 The converse of corollary 3.8.1 is that a convex function can be discontinuous on the boundary of its domain (example 3.52). This is not a mere curiosity. Economic life often takes place at the boundaries of convex sets, where the possibility of discontinuities must be taken into account. This accounts for some of the unwelcome contortions necessary in, for example, duality theory, which could otherwise be exhibited rather more elegantly.

3.7 Convex Functions

Example 3.52 (A discontinuous convex function) Let $S = \Re_+$. The function $f: S \to \Re$ defined by

$$f(x) = \begin{cases} 1, & x = 0 \\ 0, & \text{otherwise} \end{cases}$$

is convex on S but discontinuous at 0.

Exercise 3.137
Let f be a convex function on an open set S that is bounded above by M in a neighborhood of $\mathbf{x}_0$; that is, there exists an open set f containing $\mathbf{x}_0$ such that

$$f(\mathbf{x}) \leq M \quad \text{for every } \mathbf{x} \in U$$

1. Show that there exists a ball $B(\mathbf{x}_0)$ containing $\mathbf{x}_0$ such that for every $\mathbf{x} \in B(\mathbf{x}_0)$,

$$f(\alpha \mathbf{x} + (1 - \alpha)\mathbf{x}_0) \leq \alpha M + (1 - \alpha)f(\mathbf{x}_0)$$

2. Choose some $\mathbf{x} \in B(\mathbf{x}_0)$ and $\alpha \in [0, 1]$. Let $\mathbf{z} = \alpha \mathbf{x} + (1 - \alpha)\mathbf{x}_0$. Show that $\mathbf{x}_0$ can be written as a convex combination of $\mathbf{x}$, $\mathbf{x}_0$ and $\mathbf{z}$ as follows:

$$\mathbf{x}_0 = \frac{1}{1+\alpha}\mathbf{z} + \frac{\alpha}{1+\alpha}(2\mathbf{x}_0 - \mathbf{x})$$

3. Deduce that $f(\mathbf{x}_0) - f(\mathbf{z}) \leq \alpha(M - f(\mathbf{x}_0))$.
4. Show that this implies that f is continuous at $\mathbf{x}_0$.

Exercise 3.138
Let f be a convex function on an open set S which is bounded above by M in a neighborhood of $\mathbf{x}_0$. That is, there exists an open ball $B_r(\mathbf{x}_0)$ containing $\mathbf{x}_0$ such that f is bounded on $B(\mathbf{x}_0)$. Let $\mathbf{x}_1$ be an arbitrary point in S.

1. Show that there exists a number $t > 1$ such that

$$\mathbf{z} = \mathbf{x}_0 + t(\mathbf{x}_1 - \mathbf{x}_0) \in S$$

2. Define $T = \{\mathbf{y} \in X : \mathbf{y} = (1 - \alpha)\mathbf{x} + \alpha \mathbf{z}, \mathbf{x} \in B(x_0)\}$. T is a neighborhood of $\mathbf{x}_1$.
3. f is bounded above on T.

Exercise 3.139
Let f be a convex function on an open set S that is bounded at a single point. Show that f is locally bounded, that is for every $\mathbf{x} \in S$ there exists a constant M and neighborhood U containing $\mathbf{x}$ such that

$$|f(\mathbf{x}')| \leq M \quad \text{for every } \mathbf{x}' \in U$$

Exercise 3.140
Prove proposition 3.8.

Exercise 3.141
Prove corollary 3.8.1 [Hint: Use Carathéodory's theorem (exercise 1.175) and Jensen's inequality (exercise 3.122).]

Exercise 3.142 (Local convexity)
Let f be a functional on a convex open subset S of a Euclidean space X. f is convex if and only f is locally convex at every $\mathbf{x} \in S$. [Hint: Assume the contrary, and consider a cross section. Use the theorem 2.3.]

3.7.2 Quasiconcave Functions

Convex and concave functions relax the additivity requirement of linearity and dispense with homogeneity. Even this is too restrictive for many economic models, and a further generalization is commonly found. A functional f on a convex set S of a linear space X is *quasiconvex* if

$$f(\alpha \mathbf{x}_1 + (1-\alpha)\mathbf{x}_2) \leq \max\{f(\mathbf{x}_1), f(\mathbf{x}_2)\}$$

for every $\mathbf{x}_1, \mathbf{x}_2 \in S$ and $0 \leq \alpha \leq 1$

Similarly f is *quasiconcave* if

$$f(\alpha \mathbf{x}_1 + (1-\alpha)\mathbf{x}_2) \geq \min\{f(\mathbf{x}_1), f(\mathbf{x}_2)\}$$

for every $\mathbf{x}_1, \mathbf{x}_2 \in S$ and $0 \leq \alpha \leq 1$

It is *strictly quasiconcave* if the inequality is strict, that is, for every $\mathbf{x}_1 \neq \mathbf{x}_2$

$$f(\alpha \mathbf{x}_1 + (1-\alpha)\mathbf{x}_2) > \min\{f(\mathbf{x}_1), f(\mathbf{x}_2)\}, \quad 0 < \alpha < 1$$

Geometrically a function is quasiconcave if the function along a line joining any two points in the domain lies above at least one of the endpoints. In practice, quasiconcave functions are more frequently encoun-

Figure 3.12
A bell

tered than quasiconvex functions, and we will focus on the former in this section. The surface of a bell is quasiconcave (figure 3.12).

Exercise 3.143
f is quasiconcave if and only if $-f$ is quasiconvex.

Exercise 3.144
Every concave function is quasiconcave.

Exercise 3.145
Any monotone functional on $\Re$ is both quasiconvex and quasiconcave.

Example 3.53 (Power function) The general power function $f \in F(\Re_+)$

$$f(x) = x^a, \quad a \in \Re$$

is monotone, being strictly increasing if $a > 0$ and strictly decreasing $a < 0$ (exercise 2.34). Therefore (exercise 3.145), it is quasiconcave (and quasiconvex) for all a.

Recall that convex and concave functions can be characterized by convexity of associated sets (proposition 3.7). Quasiconvex and quasiconcave functions have an analogous geometric characterization in terms of their upper and lower contour sets (section 2.1.1).

Proposition 3.9 (Quasiconcavity) *A functional f is quasiconcave if and only if every upper contour set is convex; that is, $\gtrsim_f(c) = \{\mathbf{x} \in X : f(\mathbf{x}) \geq c\}$ is convex for every $c \in \Re$.*

Proof Assume that f is quasiconcave, and choose some $c \in \Re$. If $\gtrsim_f(c)$ is empty, then it is trivially convex. Otherwise, choose $\mathbf{x}_1, \mathbf{x}_2 \in \gtrsim_f(c)$. Then $f(\mathbf{x}_1) \geq c$ and $f(\mathbf{x}_2) \geq c$. Since f is quasiconcave,

$$f(\alpha \mathbf{x}_1 + (1-\alpha)\mathbf{x}_2) \geq \min\{f(\mathbf{x}_1), f(\mathbf{x}_2)\} \geq c$$

for every $0 \leq \alpha \leq 1$, and therefore $\alpha \mathbf{x}_1 + (1-\alpha)\mathbf{x}_2 \in \succsim_f(c)$. That is, $\succsim_f(c)$ is convex. Conversely, assume that $\succsim_f(c)$ is convex for every $c \in \Re$. Choose any $\mathbf{x}_1$ and $\mathbf{x}_2$ in the domain of f, and let

$$c = \min\{f(\mathbf{x}_1), f(\mathbf{x}_2)\}$$

Then $\mathbf{x}_1, \mathbf{x}_2 \in \succsim_f(c)$. Since $\succsim_f(c)$ in convex, $\alpha \mathbf{x}_1 + (1-\alpha)\mathbf{x}_2 \in \succsim_f(c)$ for every $0 \leq \alpha \leq 1$. Consequently

$$f(\alpha \mathbf{x}_1 + (1-\alpha)\mathbf{x}_2) \geq c = \min\{f(\mathbf{x}_1), f(\mathbf{x}_2)\}$$

f is quasiconcave. □

Exercise 3.146
A functional f is quasiconvex if and only if every lower contour set is convex; that is, $\precsim_f(c) = \{\mathbf{x} \in X : f(\mathbf{x}) \leq c\}$ is convex for every $a \in \Re$.

Remark 3.13 This geometric characterization highlights the sense in which quasiconcavity generalizes concavity. A function is concave if and only if its hypograph is concave. The hypograph of a function $f: X \to \Re$ is a subset of $X \times \Re$. If we think of $\Re$ forming the vertical axis, the contour sets can be thought of as horizontal cross sections of the hypograph. Clearly, a convex hypograph (concave function) will have convex cross sections. But a hypograph may have convex cross sections without itself being convex. We illustrate with examples from producer and consumer theory.

Example 3.54 (Convex technology) Suppose that the technology of a firm producing a single output y can be represented by the production function f defined by

$$y = f(\mathbf{x}) = \sup\{y : \mathbf{x} \in V(y)\}$$

The input requirement sets (example 1.8) are the upper contour sets of f, that is,

$$V(y) = \{\mathbf{x} \in \Re_+^n : f(\mathbf{x}) \geq y\}$$

The firm's technology is convex—$V(y)$ convex for every y (example 1.163)—if and only if the production f is quasiconcave. This is less restrictive than assuming that the production function f is concave, which is equivalent to the assumption that the production possibility set Y is convex (example 3.47). The assumption of a convex technology ($V(y)$ convex or f quasiconcave) is typical in economic models, since it does not preclude increasing returns to scale.

Example 3.55 (Convex preferences) Recall that a preference relation $\succsim$ is convex if and only if the upper preference sets $\succsim(\mathbf{y})$ are convex for every $\mathbf{y}$ (section 1.6). A utility function u represents a convex preference relation $\succsim$ if and only if u is quasiconcave.

Although quasiconvex functions are less commonly encountered, there is one important example of a quasiconvex function in economics.

Example 3.56 (Indirect utility function) The consumer's indirect utility function (example 2.90)

$$v(\mathbf{p}, m) = \sup_{\mathbf{x} \in X(\mathbf{p},m)} u(\mathbf{x})$$

which measures the maximum utility attainable given prices and income, is quasiconvex in prices $\mathbf{p}$.

Exercise 3.147
Show that the indirect utility function is quasiconvex. [Hint: Show that the lower contour sets $\precsim_v(c) = \{\mathbf{p} : v(\mathbf{p}, m) \leq c\}$ are convex for every c.]

Properties of Quasiconcave Functions

It is important to note that there is no counterpart to the first part of exercise 3.131—quasiconcavity is not preserved by addition (example 3.57). On the other hand, exercise 3.133 admits a significant generalization (exercise 3.148).

Example 3.57 The function $f(x) = -2x$ is concave and $g(x) = x^3 + x$ is quasiconcave, but their sum $(f + g)(x) = x^3 - x$ is neither concave nor quasiconcave.

Exercise 3.148
If f is quasiconcave and g is increasing, then $g \circ f$ is quasiconcave.

Example 3.58 (CES function) The CES function

$$f(\mathbf{x}) = (\alpha_1 x_1^\rho + \alpha_2 x_2^\rho + \cdots \alpha_n x_n^\rho)^{1/\rho}, \quad \alpha_i > 0, \rho \neq 0$$

is quasiconcave on $\mathfrak{R}^n_+$ provided that $\rho \leq 1$. To see this, let $h(\mathbf{x}) = \alpha_1 x_1^\rho + \alpha_2 x_2^\rho + \cdots \alpha_n x_n^\rho$ so that $f(\mathbf{x}) = (h(\mathbf{x}))^{1/\rho}$. From example 3.44, we know that x_i^ρ is concave if $0 < \rho \leq 1$ and convex otherwise. Therefore (exercise 3.133) h is concave if $0 < \rho \leq 1$ and convex otherwise.

There are two cases to consider:

when $0 < \rho \leq 1$, $f(\mathbf{x}) = (h(\mathbf{x}))^{1/\rho}$ is an increasing function of concave function and is therefore quasiconcave (exercise 3.148).

when $\rho < 0$,

$$f(\mathbf{x}) = (h(\mathbf{x}))^{1/\rho} = \left(\frac{1}{h(\mathbf{x})}\right)^{-1/\rho}$$

Since h is convex, $1/h$ is concave (exercise 3.135) and $-1/\rho > 0$. Again, f is an increasing function of concave function and is therefore quasiconcave (exercise 3.148).

Note that we cannot use exercise 3.133 to conclude that the CES function is concave when $0 < \rho < 1$, since $g(y) = y^{1/\rho}$ is then convex while h is concave. However, we will show later (example 3.74) that it is in fact concave when $\rho < 1$.

Exercise 3.149 (CES function)
The CES function

$$f(\mathbf{x}) = (\alpha_1 x_1^\rho + \alpha_2 x_2^\rho + \cdots \alpha_n x_n^\rho)^{1/\rho}, \qquad \alpha_i > 0, \rho \neq 0$$

is convex on $\mathfrak{R}^n_+$ if $\rho \geq 1$.

Remark 3.14 Production and utility functions are usually assumed to be quasiconcave, so as to represent convex technologies (example 3.54) and convex preferences (example 3.55) respectively. Consequently, when the CES functional form is used as a production or utility function, it is normally restricted so that $\rho < 1$.

Recall that a monotonic transformation is a strictly increasing functional on $\mathfrak{R}$. A monotonic transformation (example 2.60) of a concave function is called concavifiable. Formally a function $f \in F(\mathfrak{R})$ is *concavifiable* if there exists a strictly increasing function $g \in F(\mathfrak{R})$ such that $g \circ f$ is concave. Every concavifiable function is quasiconcave (exercise 3.148). However, the converse is not true in general. There exist quasiconcave functions that are not concavifiable.

Exercise 3.150
Any strictly increasing functional on $\Re$ is concavifiable.

The following results help in recognizing quasiconcave functions.

Exercise 3.151
Let f and g be affine functionals on a linear space X, and let $S \subseteq X$ be a convex set on which $g(\mathbf{x}) \neq 0$. The function

$$h(\mathbf{x}) = \frac{f(\mathbf{x})}{g(\mathbf{x})}$$

is both quasiconcave and quasiconvex on S. [Hint: Use exercise 3.39.]

Exercise 3.152
Let f and g be strictly positive definite functions on a convex set S with f concave and g convex. Then

$$h(\mathbf{x}) = \frac{f(\mathbf{x})}{g(\mathbf{x})}$$

is quasiconcave on S. [Hint: Consider the upper contour sets $\succsim_h(a)$.]

Exercise 3.153
Let f and g be strictly positive definite concave functions on a convex set S. Then their product

$$h(\mathbf{x}) = f(\mathbf{x})g(\mathbf{x})$$

is quasiconcave on S [Hint: Use exercise 3.135.]

The following result should be compared with exercise 3.134.

Exercise 3.154
If f is nonnegative definite,

$\log f$ concave $\Rightarrow f$ quasiconcave

Exercise 3.155
Let $f_1, f_2, \ldots, f_n$ be nonnegative definite concave functions on a convex set S. The function

$$f(\mathbf{x}) = (f_1(\mathbf{x}))^{\alpha_1}(f_2(\mathbf{x}))^{\alpha_2} \ldots (f_n(\mathbf{x}))^{\alpha_n}$$

is quasiconcave on S for any $\alpha_1, \alpha_2, \ldots, \alpha_n \in \Re_+$.

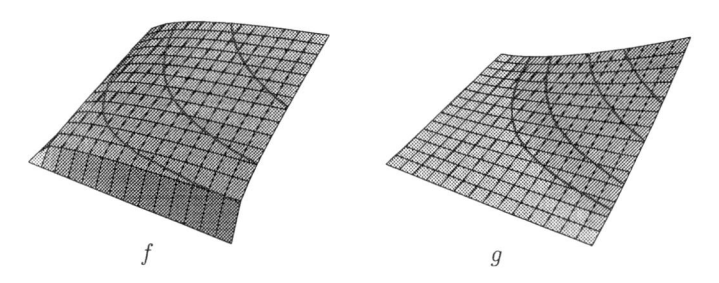

Figure 3.13
The Cobb-Douglas function is quasiconcave

Example 3.59 (Cobb-Douglas) As an immediate application of the preceding exercise, we note that the Cobb-Douglas function

$$f(\mathbf{x}) = x_1^{a_1} x_2^{a_2} \ldots x_n^{a_n}, \quad a_i > 0$$

is quasiconcave on $\Re_+^n$. Figure 3.13 illustrates two Cobb-Douglas functions

$$f(\mathbf{x}) = x_1^{1/3} x_2^{1/3} \quad \text{and} \quad g(\mathbf{x}) = x_1^{4/3} x_2^{4/3}$$

Note that f is concave but g is not. However, both are quasiconcave, as indicated by the curvature of the isoquants.

3.7.3 Convex Maximum Theorems

Convexity in optimization problems yields some useful counterparts to the maximum theorems for monotone and continuous problems (theorems 2.1 and 2.3). The most straightforward result applies to optimization problems in which the constraint is independent of the parameters.

Proposition 3.10 (Convex maximum theorem) *Assume that $f \colon X \times \Theta \to \Re$ is convex in θ. Then the value function*

$$v(\theta) = \max_{x \in X} f(\mathbf{x}, \theta)$$

is convex in θ.

Example 3.60 (Profit function) Earlier (example 3.42) we showed directly that the profit function of a competitive firm

$$\Pi(\mathbf{p}) = \sup_{\mathbf{y} \in Y} \sum_i p_i y_i$$

is convex in **p**. This is a particular case of proposition 3.10, since the objective function is linear in **p**.

To apply proposition 3.10, it is not necessary that the constraint set be completely free of parameters, only that they be free of the parameters of interest. This is illustrated in example 3.61.

Example 3.61 (Cost function) The cost minimization problem

$$\min_{\mathbf{x} \in V(y)} \mathbf{w}^T \mathbf{x}$$

is equivalent to maximization problem

$$\max_{\mathbf{x} \in V(y)} -\mathbf{w}^T \mathbf{x} \qquad (27)$$

The objective function in (27) is convex (linear) in **w**, and the constraint is independent of **w**. Therefore (27) fits the requirements of proposition 3.10, and its value function

$$v(\mathbf{w}) = \sup_{\mathbf{x} \in V(y)} -\mathbf{w}^T \mathbf{x}$$

is convex in **w**. For every output y, the cost function is

$$c(\mathbf{w}, y) = -v(\mathbf{w})$$

which is therefore concave in **w** (exercise 3.124). This duplicates a result found directly in exercise 3.126

Exercise 3.156
Prove proposition 3.10. [Hint: Adapt example 3.42.]

Many optimization problems have constraints that depend on parameters of interest, so proposition 3.10 cannot be applied. A more explicit counterpart to theorems 2.1 and 2.3 is provided by the following theorem, applicable to general constrained optimization problems with concave objectives and convex constraints.

Theorem 3.1 (Concave maximum theorem) *Consider the general constrained maximization problem*

$$\max_{\mathbf{x} \in G(\theta)} f(\mathbf{x}, \theta)$$

where X and Θ are linear spaces. Let $\Theta^* \subset \Theta$ denote the set of parameter values for which a solution exists. If

- the objective function $f: X \times \Theta \to \Re$ is quasiconcave in X and
- the constraint correspondence $G: \Theta \rightrightarrows X$ is convex-valued

then the solution correspondence $\varphi: \Theta^* \rightrightarrows X$ defined by

$$\varphi(\boldsymbol{\theta}) = \arg \max_{\mathbf{x} \in G(\boldsymbol{\theta})} f(\mathbf{x}, \boldsymbol{\theta})$$

is convex-valued. Furthermore, if

- the objective function $f: X \times \Theta \to \Re$ is (strictly) concave in $X \times \Theta$ and
- the constraint correspondence $G: \Theta \rightrightarrows X$ is convex

the value function

$$v(\boldsymbol{\theta}) = \sup_{\mathbf{x} \in G(\boldsymbol{\theta})} f(\mathbf{x}, \boldsymbol{\theta})$$

is (strictly) concave in $\boldsymbol{\theta}$.

Proof

Convexity of $\varphi(\boldsymbol{\theta})$ For any $\boldsymbol{\theta} \in \Theta^*$, let $\mathbf{x}_1, \mathbf{x}_2 \in \varphi(\boldsymbol{\theta})$. This implies that

$$f(\mathbf{x}_1, \boldsymbol{\theta}) = f(\mathbf{x}_2, \boldsymbol{\theta}) = v(\boldsymbol{\theta}) \geq f(\mathbf{x}, \boldsymbol{\theta}) \quad \text{for every } \mathbf{x} \in G(\boldsymbol{\theta})$$

Let $\bar{\mathbf{x}} = \alpha \mathbf{x}_1 + (1 - \alpha) \mathbf{x}_2$. Since f is quasiconcave,

$$f(\bar{\mathbf{x}}, \boldsymbol{\theta}) \geq \min\{f(\mathbf{x}_1, \boldsymbol{\theta}), f(\mathbf{x}_2, \boldsymbol{\theta})\} = v(\boldsymbol{\theta})$$

which implies that $\bar{\mathbf{x}} \in \varphi(\boldsymbol{\theta})$. For every $\boldsymbol{\theta}$, $\varphi(\boldsymbol{\theta})$ is convex.

Concavity of v Let $\boldsymbol{\theta}_1, \boldsymbol{\theta}_2$ belong to Θ^*. Choose any optimal solutions $\mathbf{x}_1 \in \varphi(\boldsymbol{\theta}_1)$ and $\mathbf{x}_2 \in \varphi(\boldsymbol{\theta}_2)$. Let

$$\bar{\boldsymbol{\theta}} = \alpha \boldsymbol{\theta}_1 + (1 - \alpha) \boldsymbol{\theta}_2, \quad \bar{\mathbf{x}} = \alpha \mathbf{x}_1 + (1 - \alpha) \mathbf{x}_2$$

Since $\mathbf{x}_1 \in G(\boldsymbol{\theta}_1)$, $\mathbf{x}_2 \in G(\boldsymbol{\theta}_2)$, and G is convex, $\bar{\mathbf{x}} \in G(\bar{\boldsymbol{\theta}})$. Thus $\bar{\mathbf{x}}$ is feasible for $\bar{\boldsymbol{\theta}}$ so

$$v(\bar{\boldsymbol{\theta}}) = \sup_{\mathbf{x} \in \varphi(\bar{\boldsymbol{\theta}})} f(\mathbf{x}, \bar{\boldsymbol{\theta}}) \geq f(\bar{\mathbf{x}}, \bar{\boldsymbol{\theta}}) \geq \alpha f(\mathbf{x}_1, \boldsymbol{\theta}_1) + (1 - \alpha) f(\mathbf{x}_2, \boldsymbol{\theta}_2)$$

$$= \alpha v(\boldsymbol{\theta}_1) + (1 - \alpha) v(\boldsymbol{\theta}_2)$$

v is concave.

Strict concavity of v Furthermore, if f is strictly concave,

$$v(\bar{\boldsymbol{\theta}}) = \sup_{\mathbf{x} \in \varphi(\bar{\boldsymbol{\theta}})} f(\mathbf{x}, \bar{\boldsymbol{\theta}}) \geq f(\bar{\mathbf{x}}, \bar{\boldsymbol{\theta}}) > \alpha f(\mathbf{x}_1, \boldsymbol{\theta}_1) + (1 - \alpha) f(\mathbf{x}_2, \boldsymbol{\theta}_2)$$
$$= \alpha v(\boldsymbol{\theta}_1) + (1 - \alpha) v(\boldsymbol{\theta}_2)$$

so v is strictly concave. □

The first part of this theorem, requiring only quasiconcavity of the objective function (and convex-valued constraint), is a key result in optimization (see proposition 3.16). Strict quasiconcavity leads to the following important corollary.

Corollary 3.1.1 *Let* $\Theta^* \subset \Theta$ *denote the set of parameter values for which a solution exists in the general constrained maximization problem*

$$\max_{\mathbf{x} \in G(\boldsymbol{\theta})} f(\mathbf{x}, \boldsymbol{\theta})$$

where X and Θ are linear spaces. If

- *the objective function $f \colon X \times \Theta \to \Re$ is strictly quasiconcave in X and*
- *the constraint correspondence $G \colon \Theta \rightrightarrows X$ is convex-valued*

then the solution correspondence $\varphi \colon \Theta^* \rightrightarrows X$ defined by

$$\varphi(\boldsymbol{\theta}) = \arg\max_{\mathbf{x} \in G(\boldsymbol{\theta})} f(\mathbf{x}, \boldsymbol{\theta})$$

is single-valued; that is, φ is a function from Θ^* to X.

Example 3.62 (Demand functions) Corollary 3.1.1 can be applied directly to the consumer's problem (example 2.90). If the consumer's preferences are strictly convex, the utility function is strictly quasiconcave, and corollary 3.1.1 implies a unique optimal solution for every $\mathbf{p}$ and m. That is, the consumer's demand correspondence $\mathbf{x}(\mathbf{p}, m)$ is a function.

Exercise 3.157
Prove corollary 3.1.1.

Example 3.63 (Cost function) We have previously shown (exercise 3.126) that the cost function $c(\mathbf{w}, y)$ of a competitive firm is concave in $\mathbf{w}$. Suppose in addition that the firm's production possibility set Y is convex. Then the input requirements sets

$$V(y) = \{\mathbf{x} \in \Re_+^n : (y, -\mathbf{x}) \in Y\}$$

define a convex correspondence (exercise 2.25), and the cost minimization problem

$$\min_{\mathbf{x} \in V(y)} \sum_{i=1}^n w_i x_i = \max_{\mathbf{x} \in V(y)} -\sum_{i=1}^n w_i x_i$$

satisfies the requirements of theorem 3.1 (the objective function is linear). This implies that the value function

$$c(\mathbf{w}, y) = \inf_{\mathbf{x} \in V(y)} \sum_{i=1}^n w_i x_i$$

is concave in $\mathbf{w}$ and y jointly.

Exercise 3.158 (Uniqueness of the optimal plan)
In the dynamic programming problem (example 2.32)

$$\max_{x_1, x_2, \ldots} \sum_{t=0}^\infty \beta^t f(x_t, x_{t+1})$$

subject to $\quad x_{t+1} \in G(x_t)$

$t = 0, 1, 2, \ldots, x_0 \in X$ given

Assume that

- f is bounded, continuous and strictly concave on $X \times X$.
- $G(x)$ is nonempty, compact-valued, convex-valued, and continuous for every $x \in X$
- $0 \leq \beta < 1$

We have previously shown (exercise 2.124) that an optimal policy exists under these assumptions. Show also that

1. the value function v is strictly concave
2. the optimal policy is unique

Example 3.64 (Optimal economic growth) In the optimal economic growth model (example 2.33), assume that

- u is continuous and strictly concave on $\Re_+$
- F is continuous and increasing on $\Re_+$ with $F(0) = 0$
- there exists $\bar{k} > 0$ such that $F(k) \leq k$ for every $k \geq \bar{k}$
- $0 \leq \beta < 1$

Then there exists an optimal growth policy $(k_0, k_1^*, k_2^*, \ldots)$ for every starting point k_0 (example 2.93). Furthermore

- the optimal growth policy $(k_0, k_1^*, k_2^*, \ldots)$ is unique (exercise 3.158) and
- converges monotonically to some steady state k^* (exercise 3.159).

Whether capital accumulates or decumulates under the optimal policy depends on the relationship between the limiting value k^* and the initial value k_0. If $k_0 < k^*$, then k_t^* grows increases monotonically to k^*. Conversely, if the economy starts with an oversupply of capital $k_0 > k^*$, capital will be progressively reduced.

Exercise 3.159
Assuming that u is strictly concave, show that the optimal growth model satisfies the requirements of exercise 2.126. Hence conclude that the optimal policy converges monotonically to a steady state.

Unfortunately, the requirements of the second part of theorem 3.1, joint concavity of f in $\mathbf{x}$ and $\boldsymbol{\theta}$ and convexity of the constraint correspondence, are quite stringent and often missing in practice. Example 3.65 illustrates that it is not sufficient that the constraint be convex-valued, while example 3.66 illustrates what this requirement means in the most typical setting. Often theorem 3.1 can be applied to those parts of the problem that satisfy the conditions, holding the other parameters constant. This procedure is illustrated in example 3.67, where we establish concavity of the indirect utility function in income, by holding prices constant.

Example 3.65 Let $X = \Theta = [0, 1]$, and define

$f(x, \theta) = x \quad$ for every $(x, \theta) \in X \times \Theta$

$G(\theta) = [0, \theta^2] \quad$ for every $\theta \in \Theta$

Since f is strictly increasing, the optimal solution correspondence is

$$\varphi(\theta) = \{\theta^2\} \quad \text{for every } \theta \in \Theta$$

G is convex-valued but not convex, and the value function

$$v(\theta) = \sup_{x \in [0, \theta^2]} x = \theta^2$$

is not concave on Θ.

Example 3.66 Suppose that the constraint set $G(\theta)$ in the constrained optimization problem

$$\max_{\mathbf{x} \in G(\theta)} f(\mathbf{x}, \theta)$$

is defined by a set of inequalities (example 2.40)

$$g_1(\mathbf{x}, \theta) \leq c_1$$

$$g_2(\mathbf{x}, \theta) \leq c_2$$

$$\ldots$$

$$g_m(\mathbf{x}, \theta) \leq c_m$$

where each $g(\mathbf{x}, \theta)$ is convex jointly in $\mathbf{x}$ and θ. Then the correspondence

$$G(\theta) = \{\mathbf{x} \in X : g_j(\mathbf{x}, \theta) \leq c_j, j = 1, 2, \ldots, m\}$$

is convex (exercise 2.26). Provided that the objective function $f(\mathbf{x}, \theta)$ is (strictly) concave in $\mathbf{x}$ and θ, the value function

$$v(\theta) = \sup_{\mathbf{x} \in G(\theta)} f(\mathbf{x}, \theta)$$

is (strictly) concave in θ (theorem 3.1).

Example 3.67 (Consumer theory) Theorem 3.1 cannot be applied directly to deduce general properties of the indirect utility function (example 2.90), since the budget constraint is not convex in $\mathbf{x}$ and $\mathbf{p}$ jointly. However, for given prices $\bar{\mathbf{p}}$, the budget constraint

$$X(m) = \{\mathbf{x} \in X : \bar{p}^T \mathbf{x} \leq m\}$$

is convex in m. If the utility function is concave, the consumer's problem (with constant prices)

$$\max_{\mathbf{x} \in X(m)} u(\mathbf{x})$$

satisfies the conditions of theorem 3.1. We can deduce that the indirect utility function $v(\mathbf{p}, m)$ is concave in income m. We have previously shown that the indirect utility function $v(\mathbf{p}, m)$ is quasiconvex in p (example 3.56). This is as far as we can go in deducing general properties of the indirect utility function.

3.7.4 Minimax Theorems

Let $f(x, y)$ be a continuous functional on a compact domain $X \times Y$ in a normed linear space. It is always the case (exercise 3.161) that

$$\max_{\mathbf{x} \in X} \min_{\mathbf{y} \in Y} f(\mathbf{x}, \mathbf{y}) \leq \min_{\mathbf{y} \in Y} \max_{\mathbf{x} \in X} f(\mathbf{x}, \mathbf{y}) \tag{28}$$

If (28) is satisfied as an equality

$$\max_{\mathbf{x} \in X} \min_{\mathbf{y} \in Y} f(\mathbf{x}, \mathbf{y}) = \min_{\mathbf{y} \in Y} \max_{\mathbf{x} \in X} f(\mathbf{x}, \mathbf{y})$$

so that the order of max and min does not matter, there exists a point $(\mathbf{x}^*, \mathbf{y}^*) \in X \times Y$ satisfying

$$f(\mathbf{x}, \mathbf{y}^*) \leq f(\mathbf{x}^*, \mathbf{y}^*) \leq f(\mathbf{x}^*, \mathbf{y}) \quad \text{for every } \mathbf{x} \in X \text{ and } \mathbf{y} \in Y$$

Such a point is called a *saddle point*, since it simultaneously maximizes f over X and minimizes f over Y (figure 3.14).

Exercise 3.160 (Saddle point)
Let X and Y be compact subsets of a finite-dimensional normed linear space, and let f be a continuous functional on $X \times Y$. Then

$$\max_{\mathbf{x} \in X} \min_{\mathbf{y} \in Y} f(\mathbf{x}, \mathbf{y}) = \min_{\mathbf{y} \in Y} \max_{\mathbf{x} \in X} f(\mathbf{x}, \mathbf{y})$$

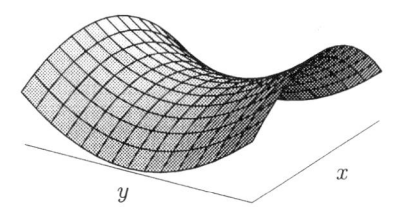

Figure 3.14
A saddle point

if and only if there exists a point $(\mathbf{x}^*, \mathbf{y}^*) \in X \times Y$ such that

$$f(\mathbf{x}, \mathbf{y}^*) \leq f(\mathbf{x}^*, \mathbf{y}^*) \leq f(\mathbf{x}^*, \mathbf{y}) \qquad \text{for every } \mathbf{x} \in X \text{ and } \mathbf{y} \in Y$$

Theorems that specify the additional conditions on f, X, and Y necessary to ensure equality in (28), and hence the existence of saddle points, are known as *minimax theorems*. The original minimax theorem (exercise 3.262) was due to von Neumann, who used it to demonstrate the existence of solutions to zero-sum games (section 3.9.4). Von Neumann's theorem applied to bilinear functions on the standard simplex in $\Re^n$. The following generalization to quasiconcave functions on convex sets is a straightforward application of Kakutani's theorem (theorem 2.7).

Proposition 3.11 (Minimax theorem) *Let X and Y be compact, convex subsets of a finite-dimensional normed linear space, and let f be a continuous functional on $X \times Y$ which is quasiconcave on X and quasiconvex on Y. Then*

$$\max_{\mathbf{x}} \min_{\mathbf{y}} f(\mathbf{x}, \mathbf{y}) = \min_{\mathbf{y}} \max_{\mathbf{x}} f(\mathbf{x}, \mathbf{y})$$

Proof Define the correspondences $\varphi \colon Y \rightrightarrows X$ and $\psi \colon X \rightrightarrows Y$ by

$$\varphi(\mathbf{y}) = \arg \max_{\mathbf{x} \in X} f(\mathbf{x}, \mathbf{y})$$

and

$$\psi(\mathbf{x}) = \arg \min_{\mathbf{y} \in Y} f(\mathbf{x}, \mathbf{y}) = \arg \max_{\mathbf{y} \in Y} -f(\mathbf{x}, \mathbf{y})$$

By the continuous maximum theorem (theorem 2.3) φ and ψ are nonempty, compact, and upper hemicontinuous. By the concave maximum theorem (theorem 3.1), φ is convex-valued. Similarly, since $(-f)$ is quasiconcave (exercise 3.143), ψ is also convex-valued.

The correspondence $\Phi \colon X \times Y \rightrightarrows X \times Y$ defined by

$$\Phi(\mathbf{x}, \mathbf{y}) = \varphi(\mathbf{y}) \times \psi(\mathbf{x})$$

is closed and convex-valued (proposition 1.2, exercises 2.107, 1.165). By Kakutani's theorem (theorem 2.7), Φ has a fixed point $(\mathbf{x}^*, \mathbf{y}^*)$ such that

$$\mathbf{x}^* \in \arg \max_{\mathbf{x} \in X} f(\mathbf{x}, \mathbf{y}^*) \quad \text{and} \quad \mathbf{y}^* \in \arg \min_{\mathbf{y} \in Y} f(\mathbf{x}^*, \mathbf{y})$$

That is,

$$f(\mathbf{x}, \mathbf{y}^*) \le f(\mathbf{x}^*, \mathbf{y}^*) \le f(\mathbf{x}^*, y) \quad \text{for every } \mathbf{x} \in X \text{ and } \mathbf{y} \in Y$$

In other words, $(\mathbf{x}^*, \mathbf{y}^*)$ is a saddle point. This implies (exercise 3.160) that

$$\max_x \min_y f(x, y) = \min_y \max_x f(x, y) \qquad \square$$

Exercise 3.161
If f is a continuous functional on a compact domain $X \times Y$,

$$\max_x \min_y f(x, y) \le \min_y \max_x f(x, y)$$

3.8 Homogeneous Functions

Concave and convex functions generalize the additivity property of linear functionals. Homogeneous functions generalize homogeneity. If S is a cone in linear space X, a functional $f \in F(S)$ is *homogeneous of degree k* if for every $\mathbf{x} \in S$,

$$f(t\mathbf{x}) = t^k f(\mathbf{x}) \qquad \text{for every } t \in \Re_{++}$$

This definition relaxes the homogeneity requirement of a linear function (and dispenses with additivity).

Example 3.68 (Power function) The general power function $f \in F(\Re_+)$ (example 2.56)

$$f(x) = x^a$$

is homogeneous of degree a, since

$$f(tx) = (tx)^a = t^a x^a = t^a f(x)$$

In fact, every homogeneous function on $\Re_+$ is a power function (exercise 3.162).

Exercise 3.162
A function $f \colon \Re_+ \to \Re_+$ is homogeneous of degree a if and only if it is (a multiple of) a power function, that is,

$$f(x) = A x^a \qquad \text{for some } A \in \Re$$

Example 3.69 (Cobb-Douglas) The Cobb-Douglas function

$$f(\mathbf{x}) = x_1^{a_1} x_2^{a_2} \ldots x_n^{a_n}$$

is homogeneous of degree $a_1 + a_2 + \cdots + a_n$, since for any $t > 0$,

$$\begin{aligned} f(t\mathbf{x}) &= (tx_1)^{a_1} (tx_2)^{a_2} \ldots (tx_n)^{a_n} \\ &= t^{a_1+a_2+\cdots+a_n} x_1^{a_1} x_2^{a_2} \ldots x_n^{a_n} \\ &= t^{a_1+a_2+\cdots+a_n} f(\mathbf{x}) \end{aligned}$$

Exercise 3.163 (CES function)
Show that the CES function

$$f(\mathbf{x}) = (a_1 x_1^\rho + a_2 x_2^\rho + \cdots a_n x_n^\rho)^{1/\rho}$$

is homogeneous of degree one.

The explicit characterization of homogeneous functions on $\Re_+$ as power functions (exercise 3.162) can help us understand the structure of homogeneous functions on more complex domains. Suppose that f is homogeneous of degree k on S. For any $\mathbf{x}_0 \in S$, the function $h \in F(\Re_+)$ defined by

$$h(t) = f(t\mathbf{x}_0), \quad t \in \Re_+$$

is also homogeneous of degree k (exercise 3.164). h provides a cross section of f along a ray $\{t\mathbf{x}_0 : \mathbf{x}_0 \in S, t > 0\}$ through the point $\mathbf{x}_0$ (see section 2.1.4). By exercise 3.162, h is a power function, that is,

$$h(t) = At^k$$

Therefore any homogeneous function looks like a power function when viewed along a ray.

Example 3.70 (Cobb-Douglas) The two-variable Cobb-Douglas function

$$f(x_1, x_2) = x_1^{a_1} x_2^{a_2}$$

is homogeneous of degree $a_1 + a_2$. Figure 2.11 illustrates this function when $a_1 + a_2 = 1$ (example 2.41). While clearly nonlinear when considered over its whole domain, the surface is linear when considered along any ray through the origin. If a ruler were laid along this Cobb-Douglas surface so that it passed through the origin, it would align with the surface along its entire length. Similarly, when $a_1 = a_2 = 1$, the Cobb-Douglas function is homogeneous of degree 2 and looks like a quadratic t^2 along any ray.

Exercise 3.164
Let $f: S \to \Re$ be homogeneous of degree k. For any $\mathbf{x}_0 \in S$, the functional defined $h(t) = f(t\mathbf{x}_0)$ is homogeneous of degree k on $\Re_+$.

Homogeneous functions arise naturally in economics. Homogeneity restricts the behavior of a function when all variables change in the same proportion, which represents two recurrent situations in economic analysis. In a production context it corresponds to changing the scale of production, leaving the relative proportions of different inputs fixed. Constant returns to scale implies that the production function is homogeneous of degree one. In a function of prices (e.g., a profit function), scaling corresponds to changing all prices in the same proportion (inflation), leaving relative prices unchanged. The degree of homogeneity k can be positive, negative, or zero. The most common examples encountered in economics are homogeneity of degree 0 and homogeneity of degree 1. Functions homogeneous of degree 0 are constant along any ray. Functions homogeneous of degree 1 are sometimes called *linearly homogeneous* functions, since they are linear along any ray.

Example 3.71 (Profit function) The profit function of a competitive firm (example 2.29)

$$\Pi(\mathbf{p}) = \sup_{\mathbf{y} \in Y} \sum_i p_i y_i$$

is homogeneous of degree one. To see this, suppose that the production plan $\mathbf{y}^*$ maximizes the firms profit at prices $\mathbf{p}$, that is,

$$\mathbf{p}^T \mathbf{y}^* \geq \mathbf{p}^T \mathbf{y} \quad \text{for every } \mathbf{y} \in Y$$

Therefore for every $t > 0$,

$$(t\mathbf{p})^T \mathbf{y}^* \geq (t\mathbf{p})^T \mathbf{y} \quad \text{for every } \mathbf{y} \in Y$$

and therefore $\mathbf{y}^*$ also maximizes the firm's profit at prices $t\mathbf{p}$. Consequently

$$\Pi(t\mathbf{p}) = (t\mathbf{p})^T \mathbf{y}^* = t \sum_i p_i y_i = t\Pi(\mathbf{p})$$

We conclude that the profit function is homogeneous of degree one. This implies that if all prices are increased in the same proportion, the firm's maximum profit increases proportionately.

Example 3.72 (Demand function) If the consumer's preferences are strictly convex, the optimal solution to the consumer's problem (examples 1.113, 2.91, and 3.67) is a set of demand functions $x_i(\mathbf{p}, m)$, each specifying the consumer's demand for commodity i as a function of prices $\mathbf{p}$ and m. One of the most important properties of the consumer demand functions is that they are homogeneous of degree zero. This means that demand is invariant to the general level of prices and income—only relative prices matter.

To verify homogeneity, we note that if commodity bundle $\mathbf{x}$ is affordable at prices $\mathbf{p}$ and income m, it is also affordable at prices $t\mathbf{p}$ and income tm for every $t > 0$, since

$$\mathbf{p}^T \mathbf{x} \leq m \Leftrightarrow (t\mathbf{p})^T \mathbf{x} \leq tm$$

Therefore the consumer's budget set is invariant to proportionate changes in prices and income

$$X(t\mathbf{p}, tm) = X(\mathbf{p}, m) \quad \text{for every } t > 0$$

which implies that the consumer's optimal choice will also be invariant to proportionate changes in prices and income.

Exercise 3.165 (Cost function)
Show that the cost function $c(\mathbf{w}, y)$ of a competitive firm (example 2.31) is homogeneous of degree one in input prices $\mathbf{w}$.

Exercise 3.166 (Cost function with constant returns to scale)
If the production function of a competitive firm is homogeneous of degree one, then the cost function $c(\mathbf{w}, y)$ is homogeneous of degree one in y, that is,

$$c(\mathbf{w}, y) = y c(\mathbf{w}, 1)$$

where $c(\mathbf{w}, 1)$ is the cost of producing one unit (unit cost).

Exercise 3.167 (Indirect utility function)
Show that the indirect utility function $v(\mathbf{p}, m)$ (example 2.90) is homogeneous of degree zero in $\mathbf{p}$ and m.

Analogous to convex functions (proposition 3.7), linearly homogeneous functions can be characterized by their epigraph.

Exercise 3.168
A function $f: S \to \Re$ is linearly homogeneous if and only if epi f is a cone.

The following useful proposition show how quasiconcavity and homogeneity combine to produce full concavity. Quasiconcavity ensures convexity of the upper contour sets, while homogeneity of degree $k \leq 1$ strengthens this to convexity of the hypograph (see remark 3.13).

Proposition 3.12 *Let f be a strictly positive definite functional that is homogeneous of degree k, $0 < k \leq 1$. Then f is quasiconcave if and only if f is concave.*

Proof The "if" part is trivial (exercise 3.144). The "only-if" part is developed in exercises 3.169 through 3.171. □

Exercise 3.169
If $f \in F(S)$ is strictly positive definite, quasiconcave, and homogeneous of degree one, then f is superadditive, that is,

$$f(\mathbf{x}_1 + \mathbf{x}_2) \geq f(\mathbf{x}_1) + f(\mathbf{x}_2) \quad \text{for every } \mathbf{x}_1, \mathbf{x}_2 \in S$$

Exercise 3.170
If $f \in F(S)$ is strictly positive definite, quasiconcave, and homogeneous of degree one, then f is concave.

Exercise 3.171
Generalize exercise 3.170 to complete the proof of proposition 3.12.

Example 3.73 (Cobb-Douglas) We have previously shown that the Cobb-Douglas function

$$f(\mathbf{x}) = x_1^{a_1} x_2^{a_2} \ldots x_n^{a_n}, \quad a_i > 0$$

is quasiconcave and homogeneous of degree $a_1 + a_2 + \cdots + a_n$. By proposition 3.12, we can conclude that the Cobb-Douglas function is concave provided $a_1 + a_2 + \cdots + a_n \leq 1$.

Example 3.74 (CES function) We have previously shown that the CES function

$$f(\mathbf{x}) = (\alpha_1 x_1^\rho + \alpha_2 x_2^\rho + \cdots \alpha_n x_n^\rho)^{1/\rho}, \quad \alpha_i > 0, \rho \neq 0$$

is quasiconcave if $\rho \leq 1$ (example 3.58) and convex if $\rho \leq 1$ (exercise 3.149). Since the CES function is positive and homogeneous of degree one (exercise 3.163), proposition 3.12 implies that the CES function is in fact concave if $\rho \leq 1$ and convex otherwise.

3.8.1 Homothetic Functions

Analogous to the generalization of concave to quasiconcave functions, there is corresponding generalization of homogeneity. A functional f defined on a convex cone S in a linear space X is *homothetic* if

$$f(\mathbf{x}_1) = f(\mathbf{x}_2) \Rightarrow f(t\mathbf{x}_1) = f(t\mathbf{x}_2)$$

for every $\mathbf{x}_1, \mathbf{x}_2 \in S$ and $t > 0$

Geometrically, if two points belong to the same contour of a homothetic function, then every scalar multiple of these points belong to a common contour. In other words, a function is homothetic if its contours are radial expansions of each other. Clearly, every homogeneous function is homothetic, but not every homothetic function is homogeneous, as is shown by the following example.

Example 3.75 The function $f \colon \Re_{++}^2 \to \Re$ defined by

$$f(x_1, x_2) = \log x_1 + \log x_2$$

is homothetic but not homogeneous, since

$$f(tx_1, tx_2) = \log tx_1 + \log tx_2 = 2\log t + \log x_1 + \log x_2$$
$$= 2\log t + f(x_1, x_2)$$

and therefore $f(\mathbf{x}^1) = f(\mathbf{x}^2)$ implies that

$$f(t\mathbf{x}^1) = 2\log t + f(\mathbf{x}^1) = 2\log t + f(\mathbf{x}^2) = f(t\mathbf{x}^2) \quad \text{for every } t > 0$$

Exercise 3.172 (Homothetic preferences)
A preference relation $\succsim$ (section 1.6) on a cone S is said to be *homothetic* if

$$\mathbf{x}_1 \sim \mathbf{x}_2 \Rightarrow t\mathbf{x}_1 \sim t\mathbf{x}_2 \quad \text{for every } \mathbf{x}_1, \mathbf{x}_2 \in S \text{ and } t > 0$$

Show that a continuous preference relation is homothetic if and only if every utility representation (example 2.58) is homothetic.

Every monotonic transformation of a homogeneous function is homothetic (exercise 3.174). For strictly increasing functions the converse is also true (exercise 3.175). This provides an equivalent characterization of homotheticity that is particularly useful in economic analysis.

Proposition 3.13 (Homotheticity) *Let f be a strictly increasing functional on a cone S in an linear space X. f is homothetic if and only if it is a monotonic transformation of a homogeneous function.*

Proof Exercises 3.174 and 3.175. □

Remark 3.15 (Equivalent definitions of homotheticity) Many texts use the characterization in proposition 3.13 to define homotheticity, stating that a function is homothetic if it is a monotonic transformation of a homogeneous function (Simon and Blume 1994, p. 500). Other texts define a homothetic function as monotonic transformation of a *linearly* homogeneous function (Varian 1992, p. 18). Clearly, these definition are equivalent to one another (exercise 3.173) and equivalent to our definition for strictly increasing functions (proposition 3.13).

Example 3.76 (Log-linear function) The log-linear function

$$f(\mathbf{x}) = a_1 \log x_1 + a_2 \log x_2 + \cdots + a_n \log x_n$$

is commonly used in empirical work. It is not homogeneous, since

$$f(t\mathbf{x}) = a_1 \log(tx_1) + a_2 \log(tx_2) + \cdots + a_n \log(t\mathbf{x}_n)$$
$$= (a_1 + a_2 + \cdots + a_n) \log t + a_1 \log x_1 + a_2 \log x_2 + \cdots + a_n \log x_n$$
$$= (a_1 + a_2 + \cdots + a_n) \log t + f(\mathbf{x})$$

It is, however, homothetic, since it is an increasing transformation of the homogeneous Cobb-Douglas function (example 3.69)

$$f(\mathbf{x}) = \log(x_1^{a_1} x_2^{a_2} \ldots x_n^{a_n}) = a_1 \log x_1 + a_2 \log x_2 + \cdots + a_n \log x_n$$

Exercise 3.173
Suppose that f is a monotonic transformation of a homogeneous function. Show that f is a monotonic transformation of a linearly homogeneous function.

Exercise 3.174
If h is a homogeneous functional on S and $g \colon \Re \to \Re$ is strictly increasing, then $f = g \circ h$ is homothetic.

Exercise 3.175
Let f be a strictly increasing homothetic functional on a cone S in an linear space X. Then there exists a linearly homogeneous function $h: S \to \Re$ and a strictly increasing function $g: \Re \to \Re$ such that $f = g \circ h$. [Hint: Define $g(\alpha) = f(\alpha \mathbf{x}_0)$ for some $\mathbf{x}_0 \in S$, and show that $h = g^{-1} \circ f$ is homogeneous of degree one.]

Exercise 3.176 (Homothetic technology)
If the production function of a competitive firm is homothetic, then the cost function is separable, that is,

$$c(\mathbf{w}, y) = \varphi(y) c(\mathbf{w}, 1)$$

where $c(\mathbf{w}, 1)$ is the cost of producing one unit (unit cost). [Hint: Use exercise 3.166.]

Exercise 3.177 (Concavifiability)
A strictly positive definite, strictly increasing, homothetic, and quasiconcave functional is concavifiable.

3.9 Separation Theorems

A hyperplane $H_f(c)$ in a linear space X divides the space into two sets $\{\mathbf{x} \in X : f(\mathbf{x}) \geq c\}$ and $\{\mathbf{x} \in X : f(\mathbf{x}) \leq c\}$ called *halfspaces*. These are the upper $\succsim_f(c)$ and lower $\precsim_f(c)$ contour sets of f respectively. The halfspaces are closed sets if f is continuous (exercise 2.77). A hyperplane is said to *separate* two sets A and B if they lie on opposite sides of the hyperplane so that each is contained in opposing halfspaces. Formally $H_f(c)$ separates A and B if

either $\quad f(\mathbf{x}) \leq c \leq f(\mathbf{y}) \quad$ or $\quad f(\mathbf{x}) \geq c \geq f(\mathbf{y})$

for every $\mathbf{x} \in A$ and $\mathbf{y} \in B$

Similarly the hyperplane $H_f(c)$ *bounds* a set S if S is wholly contained in one or other of the halfspaces, that is,

either $\quad f(\mathbf{x}) \leq c \quad$ or $\quad f(\mathbf{x}) \geq c \quad\quad$ for every $\mathbf{x} \in S$

A hyperplane $H_f(c)$ is a *supporting hyperplane* to S at $\mathbf{x}_0 \in S$ if $H_f(c)$ bounds S and contains $\mathbf{x}_0$ (figure 3.15). A surprising number of questions

3.9 Separation Theorems

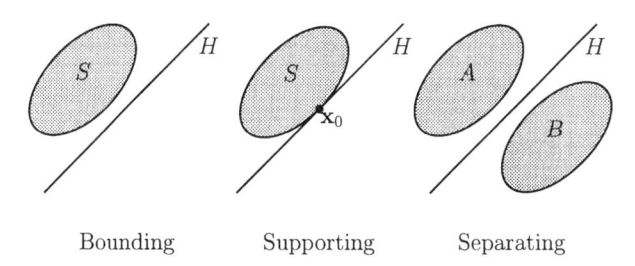

Figure 3.15
Bounding, separating, and supporting hyperplanes

in economics can be posed in terms of the existence of separating or supporting hyperplanes to appropriately defined sets.

Two sets A and B can be separated if and only if there exists a linear functional f and constant c such that

$$f(\mathbf{x}) \leq c \leq f(\mathbf{y}) \quad \text{for every } \mathbf{x} \in A \text{ and } \mathbf{y} \in B$$

That is, there exists a linear functional that values every point in A less than any point in B. The connection with optimization becomes more transparent when we rewrite the condition as asserting the existence of linear functional, which is maximized over A and minimized over B, that is,

$$\sup_{\mathbf{x} \in A} f(\mathbf{x}) \leq c \leq \inf_{\mathbf{y} \in B} f(\mathbf{y})$$

Exercise 3.178
Assume that $H_f(c)$ is a supporting hyperplane to a set S at $\mathbf{x}_0$. Show that either $\mathbf{x}_0$ maximizes f or $\mathbf{x}_0$ minimizes f on the set S.

As the illustrations in figure 3.16 suggest, the fundamental requirement for separation is convexity. This is the content of following basic separation theorem, whose proof is developed in the next section.

Theorem 3.2 (Separating hyperplane theorem) *Let A and B be nonempty, disjoint, convex subsets in a normed linear space X. Assume that* **either** *at least one of the sets has a nonempty interior* **or** *X is finite-dimensional. Then there exists a continuous linear functional $f \in X^*$ and a number c such that*

$$f(\mathbf{x}) \leq c \leq f(\mathbf{y}) \quad \text{for every } \mathbf{x} \in A \text{ and } \mathbf{y} \in B$$

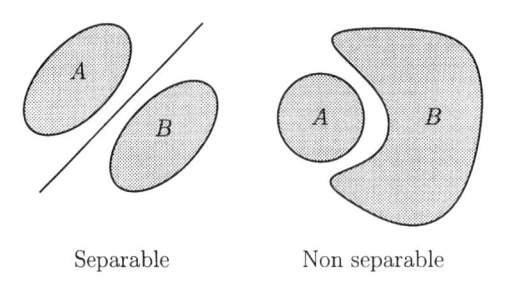

Figure 3.16
Convexity is the fundamental requirement for separation

Moreover separation is strict on the interiors of A and B, that is,

$$f(\mathbf{x}) < c < f(\mathbf{y}) \quad \text{for every } \mathbf{x} \in \text{int } A \text{ and } \mathbf{y} \in \text{int } B$$

Actually it is not necessary that the convex sets be entirely disjoint. Separation is possible if the convex sets share a common boundary, provided they have no interior points in common. In fact this is a necessary and sufficient condition for separation. Thus we have the following useful corollary.

Corollary 3.2.1 *Let A and B be nonempty, convex subsets in a normed linear space X with int $A \neq \emptyset$. Then A and B can be separated if and only if int $A \cap B = \emptyset$.*

Corollary 3.2.2 (Supporting hyperplane) *Let $\mathbf{x}_0$ be a boundary point of a convex set S in normed linear space. Assume that S has a nonempty interior. Then there exists a supporting hyperplane at $\mathbf{x}_0$; that is, there exists a continuous linear functional $f \in X^*$ such that*

$$f(\mathbf{x}_0) \leq f(\mathbf{x}) \quad \text{for every } \mathbf{x} \in S$$

In many applications one of the convex sets to be separated is a subspace; when this is the case, the separating hyperplane necessarily contains the subspace.

Corollary 3.2.3 (Subspace separation) *Let S be a convex subset of linear space X with a nonempty interior, and let Z be a subspace that is disjoint from the interior of S. Then there exists a separating hyperplane which contains Z, that is there exists a continuous linear functional $f \in X^*$ such that*

3.9 Separation Theorems

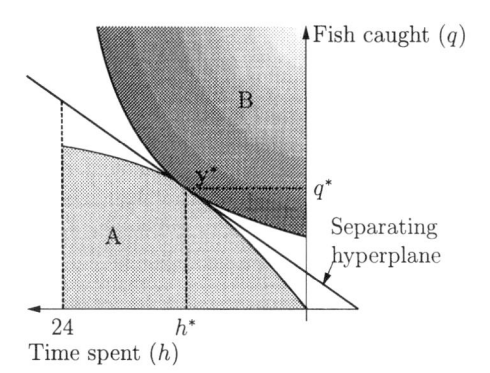

Figure 3.17
Robinson's choice of lifestyle

$$f(\mathbf{x}) \geq 0 \quad \text{for every } \mathbf{x} \in S$$

and

$$f(\mathbf{z}) = 0 \quad \text{for every } \mathbf{z} \in Z$$

The classic application of the separating hyperplane theorem in economics is the second theorem of welfare economics, which shows it is possible for decentralization to achieve Pareto optimality. The following example involving a single producer and consumer illustrates the essential idea.

Example 3.77 (Robinson Crusoe) Isolated on a desert island, Robinson Crusoe survives by catching fish. Although fish are plentiful in the lagoon, the more time he spends fishing, the more wary become the fish, and the harder they are to catch. Robinson does not like fishing—it is hard work, and he would prefer to spend his time sitting on the beach dreaming of being rescued.

Robinson's predicament is illustrated in figure 3.17. He has a single input (time) and a single output (fish). His only productive activity (fishing) exhibits diminishing returns. His production opportunities are prescribed by the convex production possibility set A. Each point in $\mathbf{y} \in A$ is a pair (h, q) specifying the time spent fishing (h) and the resulting catch of fish (q). Since fishing time is an input, h is negative (example 1.7). We assume that A is closed.

A is the set of attainable or feasible lifestyles for Robinson. His choice of the best lifestyle $(h, q) \in A$ is guided by his preferences, specifically his trade-off between food and leisure. We assume that his preferences are strictly convex, continuous, and monotonic. Since total time is limited (to 24 hours a day), the feasible set A is compact. Consequently there exists a best choice $(h^*, q^*) \in A$ that is at least as good as any other lifestyle $(h, q) \in Y$ (proposition 1.5).

Robinson fulfills two roles in our model: he is both consumer and producer. Suppose that we want to separate these roles, allowing Robinson the consumer to act independently of Robinson the producer, exchanging fish for labor at arm's-length. The separating hyperplane theorem guarantees that there exist a price of fish p and wage rate w that achieves precisely this decentralization. To see this, let B denote the set of all feasible lifestyles which are at least as good as (h^*, q^*). That is,

$$B = \succsim(h^*, q^*) = \{(h, q) : (h, q) \succsim (h^*, q^*)\}$$

B is convex. Furthermore B contains no interior points of A (exercise 3.179). Consequently (theorem 3.2, corollary 3.2.1) there is a linear functional f and number c such that

$$f(\mathbf{y}) \leq c \leq f(\mathbf{y}') \quad \text{for every } \mathbf{y} \in A \text{ and } \mathbf{y}' \in B \tag{29}$$

See figure 3.17. A and B are convex sets in $\Re^2$. Consequently (proposition 3.4) there exist numbers w and p such that $f(h, q) = wh + pq$. If Robinson the producer buys labor at wage rate w and sells fish at price p, $f(h, q)$ measures the net profit achieved from the production plan (h, q). Simultaneously $f(h, q)$ measures the net cost to Robinson the consumer of buying q fish at price p, while selling h hours of labor at wage rate w. Since (h^*, q^*) belongs to both A and B, (29) implies that

1. $wh^* + pq^* \geq wh + pq$ for every $\mathbf{y} = (h, q) \in A$ At the prices (w, p), Robinson the producer maximizes his profit $wh + pq$ at the production plan (h^*, q^*).

2. $wh^* + pq^* \leq wh + pq$ for every $\mathbf{y}' = (h, q) \in B$ At the prices (w, p), Robinson the consumer minimizes the cost $wh + pq$ of achieving a lifestyle at least as satisfying as (h^*, q^*).

Remark 3.16 (Second theorem of welfare economics) The first theorem of welfare economics (exercise 1.251) establishes the Pareto optimality of competitive markets. The second theorem of welfare economics is the converse, asserting that the achievement of Pareto optimal outcomes can be decentralized through competitive markets.

In the Robinson Crusoe economy, the existence of prices that enable the decentralization of the production and consumption sides of Robinson's existence is a straightforward application of the separating hyperplane theorem. Trading at these prices, Robinson the producer and Robinson the consumer will independently achieve a compatible outcome.

At first glance the Robinson Crusoe economy seems a very special case, since it involves a single consumer, a single producer, and only two commodities. Fortunately, these limitations are more apparent than real. Nothing in the derivation in example 3.77 hinged on there only being two commodities, and the extension to $l > 2$ commodities is trivial.

The assumption of price-taking behavior when there are only two agents is far-fetched. Fortunately the extension to multiple consumers and producers is also straightforward. Exercise 3.228 establishes the second theorem for an exchange economy with many consumers but no producers. Adding multiple producers brings no further conceptual insight, although the need to take account of the distribution of profits in the economy complicates the notational burden. For this reason we invite the reader to consult standard texts such as Mas-Colell et al. (1995), Starr (1997), and Varian (1992) for a general treatment.

The one ingredient of the Robinson Crusoe economy that cannot be dispensed with is convexity. Convexity of both technology and preferences is indispensable to ensure the separation of production and upper preference sets and hence the possibility of decentralization through markets. With many agents, the convexity requirements can be relaxed somewhat. On the production side, convexity of the aggregate production set suffices, even if the technology of individual producers in not convex. Similarly the aggregation of large numbers of consumers alleviates individual nonconvexity (Hildenbrand and Kirman 1976).

Exercise 3.179
In example 3.77 show that int $A \cap B = \emptyset$.

Exercise 3.180
In example 3.77 Robinson the producer makes a profit of $wh^* + pq^*$. This is Robinson the consumer's income, so his budget set is

$$X = \{(h, q) : wh + pq \leq wh^* + pq^*\}$$

This is halfspace below the separating hyperplane in figure 3.17. Note that $A \subseteq X$.

(h^*, q^*) is the optimal choice in the feasible set A. Show that it is also Robinson the consumer's optimal choice in the larger budget set X. Consequently (h^*, q^*) solves the consumer's problem when the prices are w and p.

Exercise 3.181 (Subgradient)
Let f be a convex function defined on a convex set S in a normed linear space X. For every $\mathbf{x}_0 \in \text{int } S$ there exists a linear functional $g \in X^*$ that bounds f in the sense that

$$f(\mathbf{x}) \geq f(\mathbf{x}_0) + g(\mathbf{x} - \mathbf{x}_0) \qquad \text{for every } \mathbf{x} \in S$$

Such a linear functional g is called a *subgradient* of f at $\mathbf{x}_0$. [Hint: Consider a supporting hyperplane to epi f at $(\mathbf{x}_0, f(\mathbf{x}_0))$.]

Example 3.78 (Profit function) Figure 3.18 shows a cross section through the profit function of a competitive firm. The straight line is the graph of a subgradient of the profit function. It shows that profit attainable by the firm if it *does not change* it production activities as the price p varies. The fact that it bounds the profit function from below shows that the firm can attain a higher profit by adjusting its production plans in response to price changes.

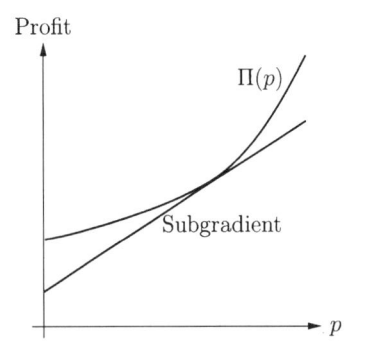

Figure 3.18
A subgradient of the profit function

3.9 Separation Theorems

Proof of the Basic Separation Theorem

The separating hyperplane theorem is one of the most intuitive results in mathematics. A few minutes drawing figures should convince you of the veracity of the separation theorem in the plane $\Re^2$. Fortunately this is not one of those occasions where our intuition leads us astray in higher dimensions, and the passage to higher dimensions introduces no major complications. However, proving this is not trivial. Indeed, a proof for an arbitrary linear space involves some sophisticated mathematics. In the special case of Euclidean space, a proof of the separating hyperplane theorem is a useful illustration of the interplay of algebraic and topological concepts in linear spaces. It is established in the following exercises.

Exercise 3.182

Let S be a nonempty, *closed*, convex set in a Euclidean space X and $\mathbf{y} \notin S$. There exists a continuous linear functional $f \in X^*$ and a number c such that

$$f(\mathbf{y}) < c \leq f(\mathbf{x}) \qquad \text{for every } \mathbf{x} \in S$$

Exercise 3.183

Let $\mathbf{y}$ be a boundary point of a nonempty, convex set S in a Euclidean space X. There exists a supporting hyperplane at $\mathbf{y}$; that is, there exists a continuous linear functional $f \in X^*$ such that

$$f(\mathbf{y}) \leq f(\mathbf{x}) \qquad \text{for every } \mathbf{x} \in S$$

[Hint: If $\mathbf{y} \in \mathbf{b}(S)$, there exists a sequence $\mathbf{y}^n \to \mathbf{y}$ with $\mathbf{y}^n \notin \overline{S}$.]

Exercise 3.184

Generalize exercise 3.182 to dispense with the assumption that S is closed. That is, let S be a nonempty, convex set in a Euclidean space X and $\mathbf{y} \notin S$. There exists a continuous linear functional $f \in X^*$ such that

$$f(\mathbf{y}) \leq f(\mathbf{x}) \qquad \text{for every } \mathbf{x} \in S$$

[Hint: Consider separately the two possible cases: $y \in \overline{S}$ and $y \notin \overline{S}$.]

Exercise 3.185

Let S be an open convex subset of a linear space X and $f \in X^*$ a nonzero linear functional on X. Then $f(S)$ is an open interval in $\Re$.

Exercise 3.186
Prove theorem 3.2 (assuming that X is Euclidean). [Hint: Apply exercise 3.184 to separate $\mathbf{0}$ from $S = B + (-A)$. Then use exercise 3.185.]

Remark 3.17 Finite dimensionality was used at two crucial stages in the derivation above. In exercise 3.72 finite dimensionality ensured the compactness of $\hat{S}$, to which we applied the Weierstrass theorem to guarantee the existence of a closest point to $\mathbf{y}$. In exercise 3.183 finite dimensionality was required to ensure the existence of a convergent subsequence of linear functionals. Holmes (1975, pp. 14–16) gives a general proof of the separating hyperplane theorem.

Exercise 3.187
Prove corollary 3.2.1.

Exercise 3.188
Prove corollary 3.2.2

Exercise 3.189
Let $H_f(c)$ be a bounding hyperplane of a cone C in a normed linear space X, that is, $f(\mathbf{x}) \geq c$ for every $\mathbf{x} \in C$. Then

$f(\mathbf{x}) \geq 0$ for every $\mathbf{x} \in C$

Exercise 3.190
Let $H_f(c)$ be a bounding hyperplane of a subspace Z of a normed linear space X, that is, $f(\mathbf{x}) \leq c$ for every $\mathbf{x} \in Z$. Then Z is contained in the kernel of f, that is,

$f(\mathbf{x}) = 0$ for every $\mathbf{x} \in Z$

Exercise 3.191
Prove corollary 3.2.3.

Separation theorems are so pervasive in mathematical economics that it is necessary to have a range of variations in the armory. In the following sections we develop some refinements of the basic separating hyperplane theorem that are useful in applications.

Strong Separation

A hyperplane $H_f(c)$ is said to *properly separate* convex sets A and B, provided that both are not contained in the hyperplane itself. This avoids

3.9 Separation Theorems

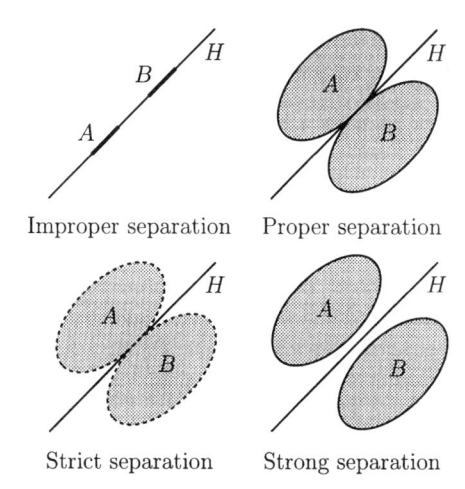

Figure 3.19
Various forms of separation

the trivial case in which $f(\mathbf{x}) = c$ for every $\mathbf{x} \in A \cup B$ (figure 3.19). Theorem 3.2 ensures proper separation whenever at least one of the sets has a nonempty interior (exercise 3.192).

Exercise 3.192 (Proper separation)
Let A and B be nonempty, convex subsets in a normed linear space X with int $A \neq \emptyset$ and int $A \cap B = \emptyset$. Then A and B can be properly separated.

Frequently stronger forms of separation are required. Two sets A and B are *strictly separated* by a hyperplane $H_f(c)$ if A and B lie in opposite *open* halfspaces defined by $H_f(c)$, that is,

$$f(\mathbf{x}) < c < f(\mathbf{y}) \quad \text{for every } \mathbf{x} \in A, \mathbf{y} \in B$$

The sets A and B are *strongly separated* by the hyperplane $H_f(c)$ if there exists some number ε such that

$$f(\mathbf{x}) < c - \varepsilon < c + \varepsilon < f(\mathbf{y}) \quad \text{for every } \mathbf{x} \in A, \mathbf{y} \in B$$

or equivalently

$$\sup_{\mathbf{x} \in A} f(\mathbf{x}) < \inf_{\mathbf{y} \in B} f(\mathbf{y})$$

Exercise 3.193 (Strict separation)
If A and B are nonempty, disjoint, convex *open* sets in a finite dimensional normed linear space X, they can be strictly separated; that is, there exists a continuous linear functional $f \in X^*$ and a numbers c such that

$$f(x) < c < f(y) \quad \text{for every } x \in A, y \in B$$

The most important variant is strong separation. The basic result is presented in the following proposition. In the following exercises we explore proofs for the finite- and infinite-dimensional cases. We then use proposition 3.14 to generalize some previous results and provide some new applications.

Proposition 3.14 (Strong separation) *Let A and B be nonempty, disjoint, convex subsets in a normed linear space X with*

- *A compact*
- *B closed.*

Then A and B can be strongly separated; that is, there exists a continuous linear functional $f \in X^*$ such that

$$\sup_{\mathbf{x} \in A} f(\mathbf{x}) < \inf_{\mathbf{y} \in B} f(\mathbf{y})$$

A straightforward proof for a finite-dimensional space is given in the following exercise.

Exercise 3.194
Prove proposition 3.14 for a finite-dimensional space X. [Hint: Apply exercise 3.182 to the set $B - A$. Compactness of A is necessary to ensure that $B - A$ is closed.]

The following exercise shows that compactness of A is essential in proposition 3.14.

Exercise 3.195
In $\Re^2$, draw the sets $A = \{x \in \Re_+^2 : x_1 x_2 \geq 1\}$ and $B = \{x \in \Re^2 : x_2 \leq 0\}$. Can these sets be strongly separated?

The essential requirement for strong separation is that the two sets be spatially disjoint. This requirement is formalized for general (infinite-dimensional) normed linear spaces in the following exercise, providing a general proof of proposition 3.14.

Exercise 3.196

1. Let A and B be nonempty, disjoint, convex subsets in a normed linear space X. A and B can be strongly separated if and only if there exists a convex neighborhood of U of 0 such that

$$(A + U) \cap B = \emptyset$$

2. Prove proposition 3.14. [Hint: Use exercise 1.208.]

Exercise 3.197

Let A and B be convex subsets in a finite-dimensional normed linear space X. A and B can be strongly separated if and only if

$$\rho(A, B) = \inf\{\|\mathbf{x} - \mathbf{y}\| : \mathbf{x} \in A, \mathbf{y} \in B\} > 0$$

Combining proposition 3.14 with corollary 3.2.3 gives the following important result, a geometric form of the Hahn-Banach theorem (proposition 3.15).

Exercise 3.198 (Geometric Hahn-Banach theorem)
Let M be a nonempty, closed, subspace of a linear space X and $y \notin M$. Then there exists a continuous linear functional $f \in X^*$ such that

$$f(\mathbf{y}) > 0 \quad \text{and} \quad f(\mathbf{x}) = 0 \quad \text{for every } \mathbf{x} \in M$$

As an application of the previous result, we use it in the following exercise to provide an alternative derivation of the Fredholm alternative (exercise 3.48). Note how a clever choice of space enables us to apply a separation theorem to derive an a straightforward proof of a fundamental theorem.

Exercise 3.199 (Fredholm alternative)
Let $g_1, g_2, \ldots, g_m$ be linear functionals on a linear space X, and let

$$S = \{\mathbf{x} \in X : g_j(\mathbf{x}) = 0, j = 1, 2, \ldots, m\} = \bigcap_{j=1}^{m} \text{kernel } g_j$$

Suppose that $f \neq 0$ is another linear functional such that such that $f(\mathbf{x}) = \mathbf{0}$ for every $\mathbf{x} \in S$. Show that

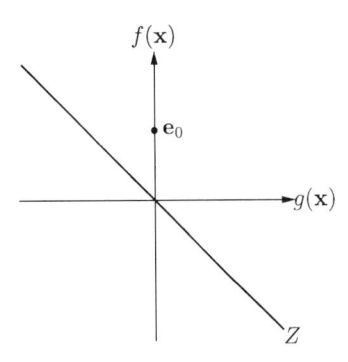

Figure 3.20
The Fredholm alternative via separation

1. The set $Z = \{f(x), -g_1(x), -g_2(x), \ldots, -g_m(x) : \mathbf{x} \in X\}$ is a subspace of $Y = \Re^{m+1}$.
2. $\mathbf{e}^0 = (1, 0, 0, \ldots, 0) \in \Re^{m+1}$ does not belong to Z (figure 3.20).
3. There exists a linear functional $\varphi \in Y^*$ such that $\varphi(\mathbf{e}^0) > 0$ and $\varphi(\mathbf{z}) = 0$ for every $\mathbf{z} \in Z$.
4. Let $\varphi(\mathbf{y}) = \lambda_\mathbf{y}^T$ where $\lambda = (\lambda_0, \lambda_1, \ldots, \lambda_m) \in Y = \Re^{(m+1)^*}$. For every $\mathbf{z} \in Z$,

$$\lambda_\mathbf{z}^T = \lambda_0 z_0 + \lambda_1 z_1 + \cdots + \lambda_m z_m = 0$$

5. $\lambda_0 > 0$.
6. $f(x) = \sum_{i=1}^m \lambda_i g_i(x)$; that is, f is linearly dependent on $g_1, g_2, \ldots, g_m$.

Exercise 3.200
Show the converse; that is, if $f(\mathbf{x}) = \sum_{i=1}^m \lambda_i g_i(\mathbf{x})$, then $f(\mathbf{x}) = 0$ for every $\mathbf{x} \in S$ where $S = \{\mathbf{x} \in X : g_j(\mathbf{x}) = 0, j = 1, 2 \ldots m\}$.

Exercise 3.201 (Gale)
Let $g_1, g_2, \ldots, g_m$ be linear functionals on a linear space X. For fixed numbers c_j, the systems of equations

$$g_j(\mathbf{x}) = c_j, \quad j = 1, 2, \ldots, m$$

is consistent if and only if

$$\sum_{j=1}^m \lambda_j g_j = 0 \Rightarrow \sum_{j=1}^m \lambda_j c_j = 0$$

for every set of numbers $\lambda_1, \lambda_2, \ldots, \lambda_m$. [Hint: Separate $\mathbf{c} = (c_1, c_2, \ldots, c_m)$ from the subspace $Z = \{g_1(x), g_2(x), \ldots, g_m(x) : x \in X\}$ in $\Re^m$.]

Exercise 3.198 can be extended to a closed convex cone when X is finite-dimensional. This result will be used in exercise 3.225.

Exercise 3.202
Let K be a closed convex cone in a finite-dimensional linear space X and M a subspace with $K \cap M = \{\mathbf{0}\}$. Then there exists a linear functional $f \in X^*$ such that

$f(\mathbf{x}) > 0 \quad$ for every $\mathbf{x} \in K \setminus \{\mathbf{0}\}$

and

$f(\mathbf{x}) = 0 \quad$ for every $\mathbf{x} \in M$

[Hint: Consider the set $\hat{K} = \{\mathbf{x} \in K : \|\mathbf{x}\|_1 = 1\}$.]

3.9.1 Hahn-Banach Theorem

Any linear functional f_0 on a subspace $Z \subset X$ can be trivially extended to a functional $f \in X^*$ on the whole space by defining

$f(\mathbf{x}) = f_0(\mathbf{z})$

where $\mathbf{x} = \mathbf{y} + \mathbf{z}$ with $\mathbf{z} \in Z$ and $\mathbf{y} \in Z^\perp$, the orthogonal complement of Z. What makes extension theorems interesting is the presence of various additional constraints which must be satisfied by the extension. The classic extension theorem is the Hahn-Banach theorem, where the extension must satisfy the additional constraint that $f(\mathbf{x}) \leq g(\mathbf{x})$ where g is convex.

Proposition 3.15 (Hahn-Banach theorem) *Let g be a convex functional on a linear space X. Suppose that f_0 is a linear functional defined on a subspace Z of X such that*

$f_0(\mathbf{x}) \leq g(\mathbf{x}) \quad$ *for every* $\mathbf{x} \in Z$

Then f_0 can be extended to a functional $f \in X^*$ such that

$f(\mathbf{x}) = f_0(\mathbf{x}) \quad$ *for every* $\mathbf{x} \in Z$

and

$f(\mathbf{x}) \leq g(\mathbf{x}) \quad$ *for every* $\mathbf{x} \in X$

Proof Exercise 3.203. □

Exercise 3.203
Suppose that f_0 is a linear functional defined on a subspace Z of X such that

$$f_0(\mathbf{x}) \leq g(\mathbf{x}) \qquad \text{for every } \mathbf{x} \in Z$$

where $g \in X^*$ is convex. Show that

1. The sets

$$A = \{(\mathbf{x}, y) : y \geq g(\mathbf{x}), \mathbf{x} \in X\}$$

and

$$B = \{(\mathbf{x}, y) : y = f_0(\mathbf{x}), \mathbf{x} \in Z\}$$

are convex subsets of the linear space $Y = X \times \Re$ (figure 3.21)

2. int $A \neq \emptyset$ and int $A \cap B = \emptyset$.

3. There exists a linear functional $\varphi \in Y^*$ with $\varphi(\mathbf{0}, 1) > 0$ such that

$$\varphi(\mathbf{x}, y) \geq 0 \qquad \text{for every } (\mathbf{x}, y) \in A$$

and

$$\varphi(\mathbf{x}, y) = 0 \qquad \text{for every } (\mathbf{x}, y) \in B$$

4. Define the functional $f \in X^*$ by $f(\mathbf{x}) = -\frac{1}{c}\varphi(\mathbf{x}, 0)$ where $c = \varphi(\mathbf{0}, y) > 0$. Then

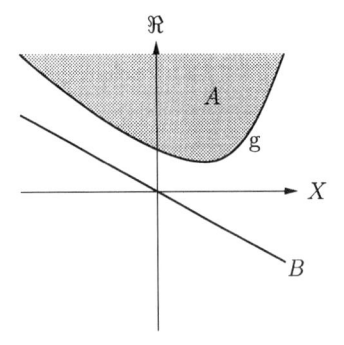

Figure 3.21
Deriving the Hahn-Banach theorem

$$f(\mathbf{x}) = -\frac{1}{c}\varphi(\mathbf{x}, y) + y \quad \text{for every } y \in \Re$$

5. f is an extension of f_0; that is $f(\mathbf{z}) = f_0(\mathbf{z})$ for every $\mathbf{z} \in Z$.
6. f is bounded by g; that is $f(\mathbf{x}) \leq g(\mathbf{x})$ for every $\mathbf{x} \in X$.

The Hahn-Banach theorem is in fact equivalent to the basic separation theorem. We established one direction of this equivalence in exercise 3.203. Luenberger (1969, p. 133) gives the reverse direction. The Hahn-Banach theorem shows that a normed linear space is well endowed with linear functionals. Some consequences are addressed in the following exercises. We will use exercise 3.205 in proposition 4.1.1 and exercise 3.206 in exercise 3.207.

Exercise 3.204
Let f_0 be a bounded linear functional on a subspace Z of a normed linear space X. Then f_0 can be extended to a linear functional on the whole space X without increasing its norm, that is,

$$\|f\|_X = \|f_0\|_Z$$

Exercise 3.205
Let $\mathbf{x}_0$ be an element of a normed linear space X. There exists a linear functional $f \in X^*$ such that $\|f\| = 1$ and $f(\mathbf{x}_0) = \|\mathbf{x}_0\|$. [Hint: Define the function $f_0(\alpha \mathbf{x}_0) = \alpha \|\mathbf{x}_0\|$ on the subspace $\mathrm{lin}\{\mathbf{x}_0\} = \{\alpha \mathbf{x}_0 : \alpha \in \Re\}$.]

Exercise 3.206
Let $\mathbf{x}_1, \mathbf{x}_2$ be distinct points in a normed linear space X. There exists a continuous linear functional $f \in X^*$ that evaluates them differently, that is, such that

$$f(\mathbf{x}_1) \neq f(\mathbf{x}_2)$$

Example 3.79 When $X = \Re^n$, the vector $\mathbf{x}_0/\|\mathbf{x}_0\|$ defines a linear functional $f \in X^*$ satisfying exercise 3.205 (exercise 3.64), namely

$$f(\mathbf{x}) = \mathbf{x}^T \left(\frac{\mathbf{x}_0}{\|\mathbf{x}_0\|} \right)$$

since

$$f(\mathbf{x}_0) = \frac{\mathbf{x}_0^T \mathbf{x}_0}{\|\mathbf{x}_0\|} = \|\mathbf{x}_0\| \quad \text{and} \quad f\left(\frac{\mathbf{x}_0}{\|\mathbf{x}_0\|}\right) = \frac{\mathbf{x}_0^T \mathbf{x}_0}{\|\mathbf{x}_0\|^2} = 1$$

By Cauchy-Schwartz inequality (exercise 3.61)

$|f(\mathbf{x})| \leq \|\mathbf{x}\|$ for every $\mathbf{x} \in X$

and therefore

$\|f\| = \sup_{\|\mathbf{x}\|=1} |f(\mathbf{x})| = 1$

Extreme Points

The following exercise is a nice illustration of the use of the Hahn-Banach theorem. Earlier (exercise 3.65) we showed that every compact set in an inner product space has an extreme point. To extend this to more general spaces requires two of our most powerful tools: Zorn's lemma and the Hahn-Banach theorem.

Exercise 3.207 (Existence of extreme points)
Let S be a nonempty compact convex subset of a normed linear space X.

1. Let $\mathfrak{F}$ be the collection of all faces of S. $\mathfrak{F}$ has a minimal element F_0.
2. F_0 contains only a single point, $\mathbf{x}_0$.
3. $\mathbf{x}_0$ is an extreme point of S

Remark 3.18 We note that convexity is unnecessary in the previous result: *every compact set in a normed linear space has an extreme point*. We assumed convexity to take advantage of the familiar concept of faces, since the result will normally be applied to convex sets (exercise 3.16). The proof of the more general result follows exactly the form outlined in the previous exercise, substituting the related concept of *extremal set* for face (Holmes 1975, p. 74).

Combining exercise 3.207 with some earlier results yields the following proposition, which is of immense practical importance for optimization because it implies that the search for an optimum can be confined to the extreme points of feasible set. In particular, this observation provides the essential rationale for the simplex algorithm in linear programming.

Proposition 3.16 (Quasiconcave functional maximized at an extreme point)
Let f be a continuous function on a compact, convex set S. If f is quasiconcave, then it attains its maximum at an extreme point of S.

Proof Let S^* be the set on which f attains its maximum, that is,

$$S^* = \{\mathbf{x}^* \in S : f(\mathbf{x}^*) \geq f(\mathbf{x}) \text{ for every } \mathbf{x} \in S\}$$

S^* is nonempty (Weierstrass theorem 2.2), compact (continuous maximum theorem 2.3), and convex (concave maximum theorem 3.1). Therefore S^* contains an extreme point (exercise 3.207). □

A particular application of this proposition is often useful.

Exercise 3.208
Let S be a compact convex set. Every supporting hyperplane to S contains an extreme point of S.

This leads to another important implication of the strong separation theorem, which is called the Krein-Milman theorem.

Exercise 3.209 (Krein-Milman theorem)
Every compact, convex set is the closed convex hull of its extreme points.

The Krein-Millman theorem underpins the standard proof of the Shapley-Folkman theorem (see exercise 3.112), which is developed in the following exercises.

Exercise 3.210
Let $\{S_1, S_2, \ldots, S_n\}$ be a collection of nonempty *compact* subsets of an m-dimensional linear space, and let $\mathbf{x} \in \text{conv} \sum_{i=1}^n S_i$. We consider the Cartesian product of the convex hulls of S_i, namely

$$P = \prod_{i=1}^n \text{conv } S_i$$

Every point in P is an n-tuple $(\mathbf{x}_1, \mathbf{x}_2, \ldots, \mathbf{x}_n)$ where each $\mathbf{x}_i$ belongs to the corresponding conv S_i. Let $P(\mathbf{x})$ denote the subset of P for which $\sum_{i=1}^n \mathbf{x}_i = \mathbf{x}$, that is,

$$P(x) = \left\{ (\mathbf{x}_1, \mathbf{x}_2, \ldots, \mathbf{x}_n) : \mathbf{x}_i \in \text{conv } S_i \text{ and } \sum_{i=1}^n \mathbf{x}_i = \mathbf{x} \right\}$$

1. Show that

a. $P(\mathbf{x})$ is compact and convex.

b. $P(\mathbf{x})$ is nonempty.

c. $P(\mathbf{x})$ has an extreme point $\mathbf{z} = (\mathbf{z}_1, \mathbf{z}_2, \ldots, \mathbf{z}_n)$ such that
- $\mathbf{z}_i \in \text{conv } S_i$ for every i
- $\sum_{i=1}^n \mathbf{z}_i = \mathbf{x}$.

2. At least $n - m$ components $\mathbf{z}_i$ of $\mathbf{z} = (\mathbf{z}_1, \mathbf{z}_2, \ldots, \mathbf{z}_n)$ are extreme points of their sets conv S_i. To show this, assume the contrary. That is, assume that there are $l > m$ components of $\mathbf{z}$ that are *not* extreme points of conv S_i. Without loss of generality, suppose that these are the first l components of $\mathbf{z}$.

a. For $i = 1, 2, \ldots, l$, there exists $\mathbf{y}_i \in X$ such that

$$\mathbf{z}_i + \mathbf{y}_i \in \text{conv } S_i \quad \text{and} \quad \mathbf{z}_i - \mathbf{y}_i \in \text{conv } S_i$$

b. There exists numbers $\alpha_1, \alpha_2, \ldots, \alpha_l$, $|\alpha_i| \leq 1$ such that

$$\alpha_1 \mathbf{y}_1 + \alpha_2 \mathbf{y}_2 + \cdots + \alpha_l \mathbf{y}_l = 0$$

c. Define

$$\mathbf{z}^+ = \begin{pmatrix} \mathbf{z}_1 + \alpha_1 \mathbf{y}_1 \\ \mathbf{z}_2 + \alpha_2 \mathbf{y}_2 \\ \ldots \\ \mathbf{z}_l + \alpha_l \mathbf{y}_l \\ \mathbf{z}_{l+1} \\ \ldots \\ \mathbf{z}_n \end{pmatrix}, \quad \mathbf{z}^- = \begin{pmatrix} \mathbf{z}_1 - \alpha_1 \mathbf{y}_1 \\ \mathbf{z}_2 - \alpha_2 \mathbf{y}_2 \\ \ldots \\ \mathbf{z}_l - \alpha_l \mathbf{y}_l \\ \mathbf{z}_{l+1} \\ \ldots \\ \mathbf{z}_n \end{pmatrix}$$

Show that $\mathbf{z}^+$ and $\mathbf{z}^-$ belong to $P(\mathbf{x})$.

d. Conclude that $\mathbf{z}$ is not an extreme point of $P(\mathbf{x})$, contradicting the assumption. Hence at least $(n - m)$ of the $\mathbf{z}_i$ are extreme points of the corresponding conv S_i.

3. $\mathbf{z}$ is the required representation of $\mathbf{x}$, that is

$$\mathbf{x} = \sum_{i=1}^n \mathbf{z}_i, \quad \mathbf{z}_i \in \text{conv } S_i$$

and $\mathbf{z}_i \in S_i$ for all but at most m indexes i.

Exercise 3.211
Let $S_1 = \{0, 1\}$, $S_2 = \{2, 3\}$, and $x = 2.5$. Illustrate the sets conv S_i, P, and $P(x)$.

Exercise 3.212 (Shapley-Folkman theorem)
Extend exercise 3.210 to allow for noncompact S_i, thus proving the Shapley-Folkman theorem. [Hint: Use Carathéodory's theorem (exercise 1.175).]

3.9.2 Duality

Exactly what an economist means by *duality* is confusing—it is an overworked term that has many different shades of meaning. Nevertheless, it is clear that the foundation of the theory of duality in economics is the following proposition, a straightforward corollary of the strong separation theorem. It establishes a correspondence between a closed, convex set and its bounding hyperplanes. Since the latter can be identified with elements of the dual space X^*, we have a correspondence between convex sets in the primal space X and elements in the dual space X^*.

Proposition 3.17 (Minkowski's theorem) *A closed, convex set in a normed linear space is the intersection of the closed halfspaces that contain it.*

Minkowski's theorem is illustrated in figure 3.22.

Exercise 3.213
Prove proposition 3.17.

Remark 3.19 This result can be strengthened to so that it is confined to halfspaces of *supporting* (as opposed to *bounding*) hyperplanes (Rockafellar 1970, p. 169), which is often how it is presented and used in economic sources (e.g., Diewert 1982, p. 547; Klein 1973, p. 327). However, it is often the weaker result that is proved (e.g., Karlin 1959, p. 398).

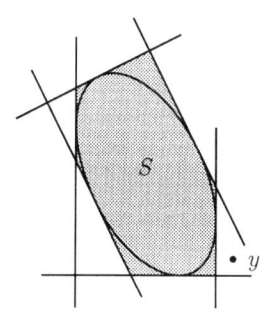

Figure 3.22
Minkowski's theorem

Example 3.80 (Recovering the technology from the cost function) The classic application of the duality of convex sets in economics involves the recovery of a production technology from cost data. To illustrate, suppose that the following data describe the behavior of a cost minimizing firm producing a fixed output y as factor prices change. x_i denotes the demand for factor i when its price is w_i.

w_1	w_2	x_1	x_2	Cost
1	2	10	5	11
2	1	6	10	11

From these data we can deduce certain facts about the technology of the firm. The isocost lines corresponding to the two price vectors are bounding hyperplanes $H_{(1,2)}(11)$ and $H_{(2,1)}(11)$ to the input requirement set $V(y)$ (figure 3.23). We can rule out certain production plans as infeasible. For example, the input combination $(8, 5)$ cannot be a feasible way of producing y ($(8, 5) \notin V(y)$), since it lies on the wrong side of the hyperplane $H_{(2,1)}(11)$. If we had more data, we could refine our knowledge of the technology.

If we know the cost function $c(\mathbf{w}, y)$ (example 2.31) of an unknown technology $V(y)$, we potentially have an infinite supply of data. Extending the previous line of reasoning, define the set

$$V^*(y) = \{\mathbf{x} : \mathbf{w}^T \mathbf{x} \geq c(\mathbf{w}, y) \text{ for every } \mathbf{w} \geq \mathbf{0}\}$$

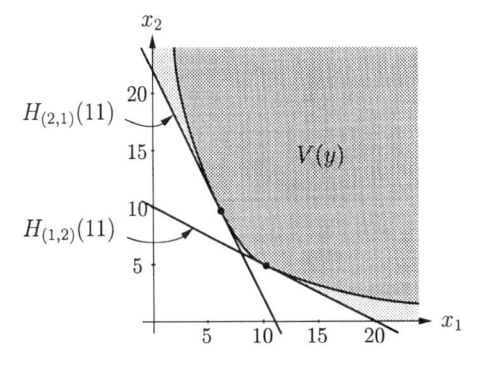

Figure 3.23
Recovering the technology from the cost function

$V^*(Y)$ is the set which is bounded by all conceivable isocost lines. It is a closed convex set that approximates the true technology $V(y)$. Furthermore, by virtue of proposition 3.17, if the technology is convex, the approximation is exact, that is, $V^*(y) = V(y)$. The practical importance of this derivation is that an analyst does not need to begin with detailed knowledge of the technology. She can concentrate on estimating the cost function, using market prices **w** that are easily observable and exogenous, and then recover the technology from the estimated cost function.

Exercise 3.214
Show that

1. $V^*(y)$ is a closed convex set containing $V(y)$
2. if $V(y)$ is convex and monotonic, then $V^*(y) = V(y)$

Example 3.81 (Leontief technology) Suppose that the cost function is linear in factor prices, that is,

$$c(\mathbf{w}, y) = \left(\sum_{i=1}^{n} b_i w_i \right) y$$

Then

$$V^*(y) = \{\mathbf{x} : \mathbf{w}^T \mathbf{x} \geq c(\mathbf{w}, y) \text{ for every } \mathbf{w} \geq \mathbf{0}\}$$

$$= \left\{ \mathbf{x} : \sum_{i=1}^{n} w_i x_i \geq \left(\sum_{i=1}^{n} b_i w_i \right) y \text{ for every } \mathbf{w} \geq \mathbf{0} \right\}$$

Equivalently, $\mathbf{x} \in V^*(y)$ if it satisfies the inequality $\sum_{i=1}^{n} w_i(x_i - b_i y) \geq 0$ for every $\mathbf{w} \geq \mathbf{0}$. This requires that $x_i \geq b_i y$ for $i = 1, 2, \ldots, n$. Equivalently

$$y = \min\left\{ \frac{x_1}{b_1}, \frac{x_2}{b_2}, \ldots, \frac{x_n}{b_n} \right\}$$

or letting $a_i = 1/b_i$,

$$y = \min\{a_1 x_1, a_2 x_2, \ldots, a_n x_n\}$$

which is known as the *Leontief production function*.

Exercise 3.215
Assume that the cost function for a convex, monotonic technology $V(y)$ is linear in y, that is, $c(\mathbf{w}, y) = \hat{c}(\mathbf{w}) y$. Then the technology exhibits constant returns to scale. [Hint: Consider $V^*(y)$.]

Polyhedral Sets

By Minkowski's theorem (proposition 3.17), any convex set is the intersection of a (possibly infinite) collection of closed halfspaces. If a finite number of halfspaces is sufficient to determine a convex set, then the set is called a *polyhedral set* or simply a *polyhedron*. Formally a set S in a normed linear space is *polyhedral* if there exist linear functionals $g_1, g_2, \ldots, g_m \in X^*$ and numbers c_1, c_2, c_m such that $S = \{\mathbf{x} \in X : g_i(\mathbf{x}) \leq c_i, i = 1, 2, \ldots, m\}$.

Exercise 3.216
Every polyhedral set is closed and convex.

Exercise 3.217
Let A be an $m \times n$ matrix. The set of solutions to the system of linear inequalities $A\mathbf{x} \leq \mathbf{c}$ is a polyhedron in $\Re^n$.

Example 3.82 (Core of a TP-coalitional game) The core of a TP-coalitional game (N, w) is the set of allocations $\mathbf{x} \in \Re^n$ for which

$$\sum_{i \in S} x_i \geq w(S) \qquad \text{for every } S \subset N$$

and

$$\sum_{i \in N} x_i = w(N)$$

The equation can be represented as a pair of inequalities

$$\sum_{i \in N} x_i \leq w(N), \quad -\sum_{i \in N} x_i \leq -w(N)$$

So the core of a TP-coalitional game is the solution to a system of linear inequalities and therefore a polyhedron (exercise 3.217). Geometrically, each coalitional constraint $\sum_{i \in S} x_i \geq v(S)$ defines a closed halfspace in $\Re^n$ bounded by the hyperplane $\sum_{i \in S} x_i = v(S)$. The core is the intersection of these 2^n closed halfspaces. Since each coalitional constraint is defined by a hyperplane, TP-coalitional games are sometimes called *hyperplane games*.

Exercise 3.218
The core of a TP-coalitional game is compact.

A polyhedral set which is nonempty and compact is called a *polytope*. Earlier (section 1.4.3) we defined a polytope as the convex hull of finite set of points. The equivalence of these two definitions follows from the Krein-Milman theorem.

Exercise 3.219
Show that a polytope can be defined alternatively as

- the convex hull of a finite set of points
- a nonempty compact polyhedral set

That is, show the equivalence of these two definitions.

Example 3.83 (Core of a TP-coalitional game) Provided it is nonempty, the core of a TP-coalitional game is a polytope (exercise 3.218). This means that the core can be represented as the convex hull of its extreme points, an alternative representation that is often more revealing than the corresponding system of inequalities.

Example 3.84 (Three way market) The characteristic function of the three way market game (example 3.7) is

$$w(\{f\}) = 1, \quad w(\{m\}) = 0, \quad w(\{s\}) = 0$$
$$w(\{f,m\}) = 2, \quad w(\{f,s\}) = 3, \quad w(\{m,s\}) = 0$$
$$w(N) = 3$$

So its core is

$$\text{core} = \{\mathbf{x} \in \Re^3 : x_1 \geq 1, x_2 \geq 0, x_3 \geq 0, x_1 + x_2 \geq 2, x_1 + x_3 \geq 3,$$
$$x_1 + x_2 + x_3 = 3\}$$

The extreme points of the core are $\{(3,0,0),(2,0,1)\}$, so the core can be represented as $\text{core} = \alpha(3,0,0) + (1-\alpha)(2,0,1)$ for $0 \leq \alpha \leq 1$, or alternatively,

$$\text{core} = \{(3-\alpha, 0, \alpha) : 0 \leq \alpha \leq 1\}$$

This expression highlights the two features of the core in this game:

- the zero payoff to player 2
- the discretion over the division of 1 unit between players 1 and 3.

382   Chapter 3   Linear Functions

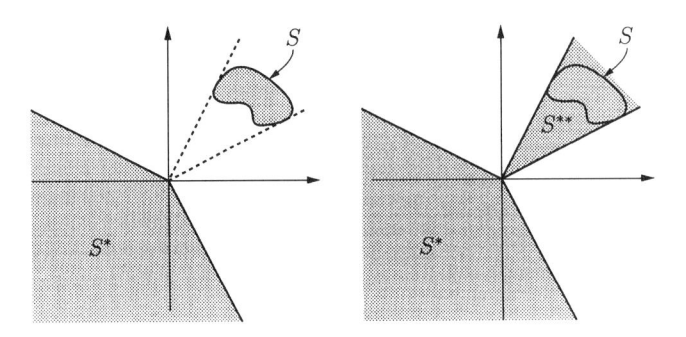

Figure 3.24
Polar cones in $\Re^2$

Duality for Cones

Given any subset S of a normed linear space X, the *polar* or *dual* cone of S is defined to be

$$S^* = \{f \in X^* : f(\mathbf{x}) \leq 0 \text{ for every } x \in S\}$$

The polar cone of S^* is called the *bipolar* of S. It is defined by

$$S^{**} = \{\mathbf{x} \in X : f(\mathbf{x}) \leq 0 \text{ for every } f \in S^*\}$$

The polar cone is a generalization of the orthogonal complement of a subspace (see figure 3.24). Polar cones provide elegant proofs of a number of results. Their basic properties are summarized in the following exercise.

Exercise 3.220 (Polar cones)
Let S be a nonempty set in a normed linear space X. Show

1. S^* is a closed convex cone in X^*
2. S^{**} is a closed convex cone in X
3. $S \subseteq S^{**}$
4. $\bar{S} \subseteq S^{**}$

Exercise 3.221
Let S_1, S_2 be nonempty sets in a normed linear space.

$$S_1 \subseteq S_2 \Rightarrow S_2^* \subseteq S_1^*$$

Whatever the nature of the set S, S^* and S^{**} are always cones (in X^* and X respectively) and $S \subseteq S^{**}$. Under what conditions is S identical to its bipolar S^{**} (rather than a proper subset)? Clearly, a necessary condition is that S is a convex cone. The additional requirement is that S is closed. This is another implication of the strong separation theorem.

Exercise 3.222 (Duality theorem for convex cones)
Let S be a nonempty convex cone in a normed linear space X. Then $S = S^{**}$ if and only if S is closed.

The duality theory of convex cones provides an elegant proof of the Farkas lemma that extends a classic result on linear equations—the Fredholm alternative (exercises 3.48 and 3.199)—to systems of linear inequalities. The Farkas lemma is an equivalent formulation of the basic separation theorem that is particularly useful for applications. Among other things, it implies the duality theorem in linear programming, the minimax and Bondareva-Shapley theorems (propositions 3.20 and 3.21) of game theory, and the Kuhn-Tucker theorem (theorem 5.3) of nonlinear programming.

Proposition 3.18 (Farkas lemma) Let $g_1, g_2, \ldots, g_m$ be linear functionals on a (reflexive) normed linear space X, and let

$$S = \{\mathbf{x} \in X : g_j(\mathbf{x}) \leq 0, j = 1, 2, \ldots, m\}$$

Suppose that $f \neq 0$ is another linear functional such that $f(\mathbf{x}) \leq 0$ for every $\mathbf{x} \in S$. Then $f \in cone\{g_1, g_2, \ldots, g_m\}$, the conic hull of the g_j. That is, there exist nonnegative constants λ_j such that

$$f(x) = \sum_{j=1}^{m} \lambda_j g_j(x)$$

Proof Let $K = cone\{g_1, g_2, \ldots, g_m\}$. Since X is reflexive (remark 3.5)

$$K^* = \{\mathbf{x} \in X : g(\mathbf{x}) \leq 0 \text{ for every } g \in K\} = S$$

We need to prove that $f \in K$. Since K is a closed convex cone, $K = K^{**}$ (exercise 3.222), and

$$K^{**} = \{g \in X^* : g(\mathbf{x}) \leq 0 \text{ for every } \mathbf{x} \in K^*\}$$
$$= \{g \in X^* : g(\mathbf{x}) \leq 0 \text{ for every } \mathbf{x} \in S\}$$

$f \in K$ if and only if $f \in K^{**}$. By assumption, $f \in K^{**}$, which implies that $f \in K$. □

Remark 3.20 To exploit the duality theory of convex cones, our proof of the Farkas lemma presumes a reflexive normed linear space (see remark 3.5). Reflexivity is not necessary, although establishing the theorem in a general linear space requires a more elaborate proof (Fan 1956, thm. 4, p. 108). Braunschweiger and Clark (1962) discuss the necessity of reflexivity. The Farkas lemma is most often encountered in $\Re^n$ or other Hilbert space, where a more direct proof can be given (exercise 3.223).

Exercise 3.223 (Farkas lemma in Hilbert space)
Use proposition 3.14 directly to prove the Farkas lemma when X is a Hilbert space. [Hint: Use the Riesz representation theorem (exercise 3.75).]

A major application of the traditional theory of linear algebra addresses the consistency of linear models; that is, it specifies necessary and sufficient conditions for the existence of solutions to a systems of linear equations or inequalities (section 3.6). In many applications we also require that the solution of the model be nonnegative. This is the most frequent application of the Farkas lemma, which provides necessary and sufficient conditions for the existence of a *nonnegative* solution to a system of linear equations, as detailed in the following exercise.

Exercise 3.224
Let A be an $m \times n$ matrix. A necessary and sufficient condition for the system of linear equations $A^T \mathbf{y} = \mathbf{c}$ to have a nonnegative solution $\mathbf{y} \in \Re^m_+$ is that $\mathbf{c}^T \mathbf{x} \leq 0$ for every $\mathbf{x} \in \Re^n$ satisfying $A\mathbf{x} \leq \mathbf{0}$.

Separation with Nonnegative Normal

In economic applications the underlying space X is usually a set of commodity bundles or allocations of resources, and the linear functionals are valuations or equivalently prices. It is often appropriate that the prices be nonnegative. The following corollary of the separating hyperplane theorem is particularly useful in such cases. We give the result for $\Re^n$, since its application in the more general spaces requires the specification of an order structure for the space.

3.9 Separation Theorems

Exercise 3.225 (Nonnegative normal)
Let S be convex set in $\Re^n$ which contains no interior points of the nonnegative orthant $\Re^n_+$. Then there exists a hyperplane with nonnegative normal $\mathbf{p} \gneq \mathbf{0}$ (that is $\mathbf{p} \neq \mathbf{0}$, $p_i \geq 0$) such that $\mathbf{p}^T\mathbf{x} \leq 0$ for every $\mathbf{x} \in S$.

Earlier (exercise 3.42) we showed that profit maximization implies efficiency. Now we establish the converse provided the production set is convex. This is another version of the second theorem of welfare economics (example 3.77).

Exercise 3.226
Suppose that the production possibility set Y is convex. Then every efficient production plan $\mathbf{y} \in Y$ is profit maximizing for some nonzero price system $\mathbf{p} \geq \mathbf{0}$.

The mirror image of exercise 3.225 is useful in some applications (exercises 3.228, 3.259).

Exercise 3.227
Let S be convex set in $\Re^n$ that contains no interior points of the nonpositive orthant $\Re^n_-$. Then there exists a hyperplane with nonnegative normal $\mathbf{p} \gneq \mathbf{0}$ such that

$$\mathbf{p}^T\mathbf{x} \geq 0 \quad \text{for every } \mathbf{x} \in S$$

Exercise 3.228 (Exchange economy)
Suppose that $\underline{\mathbf{x}}^* = (\mathbf{x}_1^*, \mathbf{x}_2^*, \ldots, \mathbf{x}_n^*)$ is a Pareto efficient allocation in an exchange economy with l commodities and n consumers (example 1.117). Assume that

- individual preferences are convex, continuous and strongly monotonic.
- $\mathbf{x}^* = \sum_{i=1}^n \mathbf{x}_i^* > \mathbf{0}$

Show that

1. The set $\succsim(\mathbf{x}^*) = \sum_{i=1}^n \succsim_i(\mathbf{x}_i^*)$ is the set of all aggregate commodity bundles that can be distributed so as to make all the consumers at least as well off as at the allocation $\underline{\mathbf{x}}^*$.

2. $S = \succsim(\mathbf{x}^*) - \mathbf{x}^*$ is nonempty, convex and contains no interior points of the nonpositive orthant $\Re^l_-$.

3. There exist prices $\mathbf{p}^* \in \Re_+^l$ such that $(\mathbf{p}^*)^T \mathbf{x} \geq (\mathbf{p}^*)^T \mathbf{x}^*$ for every $\mathbf{x} \in \succsim(\mathbf{x}^*)$.

4. For every consumer i, $(\mathbf{p}^*)^T \mathbf{x}_i \geq (\mathbf{p}^*)^T \mathbf{x}_i^*$ for every $\mathbf{x}_i \in \succ_i(\mathbf{x}_i^*)$.

5. $(\mathbf{p}^*, \underline{\mathbf{x}}^*)$ is a competitive equilibrium with endowments $\mathbf{w}_i = \mathbf{x}_i^*$

Exercise 3.229 (Second theorem of welfare economics)
Suppose that $\underline{\mathbf{x}}^* = (\mathbf{x}_1^*, \mathbf{x}_2^*, \ldots, \mathbf{x}_n^*)$ is a Pareto-efficient allocation in an exchange economy (example 1.117) in which each of the n consumers has an endowment $\mathbf{w}_i \in \Re_+^l$ of the l commodities. Assume that

- individual preferences $\succsim_i$ are convex, continuous and strongly monotonic
- $\underline{\mathbf{x}}^*$ is a feasible allocation, that is $\sum_i \mathbf{x}_i^* = \sum_i \mathbf{w}_i > 0$

Show that there exists

- a list of prices $\mathbf{p}^* \in \Re_+^l$ and
- a system of lump-sum taxes and transfers $\mathbf{t} \in \Re^n$ with $\sum_i t_i = 0$

such that $(\mathbf{p}^*, \underline{\mathbf{x}}^*)$ is a competitive equilibrium in which each consumer's after-tax wealth is $m_i = (\mathbf{p}^*)^T \mathbf{w}_i + t_i$.

Example 3.85 Let $S = \{(x, x, 0) : x \in \Re\}$. S is a subspace of $\Re^3$. There exists a hyperplane (improperly) separating S from $\Re_+^3$, namely the hyperplane orthogonal to the x_3 axis that has normal $\mathbf{p} = (0, 0, p)$. Note that $S \cap \Re_+^3 = \{(x, x, 0) : x \in \Re_+\} \neq \mathbf{0}$ and $\mathbf{p}^T \mathbf{x} = 0$ for all $\mathbf{x} = (x_1, x_2, 0) \in \Re^3$.

In general, a separating hyperplane with a strictly positive normal $(\mathbf{p} > 0)$ cannot be guaranteed, as the preceding example illustrates. However, exercise 3.225 can be strengthened to ensure strictly positive prices $(p_i > 0)$ if S is a subspace which intersects the nonnegative orthant $\Re_+^n$ precisely at $\mathbf{0}$ (exercise 3.230).

Exercise 3.230 (Positive normal)
Let S be a subspace that intersects the nonnegative orthant at $\mathbf{0}$, that is $S \cap \Re_+^n = \{\mathbf{0}\}$. Then there exists a hyperplane with positive normal $\mathbf{p} > 0$ (i.e., $p_i > 0$ for every i) such that

$$\mathbf{p}^T \mathbf{x} = 0 \quad \text{for every } \mathbf{x} \in S$$

and

$$\mathbf{p}^T\mathbf{x} > 0 \quad \text{for every } \mathbf{x} \in \Re_+^n \setminus \{\mathbf{0}\}$$

Using the duality theory of cones, we can elegantly extend this result to closed convex cones.

Exercise 3.231 (Positive normal)
Let K be a closed convex cone that intersects the nonnegative orthant at $\mathbf{0}$, that is, $K \cap \Re_+^n = \{\mathbf{0}\}$. Then there exists a hyperplane with positive normal $\mathbf{p} > 0$ (i.e., $p_i > 0$ for every i) such that $\mathbf{p}^T\mathbf{x} \leq 0$ for every $\mathbf{x} \in K$. [Hint: Show $K \cap \Re_+^n = \{\mathbf{0}\} \Rightarrow K^* \cap \Re_{++}^n \neq \emptyset$.]

Example 3.86 (No arbitrage theorem) In a simple model of portfolio investment, we showed earlier (example 3.37) that where there exists a full set of Arrow-Debreu securities, the equilibrium price of asset a must be given by

$$p_a = \sum_{s=1}^{S} r_{sa} \pi_s$$

where π_s is the price of the s Arrow-Debreu security, which guarantees a payoff of \$1 in state s and zero in every other state. In reality the hypothetical Arrow-Debreu securities do not exist, and the number of states S vastly exceeds the number of assets A. It is a surprising consequence of the separating hyperplane theorem that a similar result holds even when there are fewer assets than states of the world.

Suppose that there are A assets with prices $\mathbf{p} = (p_1, p_2, \ldots, p_A)$ and return matrix

$$R = \begin{pmatrix} r_{11} & r_{12} & \cdots & r_{1A} \\ r_{21} & r_{22} & \cdots & r_{2A} \\ \cdots & \cdots & \cdots & \cdots \\ r_{S1} & r_{S2} & \cdots & r_{SA} \end{pmatrix}$$

The value of any portfolio $\mathbf{x}$ is $\mathbf{p}^T\mathbf{x} = \sum_{a=1}^{A} p_a x_a$.

Recall that a portfolio $\mathbf{x}$ can have negative as well as positive components. A negative component x_a indicates a *short position* in the asset a. Given a system of asset prices $\mathbf{p} = (p_1, p_2, \ldots, p_A)$, an *arbitrage* is a portfolio $\mathbf{x}$ such that $\mathbf{p}^T\mathbf{x} \leq 0$ and $R\mathbf{x} \geq 0$. It is a portfolio that has zero (or negative cost) and that guarantees a nonnegative return in every state and a positive return in at least one state. An arbitrage is the financial equivalent of a "free lunch," a sure way to make money. We expect an efficient market to eliminate any arbitrage possibilities very rapidly. The no arbitrage condition precludes the existence of any arbitrage opportunities in equilibrium. Formally the *no arbitrage condition* requires any portfolio that guarantees nonnegative returns in every state to have a nonnegative value, that is,

$$R\mathbf{x} \geq 0 \Rightarrow \mathbf{p}^T\mathbf{x} \geq 0$$

This simple equilibrium condition has surprisingly deep implications.

In particular, the no arbitrage condition implies the existence of positive *state prices* $\pi = (\pi_1, \pi_2, \ldots, \pi_S)$ such that the price of any security a is given by

$$p_a = \sum_{s=1}^{S} r_{as}\pi_s$$

The state price $\pi_s > 0$ measures the value of \$1 to be received in state s. It reflects the likelihood that state s occurs. The state prices are the implicit prices of the nonexistent Arrow-Debreu securities (example 3.37). Even in the absence of such securities, financial market acts as though they exist in the sense that the equilibrium prices of all existing securities are compounds of the implicit prices of notional basic securities. The fundamental theorem of financial economics states that the no arbitrage condition is a necessary and sufficient condition for the existence of state prices. This has many implications.

Exercise 3.232 *(No arbitrage theorem)*
There is no arbitrage if and only if there exist state prices. [Hint: Apply exercise 3.230 to the set $Z = \{(-\mathbf{p}^T\mathbf{x}, R\mathbf{x}) : \mathbf{x} \in \Re^n\}$.]

3.9.3 Theorems of the Alternative

We saw earlier that the Farkas lemma provides necessary and sufficient conditions for the existence of a nonnegative solution to a system of linear equations (exercise 3.224). Specifically, if A is an $m \times n$ matrix, a necessary and sufficient condition for the system of linear equations $A^T\mathbf{y} = \mathbf{c}$ to have a nonnegative solution $\mathbf{y} \in \Re_+^m$ is that $\mathbf{c}^T\mathbf{x} \leq 0$ for every $\mathbf{x} \in \Re^n$, satisfying $A\mathbf{x} \leq \mathbf{0}$. Therefore, if $\mathbf{c} = A^T\mathbf{y}$ for some $\mathbf{y} \in \Re_+^m$, there is no $\mathbf{x} \in \Re^n$ with $A\mathbf{x} \leq \mathbf{0}$ and $\mathbf{c}^T\mathbf{x} > 0$. In other words, the Farkas lemma asserts that one and only one of these two linear systems is consistent. In this form the Farkas lemma is the archetypical *theorem of the alternative*, a collection of results which provide versatile tools in applications.

Proposition 3.19 (Farkas alternative) *Let A be an $m \times n$ matrix and $\mathbf{c}$ be a nonzero vector in $\Re^n$. Then exactly one of the following two alternatives holds:*

Either I *there exists $\mathbf{x} \in \Re^n$ such that $A\mathbf{x} \leq \mathbf{0}$ and $\mathbf{c}^T\mathbf{x} > 0$*

or II *there exists $\mathbf{y} \in \Re_+^m$ such that $A^T\mathbf{y} = \mathbf{c}$.*

Proof Assume that alternative I does not hold. That is, $\mathbf{c}^T\mathbf{x} \leq 0$ for every $\mathbf{x}$ such that $A\mathbf{x} \leq \mathbf{0}$. Then there exists $\mathbf{y} \in \Re_+^m$ such that $\mathbf{c} = A^T\mathbf{y}$ (exercise 3.224). This is alternative II. Conversely, suppose that alternative II holds. Then $\mathbf{c}^T\mathbf{x} \leq 0$ for every x such that $A\mathbf{x} \leq \mathbf{0}$ (exercise 3.224). This precludes alternative I. □

The Farkas alternative is illustrated in figure 3.25. Suppose that $m = 3$. Each inequality defines a closed halfspace, which is depicted in the diagram by a hyperplane and its associated normal. For example, the set of all $\mathbf{x}$ satisfying $\mathbf{a}_1^T\mathbf{x} \leq 0$ is given by the area below and to the left of the hyperplane $H_{\mathbf{a}_1}(0)$. The set of all $\mathbf{x}$ satisfying the three inequalities is the cone POQ in figure 3.25a. POQ is the polar cone to the cone $\mathbf{a}_1 0 \mathbf{a}_3$ generated by the normals to the hyperplanes.

There are only two possibilities for any other linear functional $\mathbf{c}$. Either the hyperplane $H_{\mathbf{c}}(0)$ intersects the cone POQ or it does not. If the hyperplane $H_{\mathbf{c}}(0)$ intersects the cone POQ (figure 3.25b), there exists a point $\hat{\mathbf{x}}$ that belongs to both $H^+ = \{\mathbf{x} : \mathbf{c}^T\mathbf{x} > 0\}$ and POQ, so that

$$A\hat{\mathbf{x}} \leq 0 \ (\hat{\mathbf{x}} \in POQ) \quad \text{and} \quad \mathbf{c}^T\hat{\mathbf{x}} > 0 \ (\hat{\mathbf{x}} \in H_{\mathbf{c}}^+)$$

$\hat{\mathbf{x}}$ satisfies alternative 1.

Alternatively, if the hyperplane $H_{\mathbf{c}}(0)$ does not intersect the cone POQ, the positive halfspace $H_{\mathbf{c}}^+$ must be disjoint from POQ (figure 3.25c). Then $\mathbf{c}^T\mathbf{x} \leq 0$ for $\mathbf{x}$ in POQ and the normal $\mathbf{c}$ must lie in the cone $\mathbf{a}_1 O \mathbf{a}_3$. Since $\mathbf{c} \in \text{cone}\{\mathbf{a}_1, \mathbf{a}_2, \mathbf{a}_3\}$, there exists $\mathbf{y} \geq \mathbf{0}$ such that $\mathbf{c} = \mathbf{y}^T A = A^T \mathbf{y}$.

□

Example 3.87 (Markov chains) As an application of the Farkas lemma, we show that every Markov chain (section 3.6.4) has a stationary distribution. Let T be the $n \times n$ transition matrix of finite Markov process. A

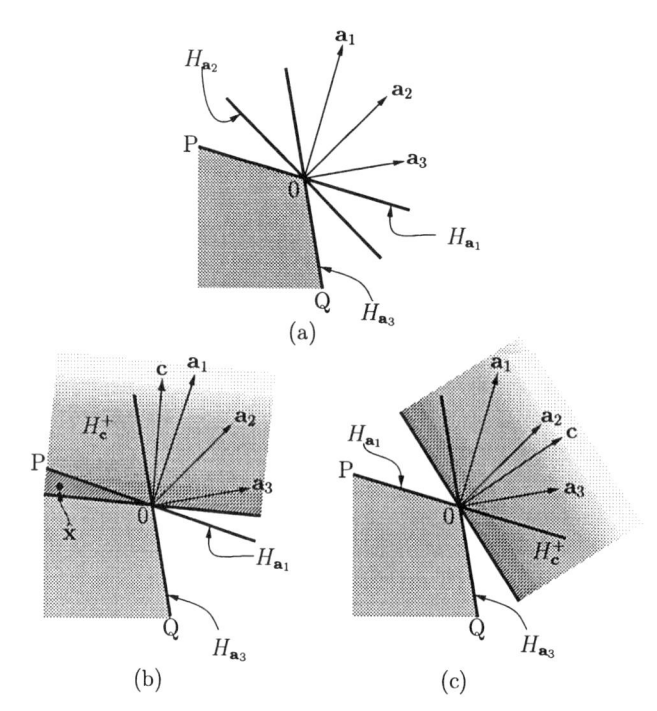

Figure 3.25
The Farkas lemma

stationary distribution **p** is a solution to the system

$$T\mathbf{p} = \mathbf{p}, \quad \sum_i p_i = 1 \tag{30}$$

which is equivalent to

$$T\mathbf{p} - I\mathbf{p} = \mathbf{0}, \quad \mathbf{1}\mathbf{p} = 1$$

where $\mathbf{1} = (1, 1, \ldots, 1) \in \Re^n$. This can be considered alternative II of the Farkas lemma with

$$A^T = \begin{pmatrix} T^T - I \\ \mathbf{1} \end{pmatrix}, \quad \mathbf{c} = \begin{pmatrix} \mathbf{0} \\ 1 \end{pmatrix}$$

Thus system (30) has a solution provided the corresponding alternative I has no solution. That is, there is no $\mathbf{x} \in \Re^{n+1}$ such that $A\mathbf{x} \le \mathbf{0}$ and

$\mathbf{cx} > 0$. Suppose, to the contrary, that this system has a solution. A is the matrix $A = (T^T - I, \mathbf{1}^T)$. Partition $\mathbf{x}$ so that

$$\mathbf{x} = \begin{pmatrix} \hat{\mathbf{x}} \\ x_{n+1} \end{pmatrix}$$

Then alternative II is the system

$$T^T \mathbf{x} - \mathbf{x} + x_{n+1} \mathbf{1}^T \leq \mathbf{0} \quad \text{and} \quad x_{n+1} > 0$$

which implies that $T^T \mathbf{x} < \mathbf{x}$ contradicting the fact that T is a stochastic matrix, with $t_{1j} + t_{2j} + \cdots + t_{nj} = 1$ for every $j = 1, 2, \ldots, n$. Therefore alternative I has no solution, which implies that alternative II has a solution $\mathbf{p} \in \Re_+^n$. This solution is a stationary distribution for the Markov process. This duplicates a result we obtained in example 2.94 using Brouwer's fixed point theorem (theorem 2.6).

The Farkas alternative is the best known of a host of similar results known as *theorems of the alternative*, because they assert that one and only one of two related linear systems will have a solution. Many of these theorems are equivalent to the Farkas lemma. We explore some variants in the following exercises.

One variant simply reverses the sense of the inequalities.

Exercise 3.233
Let A be an $m \times n$ matrix and $\mathbf{c}$ be a nonzero vector in $\Re^n$. Then exactly one of the following systems has a nonnegative solution:

Either I $A\mathbf{x} \geq \mathbf{0}$ and $\mathbf{cx} < 0$ for some $\mathbf{x} \in \Re^n$
or II $A^T \mathbf{y} = \mathbf{c}$ for some $\mathbf{y} \in \Re_+^m$.

Other variants of the Farkas alternative theorem can be obtained by using *slack variables* to transform inequalities to equations. The following example is typical.

Exercise 3.234 (Gale alternative)
Let A be an $m \times n$ matrix and $\mathbf{c}$ be a nonzero vector in $\Re^n$. Then exactly one of the following systems has a *nonnegative* solution:

Either I $A\mathbf{x} \leq \mathbf{0}$ and $\mathbf{cx} > 0$ for some $\mathbf{x} \in \Re_+^n$
or II $A^T \mathbf{y} \geq \mathbf{c}$ for some $\mathbf{y} \in \Re_+^m$.

Still other variants can be derived by creative reformulations of the alternative systems.

Exercise 3.235

Let A be an $m \times n$ matrix and $\mathbf{c}$ be a nonzero vector in $\Re^n$. Then exactly one of the following systems has a solution:

Either I $A\mathbf{x} = \mathbf{0}$ and $\mathbf{c}^T\mathbf{x} > 0$ for some $\mathbf{x} \in \Re^n_+$

or II $A^T\mathbf{y} \geq \mathbf{c}$ for some $\mathbf{y} \in \Re^m$.

[Hint: Show that system I is equivalent to $B\mathbf{x} = \mathbf{b}$, where $B = \begin{pmatrix} -A \\ \mathbf{c} \end{pmatrix}$, $\mathbf{b} = \begin{pmatrix} \mathbf{0} \\ 1 \end{pmatrix}$, $\mathbf{0} \in \Re^m$, and apply the Farkas alternative.]

Exercise 3.235 is often found in an equivalent form, known as Fan's condition. We will use this version to prove the Bondareva-Shapley theorem (proposition 3.21) for coalitional games.

Exercise 3.236 (Fan's condition)

Let $g_1, g_2, \ldots, g_m$ be linear functionals on $\Re^n$. For fixed numbers $c_1, c_2, \ldots, c_m$, the system of inequalities

$$g_j(\mathbf{x}) \geq c_j, \quad j = 1, 2, \ldots, m \tag{31}$$

is consistent for some $\mathbf{x} \in \Re^n$ if and only if $\sum_{j=1}^m \lambda_j g_j = \mathbf{0}$ implies that $\sum_{j=1}^m \lambda_j c_j \leq 0$ for every set of nonnegative numbers $\lambda_1, \lambda_2, \ldots, \lambda_m$.

These theorems of the alternative are all generalizations of a classic result in the theory of linear equations (the Fredholm alternative), which we have already encountered in our discussion of linear functionals (exercise 3.48) and as an application of the separation theorem (exercise 3.199).

Exercise 3.237 (Fredholm alternative)

Let A be an $m \times n$ matrix and $\mathbf{c}$ be a nonzero vector in $\Re^n$. Then exactly one of the following systems has a solution:

Either I $A\mathbf{x} = \mathbf{0}$ and $\mathbf{cx} > 0$

or II $A^T\mathbf{y} = \mathbf{c}$

[Hint: The system of equations $A\mathbf{x} = \mathbf{0}$ is equivalent to the system of inequalities $A\mathbf{x} \leq \mathbf{0}$, $-A\mathbf{x} \leq \mathbf{0}$.]

Table 3.2
Variants of the Farkas alternative

System I	System II	
$A\mathbf{x} = \mathbf{0}, \mathbf{c}^T\mathbf{x} > 0$	$A^T\mathbf{y} = \mathbf{c}$	Fredholm
$A\mathbf{x} \leq \mathbf{0}, \mathbf{c}^T\mathbf{x} > 0$	$A^T\mathbf{y} = \mathbf{c}, \mathbf{y} \geq \mathbf{0}$	Farkas
$A\mathbf{x} \leq \mathbf{0}, \mathbf{c}^T\mathbf{x} > 0, \mathbf{x} \geq \mathbf{0}$	$A^T\mathbf{y} \geq \mathbf{c}, \mathbf{y} \geq \mathbf{0}$	Gale
$A\mathbf{x} = \mathbf{0}, \mathbf{c}^T\mathbf{x} > 0, \mathbf{x} \geq \mathbf{0}$	$A^T\mathbf{y} \geq \mathbf{c}$	Fan

Exercise 3.238
Derive the Fredholm alternative (exercise 3.237) directly from exercise 3.199.

These variants of the Farkas alternative are summarized in table 3.2, which highlights the interplay between inequalities in one system and nonnegativity restrictions in the other. The Fredholm alternative is a standard result in linear algebra. The Farkas alternative theorem generalizes system I from homogeneous equations to inequalities, requiring nonnegativity in system II as a consequence. The Gale alternative gives existence conditions for consistency of two systems of inequalities. Compared to the Farkas lemma, it generalizes system II from nonhomogeneous equations to inequalities; it requires nonnegativity in system I as a consequence. The final variant dispenses with the nonnegativity condition in system II, by requiring that system I be strengthened to a system of homogeneous equations. This final variant provides an interesting contrast to the Fredholm alternative that heads the table.

Another useful alternative theorem, known as *Gordan's theorem*, deals with the consistency of *homogeneous* equations and inequalities. We first derive Gordan's theorem independently using an appropriate separation theorem, and present some of its variants. Later we show that Gordan's theorem is equivalent to the Farkas alternative and then use it to derive the basic separation theorem, completing a circle of equivalences.

Exercise 3.239 (Gordan's theorem)
Let A be an $m \times n$ matrix. Then exactly one of the following systems as a solution:

Either I $A\mathbf{x} > \mathbf{0}$ for some $\mathbf{x} \in \Re^n$

or II $A^T\mathbf{y} = \mathbf{0}, \mathbf{y} \gneq \mathbf{0}$ for some $\mathbf{y} \in \Re^m$

[Hint: Apply exercise 3.225 to the set $S = \{\mathbf{z} : \mathbf{z} = A\mathbf{x}, \mathbf{x} \in \Re\}$.]

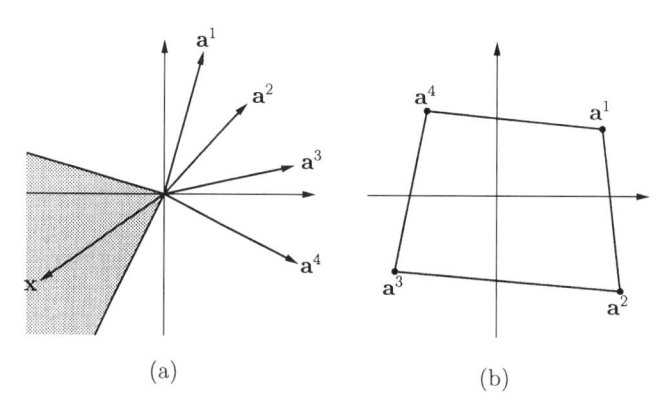

Gordan's theorem is illustrated in figure 3.26. For every matrix A, either (**I**) there exists a vector **x** that makes an obtuse angle with every row of A (figure 3.26a) or (**II**) the origin is in the convex hull of the row vectors (linear functionals) of A (figure 3.26b).

Exercise 3.240
Derive Gordan's theorem from the Farkas alternative. [Hint: Apply the Farkas alternative to the matrix $B = (A, \mathbf{1})$, the matrix B augmented with a column of ones.]

Gordan's theorem has a natural geometric interpretation which is given in the following exercise.

Exercise 3.241
Let S be a subspace in $\Re^n$ and $S^\perp$ its orthogonal complement.

Either I S contains a positive vector $\mathbf{y} > \mathbf{0}$

or II $S^\perp$ contains a nonnegative vector $\mathbf{y} \geqq \mathbf{0}$

Interchanging the role of S and $S^\perp$ in the previous exercise provides the following variant of Gordan's theorem.

Exercise 3.242 (Stiemke's theorem)
Let A be an $m \times n$ matrix. Then exactly one of the following systems has a solution:

Either I $A\mathbf{x} \geqq 0$ for some $\mathbf{x} \in \Re^n$
or II $A^T\mathbf{y} = 0, \mathbf{y} > \mathbf{0}$ for some $\mathbf{y} \in \Re^m_{++}$

Exercise 3.243
Deduce Stiemke's theorem directly from exercise 3.230.

Analogous to the Farkas lemma (exercise 3.234), system II in Gordan's theorem can be generalized to an inequality, requiring a nonnegativity restriction on system I. This gives a theorem of the alternative attributed to von Neumann.

Exercise 3.244 (von Neumann alternative I)
Let A be an $m \times n$ matrix. Then exactly one of the following systems has a *nonnegative* solution:

Either I $A\mathbf{x} > 0, \mathbf{x} > \mathbf{0}$ for some $\mathbf{x} \in \Re^n$
or II $A^T\mathbf{y} \leq 0, \mathbf{y} \geqq \mathbf{0}$ for some $\mathbf{y} \in \Re^m$.

[Hint: Use slack variables as in exercise 3.234.]

We will use the following variation to prove the fundamental minimax theorem (proposition 3.20) of game theory.

Exercise 3.245 (von Neumann alternative II)
Let A be an $m \times n$ matrix. Then exactly one of the following systems has a *nonnegative* solution:

Either I $A^T\mathbf{x} > \mathbf{0}, \mathbf{x} \geqq \mathbf{0}$ for some $\mathbf{x} \in \Re^m$
or II $A\mathbf{y} \leq \mathbf{0}, \mathbf{y} \geqq \mathbf{0}$ for some $\mathbf{y} \in \Re^n$.

[Hint: Apply the Gale alternative (exercise 3.234) to the system $A\mathbf{y} \leq \mathbf{0}$, $\mathbf{1}^T\mathbf{y} \geq 1$.]

Table 3.3 summarizes the different versions of Gordan's theorem. Note again the interplay between inequalities in one system and nonnegativity conditions in the other.

Gordan's and Stiemke's theorems provide the two extreme cases of the phenomenom called *complementary slackness*, which is of practical importance in the theory of optimization. The two homogeneous linear systems

$$A\mathbf{x} \geq 0 \quad \text{and} \quad A^T\mathbf{y} = \mathbf{0}, \quad \mathbf{y} \geq 0$$

Table 3.3
Variants of Gordan's theorem

System I	System II	
$A\mathbf{x} > \mathbf{0}$	$A^T\mathbf{y} = \mathbf{0}, \mathbf{y} \geqslant \mathbf{0}$	Gordan
$A\mathbf{x} \geqslant \mathbf{0}$	$A^T\mathbf{y} = \mathbf{0}, \mathbf{y} > \mathbf{0}$	Stiemke
$A\mathbf{x} > \mathbf{0}, \mathbf{x} > \mathbf{0}$	$A^T\mathbf{y} \leq \mathbf{0}, \mathbf{y} \geqslant \mathbf{0}$	von Neumann

are said to form a *dual pair*. A fundamental theorem of Tucker guarantees the existence of a pair of solutions **x** and **y** to every dual pair such that

$$A\mathbf{x} + \mathbf{y} > \mathbf{0} \tag{32}$$

Let $(A\mathbf{x})_j$ denote the *j*th element of the list $A\mathbf{x}$. Inequality (32) requires that $(A\mathbf{x})_j + y_j$ is positive for every *j*. In other words, for every $j = 1, 2, \ldots, m$ either $y_j > 0$ or the corresponding *j*th element of $A\mathbf{x}$ is positive. For each coordinate *j*, only one of the constraints $y_j \geq 0$ and $(A\mathbf{x})_j \geq 0$ is binding. Tucker's theorem follows quite readily from Gordan's and Stiemke's theorem, using the result in the following exercise.

Exercise 3.246
Assume the system $A\mathbf{x} \geqslant \mathbf{0}$ has a solution, while $A\mathbf{x} > \mathbf{0}$ has no solution. Then A can be decomposed into two consistent subsystems

$$B\mathbf{x} > \mathbf{0} \quad \text{and} \quad C\mathbf{x} = \mathbf{0}$$

such that $C\mathbf{x} \geqslant \mathbf{0}$ has no solution.

Exercise 3.247 (*Tucker's theorem*)
Let A be an $m \times n$ matrix. The dual pair

$$A\mathbf{x} \geq \mathbf{0} \quad \text{and} \quad A^T\mathbf{y} = \mathbf{0}, \quad \mathbf{y} \geq \mathbf{0}$$

possess a pair of solutions $\mathbf{x} \in \Re^n$ and $\mathbf{y} \in \Re^m$ such that $A\mathbf{x} + \mathbf{y} > \mathbf{0}$.

Once again, an equation can be relaxed by including a nonnegativity condition.

Exercise 3.248 (*von Neumann*)
Let A be an $m \times n$ matrix. The dual pair

$$A\mathbf{x} \geq \mathbf{0}, \mathbf{x} \geq \mathbf{0} \quad \text{and} \quad A^T\mathbf{y} \leq \mathbf{0}, \quad \mathbf{y} \geq \mathbf{0}$$

possess a pair of solutions $\mathbf{x} \in \Re^n$ and $\mathbf{y} \in \Re^m$ such that

$A\mathbf{x} + \mathbf{y} > \mathbf{0}$ and $\mathbf{x} - A^T\mathbf{y} > \mathbf{0}$

[Hint: Apply Tucker's theorem to the $2m \times n$ matrix $\begin{pmatrix} A \\ I \end{pmatrix}$.]

Tucker's theorem was described as the *key theorem* by Good (1959), since so many results can be derived from it. Gordan's and Stiemke's theorems are obvious corollaries. The Farkas lemma is another corollary. It can also provide theorems that are apparently more general, such as Motzkin's theorem.

Exercise 3.249
Show that Gordan's and Stiemke's theorems are special cases of Tucker's theorem.

Exercise 3.250
Derive the Farkas lemma from Tucker's theorem.

Exercise 3.251 (Motzkin's theorem)
Let A, B, and C be matrices of order $m_1 \times n$, $m_2 \times n$ and $m_3 \times n$ respectively with A nonvacuous. Then either

Either I $A\mathbf{x} > \mathbf{0}, B\mathbf{x} \geq \mathbf{0}, C\mathbf{x} = \mathbf{0}$ has a solution $\mathbf{x} \in \Re^n$

or II $A^T\mathbf{y}_1 + B^T\mathbf{y}_2 + C^T\mathbf{y}_3 = \mathbf{0}$ has a solution $\mathbf{y}_1 \in \Re^{m_1}$, $\mathbf{y}_2 \in \Re^{m_2}$, $\mathbf{y}_3 \in \Re^{m_3}$ with $\mathbf{y}_1 \gneq \mathbf{0}, \mathbf{y}_2 \geq \mathbf{0}$.

Exercise 3.250 completes the cycle

Farkas lemma $\Rightarrow$ Gordan's theorem $\Rightarrow$ Tucker's theorem

$\Rightarrow$ Farkas lemma

establishing the mutual equivalence of the basic theorems of the alternative. These theorems were in turn derived from appropriate separation theorems. In the following exercise we reverse this process, deriving the basic separation theorem from Gordan's theorem. This establishes the fundamental equivalence between the separation theorems and theorems of the alternative in $\Re^n$.

Exercise 3.252
Let S be a nonempty convex set in $\Re^n$ with $\mathbf{0} \notin S$.

1. For every point $\mathbf{a} \in S$, define the polar set $S_\mathbf{a}^* = \{\mathbf{x} \in \Re^n : \|\mathbf{x}\| = 1, \mathbf{x}^T\mathbf{a} \geq 0\}$. $S_\mathbf{a}^*$ is a nonempty closed set.

2. For any finite set of points $\{\mathbf{a}_1, \mathbf{a}_2, \ldots, \mathbf{a}_m\}$ in S, let A be the $m \times n$ matrix whose rows are $\mathbf{a}_i$. The system

$$A^T \mathbf{y} = \sum_{i=1}^m y_i \mathbf{a}_i = \mathbf{0}$$

has no solution $\mathbf{y} \in \Re_+^m$.

3. The system $A\mathbf{x} > 0$, $i = 1, 2 \ldots, m$ has a solution $\bar{\mathbf{x}} \in \Re^n$, $\bar{\mathbf{x}} \neq \mathbf{0}$.
4. $\bar{\mathbf{x}} \in \bigcap_{i=1}^m S_{\mathbf{a}_i}^*$.
5. $\bigcap_{\mathbf{a} \in S} S_{\mathbf{a}}^* \neq \emptyset$.
6. There exists a hyperplane $f(\mathbf{a}) = \mathbf{p}^T \mathbf{a}$ that separates S from $\mathbf{0}$ such that $\mathbf{p}^T \mathbf{a} \geq 0$ for every $\mathbf{a} \in S$.

3.9.4 Further Applications

We have already presented a number of applications of the separation theorems or the equivalent theorems of the alternative. The classic application in economics is the second theorem of welfare economics (example 3.228). Another important application in economics is the theory of duality in consumer and producer theory (example 3.80). In finance, the no arbitrage theorem is a straightforward application of the Farkas lemma. In chapter 5 we will use the Farkas lemma again to prove the Kuhn-Tucker theorem (theorem 5.3) for constrained optimization. We will use the separating hyperplane theorem directly to derive a stronger result for concave programming. To complete this survey of applications, we now derive two fundamental theorems of game theory. These applications are summarized in figure 3.27.

Zero-Sum Games and the Minimax Theorem

In 1953 an intellectual paternity dispute was aired in the pages of *Econometrica*. The famous French mathematician Maurice Fréchet claimed that the even more famous French mathematician Emile Borel should be credited with initiating the formal study of game theory. In particular, Borel proposed the concept of a strategy, and suggested resort to mixed strategies to avoid loss and maximize expected payoff. John von Neumann, the acknowledged father of game theory, demurred. He suggested that the importance of Borel's contribution was diminished because the latter had not formulated the essential minimax theorem, and indeed had

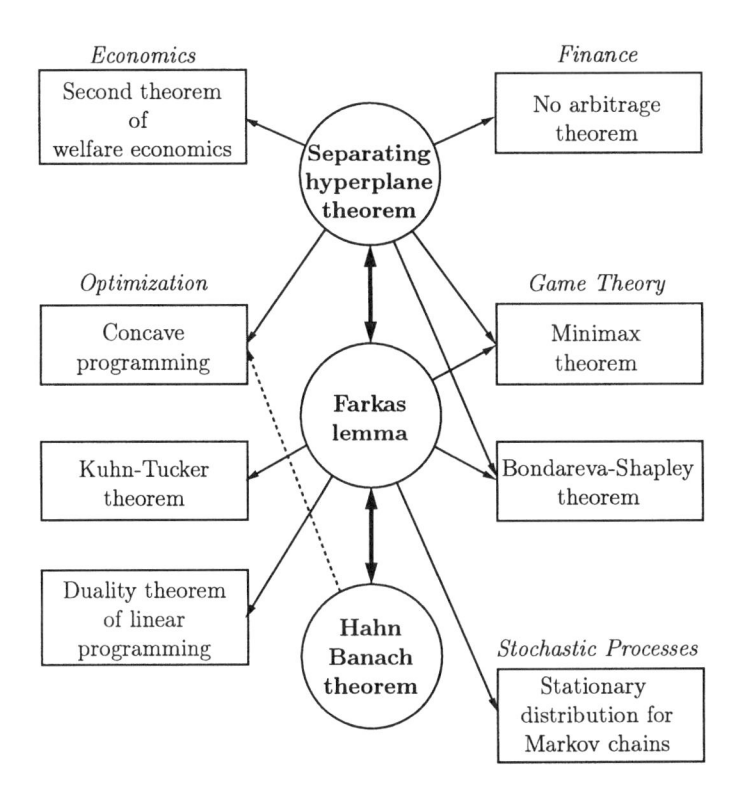

Figure 3.27
Applications of the separating hyperplane theorem

doubted it would hold in general. "Throughout the period in question," wrote von Neumann, "I felt there was nothing worth publishing until the minimax theorem was proved" (see von Neumann 1953).

What is this fundamental theorem of game theory without which there is nothing worth publishing? The minimax theorem establishes the existence of optimal strategies in games in which the interests of the players are diametrically opposed. Von Neumann proved the minimax theorem in 1928 using a fixed point argument. Subsequent authors devised more elementary proofs based on the separation theorem or equivalent theorems of the alternative.

A two person zero-sum game (example 2.40) comprises:

- two players 1 and 2

- for each player $i = 1, 2$ a nonempty set of pure strategies S_i
- a function $u: S_1 \times S_2 \to \Re$ representing the payoff (from player 2 to 1)

The function $u(s_i^1, s_j^2)$ denotes the payoff to player 1 if he chooses strategy s_i^1 and his opponent chooses s_j^2. Player 1 seeks to maximize $u(s_i^1, s_j^2)$ while player 2 seeks to minimize it.

A game is *finite* if each player has only a finite number of pure strategies. Let $m = |S_1|$ and $n = |S_2|$ be the number of strategies of players 1 and 2 respectively. For any finite two person zero-sum game, there exists an $m \times n$ matrix A that represents u, in the sense that its elements $a_{ij} = u(s_i^1, s_j^2)$ represent the payment from 2 to 1 when the players choose strategies i and j respectively.

Let $u(\mathbf{p}, j)$ denote the expected payoff if player 1 adopts the mixed strategy $\mathbf{p} = (p_1, p_2, \ldots, p_m)$ while player 2 plays her j pure strategy. That is, define

$$u(\mathbf{p}, j) = \sum_{i=1}^{m} p_i a_{ij}$$

The worst that can happen to player 1 when he plays the mixed strategy $\mathbf{p}$ is

$$v_1(\mathbf{p}) = \min_{j=1}^{n} u(\mathbf{p}, j)$$

We call this player 1's *security level* when playing the strategy $\mathbf{p}$. Similarly, if

$$u(i, \mathbf{q}) = \sum_{j=1}^{n} q_i a_{ij}$$

denotes the expected outcome when player 2 plays the mixed strategy $\mathbf{q}$ and player 1 plays strategy i, the worst outcome from the viewpoint of player 2 is

$$v_2(\mathbf{q}) = \max_{i=1}^{n} u(i, \mathbf{q})$$

Note that player 1 wishes to maximize the payoff and succeeds in the game if $v_1(\mathbf{p}) \geq 0$. On the other hand, player 2 seeks to minimize the payoff (from 2 to 1), and succeeds if $v_2(\mathbf{q}) \leq 0$. Surprisingly, in any zero-sum game, at least one of the players has a strategy that guarantees she cannot lose. That is, *either* player 1 has a strategy **p** that ensures an expected outcome favorable to him ($v_1(\mathbf{p}) \geq 0$) *or* player 2 has a strategy **q** that ensures an outcome favorable to her ($v_2(\mathbf{q}) \leq 0$).

Exercise 3.253
In any finite two person zero-sum game, at least one of the players has a mixed strategy that guarantees she cannot lose. That is, either $v_1(\mathbf{p}) \geq 0$ or $v_2(\mathbf{q}) \leq \mathbf{0}$. [Hint: Use the von Neumann alternative (exercise 3.245).]

Example 3.88 (Chess) Chess can be considered a zero-sum game in which a payoff of 1 is assigned for a win, -1 for a loss and 0 for a draw. Each players has a finite (though extremely large) set of possible strategies. (A game is declared a draw if the same position is repeated three times.) The preceding proposition asserts that at least one of the players, White or Black, has a mixed strategy that will ensure that she can expect not to lose on average. In fact it can be shown that one of the players has a pure strategy that ensures she will never the lose. The only difficulty is that no one knows which player, White or Black, has the winning strategy, let alone the specification of that strategy.

This example illustrates both the power and the limitation of abstraction. On the one hand, the apparently innocuous proposition (exercise 3.253), based on the separation of convex sets, has enormous ramifications for this classic strategic game. Without abstraction, a study of the rules and possibilities of chess would have proved overwhelmingly too complicated to yield the conclusion. On the other hand, because of the high degree of abstraction, the proposition shields little light on optimal behavior in real chess games, and a future for chess masters remains assured.

In seeking the most propitious outcome, player 1 would do well to choose his strategy **p** in order to maximize his security level $v(\mathbf{p})$. If he does this, the value (expected payoff) of the game to player 1 is

$$v_1 = \max_{\mathbf{p}} v_1(\mathbf{p}) = \max_{\mathbf{p}} \min_j u(\mathbf{p}, j)$$

Similarly the value of the game to player 2 is

$$v_2 = \min_{\mathbf{q}} v_2(\mathbf{q}) = \min_{\mathbf{q}} \max_i u(i, \mathbf{q})$$

This also is the best that player 1 can hope to achieve from the game. By the previous result, either $v_1 \geq 0$ or $v_2 \leq 0$. The minimax theorem goes a step further and asserts that in fact $v_1 = v_2 = v$, which is called the *value* of the game.

Proposition 3.20 (Minimax theorem) *For any two person zero-sum game, $v_1 = v_2$. Every game has a value.*

Exercise 3.254
Prove proposition 3.20 by showing

1. for every $c \in \Re$, *either* $v_1 \geq c$ *or* $v_2 \leq c$
2. $v_1 = v_2$

[Hint: Apply the previous exercise to the game $\hat{u}(s^1, s^2) = u(s^1, s^2) - c$.]

Example 3.89 (Rock–Scissors–Paper) The payoff matrix for the game Rock–Scissors–Paper (example 2.34) is

		Chris	
	Rock	Scissors	Paper
Jenny Rock	0	1	−1
Scissors	−1	0	1
Paper	1	−1	0

If Jenny adopts the mixed strategy $\mathbf{p} = (\frac{1}{3}, \frac{1}{3}, \frac{1}{3})$, her expected payoff is zero regardless of the Chris's choice. That is,

$$u(\mathbf{p}, \text{Rock}) = u(\mathbf{p}, \text{Scissors}) = u(\mathbf{p}, \text{Paper}) = 0$$

Therefore

$$v_1(\mathbf{p}) = \min\{u(\mathbf{p}, \text{Rock}), u(\mathbf{p}, \text{Scissors}), u(\mathbf{p}, \text{Paper})\} = 0$$

which implies that $v_1 = \max_{\mathbf{p}} v_1(\mathbf{p}) \geq 0$. If Chris adopts a similar strategy $\mathbf{q} = (\frac{1}{3}, \frac{1}{3}, \frac{1}{3})$, $v_2(\mathbf{q}) = 0$ and therefore $v_2 = \min_{\mathbf{q}} v_2(\mathbf{q}) \leq 0$. We conclude that $v = 0$. This value of this game is zero, since it is symmetric.

Exercise 3.255
Let v denote the value of a two person zero-sum game. Show that

1. Player 1 has an optimal strategy $\mathbf{p}^*$ that achieves the value v. Similarly player 2 has an optimal strategy $\mathbf{q}^*$.
2. The strategy pair $(\mathbf{p}^*, \mathbf{q}^*)$ constitutes a Nash equilibrium of the game.

Exercise 3.256
Every finite two-person zero-sum game has a Nash equilibrium.

Example 3.90 Consider a zero-sum game in which player 1 has two strategies $\{s_1, s_2\}$, while player 2 has 5 strategies $\{t_1, t_2, t_3, t_4, t_5\}$. The payoffs to player 1 are

		t_1	t_2	t_3	t_4	t_5
Player 1	s_1	−1	1	3	1	1
	s_2	2	4	2	1	−2

The set of feasible outcomes is illustrated in Figure 3.28, where each labeled point represents the pair of outcomes corresponding to a particular pure strategy of player 2. For example, if player 2 selects t_1, the outcome will be either −1 or 2 depending on the choice of player 1. This is the point labeled t_1. By using mixed strategies, player 2 in effect chooses a point in the convex hull Z of these primary points. Player 1 chooses which coordinate determines the outcome.

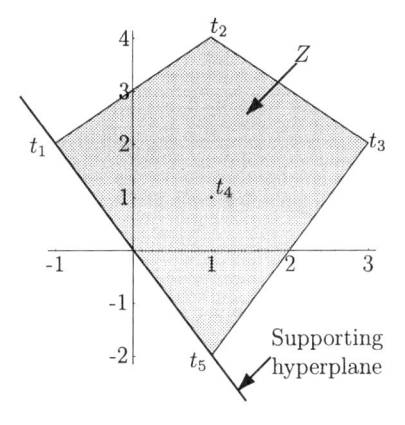

Figure 3.28
The feasible payoffs

Player 1 seeks to maximize the payoff and player 2 to minimize it. In particular, player 2 would like to achieve a negative outcome, which results in a payment from 1 to 2. Unfortunately, she cannot to this. The best that she can do is to play a 50–50 mixture of t_1 and t_5, which guarantees her a payoff of zero *irrespective* of the choice of player 1. This is her value v_2 which is achieved with the mixed strategy $\mathbf{q} = (\frac{1}{2}, 0, 0, 0, \frac{1}{2})$. If she attempts to secure a better outcome (by favoring t_1 or t_5), player 1 can counteract by choosing the appropriate coordinate. Resorting to any of her other strategies t_2, t_3, and t_4 can only make things worse. Breaking even, $v_2 = 0$, is the best she can hope to achieve in this game.

We can illustrate the strategic choice of player 1 in a similar diagram. Every mixed strategy $\mathbf{p}$ for player 1 defines a linear functional $f_\mathbf{p}$ on the set Z of feasible outcomes defined by

$$f_\mathbf{p}(\mathbf{z}) = p_1 z_1 + p_2 z_2$$

$f_\mathbf{p}(\mathbf{z})$ measures the expected payoff when player 1 chooses $\mathbf{p}$ and player 2 chooses $\mathbf{z}$. The linear functional $f_\mathbf{p}$ corresponding to the mixed strategy $\mathbf{p}$ can be illustrated by the set of hyperplanes representing the contours of $f_\mathbf{p}$. Each hyperplane traces out the various choices in Z that generate the same expected payoff, given that player 1 chooses the mixed strategy $\mathbf{p}$.

For example, if player 1 adopts the mixed strategy $(\frac{1}{2}, \frac{1}{2})$ and player 2 responds with the t_1, the expected payoff is

$$f((-1, 2)) = \tfrac{1}{2}(-1) + \tfrac{1}{2}(2) = \tfrac{1}{2}$$

Other choices of player 2 that also generate an expected payoff of $\frac{1}{2}$ lie along the hyperplane through t_1. Clearly, the best response of player 2 to the mixed strategy $(\frac{1}{2}, \frac{1}{2})$ is t_5 which achieves an expected payoff

$$f((-1, 2)) = \tfrac{1}{2}(1) + \tfrac{1}{2}(-2) = -\tfrac{1}{2}$$

This is the security level $v_1(\mathbf{p})$ of the strategy $(\frac{1}{2}, \frac{1}{2})$.

Different mixed strategies $\mathbf{p}$ give rise to hyperplanes of different slopes. Player 1's choice of mixed strategy determines the slope of the set of hyperplanes of equal expected payoff. Player 2's choice of an element in Z selects which particular hyperplane determines the outcome. The expected payoff associated with any particular hyperplane can be read off its intersection with the 45 degree line.

The security level $v_1(\mathbf{p})$ of any mixed strategy $\mathbf{p}$ is given by the expected payoff the supporting hyperplane which bounds Z from below. As already stated, the security level $v_1(\mathbf{p})$ of the strategy $(\frac{1}{2}, \frac{1}{2})$ that supports Z at t_5 is $-\frac{1}{2}$. Player 1 should seek that mixed strategy whose supporting hyperplane has the highest possible intersection with the 45 degree line. In this example, the optimal strategy for player 1 will be that associated with the hyperplane that is aligned with the southwest boundary of the set Z. In this case the optimal strategy for player 1 is the mixed strategy $(\frac{2}{3}, \frac{1}{3})$. The supporting hyperplane corresponding to this strategy passes through the origin and has a security level of 0. The supporting hyperplane corresponding to any other mixed strategy will intersect the 45 degree below the origin, indicating a negative security level.

We conclude that 0 is the best that player 1 can guarantee, which is therefore the value v_1 of the game to player 1. This is equal to the best that player 2 can guarantee, and therefore $v_2 = v_1$.

Exercise 3.257
In the context of the previous example, prove that $v_1(\mathbf{p}) < 0$ for every $\mathbf{p} \neq (\frac{2}{3}, \frac{1}{3})$ and therefore that $v_1 = 0$.

Building on the insight of the previous example, we sketch an alternative proof of the minimax theorem which makes explicit use of a separation theorem. Consider any finite two-person zero-sum game in which player 1 has m pure strategies and player 2 has n strategies. Let A be the $m \times n$ matrix that represents the payoff function, that is, $a_{ij} = u(s_i^1, s_j^2)$. The set of feasible outcomes

$$Z = \{\mathbf{z} = A\mathbf{q} : \mathbf{q} \in \Delta^{m-1}\}$$

is the convex hull of the columns of A. Player 2's choice of $\mathbf{q}$ selects a particular point of $\mathbf{z} \in Z$, a convex polyhedron in $\Re^m$. Player 1's strategy defines a linear functional $f_\mathbf{p}$ which evaluates $\mathbf{z} \in Z$ to determine the expected payoff.

Assume initially that the value of the game to player 2 is zero, that is, $v_2 = 0$. This implies that there exists some $\mathbf{z} \in Z$ with $\mathbf{z} \leq \mathbf{0}$. Furthermore player 2 can only improve on this outcome if there exists an element $\mathbf{z} \in Z$ with $\mathbf{z} < \mathbf{0}$. That is, an expected payoff of zero is the best that player 2 can hope to achieve if Z contains no interior point of the negative orthant $\Re^m_-$. If Z is disjoint from the negative orthant, the separating hyperplane theorem guarantees the existence of linear functional such that $f_\mathbf{p}$ such that $f_\mathbf{p}(\mathbf{z}) \geq 0$ for every $\mathbf{z} \in Z$. In other words, there exists a mixed strategy

$\mathbf{p} \in \Delta^{m-1}$ with security level $v_1(\mathbf{p}) \geq 0 = v_2$. Since $v_1 \leq v_2$, we conclude that $v_1 = v_2$.

Consider now an arbitrary two person zero-sum game in which $v_2 = c \neq 0$. By subtracting c from every outcome, this game can be transformed into a strategically equivalent game with $v_2 = 0$. By the preceding argument, there exists an optimal strategy $\mathbf{p}^*$ for player 1 that ensures an expected payoff of zero. This same strategy ensures a payoff of c to player 1 in the original game. We conclude that $v_1 = v_2$. The argument is made precise in the following exercises.

Exercise 3.258
Let A be a $m \times n$ matrix which represents (exercise 3.253) the payoff function of a two-person zero-sum game in which player 1 has m pure strategies and player 2 has n strategies. Let Z be the convex hull of the columns of A, that is, $Z = \{\mathbf{z} = A\mathbf{q} : \mathbf{q} \in \Delta^{n-1}\}$. Assume that $v_2 = 0$. Show that

1. $Z \cap \Re^n_- \neq \emptyset$.
2. $Z \cap \text{int } \Re^n_- = \emptyset$.
3. There exists $\mathbf{p}^* \in \Delta^{m-1}$ such that $f_{\mathbf{p}^*}(\mathbf{z}) \geq 0$ for every $\mathbf{z} \in S$.
4. $v_1 = 0 = v_2$.

Exercise 3.259
Prove the minimax theorem by extending the previous exercise to an arbitrary two-person zero-sum game with $v_2 = c \neq 0$.

Exercise 3.260
Show that the set of optimal strategies for each player is convex.

Exercise 3.261
Let f be a bilinear functional on the product of two simplices, that is $f : \Delta^m \times \Delta^n \to \Re$. Then

$$\max_{\mathbf{x}} \min_{\mathbf{y}} f(\mathbf{x}, \mathbf{y}) = \min_{\mathbf{y}} \max_{\mathbf{x}} f(\mathbf{x}, \mathbf{y})$$

The Core of a TP-Coalitional Game

It is highly desirable that any proposed solution to a coalitional game belong to the core. Unfortunately, in some games the core is empty. It would be extremely useful to have a set of necessary and sufficient conditions for the core of a game to be nonempty. A necessary condition for the existence of core allocations in a game (N, w) is that the game be cohesive, that is,

$$w(N) \geq \sum_{k=1}^{K} w(S_k) \tag{33}$$

for every partition $\{S_1, S_2, \ldots, S_K\}$ of N. However, cohesivity is not sufficient to guarantee the existence of outcomes in the core (exercise 3.262). The appropriate characterization is obtained by extending the concept of cohesivity to more general families of coalitions, called balanced families.

Exercise 3.262 (Three-person majority game)
A classic example in coalitional game theory is the three-person majority game, in which the allocation of \$1 among three persons is decided by majority vote. Let $N = \{1, 2, 3\}$. The characteristic function is

$w(\{i\}) = 0, \quad i = 1, 2, 3$

$w(\{i, j\}) = 1, \quad i, j \in N, i \neq j$

$w(N) = 1$

1. Show that the three-person majority game is cohesive.
2. Show that its core is empty.

Exercise 3.263
Show that cohesivity is necessary, but not sufficient, for the existence of a core.

A partition $\{S_1, S_2, \ldots, S_K\}$ is a family of coalitions in which each player I belongs to one and only one of the coalitions. A *balanced family* of coalitions $\mathcal{B}$ is a set of coalitions $\{S_1, S_2, \ldots, S_K\}$ together with a corresponding set of positive numbers $\lambda_1, \lambda_2, \ldots, \lambda_K$ called *weights* such that for every player $i \in N$, the sum of the weights of the coalitions to which player i belongs sum to one. That is, for every player $i \in N$,

$$\sum_{k:\, S_k \ni i} \lambda_k = 1, \quad \text{for every } i \in N$$

Compared to a simple partition, the coalitions in a balanced family do not require the exclusive allegiance of their members.

In one possible interpretation of the concept of a balanced family, each player allocates her time among the various coalitions in $\mathscr{B}$ to which she belongs. The weight λ_S of coalition S represents the proportion of the available time which each member spends in that coalition. The family is balanced with weights λ_S if each player's time is fully allocated to coalitions in the family $\mathscr{B}$.

Example 3.91 Any partition $\mathscr{P} = \{S_1, S_2, \ldots, S_K\}$ of the player set N is a balanced family with weights

$$\lambda_S = 1, \quad \text{for every } S \in \mathscr{P}$$

This confirms that a balanced family is a generalized partition.

Example 3.92 The set of all coalitions in a three-player game ($N = \{1, 2, 3\}$) is

$$\mathscr{N} = \{\varnothing, \{1\}, \{2\}, \{3\}, \{1,2\}, \{1,3\}, \{2,3\}, \{1,2,3\}\}$$

The family of two-player coalitions $\mathscr{B} = \{\{1,2\}, \{1,3\}, \{2,3\}\}$ together with weights $\lambda_S = \frac{1}{2}$ for every $S \in \mathscr{B}$ is a balanced family of coalitions.

Another balanced family of coalitions is $\mathscr{B} = \{\{1,2\}, \{1,3\}, \{2,3\}, \{1,2,3\}\}$ with weights

$$\lambda_S = \begin{cases} \frac{1}{2}, & S = N \\ \frac{1}{4}, & S \in \mathscr{B}, S \neq N \end{cases}$$

To verify this, calculate the allegiance of player 1,

$$\sum_{S \ni 1} \lambda_S = \lambda_{\{1,2\}} + \lambda_{\{1,3\}} + \lambda_{\{1,2,3\}} = \tfrac{1}{4} + \tfrac{1}{4} + \tfrac{1}{2} = 1$$

and similarly for the other two players.

Exercise 3.264
List three other balanced families of coalitions for the three-player game.

Exercise 3.265
Find a nontrivial (i.e., not a partition) balanced family of coalitions for the four-player game with $N = \{1, 2, 3, 4\}$.

The following exercise presents an alternative representation of a balanced collection, which will be useful below.

Exercise 3.266
A collection $\mathscr{B}$ of coalitions is balanced if and only if there exists positive weights $\{\lambda_S : S \in \mathscr{B}\}$ such that

$$\mathbf{e}_N = \sum_{S \in \mathscr{B}} \lambda_S \mathbf{e}_S \quad \text{and} \quad g_N = \sum_{S \in \mathscr{B}} \lambda_S g_S$$

where $\mathbf{e}_S$ is the characteristic vector of the coalition S (example 3.19) and $g_S \colon X \to \Re$ represents the share of coalition S at the allocation $\mathbf{x}$ (example 3.16).

In a natural extension of cohesivity (33) a cooperative game $G = (N, w)$ is *balanced* if

$$w(N) \geq \sum_{S \in \mathscr{B}} \lambda_S w(S)$$

for every balanced family of coalitions $\mathscr{B}$. Balance is the appropriate necessary and sufficient condition for nonemptyness of the core.

Proposition 3.21 (Bondareva-Shapley theorem) *A TP-coalitional game has a nonempty core if and only if it is balanced.*

Proof $\mathbf{x} \in X$ belongs to the core if and only if it satisfies the system of inequalities (example 3.82)

$$g_S(\mathbf{x}) \geq w(S) \quad \text{for every } S \subseteq N$$
$$-g_N(\mathbf{x}) \geq -w(N)$$

where $g_S(\mathbf{x}) = \sum_{i \in S} x_i$ measures the share of coalition S at the allocation $\mathbf{x}$. Therefore a core allocation exists if and only if the preceding system of inequalities is consistent. Applying Fan's condition (exercise 3.236), a necessary and sufficient condition for consistency is, for every set of nonnegative scalars $\{\lambda_S : S \subseteq N\}$ and μ,

$$\sum_{S \subseteq N} \lambda_S g_S - \mu g_N = \mathbf{0}$$

$$\Rightarrow \sum_{S \subseteq N} \lambda_S w(S) - \mu w(N) \leq 0$$

Without loss of generality, we can normalize so that $\mu = 1$ (exercise 3.267). Consistency requires that whenever

$$\sum_{S \subseteq N} \lambda_S g_S = g_N \tag{34}$$

then

$$\sum_{S \subseteq N} \lambda_S w(S) \leq w(N) \tag{35}$$

Let $\mathscr{B}$ denote the set of coalitions with positive weight, that is,

$$\mathscr{B} = \{ S \subseteq N \mid \lambda_S > 0 \}$$

Then (34) requires that $\mathscr{B}$ be balanced by the weights λ (exercise 3.266). Therefore (35) must hold for every balanced collection. That is, the game must be balanced. We conclude that the inequalities defining the core are consistent if and only if the game is balanced. □

Exercise 3.267
Why can we normalize so that $\mu = 1$ in the previous proof?

Exercise 3.268
The set of balanced games forms a convex cone in the set of all TP-coalitional games $\mathscr{G}^N$.

Remark 3.21 (Finding the right space) The preceding proof of the Bondareva-Shapley theorem is elegant, but it is does not yield much insight into why it works. We know that it is based on Fan's condition, which in turn is derived from the Farkas lemma which is equivalent to the basic separation theorem. But the simple icon of the separation of convex sets is not transparent.

Yet the Bondareva-Shapley theorem seems tailor-made for a separation argument. Let Y be the set of allocations which satisfies the claims of all the coalitions, that is

$$Y = \{ \mathbf{x} \in \Re^n : g_S(\mathbf{x}) \geq w(S) \text{ for every } S \subseteq N \}$$

The core is empty precisely if Y is disjoint from the set of feasible outcomes X. Clearly, both X and Y are convex and nonempty. This perspective seems to demand the application of a separation theorem. Unfortunately, this will not yield the desired result. The reason is the X and Y live in the wrong space. X and Y are subsets of $\Re^n$, whereas we are looking for a result applying to the set of coalitions, which live in a

3.9 Separation Theorems

higher-dimensional space. To achieve our goal with a separation argument requires that we translate these sets to a higher-dimensional space. Exercise 3.269 presents one method of doing this.

These observations exemplify a general point, namely that successful application of separation arguments requires formulation of the problem in an appropriate linear space, which may not necessarily be the one in which the problem originates. Exercise 3.199 provides another example where using the right space is crucial.

Exercise 3.269
To construct a suitable space for the application of a separation argument, consider the set of points $A^0 = \{(\mathbf{e}_S, w(S)) : S \subseteq N\}$, where $\mathbf{e}_S$ is characteristic vector of the coalition S (example 3.19) and $w(S)$ is its worth. Let A be the conic hull of A^0, that is,

$$A = \left\{ \mathbf{y} \in \Re^{n+1} : \mathbf{y} = \sum_{S \subseteq N} \lambda_S (\mathbf{e}_S, w(S)), \lambda_S \geq 0 \right\}$$

Let B be the interval $B = \{(\mathbf{e}_N, w(N) + \varepsilon) : \varepsilon > 0\}$. Clearly, A and B are convex and nonempty.

We assume that the game is balanced and construct a payoff in the core.

1. Show that A and B are disjoint if the game is balanced. [Hint: Show that $A \cap B \neq \emptyset$ implies that the game is unbalanced.]

2. Consequently there exists a hyperplane that separates A and B. That is, there exists a nonzero vector $(\mathbf{z}, z_0) \in \Re^n \times \Re$ such that

$$(\mathbf{z}, z_0)^T \mathbf{y} \geq c > (\mathbf{z}, z_0)^T (\mathbf{e}_N, w(N) + \varepsilon) \tag{36}$$

for all $\mathbf{y} \in A$ and all $\varepsilon > 0$. Show that

a. $(\mathbf{e}_\emptyset, 0) \in A$ implies that $c = 0$.

b. $(\mathbf{e}_N, w(N)) \in A$ implies that $z_0 < 0$. Without loss of generality, we can normalize so that $z_0 = -1$.

3. Show that (36) implies that the payoff vector $\mathbf{z}$ satisfies the inequalities

$$\mathbf{e}_S^T \mathbf{z} \geq w(S) \quad \text{for every } S \subseteq N \text{ and } \mathbf{e}_N^T \mathbf{z} \leq w(N)$$

Therefore $\mathbf{z}$ belongs to the core.

Another alternative proof of the Bondareva-Shapley theorem makes ingenious use of the minimax theorem. This approach is developed in the following exercises.

Exercise 3.270 (0–1 Normalization)
A TP-coalitional game (N, w) is called *0–1 normalized* if

$$w(\{i\}) = 0 \quad \text{for every } i \in N \text{ and } w(N) = 1$$

Show that

1. to every essential game (N, w) there is a corresponding 0–1 normalized game (N, w^0)
2. $\text{core}(N, w) = \alpha \, \text{core}(N, w^0) + \mathbf{w}$, where $\mathbf{w} = (w_1, w_2, \ldots, w_n)$, $w_i = w(\{i\})$, $\alpha = w(N) - \sum_{i \in N} w_i$
3. $\text{core}(N, w) = \emptyset \Leftrightarrow \text{core}(N, w^0) = \emptyset$

Exercise 3.271
Let (N, w) be a 0–1 normalized TP-coalitional game, and let $\mathscr{A}$ be the set of all nontrivial coalitions $\mathscr{A} = \{S \subseteq N : w(S) > 0\}$. Consider the following two-player zero-sum game. Player 1 chooses a player $i \in N$ and player 2 chooses a coalition $S \in \mathscr{A}$. The payoff (from 2 to 1) is

$$u(i, S) = \begin{cases} \dfrac{1}{w(S)}, & i \in S \\ 0 & i \notin S \end{cases}$$

Let δ be the value of the two-person zero-sum game. Show that

$\mathbf{x} \in \text{core}(N, w) \Leftrightarrow \delta \geq 1$

[Hint: Consider any $\mathbf{x} \in \text{core}(N, w)$ as a mixed strategy for player 1 in the two-person zero-sum game.]

Exercise 3.272
Let (N, w) be a 0–1 normalized TP-coalitional game, and let G be the corresponding two-person zero-sum game described in the previous exercise with value δ. Show that $\delta \geq 1$ if (N, w) is balanced. [Hint: Consider a mixed strategy for player 2.]

Exercise 3.273
$\text{Core}(N, w) \neq \emptyset \Rightarrow (N, w)$ balanced.

3.9 Separation Theorems

Combining the three previous exercises establishes the cycle of equivalences

(N, w) balanced $\Rightarrow \delta \geq 1 \Rightarrow (\text{core}(N, w) \neq \emptyset) \Rightarrow (N, w)$ balanced

and establishes once again the Bondareva-Shapley theorem.

The Bondareva-Shapley theorem can be used both positively to establish that game has a nonempty core and negatively to prove that a game has an empty core. To establish that a game has an empty core, the theorem implies that it is sufficient to find a single-balanced family of coalitions $\mathscr{B}$ for which

$$\sum_{S \in \mathscr{B}} \lambda_S w(S) > w(N)$$

On the other hand, to show that a game has a nonempty core, we have to show that it is balanced, that is,

$$\sum_{S \in \mathscr{B}} \lambda_S w(S) \leq w(N)$$

for every balanced family of coalitions $\mathscr{B}$. We give an example of each usage.

Example 3.93 The four player game with $N = \{1, 2, 3, 4\}$ and characteristic function

$$w(S) = \begin{cases} 1 & \text{if } S = N \\ \frac{3}{4} & \text{if } S \in \mathscr{B} \\ 0 & \text{otherwise} \end{cases}$$

where $\mathscr{B} = \{\{1,2,3\}, \{1,4\}, \{2,4\}, \{3,4\}\}$ has an empty core. First observe that $\mathscr{B}$ is a balanced collection of coalitions with weights $\lambda_{\{1,2,3\}} = \frac{2}{3}$ and $\lambda_{\{1,4\}} = \lambda_{\{2,4\}} = \lambda_{\{3,4\}} = \frac{1}{3}$ (exercise 3.265). However, since

$$\frac{2}{3} w(1,2,3) + \frac{1}{3} w(1,4) + + \frac{1}{3} w(2,4) + \frac{1}{3} w(3,4)$$

$$= \left(\frac{2}{3} + \frac{1}{3} + \frac{1}{3} + \frac{1}{3} \right) \frac{3}{4} = \frac{5}{4} > w(N)$$

the game is unbalanced. Applying the Bondareva-Shapley theorem (proposition 3.21), we conclude that this game has an empty core.

Example 3.94 (Market game) Assume that the preferences of each player i in a market game (example 1.118) can be represented by a concave utility function u_i. We can formulate an exchange economy as a TP-coalitional game by assuming that the worth of any coalition S as the maximum aggregate utility which it can attain by trading amongst themselves. Formally

$$w(S) = \max_{\mathbf{x}} \left(\sum_{i \in S} u_i(\mathbf{x}_i) : \sum_{i \in S} \mathbf{x}_i = \sum_{i \in S} \omega_i \right) \tag{37}$$

We show that this game is balanced.

Consider an arbitrary balanced collection of coalitions, $\mathscr{B}$, with weights λ_S. For every $S \in \mathscr{B}$, let $\mathbf{x}^S = \{\mathbf{x}_i : i \in S\}$ be an allocation to S that achieves $w(S)$, that is,

$$\sum_{i \in S} u_i(\mathbf{x}_i^S) = \max \left\{ \sum_{i \in S} u_i(\mathbf{x}_i) : \sum_{i \in S} \mathbf{x}_i = \sum_{i \in S} \omega_i \right\}$$

$\mathbf{x}^S$ is the best that coalition S can do with its own resources.

Now construct a new allocation $\bar{\mathbf{x}}$ which is a convex combination of the allocation $\mathbf{x}^S$ with weights λ_S, that is,

$$\bar{\mathbf{x}}_i = \sum_{\substack{S \in \mathscr{B} \\ S \ni i}} \lambda_S \mathbf{x}_i^S$$

We will show that $\bar{\mathbf{x}}$ is a feasible allocation for the grand coalition, which is at least as good as $\mathbf{x}^S$ for every player. To show that $\bar{\mathbf{x}}$ is feasible, we sum the individual allocations.

$$\sum_{i \in N} \bar{\mathbf{x}}_i = \sum_{i \in N} \left(\sum_{S \in \mathscr{B}, S \ni i} \lambda_S x_i^S \right) = \sum_{S \in \mathscr{B}} \lambda_S \left(\sum_{i \in S} x_i^S \right)$$

$$= \sum_{S \in \mathscr{B}} \lambda_S \left(\sum_{i \in S} \omega_i \right) = \sum_{i \in N} \omega_i \sum_{S \in \mathscr{B}_i} \lambda_S = \sum_{i \in N} \omega_i$$

where the last equality utilizes the fact that $\mathscr{B}$ is balanced. Since the players' utility functions are concave, we have

$$u_i(\bar{\mathbf{x}}_i) = u_i \left(\sum_{S \in \mathscr{B}_i} \lambda_S \mathbf{x}_i^S \right) \geq \sum_{S \in \mathscr{B}_i} \lambda_S u_i(\mathbf{x}_i^S)$$

Summing over all the players, we have

$$\sum_{i \in N} u_i(\bar{\mathbf{x}}_i) \geq \sum_{i \in N} \sum_{S \in \mathcal{B}_i} \lambda_S u_i(\mathbf{x}_i^S) = \sum_{S \in \mathcal{B}} \lambda_S \sum_{i \in S} u(\mathbf{x}_i^S) = \sum_{S \in \mathcal{B}} \lambda_S w(S)$$

Since $\bar{\mathbf{x}}$ is feasible for the grand coalition, we must have

$$w(N) \geq \sum_{i \in N} u_i(\bar{\mathbf{x}}_i) \geq \sum_{S \in \mathcal{B}} \lambda_S w(S)$$

Since $\mathcal{B}$ was an arbitrary balanced collection, this must be true for all balanced collections.

Therefore we have demonstrated that the pure exchange economy game is balanced. Applying the Bondareva-Shapley theorem (proposition 3.21), we conclude that there exist core allocations in this game.

3.9.5 Concluding Remarks

It is remarkable that so many fundamental results in economics, optimization, and finance depend essentially on separation theorems or theorems of the alternative. It is also worth noting that there are other key results for which separation arguments are not sufficient. The existence of competitive equilibria in market economies and Nash equilibria in strategic games and the nonemptiness of the core in general coalitional games are more profound, requiring less intuitive fixed point arguments for their derivation (section 2.4.4).

3.10 Notes

The first half of this chapter is the central core of linear algebra, for which there are a host of suitable references. Halmos (1974) is a classic text. Smith (1998) is an elementary exposition with many examples. Janich (1994) is concise and elegant, but more advanced. Simon and Blume (1994) is thorough and written with economic applications in mind. Example 3.7 is due to Shubik (1982). Magill and Quinzii (1996, pp. 108–13) provide an instructive application of the adjoint transformation in general equilibrium theory.

Roberts and Varberg (1973) provide a comprehensive account of the mathematics of convex functions. Madden (1986) is written for economists, and deals with both convex and homogeneous functions. The relationship between convex and quasiconvex functions is thoroughly explored in Greenberg and Pierskalla (1971). Other generalization and variations found the economics literature are surveyed in Diewert et al. (1981). Exercise 3.127 is adapted from Sydsaeter and Hammond (1995). Example 3.65 is from Sundaram (1996).

A slight generalization (with f semicontinuous) of the minimax theorem (3.11) is sometimes known as Sion's theorem. Our proof follows Kakutani (1941), who in fact developed his fixed point theorem in the course of establishing a generalization of von Neumann's minimax theorem (exercise 3.261). While Kakutani's theorem provides an elegant proof, the minimax theorem does not require a fixed point argument. An elementary proof can be found in Karlin (1959, pp. 29–30). Berge (1963, p. 210) proves Sion's theorem using the separation theorem. Some variations can be found in Border (1985, pp. 74–77).

Section 3.9 was assimilated from many sources. Koopmans (1957) is an insightful account of their role in decentralization. Bosch and Smith (1998) present an interesting use of separating hyperplanes to assign authorship to the disputed Federalist papers advocating ratification of the U.S. Constitution. Exercise 3.112 is adapted from Zhou (1993). Exercise 3.210, which is the standard proof of the Shapley-Folkman theorem, is adapted from Green and Heller (1981) and Ellickson (1993). Although we prove Fan's condition (exercise 3.236) for $\Re^n$, it in fact applies in arbitrary linear spaces (Fan 1956; Holmes 1975). Theorems of the alternative are discussed by Gale (1960) and Ostaszewski (1990). Comprehensive treatments are given by Mangasarian (1994) and Panik (1993).

The exchange between Fréchet and von Neumann regarding Borel's contribution to the theory of games begins with Fréchet (1953). There follow English translations of the three notes of Borel, a commentary by Fréchet, and the response from von Neumann (1953).

Proposition 3.21 was discovered independently by Olga Bondareva and Lloyd Shapley, who also proposed the Shapley value (example 3.6). Our proof follows Moulin (1986). The alternative proof (exercise 3.269) using the separation theorem is due to Osborne and Rubinstein (1994). The third approach via the minimax theorem (exercises 3.271 to 3.273) is from a lecture by Sergiu Hart.

4 Smooth Functions

All science is dominated by the idea of approximation.
—Bertrand Russell

The last chapter explored the powerful structure of linear functions. We now extend this analysis to nonlinear functions. By approximating a nonlinear function by a linear function, we can exploit the precise structure of the linear approximation to obtain information about the behavior of the nonlinear function in the area of interest. Functions that can be approximated locally by linear functions are called *smooth functions*. A smooth function is one whose surface or graph has no sharp edges or points. Among other things, economists use linear approximation of smooth functions to characterize the solution of a constrained optimization problem (chapter 5) and to derive properties of the optimal solution (chapter 6).

The content of this chapter is usually called multivariate calculus. In elementary (univariate) calculus, the derivative of a function from $\Re \to \Re$ is defined as the limit of the ratio of two numbers (an element of $\Re$). This concept of the derivative does not generalize to functions on higher-dimensional spaces. Therefore it is conventional to modify the definition of the derivative when progressing from elementary to multivariate calculus. We introduce the more appropriate definition in the next section by working through some examples.

4.1 Linear Approximation and the Derivative

Table 4.1 compares the values of the functions

$$f(x) = 10x - x^2 \quad \text{and} \quad g(x) = 9 + 4x$$

We observe that the functions attain the same value at $x = 3$, and have similar values in a neighborhood of 3. We say that the affine function g approximates the nonlinear function f.

The accuracy of the approximation deteriorates as we move away from their intersection at $x = 3$. Table 4.1 gives three different measures of this approximation error. Most obvious is the *actual error*, the numerical difference between the two functions, $f(x) - g(x)$ (column 4). More useful is the *percentage error* (column 5), which expresses the error as a percentage of the true value $f(x)$, that is,

$$\text{percentage error} = \frac{f(x) - g(x)}{f(x)} \times \frac{100}{1}$$

We see from the table that g approximates f quite well in the region of $x = 3$. Even at $x = 3.5$, which involves a 17 percent change in x, the approximation error is only 1.1 percent.

This suggests yet another measure of the approximation error, the *relative error* (column 6), which gives the ratio of the absolute error to the deviation in x from the point of intersection, $x = 3$, that is,

$$\text{relative error} = \frac{f(x) - g(x)}{x - 3}$$

We observe from table 4.1 that the relative error decreases uniformly as we approach $x = 3$ from either above or below. This is a more demanding criterion than decreasing absolute error or percentage error. We will show below that this is precisely the requirement of a "good" approximation.

Geometrically the similarity between f and g in the neighborhood of $x = 3$ is reflected in the fact that the graphs of the two functions are tangential at $(3, 21)$. That is, their graphs intersect at the point $(3, 21)$ and are barely distinguishable over the surrounding neighborhood (figure 4.1).

Table 4.1
Approximating a quadratic function

x	$f(x)$	$g(x)$	Actual	Percentage	Relative
1.000	9.000	13.000	−4.000	−44.444	−2.000
2.000	16.000	17.000	−1.000	−6.250	−1.000
2.500	18.750	19.000	−0.250	−1.333	−0.500
2.900	20.590	20.600	−0.010	−0.049	−0.100
2.990	20.960	20.960	−0.000	−0.000	−0.010
2.999	20.996	20.996	−0.000	−0.000	−0.001
3.000	21.000	21.000	0.000	0.000	NIL
3.001	21.004	21.004	−0.000	−0.000	0.001
3.010	21.040	21.040	−0.000	−0.000	0.010
3.100	21.390	21.400	−0.010	−0.047	0.100
3.500	22.750	23.000	−0.250	−1.099	0.500
4.000	24.000	25.000	−1.000	−4.167	1.000
5.000	25.000	29.000	−4.000	−16.000	2.000

Approximation error column groups: Actual, Percentage, Relative.

4.1 Linear Approximation and the Derivative

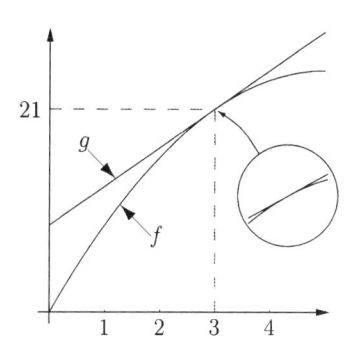

Figure 4.1
The tangency of f and g

Table 4.2 analyzes another approximation for f. The function $h(x) = 12 + 3x$ provides a rough approximation of f, but it is a less satisfactory approximation than g. Although the actual and percentage errors decline as we approach $x = 3$, the relative error does not. This is illustrated in figure 4.2, which shows that the graphs of f and h are close but not tangential.

Exercise 4.1
Show that the function $f(x) = 10x - x^2$ represents the total revenue function for a monopolist facing the market demand curve

$$x = 10 - p$$

where x is the quantity demanded and p is the market price. In this context, how should we interpret $g(x) = 9 + 4x$?

Exercise 4.2 (Growth rates)
Suppose that nominal GDP rose 10 percent in your country last year, while prices rose 5 percent. What was the growth rate of real GDP?

Moving to higher dimensions, table 4.3 shows how the linear function

$$g(x_1, x_2) = \tfrac{1}{3}x_1 + \tfrac{2}{3}x_2$$

approximates the Cobb-Douglas function

$$f(x_1, x_2) = x_1^{1/3} x_2^{2/3}$$

420   Chapter 4   Smooth Functions

Table 4.2
Another approximation for the quadratic function

			Approximation error		
x	$f(x)$	$h(x)$	Absolute	Percentage	Relative
1.000	9.000	15.000	−6.000	−66.667	−3.000
2.000	16.000	18.000	−2.000	−12.500	−2.000
2.500	18.750	19.500	−0.750	−4.000	−1.500
2.900	20.590	20.700	−0.110	−0.534	−1.100
2.990	20.960	20.970	−0.010	−0.048	−1.010
2.999	20.996	20.997	−0.001	−0.005	−1.001
3.000	21.000	21.000	0.000	0.000	NIL
3.001	21.004	21.003	0.001	0.005	−0.999
3.010	21.040	21.030	0.010	0.047	−0.990
3.100	21.390	21.300	0.090	0.421	−0.900
3.500	22.750	22.500	0.250	1.099	−0.500
4.000	24.000	24.000	0.000	0.000	0.000
5.000	25.000	27.000	−2.000	−8.000	1.000

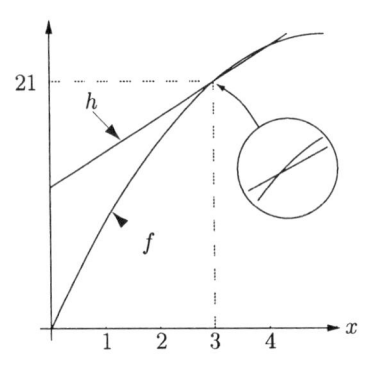

Figure 4.2
f and h are not tangential

4.1 Linear Approximation and the Derivative

Table 4.3
Approximating the Cobb-Douglas function

x	$\mathbf{x}_0 + \mathbf{x}$	$f(\mathbf{x}_0 + \mathbf{x})$	$g(\mathbf{x}_0 + \mathbf{x})$	Approximation error Percentage	Relative
At their intersection:					
(0.0, 0.0)	(8.0, 8.0)	8.0000	8.0000	0.0000	NIL
Around the unit circle:					
(1.0, 0.0)	(9.0, 8.0)	8.3203	8.3333	−0.1562	−0.0130
(0.7, 0.7)	(8.7, 8.7)	8.7071	8.7071	−0.0000	−0.0000
(0.0, 1.0)	(8.0, 9.0)	8.6535	8.6667	−0.1522	−0.0132
(−0.7, 0.7)	(7.3, 8.7)	8.2076	8.2357	−0.3425	−0.0281
(−1.0, 0.0)	(7.0, 8.0)	7.6517	7.6667	−0.1953	−0.0149
(−0.7, −0.7)	(7.3, 7.3)	7.2929	7.2929	0.0000	0.0000
(0.0, −1.0)	(8.0, 7.0)	7.3186	7.3333	−0.2012	−0.0147
(0.7, −0.7)	(8.7, 7.3)	7.7367	7.7643	−0.3562	−0.0276
Around a smaller circle:					
(0.10, 0.00)	(8.1, 8.0)	8.0332	8.0333	−0.0017	−0.0014
(0.07, 0.07)	(8.1, 8.1)	8.0707	8.0707	0.0000	0.0000
(0.00, 0.10)	(8.0, 8.1)	8.0665	8.0667	−0.0017	−0.0014
(−0.07, 0.07)	(7.9, 8.1)	8.0233	8.0236	−0.0035	−0.0028
(−0.10, 0.00)	(7.9, 8.0)	7.9665	7.9667	−0.0018	−0.0014
(−0.07, −0.07)	(7.9, 7.9)	7.9293	7.9293	0.0000	0.0000
(0.00, −0.10)	(8.0, 7.9)	7.9332	7.9333	−0.0018	−0.0014
(0.07, −0.07)	(8.1, 7.9)	7.9762	7.9764	−0.0035	−0.0028
Parallel to the x_1 axis:					
(−4.0, 0.0)	(4.0, 8.0)	6.3496	6.6667	−4.9934	−0.0793
(−2.0, 0.0)	(6.0, 8.0)	7.2685	7.3333	−0.8922	−0.0324
(−1.0, 0.0)	(7.0, 8.0)	7.6517	7.6667	−0.1953	−0.0149
(−0.5, 0.0)	(7.5, 8.0)	7.8297	7.8333	−0.0460	−0.0072
(−0.1, 0.0)	(7.9, 8.0)	7.9665	7.9667	−0.0018	−0.0014
(0.0, 0.0)	(8.0, 8.0)	8.0000	8.0000	0.0000	NIL
(0.1, 0.0)	(8.1, 8.0)	8.0332	8.0333	−0.0017	−0.0014
(0.5, 0.0)	(8.5, 8.0)	8.1633	8.1667	−0.0411	−0.0067
(1.0, 0.0)	(9.0, 8.0)	8.3203	8.3333	−0.1562	−0.0130
(2.0, 0.0)	(10.0, 8.0)	8.6177	8.6667	−0.5678	−0.0245
(4.0, 0.0)	(12.0, 8.0)	9.1577	9.3333	−1.9177	−0.0439
Parallel to the x_2 axis:					
(0.0, −4.0)	(8.0, 4.0)	5.0397	5.3333	−5.8267	−0.0734
(0.0, −2.0)	(8.0, 6.0)	6.6039	6.6667	−0.9511	−0.0314
(0.0, −1.0)	(8.0, 7.0)	7.3186	7.3333	−0.2012	−0.0147
(0.0, −0.5)	(8.0, 7.5)	7.6631	7.6667	−0.0466	−0.0071
(0.0, −0.1)	(8.0, 7.9)	7.9332	7.9333	−0.0018	−0.0014
(0.0, 0.0)	(8.0, 8.0)	8.0000	8.0000	0.0000	NIL
(0.0, 0.1)	(8.0, 8.1)	8.0665	8.0667	−0.0017	−0.0014
(0.0, 0.5)	(8.0, 8.5)	8.3300	8.3333	−0.0406	−0.0068
(0.0, 1.0)	(8.0, 9.0)	8.6535	8.6667	−0.1522	−0.0132
(0.0, 2.0)	(8.0, 10.0)	9.2832	9.3333	−0.5403	−0.0251
(0.0, 4.0)	(8.0, 12.0)	10.4830	10.6667	−1.7524	−0.0459

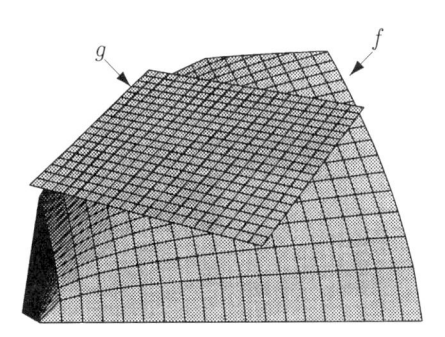

Figure 4.3
The tangency of f and g

in the neighborhood of the point $(8, 8)$. To facilitate future exposition, we have expressed the values of **x** relative to the point of intersection, $\mathbf{x}_0 = (8, 8)$.

One consequence of moving from one dimension to two is that there is now an infinity of directions along which f and g can be compared, making the evaluation of the closeness of the approximation more complicated. In the first half of table 4.3, we have evaluated the approximation around a circle of unit radius centered at the point $\mathbf{x}_0$ and also around a circle of radius 0.1. We observe that the approximation error varies with the direction of evaluation but declines as we approach the point $\mathbf{x}_0$. The bottom half of table 4.3 evaluates the approximation error over a wider interval along two particular directions, namely parallel to the x_1 and x_2 axes. Here the results parallel the approximation of a univariate function in table 4.1. The relative error declines uniformly as we approach the point of intersection, $\mathbf{x}_0$. The Cobb-Douglas function f and its linear approximation g are illustrated in figure 4.3, where we see that the graph of g is a plane that is tangential to the curved surface of f at the point $\mathbf{x}_0$.

Example 4.1 Suppose that a production process requires two inputs, k and l which we will call "capital" and "labor" respectively. The quantity of output depends upon the inputs according to the production function f,

$$f(k, l) = k^{1/3} l^{2/3}$$

Suppose that 8 units of each input are used currently, and will produce an output of 8 units. We have just shown how we can approximate the behavior of the production function f in the neighborhood of the input mix $(8,8)$ by the linear function

$$g(k,l) = \tfrac{1}{3}k + \tfrac{2}{3}l$$

This linear approximation enables us to estimate the impact of any change in the input mix.

For example, if we add an additional unit of labor l to the existing quantity of capital k, we can see from table 4.3 that the output $f(8,9)$ increases to 8.6523, an increase of 0.6523 units of output. Alternatively, we can estimate the new output using g

$$g(8,9) = \tfrac{1}{3}8 + \tfrac{2}{3}9 = 8\tfrac{2}{3}$$

The coefficients of k and l in g are known as the *marginal products* of capital and labor respectively (example 4.5). g can be thought of as an approximate linear production function, in which the contribution of each of the inputs x_1 and x_2 is evaluated by their respective marginal products $1/3$ and $2/3$ at the input combination $(8,8)$.

To facilitate the estimation of changes in output, g can be rewritten as

$$g(k,l) = 8 + \tfrac{1}{3}(k-8) + \tfrac{2}{3}(l-8)$$

If we increase both capital and labor by one unit, total output can be estimated as

$$g(9,9) = 8 + \tfrac{1}{3} \times 1 + \tfrac{2}{3} \times 1 = 9$$

which in this case equals the actual output $f(9,9)$.

The examples illustrate the sense in which a nonlinear function can be approximated by a linear function. Now let us generalize these examples. Let $f\colon X \to Y$ be a function between normed linear spaces X and Y. We say that f is differentiable at some point $\mathbf{x}_0$ in X if it has a good linear approximation at $\mathbf{x}_0$, that is there exists a linear function $g\colon X \to Y$ such that

$$f(\mathbf{x}_0 + \mathbf{x}) \approx f(\mathbf{x}_0) + g(\mathbf{x}) \qquad (1)$$

for every $\mathbf{x}$ in a neighborhood of $\mathbf{x}_0$ where $\approx$ means "is approximately equal to."

How do we know when we have a good approximation? Is there a best linear approximation? In the preceding examples we showed that the *relative error* decreased as we approached the point $\mathbf{x}_0$, and this was the case no matter from which direction we approached. This is the property that we want the approximation to possess. We now make this precise.

For any function $f: X \to Y$ and *any* linear function $g: X \to Y$, the error at $\mathbf{x}$ in approximating f by $f(\mathbf{x}_0) + g(\mathbf{x})$ is given by

$$\varepsilon(\mathbf{x}) = f(\mathbf{x}_0 + \mathbf{x}) - f(\mathbf{x}_0) - g(\mathbf{x})$$

$\varepsilon(\mathbf{x}) \in Y$ is the *actual error* as defined above. We can decompose $\varepsilon(\mathbf{x})$ into two components

$$\varepsilon(\mathbf{x}) = \eta(\mathbf{x}) \|\mathbf{x}\|$$

where

$$\eta(\mathbf{x}) = \frac{\varepsilon(\mathbf{x})}{\|\mathbf{x}\|} = \frac{f(\mathbf{x}_0 + \mathbf{x}) - f(\mathbf{x}_0) - g(\mathbf{x})}{\|\mathbf{x}\|}$$

is the *relative error*. If for a particular linear function g, the relative error $\eta(\mathbf{x})$ gets smaller and smaller as $\mathbf{x}$ gets smaller, then their product $\|\mathbf{x}\|\eta(\mathbf{x})$ becomes negligible and we consider g a good approximation of f in the neighborhood of $\mathbf{x}_0$.

Formally the function f is *differentiable* at $\mathbf{x}_0 \in X$ if there is a (continuous) linear function $g: X \to Y$ such that for all $\mathbf{x} \in X$,

$$f(\mathbf{x}_0 + \mathbf{x}) = f(\mathbf{x}_0) + g(\mathbf{x}) + \eta(\mathbf{x})\|\mathbf{x}\| \tag{2}$$

with $\eta(\mathbf{x}) \to \mathbf{0}_Y$ as $\mathbf{x} \to \mathbf{0}_X$. We call the linear function g the *derivative* of f at $\mathbf{x}_0$ and denote it by $Df[\mathbf{x}_0]$ or $f'[\mathbf{x}_0]$. The derivative is the *best linear approximation* to the function f at $\mathbf{x}_0$ in terms of minimizing the relative error.

Three points deserve emphasis:

• There can be at most one linear function g that satisfies (2). The derivative of a function is unique (exercise 4.4).

• Every linear function can be approximated by itself. Therefore every continuous linear function is differentiable (exercise 4.6).

• Not every function is differentiable. For a given function f and point $\mathbf{x}_0$, there may be no continuous linear function that satisfies (2) (example 4.7).

4.1 Linear Approximation and the Derivative

Remark 4.1 (Differentiability and continuity) The adjective "continuous" in the definition of a differentiable function is redundant in finite-dimensional spaces, since every linear function on a finite-dimensional space is continuous (exercise 3.31). It is not redundant in infinite-dimensional spaces, and it implies that every differentiable function is continuous (exercise 4.5). Some authors (e.g., Chillingworth 1976, p. 53; Dieudonne 1960, p. 143) proceed in the other direction. They define differentiabilty only for continuous functions, which implies that the derivative (if it exists) must be continuous.

Remark 4.2 (Affine or linear) We must be wary of some semantic confusion here. Strictly speaking, the "linear approximation" $f(\mathbf{x}_0) + g(\mathbf{x})$ is an affine function (section 3.2). The derivative $g = Df[\mathbf{x}_0]$ is its linear component (exercise 3.39). Most authors do not make this distinction, using the term linear to refer to the affine function $f(\mathbf{x}_0) + Df[\mathbf{x}_0]$. (Recall example 3.10.) To avoid being pedantic, we will not slavishly adhere to the distinction between affine and linear maps and follow the customary usage when referring to the derivative.

Exercise 4.3
Show that the definition (2) can be equivalently expressed as

$$\lim_{\mathbf{x} \to \mathbf{0}_X} \eta(\mathbf{x}) = \lim_{\mathbf{x} \to \mathbf{0}_X} \frac{f(\mathbf{x}_0 + \mathbf{x}) - f(\mathbf{x}_0) - g(\mathbf{x})}{\|\mathbf{x}\|} = \mathbf{0}_Y \tag{3}$$

Exercise 4.4 (Derivative is unique)
The derivative of a function is unique.

Exercise 4.5 (Differentiable implies continuous)
If $f: X \to Y$ is differentiable at $\mathbf{x}_0$, then f is continuous at $\mathbf{x}_0$.

Exercise 4.6 (Linear functions)
Every continuous linear function is differentiable with $Df[\mathbf{x}] = f$.

Exercise 4.7 (Constant functions)
A constant function $f: X \to Y$ is differentiable with $Df[\mathbf{x}] = \mathbf{0}$. That is, the derivative of a constant function is the zero map in $L(X, Y)$.

Example 4.2 (Derivative in $\Re$) It is easy to show that (2) is equivalent to the familiar definition of the derivative of univariate functions. Recall from elementary calculus that a function $f: \Re \to \Re$ is differentiable at x_0 if the limit

$$\lim_{x \to 0} \frac{f(x_0 + x) - f(x_0)}{x} \tag{4}$$

exists. This limit, called the derivative of f at x_0, is a real number, say α. It satisfies the equation

$$\lim_{x \to 0} \frac{f(x_0 + x) - f(x_0)}{x} - \alpha = 0$$

or

$$\lim_{x \to 0} \frac{f(x_0 + x) - f(x_0) - \alpha x}{x} = 0$$

The product αx defines a linear functional $g(x) = \alpha x$ on $\Re$, which is the linear function of equation (3), that is,

$$\lim_{x \to 0} \frac{f(x_0 + x) - f(x_0) - g(x)}{x} = 0$$

Any linear functional on $\Re$ necessarily takes the form αx for some $\alpha \in \Re$ (proposition 3.4). Elementary calculus identifies the derivative (the linear function αx) with its dual representation α (a number). In geometric terms, the derivative of a function is the linear function whose graph is tangential to the function at a particular point. Elementary calculus *represents* that tangent (the graph of a function) by its slope.

Exercise 4.8
Evaluate the error in approximating the function

$$f(x_1, x_2) = x_1^{1/3} x_2^{2/3}$$

by the linear function

$$g(x_1, x_2) = \tfrac{1}{3} x_1 + \tfrac{2}{3} x_2$$

at the point $(2, 16)$. Show that the linear function

$$h(x_1, x_2) = \tfrac{4}{3} x_1 + \tfrac{1}{3} x_2$$

is a better approximation at the point $(2, 16)$.

Remark 4.3 (Notation) As exercise 4.8 demonstrates, the particular linear function g that best approximates a given function f depends on the point $\mathbf{x}_0$ at which it is approximated. Different $\mathbf{x}_0$ in X require different linear approximating functions. To emphasize that the derivative is a function, we will use square brackets to denote the point of approximation, so as to distinguish this from the point of evaluation of the function. That is, $Df[\mathbf{x}_0]$ represents the derivative of the function f at the point $\mathbf{x}_0$, and $Df[\mathbf{x}_0](\mathbf{x})$ denotes the value of the approximating function $Df[\mathbf{x}_0]$ at the point $\mathbf{x}$. Sometimes, particularly when $X \subseteq \Re$, we will use the synonymous notation $f'[\mathbf{x}_0]$ for the derivative function, and $f'[\mathbf{x}_0](\mathbf{x})$ for its value at a point. Where there is no ambiguity, we may even omit explicit mention of the point of approximation, so that f' represents the derivative function (at some implicit point of approximation) and $f'(\mathbf{x})$ denotes the value of this function at $\mathbf{x}$.

A function $f: X \to Y$ is *differentiable* if it is differentiable at all $\mathbf{x}_0 \in X$. In that case the derivative defines another function Df on X, a function that takes its values in $BL(X, Y)$ rather than in Y. In other words, the derivative Df defines a function from X to $BL(X, Y)$, the space of all continuous linear functions from X to Y. If this function $Df: X \to BL(X, Y)$ is continuous, we say that the original function f is *continuously differentiable*. This means that nearby points in X give rise to nearby functions in $BL(X, Y)$. The set of all continuously differentiable functions between spaces X and Y is denoted $C^1(X, Y)$. An individual continuously differentiable function is said to be C^1.

Example 4.3 (Tangent hyperplane) The graph of the affine approximation $f(\mathbf{x}_0) + Df[\mathbf{x}_0]$ to a differentiable function on $\Re^n$ defines a hyperplane in $\Re^{n+1}$ (exercise 3.40) which is known as the *tangent hyperplane* to f at $\mathbf{x}_0$ (figure 4.3). Intuitively, for there to exist a well-defined tangent plane at any point, the surface must be smooth, that is, have no sharp edges or points. A function is differentiable if has a tangent hyperplane everywhere.

If we think of $\mathbf{dx}$ as a "small" (i.e., close to zero) vector in X, and define df to be the change in the function f in a neighborhood of $\mathbf{x}_0$, then (1) can be rewritten as

$$df \equiv f(\mathbf{x}_0 + \mathbf{dx}) - f(\mathbf{x}_0) \approx \underbrace{Df[\mathbf{x}_0]}_{\text{linear function}} \underbrace{(\mathbf{dx})}_{\text{small vector}} \qquad (5)$$

This is sometimes known as the *total differential* of f.

Thinking about approximation geometrically, we perceive that one function will approximate another when their graphs almost coincide. More precisely, their graphs must intersect at some point and not diverge too dramatically in the neighborhood of that point. We say that the graphs are *tangent* at the point of intersection. The graph of a linear function is characterized by the absence of curvature. Therefore, to be tangent to linear function, a nonlinear function must not exhibit excessive curvature. It must be *smooth* without sharp corners or edges. We will give a precise definition of smoothness in a later section.

Example 4.4 (Tangent functions) Two function $f, h: X \to Y$ are *tangent* at $\mathbf{x}_0 \in X$ if

$$\lim_{\mathbf{x} \to 0} \frac{f(\mathbf{x}_0 + \mathbf{x}) - h(\mathbf{x}_0 + \mathbf{x})}{\|\mathbf{x}\|} = \mathbf{0}$$

Exercise 4.4 shows that among all the functions that are tangent to a function $f(\mathbf{x})$ at a point $\mathbf{x}_0$, there is most one affine function of the form $f(\mathbf{x}_0) + h(\mathbf{x})$ where $h(\mathbf{x})$ is linear. This is the derivative.

Remark 4.4 (Differentiation) The process of finding the best linear approximation to a given function is called *differentiation*. This process may be view abstractly as follows. In the preceding chapter we showed (exercise 3.33) that the set $BL(X, Y)$ of all continuous linear functions between two linear spaces X and Y is itself a linear space, and hence a subspace of the set of all functions between X and Y. Given an arbitrary function $f: X \to Y$ and point $\mathbf{x}_0 \in X$, imagine searching through the subspace of all continuous linear functions $BL(X, Y)$ for a function g to approximate f in the sense of (2). We may be unable to find such a function. However, if we do find such a function, we can give up the search. There can be no better approximation in the set $BL(X, Y)$.

Remark 4.5 (Econometric estimation) There is an analogy here with the procedures of econometrics, which may be instructive to some readers. In econometrics we are presented with some relationship between a dependent variable y and a set of independent variables $\mathbf{x}$ that can be represented by a function f, that is,

$$y = f(\mathbf{x})$$

The function f is unknown, although there is a available of set of observations or points $\{(\mathbf{x}_t, y_t)\}$. In the technique of linear regression, it is supposed that the function f can be represented as a (continuous) linear function of $\mathbf{x}$ plus a random error, ε, that is,

$$y = f(\mathbf{x}_0) + g(\mathbf{x}) + \varepsilon$$

$f(\mathbf{x}_0)$ is called the constant term.

Estimation involves choosing an element of the set $BL(X, Y)$ to minimize some measure of the error depending on the presumed statistical properties of ε. For example, in ordinary least squares estimation, a linear function $g \in BL(X, Y)$ is chosen so as to minimize the squared absolute error summed over the set of observations $\{(\mathbf{x}_t, y_t)\}$.

In calculus, on the other hand, the function f is known, and the linear function $g \in BL(X, Y)$ is chosen so as to minimize the relative error in the neighborhood of some point of interest $\mathbf{x}_0$. The precise way in the which the function g is found (the estimation procedure) is the process of differentiation. It will be discussed in the next two sections.

Exercise 4.9 (The rank order function)
A subtle example that requires some thought is provided by the rank order function, which sorts vectors into descending order. Let $r: \Re^n \to \Re^n$ be the function that ranks vectors in descending order; that is, $y = r(\mathbf{x})$ is defined recursively as

$$y_1 = \max_i x_i, \quad y_2 = \max_{x_i \neq y_1} x_i, \ldots, \quad y_n = \min_i x_i$$

For example, $r(1, 2, 3, 4, 5) = (5, 4, 3, 2, 1)$ and $r(66, 55, 75, 81, 63) = (81, 75, 67, 63, 55)$. Show that r is nonlinear and differentiable for all $\mathbf{x}$ such that $x_i \neq x_j$

Before analyzing the properties of the derivative, we show how the derivative of a function can be computed in applications of practical interest.

4.2 Partial Derivatives and the Jacobian

Most functions that we encounter in economic models are functionals on $\Re^n$, which can be decomposed into simpler functions (section 2.1.3). For example, given a functional $f: \Re^n \to \Re$, define the function $h: \Re \to \Re$ obtained by allowing only one component of $\mathbf{x}$ to vary while holding the others constant, that is,

$$h(t) = f(x_1^0, x_2^0, \ldots, x_{i-1}^0, t, x_{i+1}^0, \ldots, x_n^0)$$

Geometrically the graph of h corresponds to a cross section of the graph of f through $\mathbf{x}^0$ and parallel to the x_i axis (section 2.1.4). If the limit

$$\lim_{t \to 0} \frac{h(\mathbf{x}_i^0 + t) - h(\mathbf{x}_i^0)}{t}$$

exists, then we call it the ith *partial derivative* of f at $\mathbf{x}^0$, denoted $D_{x_i}f[\mathbf{x}^0]$ or $\partial f[\mathbf{x}^0]/\partial x_i$ for even $f_{x_i}[\mathbf{x}^0]$. Geometrically the partial derivative measures the slope of the tangent to the function h, which in turn is the slope of the cross section of f through $\mathbf{x}^0$. Throughout this section, X is a subspace of $\Re^n$.

Example 4.5 (Marginal product) Recall example 4.1 where total output was specified by the production function

$$f(k, l) = k^{1/3} l^{2/3}$$

Suppose that one input k ("capital") is fixed at 8 units. The restricted production function

$$h(l) = f(8, l) = 2l^{2/3}$$

specifies attainable output as a function of the variable input l ("labor"), given a fixed capital of 8 units (figure 4.4). It represents a cross section through the production surface $f(k, l)$ at $(8, 8)$ parallel to the labor axis. To analyze the effect of small changes in l, we can use

$$\lim_{t \to 0} \frac{h(8 + t) - h(8)}{t}$$

which is precisely the partial derivative of $f(k, l)$ with respect to l at the point $(8, 8)$. The result is called the *marginal product of labor*, and it measures the change in output resulting from a small change in l while holding k constant. Geometrically it measures the slope of the tangent to the restricted production function (figure 4.4) at the point $l = 8$. Similarly the partial derivative of f with respect to k is the marginal product of capital.

Example 4.6 (Directional derivatives) Recall that one of the complications of moving to a multivariate function is the infinity of directions

4.2 Partial Derivatives and the Jacobian

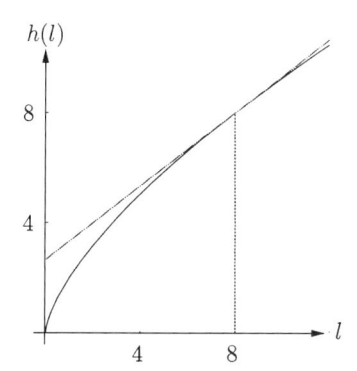

Figure 4.4
The restricted production function

in which it is possible to move from any point. Consider the behavior of a function $f: X \to \Re$ in the neighborhood of some point $\mathbf{x}^0$. Any other point $\mathbf{x} \in X$ defines a direction of movement, and the function

$$h(t) = f(\mathbf{x}^0 + t\mathbf{x})$$

can be considered a function of the single variable t. Geometrically the graph of h is a vertical cross section of the graph of f through $\mathbf{x}^0$ and $\mathbf{x}$. If h is differentiable at 0 so that the limit

$$\lim_{t \to 0} \frac{h(t) - h(0)}{t} = \lim_{t \to 0} \frac{f(\mathbf{x}^0 + t\mathbf{x}) - f(\mathbf{x}^0)}{t}$$

exists, it is called *the directional derivative* of f at $\mathbf{x}^0$ in the direction $\mathbf{x}$. It is denoted $\vec{D}_\mathbf{x} f[\mathbf{x}^0]$.

Exercise 4.10
Let $f: X \to Y$ be differentiable at $\mathbf{x}^0$ with derivative $Df[\mathbf{x}^0]$, and let $\mathbf{x}$ be a vector of unit norm ($\|\mathbf{x}\| = 1$). Show that the directional derivative of f at $\mathbf{x}^0$ in the direction $\mathbf{x}$ is the *value* of the linear function $Df[\mathbf{x}^0]$ at $\mathbf{x}$, that is,

$$\vec{D}_\mathbf{x} f[\mathbf{x}^0] = Df[\mathbf{x}^0](\mathbf{x})$$

Exercise 4.11
Show that the ith partial derivative of the function $f: \Re^n \to \Re$ at some point $\mathbf{x}^0$ corresponds to the directional derivative of f at $\mathbf{x}^0$ in the direc-

tion $\mathbf{e}_i$, where

$$\mathbf{e}_i = (0, 0, \ldots, 1, \ldots, 0)$$

is the i unit vector. That is,

$$D_{x_i} f[\mathbf{x}^0] = \vec{D}_{\mathbf{e}_i} f[\mathbf{x}^0]$$

Exercise 4.12
Calculate the directional derivative of the function

$$f(x_1, x_2) = x_1^{1/3} x_2^{2/3}$$

at the point $(8, 8)$ in the direction $(1, 1)$.

We now relate partial derivatives to the derivative defined in the previous section. If $f: X \to \Re$ is differentiable at $\mathbf{x}^0$, its derivative $Df[\mathbf{x}^0]$ is a *linear* functional on X. If the domain X is $\Re^n$, there exists (proposition 3.4) a vector $\mathbf{p} = (p_1, p_2, \ldots, p_n)$ that represents $Df[\mathbf{x}^0]$ with respect to the standard basis in the sense that

$$Df[\mathbf{x}^0](\mathbf{x}) = p_1 x_1 + p_2 x_2 + \cdots + p_n x_n$$

where $\mathbf{x} = (x_1, x_2, \ldots, x_n)$. This vector is called the *gradient* of f at $\mathbf{x}^0$. It is denoted grad $f(\mathbf{x}^0)$ or $\nabla f(\mathbf{x}^0)$. Moreover the components of $\mathbf{p}$ are precisely the *partial derivatives* of f at $\mathbf{x}^0$ (exercise 4.13), that is,

$$\nabla f(\mathbf{x}^0) = (p_1, p_2, \ldots p_n) = (D_{x_1} f[\mathbf{x}^0], D_{x_2} f[\mathbf{x}^0], \ldots, D_{x_n} f[\mathbf{x}^0])$$

so

$$Df[\mathbf{x}^0](\mathbf{x}) = \sum_{i=1}^n D_{x_i} f[\mathbf{x}^0] x_i \qquad (6)$$

The gradient of a differentiable function has an important geometric interpretation: it points in the direction in which the function increases most rapidly (exercise 4.16). For example, on a hill, the gradient at each point indicates the direction of steepest ascent.

Exercise 4.13
Show that the gradient of a differentiable functional on $\Re^n$ comprises the vector of its partial derivatives, that is,

$$\nabla f(\mathbf{x}^0) = (D_{x_1} f[\mathbf{x}^0], D_{x_2} f[\mathbf{x}^0], \ldots, D_{x_n} f[\mathbf{x}^0])$$

Exercise 4.14
Show that the derivative of a functional on $\Re^n$ can be expressed as the inner product

$$Df[\mathbf{x}^0](\mathbf{x}) = \nabla f(\mathbf{x}^0)^T \mathbf{x}$$

Exercise 4.15 (Nonnegative gradient)
If a differentiable functional f is increasing, then $\nabla f(\mathbf{x}) \geq \mathbf{0}$ for every $\mathbf{x} \in X$; that is, every partial derivative $D_{x_i} f[\mathbf{x}]$ is nonnegative.

Exercise 4.16
Show that the gradient of a differentiable function f points in the direction of greatest increase. [Hint: Use exercise 3.61.]

Partial derivatives provide a practical means for computing the derivative of a differentiable function. A fundamental result of chapter 3 was that any linear mapping is completely determined by its action on a basis for the domain. This implies that the derivative of a functional on $\Re^n$ is fully summarized by the n partial derivatives, which can be readily calculated for any given function. Formula (6) then gives a mechanism for computing the value of the derivative $Df[\mathbf{x}^0](\mathbf{x})$ for arbitrary $\mathbf{x}$. The derivative approximates a nonlinear function by combining linearly the separate effects of marginal changes in each of the variables.

Remark 4.6 (Differentiability) The preceding discussion presumes that the function f is differentiable, which implies the existence of partial derivatives. The converse is not necessarily true. We cannot infer from the existence of its partial derivatives that the function is differentiable. The reason is that the partial derivatives only approximate a function parallel to the axes, whereas the derivative approximates the function in all directions. It is possible for a function to have a linear approximation in certain directions and not in others (example 4.7). Later (exercise 4.37) we will show that continuity of the partial derivatives is sufficient to guarantee linear approximation in all directions and hence differentiability

Differentiability is seldom an issue in economics. Commonly an explicit functional form is not imposed on an economic model. Rather, the functions are simply assumed to have the required properties, such as differentiability or convexity. The results derived then apply to all functions that exhibit the assumed properties.

Example 4.7 (A nondifferentiable function) Let $f: \Re^2 \to \Re$ be defined by

$$f(x_1, x_2) = \begin{cases} 0 & \text{if } x_1 = x_2 = 0 \\ \dfrac{x_1 x_2}{x_1^2 + x_2^2} & \text{otherwise} \end{cases}$$

The partial derivatives of f exist for all $(x_1, x_2) \in \Re^2$ with $D_{x_1}f[(0,0)] = D_{x_2}f[(0,0)] = 0$. But f is not continuous at $(0,0)$, and hence not differentiable there.

Example 4.8 (Marginal product again) We showed earlier (example 4.1) that the Cobb-Douglas production function

$$f(k, l) = k^{1/3} l^{2/3}$$

could be approximated by the linear function

$$g(k, l) = \tfrac{1}{3}k + \tfrac{2}{3}l$$

We will show later that the partial derivatives of f at the point $(8, 8)$ are

$$D_k f[(8,8)] = \tfrac{1}{3} \quad \text{and} \quad D_l f[(8,8)] = \tfrac{2}{3}$$

Therefore the derivative of f at $(8,8)$ is, from (6),

$$Df[(8,8)](k, l) = D_k f[(8,8)]k + D_l f[(8,8)]l = \tfrac{1}{3}k + \tfrac{2}{3}l$$

which is precisely the approximating function g. We now recognize that the coefficients of k and l in this function are the marginal products of k and l respectively (example 4.5). Therefore g could be rewritten as

$$g(k, l) = \text{MP}_k k + \text{MP}_l l$$

where MP_k and MP_l denote the marginal products of k and l respectively evaluated at $(8, 8)$. Output at the input combination (k, l) can be approximated by adding the contributions of k and l, where each contribution is measured by the quantity used times its marginal product at $(8, 8)$.

The accuracy of the approximation provided by g will depend on the degree to which the marginal products of f vary as the quantities of k and l change. The approximation would be perfect if the marginal products were constant, but this will be the case if and only if the production function is linear or affine. It is precisely because economic realism requires nonconstant marginal products that we need to resort to nonlinear functions to model production.

Example 4.9 (Marginal utility) Suppose that a consumer's preferences can be represented (example 2.58) by the utility function $u: X \to \Re$ where $X \subset \Re^n$ is the consumption set. The consumer is currently consuming a bundle of goods and services $\mathbf{x}^0 \in X$. It is desired to measure impact on utility of small changes in the composition of this bundle. Let $\mathbf{dx} \in X$ denote the changes in the consumption bundle. Note that some components of $\mathbf{dx}$ may be positive and some negative. If the utility function is differentiable, we can estimate the resulting change in utility du by the total differential (5)

$$du \equiv u(\mathbf{x}^0 + \mathbf{dx}) - u(\mathbf{x}^0) \approx Du[\mathbf{x}^0](\mathbf{dx})$$

Using (6) this can be expressed as

$$du \approx \sum_{i=1}^{n} D_{x_i} u[\mathbf{x}^0] \, dx_i \tag{7}$$

where $\mathbf{dx} = (dx_1, dx_2, \ldots, dx_n)$.

The partial derivative $D_{x_i} u[\mathbf{x}^0]$ is called the *marginal utility* of good i. It estimates the added utility obtained from consuming an additional unit of good i holding the consumption of all other goods constant. Using (7), we approximate the change in utility in moving from consumption bundle $\mathbf{x}^0$ to $\mathbf{x}^0 + \mathbf{dx}$ by estimating the change arising from each commodity separately ($D_{x_i} u[\mathbf{x}^0] \, dx_i$) and then summing these effects. This estimate of the total change in utility may be inaccurate for two reasons:

- It presumes that the marginal utility of each good is constant as the quantity consumed is changed.
- It ignores interactions between different goods.

The extent of the inaccuracy will depend upon the utility function u, the point $\mathbf{x}$ and the degree of change $\mathbf{dx}$. However, for a differentiable function, we know that the inaccuracy becomes negligible for sufficiently small changes.

Example 4.10 (Marginal rate of substitution) We suggested in chapter 3 that any linear functional could be thought of as valuation function, with its components as prices (example 3.11). This applies to the gradient of the utility function, where the marginal utilities measure the maximum amount that the consumer would be willing to exchange for an additional unit of one commodity, holding constant the consumption of the other commodities. These are usually referred to as *subjective* or *reservation* prices.

The absolute value of these reservation prices is of limited significance, since it depends on the particular function used to represent the consumer's preferences. However, the implied relative prices between goods are significant because they are invariant to different representations of the utility function (example 4.17). At any consumption bundle the relative reservation price between any two goods i and j is called the *marginal rate of substitution* that is

$$\text{MRS}_{ij} = \frac{D_{x_i} u[\mathbf{x}^0]}{D_{x_j} u[\mathbf{x}^0]}$$

It measures the quantity of good j the consumer is willing to forgo for one additional unit of good i. As we will see in the next chapter (example 5.15), choosing an optimal consumption bundle amounts to adjusting the consumption bundle until these relative prices align with the market prices.

Economists use contours to help analyze functions in higher-dimensional spaces (section 2.1.4). Given a functional f on a set $X \subseteq \Re^n$, the contour of f through $c = f(\mathbf{x}^0)$ is

$$f^{-1}(c) = \{\mathbf{x} \in X : f(\mathbf{x}) = c\}$$

Familiar examples include isoquants and indifference curves, which are contours of production and utility functions respectively. If f is differentiable at $\mathbf{x}^0$, the derivative $Df[\mathbf{x}^0]$ defines a hyperplane (exercise 3.49) in X,

$$H = \{\mathbf{x} \in X : Df[\mathbf{x}^0](\mathbf{x}) = 0\}$$

which is tangential to the contour through $f(\mathbf{x}^0)$ (figure 4.5), and orthogonal to the gradient of f at $\mathbf{x}^0$ (exercise 4.17). This hyperplane is a linear approximation to the contour.

Example 4.11 (Slope of a contour) Suppose that $f : \Re^2 \to \Re$ is differentiable at $\mathbf{x}^0$. Then there exists a hyperplane

4.2 Partial Derivatives and the Jacobian

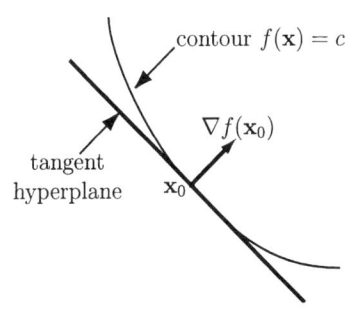

Figure 4.5
A contour and its tangent hyperplane

$$H = \{\mathbf{x} \in X : Df[\mathbf{x}^0](\mathbf{x}) = 0\}$$

that is tangential to the contour through $f(\mathbf{x}^0)$. Using (6), we see that H is a line defined by the equation

$$D_{x_1}f[\mathbf{x}^0]x_1 + D_{x_2}f[\mathbf{x}^0]x_2 = 0$$

or

$$D_{x_1}f[\mathbf{x}^0]x_1 = -D_{x_2}f[\mathbf{x}^0]x_2$$

The slope of the line is

$$\frac{x_2}{x_1} = -\frac{D_{x_1}f[\mathbf{x}^0]}{D_{x_2}f[\mathbf{x}^0]}$$

the ratio of the partial derivatives of f at $\mathbf{x}^0$.

If for example f is a utility function representing preferences over two goods, the marginal rate of substitution between the goods (example 4.10) measures the slope the indifference curve.

Exercise 4.17 (Gradient orthogonal to contour)
Let f be a differentiable functional and

$$H = \{\mathbf{x} \in X : Df[\mathbf{x}^0](\mathbf{x}) = 0\}$$

be the hyperplane tangent to the contour through $f(\mathbf{x}^0)$. Then $\nabla f(\mathbf{x}^0)$ is orthogonal to H.

Remark 4.7 (Notation) Variety is abundant in the notation employed for partial derivatives. Most common is the old-fashioned $\partial f(\mathbf{x}^0)/\partial x_i$. This is sometimes abbreviated as $f_i(\mathbf{x}^0)$ or even $f^i(\mathbf{x}^0)$. Marginal utility in this notation is $\partial u(\mathbf{x}^0)/\partial x_i$ or $u_i(\mathbf{x}^0)$. This often leads to confusion. For example, when dealing with many consumers, the marginal utility of consumer l for good i becomes $u_i^l(\mathbf{x}^0)$. Or is that the marginal utility of consumer i for good l? The marginal rate of substitution is the inelegant

$$\text{MRS}_{ij} = \frac{\partial u(\mathbf{x}^0)/\partial x_i}{\partial u(\mathbf{x}^0)/\partial x_j}$$

It is more consistent to use $D_{x_i} f[\mathbf{x}^0]$ for the partial derivative of f with respect to $\mathbf{x}_i$. When there is no ambiguity with regard to the independent variable, this is abbreviated to $D_i f[\mathbf{x}^0]$ for the ith partial derivative. Then the marginal utility of consumer l for good i is unambiguously $D_i u^l[\mathbf{x}^0]$. The marginal rate of substitution between goods i and j is $D_i u[\mathbf{x}^0]/D_j u[\mathbf{x}^0]$. This is the convention we will follow in this book.

It is also common to omit the point of evaluation $\mathbf{x}^0$ when there is ambiguity. The partial derivative $D_i f$, marginal utility $D_i u$ and marginal rate of substitution $D_i u / D_j u$ are all understood to be evaluated at a particular point, which should be obvious from the context.

The Jacobian

So far in this section, we have focused on single real-valued functions or functionals on $\Re^n$. However, we often encounter systems of functionals. Examples include systems of demand functions, payoff functions in noncooperative games and the IS-LM model in macroeconomics. Sometimes it is useful to study the components individually, where the preceding discussion applies. At other times it useful to consider the system of functions as a whole (section 2.1.3).

A system of m functionals $f_1, f_2, \ldots, f_m$ on $X \subseteq \Re^n$ defines a function $\mathbf{f}: X \to \Re^m$ where

$$\mathbf{f}(\mathbf{x}) = (f_1(\mathbf{x}), f_2(\mathbf{x}), \ldots, f_m(\mathbf{x})) \quad \text{for every } \mathbf{x} \in X$$

The function $\mathbf{f}$ is differentiable at $\mathbf{x}_0 \in X$ if and only if each component f_j is differentiable at $\mathbf{x}_0$ (exercise 4.18). The derivative $D\mathbf{f}[\mathbf{x}_0]$ is linear function between finite-dimensional spaces which can be represented by a matrix (proposition 3.1). The matrix representing the derivative $D\mathbf{f}[\mathbf{x}_0]$ of a differentiable function (with respect to the standard basis) is called the *Jacobian* of $\mathbf{f}$ at $\mathbf{x}_0$. The elements of the Jacobian J are the partial derivatives of the components f_j evaluated at $\mathbf{x}_0$:

$$J_{\mathbf{f}}(\mathbf{x}_0) = \begin{pmatrix} D_{x_1}f_1[\mathbf{x}_0] & D_{x_2}f_1[\mathbf{x}_0] & \cdots & D_{x_n}f_1[\mathbf{x}_0] \\ D_{x_1}f_2[\mathbf{x}_0] & D_{x_2}f_2[\mathbf{x}_0] & \cdots & D_{x_n}f_2[\mathbf{x}_0] \\ \vdots & \vdots & \ddots & \vdots \\ D_{x_1}f_m[\mathbf{x}_0] & D_{x_2}f_m[\mathbf{x}_0] & \cdots & D_{x_n}f_m[\mathbf{x}_0] \end{pmatrix}$$

Exercise 4.18
$\mathbf{f}: X \to \Re^m$, $X \subset \Re^n$ is differentiable $\mathbf{x}_0$ if and only if each component f_j is differentiable at $\mathbf{x}_0$. The matrix representing the derivative, the Jacobian, comprises the partial derivatives of the components of $\mathbf{f}$.

The Jacobian encapsulates all the essential information regarding the linear function that best approximates a differentiable function at a particular point. For this reason it is the Jacobian which is usually used in practical calculations with the derivative (example 4.16). To obtain the derivative of a function from $\Re^n$ to $\Re^m$, we take each of the components in turn and obtain its partial derivatives. In effect we reduce the problem of calculating the derivative of a multivariate function to obtaining the partial derivatives of each of its component functions. The Jacobian encapsulates the link between univariate and multivariate calculus.

Remark 4.8 By analogy with computer programming, our approach to multivariate calculus could be called the top–down approach, in which we start with general functions between arbitrary spaces and reduce it to the simpler familiar problem, real-valued functions of a single variable (the elements of the Jacobian). This contrast with the conventional bottom-up approach, which starts with the familiar univariate functions, and builds up to multivariate functions and then systems of such functions.

Example 4.12 (IS-LM model) The following model has formed the foundation of macroeconomic theory for the last 30 years. It consists of two equations. The IS curve describes equilibrium in the goods market

$$y = C(y, T) + I(r) + G$$

with national income y equal to aggregate output, comprising consumption C, investment I, and government spending G. Consumption $C(y, T)$ depends on income y and taxes T, while investment $I(r)$ is a function of the interest rate. The LM curve describes equilibrium in the money market

$$L(y,r) = \frac{M}{P}$$

when the demand for money L is equal to the supply M/P, where the demand for money depends on national income y and the interest rate r. Government spending G, taxes T, and the nominal money supply M are exogenous parameters. For simplicity we also assume that the price level P is fixed. Analysis of this model consists of describing the relationship between the dependent variables r, y and the three exogenous parameters G, T, M.

Designating two quantities y and r as variables and the remaining quantities G, M, and T as parameters, we can analyze the IS-LM model as a function from $\Re_+^2$ to $\Re^2$, namely

$$f_1(r, y) = y - C(y, T) - I(r) - G = 0$$

$$f_2(r, y) = L(y, r) - \frac{M}{P} = 0$$

The Jacobian of this system of equations is

$$J = \begin{pmatrix} D_r f_1 & D_y f_1 \\ D_r f_2 & D_y f_2 \end{pmatrix} = \begin{pmatrix} -D_r I & 1 - D_y C \\ D_r L & D_y L \end{pmatrix}$$

Regular Functions

A point $\mathbf{x}_0$ in the domain of a C^1 function $f \colon X \to Y$ is called a *regular point* if the derivative $Df[\mathbf{x}_0]$ has full rank (section 3.1). Otherwise, $\mathbf{x}_0$ is called a *critical point*. A function is *regular* if it has no critical points.

Two special cases occur most frequently. When f is an operator ($X = Y$), $\mathbf{x}_0$ is a regular point if and only if the Jacobian is nonsingular. When f is a functional ($Y = \Re$), $\mathbf{x}_0$ is a regular point if and only if the gradient is nonzero. If every point on a contour $\{\mathbf{x} \colon f(\mathbf{x}) = c\}$ of a functional is a regular point, then the contour is called a regular surface or $n - 1$ dimensional *manifold*.

Exercise 4.19
A point $\mathbf{x}_0$ is a regular point of a C^1 operator if and only $\det J_f(\mathbf{x}_0) \neq 0$.

Exercise 4.20
A point $\mathbf{x}_0$ is a critical point of a C^1 functional if and only if $\nabla f(\mathbf{x}_0) = \mathbf{0}$; that is, $D_{x_i} f[\mathbf{x}_0] = 0$ for each $i = 1, 2, \ldots, n$.

Critical points of smooth functionals will be the focus of our attention in the next chapter.

4.3 Properties of Differentiable Functions

In the first part of this section, we present some elementary properties of differentiable functions. These properties form the basis of the rules of differentiation, which enable the derivatives of complicated functions to be computed from the derivatives of their components. It is these rules which form the focus of most introductory treatments of calculus. In the second half of this section, we present one of the most fundamental results of differential calculus, the mean value theorem, which forms the foundation of the rest of the chapter.

4.3.1 Basic Properties and the Derivatives of Elementary Functions

Most of the standard properties of differentiable functions follow directly from definition of the derivative. We summarize them in exercises 4.21 to 4.30. Unless otherwise specified, X and Y are arbitrary normed linear spaces.

The first two results are fundamental in computing with derivatives. The first states the derivative of a sum is equal to the sum of the derivatives. That is, the differentiation is a linear operator. Exercise 4.22 shows that the derivative of a composition is the composition of their derivatives.

Exercise 4.21 (Linearity)
If $f, g: X \to Y$ are differentiable at $\mathbf{x}$, then $f + g$ and αf are differentiable at $\mathbf{x}$ with

$D(f + g)[\mathbf{x}] = Df[\mathbf{x}] + Dg[\mathbf{x}]$

$D(\alpha f[\mathbf{x}]) = \alpha Df[\mathbf{x}] \qquad \text{for every } \alpha \in \Re$

Exercise 4.22 (Chain rule)
If $f: X \to Y$ is differentiable at $\mathbf{x}$ and $g: Y \to Z$ is differentiable at $f[\mathbf{x}]$, then the composite function $g \circ f: X \to Z$ is differentiable at $\mathbf{x}$ with

$D(g \circ f)[\mathbf{x}] = Dg[f(\mathbf{x})] \circ Df[\mathbf{x}]$

Derivatives in Euclidean space can be represented by the Jacobian matrix, and composition of linear functions corresponds to matrix multiplication. Therefore the chain rule can be expressed as

$$J_{f \circ g}(\mathbf{x}) = J_g(\mathbf{y})J_f(\mathbf{x})$$

where $\mathbf{y} = f(\mathbf{x})$. Letting n, m and l denote the dimensions of X, Y, and Z respectively, the partial derivative of g_j with respect to x_i is can be computed using the familiar formula for matrix multiplication

$$D_{x_i}g_j[\mathbf{x}] = \sum_{k=1}^{m} D_{y_k}g_j[\mathbf{y}]D_{x_i}f_k[\mathbf{x}] \tag{8}$$

The reason for the name "chain rule" becomes clearer when we express (8) in the alternative notation

$$\frac{\partial g_j}{\partial x_i} = \sum_{k=1}^{m} \frac{\partial g_j}{\partial y_k} \times \frac{\partial y_k}{\partial x_i}$$

This is one instance where the alternative notation can be more transparent, provided there is no ambiguity in meaning of the symbols (see remark 4.7).

The derivative of a bilinear function is the sum of the two partial functions evaluated at the point of differentiation.

Exercise 4.23 (Bilinear functions)
Every continuous bilinear function $f: X \times Y \to Z$ is differentiable with

$$Df[\mathbf{x}, \mathbf{y}] = f(\mathbf{x}, \cdot) + f(\cdot, \mathbf{y})$$

that is,

$$Df[\mathbf{x}_0, \mathbf{y}_0](\mathbf{x}, \mathbf{y}) = f(\mathbf{x}_0, \mathbf{y}) + f(\mathbf{x}, \mathbf{y}_0)$$

Combining exercises 4.22 and 4.23 enables as to derive one of the most useful properties which is known as the *product rule*. The derivative of a the product of two functions is equal to the first times the derivative of the second plus the second times the derivative of the first. Formally we have

Exercise 4.24 (Product rule)
Let $f: X \to \Re$ be differentiable at $\mathbf{x}$ and $g: Y \to \Re$ be differentiable at $\mathbf{y}$. Then their product $fg: X \times Y \to \Re$ is differentiable at $(\mathbf{x}, \mathbf{y})$ with derivative

$$Dfg[\mathbf{x}, \mathbf{y}] = f(\mathbf{x})Dg[\mathbf{y}] + g(\mathbf{y})Df[\mathbf{x}]$$

4.3 Properties of Differentiable Functions

Exercise 4.25 (Power function)
The power function (example 2.2)

$$f(x) = x^n, \quad n = 1, 2, \ldots$$

is differentiable with derivative

$$Df[x] = f'[x] = nx^{n-1}$$

Example 4.13 (Exponential function) The exponential function (Example 2.10)

$$e^x = 1 + \frac{x}{1} + \frac{x^2}{2!} + \frac{x^3}{3!} + \cdots$$

is the limit of a series of power functions

$$f^n(x) = 1 + \frac{x}{1} + \frac{x^2}{2!} + \frac{x^3}{3!} + \cdots + \frac{x^n}{n!}$$

that is,

$$e^x = \lim_{n \to \infty} f^n(x)$$

By exercises 4.21 and 4.25, each f^n is differentiable with

$$Df^n[x] = 1 + \frac{x}{1} + \frac{x^2}{2!} + \frac{x^3}{3!} + \cdots + \frac{x^{n-1}}{(n-1)!} = f^{n-1}(x)$$

and therefore

$$\lim_{n \to \infty} Df^n[x] = \lim_{n \to \infty} f^{n-1}(x) = e^x$$

As we will show later (example 4.20), this implies that the exponential function is differentiable with derivative

$$De^x = e^x$$

Exercise 4.26
Assume that the inverse demand function for some good is given by

$$p = f(x)$$

where x is the quantity sold. Total revenue is given by

$$R(x) = f(x)x$$

Find the marginal revenue at x_0.

Recall that a homeomorphism (remark 2.12) is a continuous function that has a continuous inverse. The next result shows that a differentiable homeomorphism (with a nonsingular derivative) is differentiable. The derivative of the inverse is the inverse of the derivative of the original function. A differentiable map with a differentiable inverse is called a *diffeomorphism*.

Exercise 4.27 (Inverse function rule)
Suppose that $f \colon X \to Y$ is differentiable at **x** and its derivative is nonsingular. Suppose further that f has an inverse $f^{-1} \colon Y \to X$ that is continuous (i.e., f is a homeomorphism). Then f^{-1} is differentiable at **x** with

$$Df^{-1}[\mathbf{x}] = (Df[\mathbf{x}])^{-1}$$

Example 4.14 (Log function) The log function (example 2.55) is the inverse of the exponential function, that is,

$$y = \log x \Leftrightarrow x = e^y$$

The exponential function is differentiable and its derivative is nonsingular (example 4.13). Applying the inverse function rule (exercise 4.27), the log function is differentiable with derivative

$$D_x \log(x) = \frac{1}{D_y e^y} = \frac{1}{e^y} = \frac{1}{x}$$

Example 4.15 (General power function) The general power function is defined by (example 2.56)

$$f(x) = x^a = e^{a \log x}$$

Using the chain rule, we conclude that the general power function is differentiable with

$$D_x f(x) = D \exp(a \log x) D(a \log x) = \exp(a \log x) a \frac{1}{x} = x^a a \frac{1}{x} = a x^{a-1}$$

Exercise 4.28 (General exponential function)
When the roles are reversed in the general power function, we have the general exponential function defined as

$$f(x) = a^x$$

where $a \in \Re_+$. Show that the general exponential function is differentiable with derivative

$$D_x f(x) = a^x \log a$$

Example 4.15 provides us with two more useful rules of differentiation.

Exercise 4.29 (Reciprocal function)
Let $f: X \to \Re$ be differentiable at $\mathbf{x}$ where $f(\mathbf{x}) \neq 0$, then $1/f$ is differentiable with derivative

$$D\frac{1}{f}[\mathbf{x}] = \frac{Df[\mathbf{x}]}{(f(\mathbf{x}))^2}$$

Exercise 4.30 (Quotient rule)
Let $f: X \to \Re$ be differentiable at $\mathbf{x}$ and $g: Y \to \Re$ be differentiable at $\mathbf{y}$ with $f(\mathbf{y}) \neq 0$. Then their quotient $f/g: X \times Y \to \Re$ is differentiable at $(\mathbf{x}, \mathbf{y})$ with derivative

$$D\frac{f}{g}[\mathbf{x}, \mathbf{y}] = \frac{g(\mathbf{y})Df[\mathbf{x}] - f(\mathbf{x})Dg[\mathbf{y}]}{(g(\mathbf{y}))^2}$$

We summarize the derivatives of elementary functions in in table 4.4. Since most functional forms comprise linear combinations, products, and compositions of these elementary functions, the derivative of almost any function can be obtained by repeated application of the preceding rules to these elementary functions. We give some examples and provide other examples as exercises.

Table 4.4
Derivatives of elementary functions

Function	Derivative
$f(x)$	$f'[x_0]$
ax	a
x^a	ax_0^{a-1}
e^x	e^{x_0}
a^x	$a^{x_0} \log a$
$\log x$	$1/x_0$

Example 4.16 (Cobb-Douglas) Consider the Cobb-Douglas function

$$f(x_1, x_2) = x_1^{a_1} x_2^{a_2}$$

Holding x_1 constant, the partial function $h_2(x_2) = A x_2^{a_2}$ (where $A = x_1^{a_1}$ is a constant) is a power function whose derivative is

$$Dh_2[x_2] = A a_2 x_2^{a_2 - 1}$$

Therefore the partial derivative of f with respect to x_2 is

$$D_{x_2} f[x_1, x_2] = a_2 x_1^{a_1} x_2^{a_2 - 1}$$

Similarly the partial derivative with respect to x_1 is

$$D_{x_1} f[x_1, x_2] = a_1 x_1^{a_1 - 1} x_2^{a_2}$$

The derivative can be represented by the gradient

$$\nabla f(\mathbf{x}) = (a_1 x_1^{a_1 - 1} x_2^{a_2}, a_2 x_1^{a_1} x_2^{a_2 - 1})$$

Note that the derivative depends both on the exponents a_1 and a_2 and on the point of evaluation $\mathbf{x}^0$.

Exercise 4.31
Calculate the value of the partial derivatives of the function

$$f(k, l) = k^{2/3} l^{1/3}$$

at the point $(8, 8)$

Exercise 4.32
Show that gradient of the Cobb-Douglas function

$$f(\mathbf{x}) = x_1^{a_1} x_2^{a_2} \ldots x_n^{a_n}$$

can be expressed as

$$\nabla f(\mathbf{x}) = \left(\frac{a_1}{x_1}, \frac{a_2}{x_2}, \ldots, \frac{a_n}{x_n} \right) f(\mathbf{x})$$

Exercise 4.33
Compute the partial derivatives of the CES function (exercise 2.35)

$$f(\mathbf{x}) = (a_1 x_1^\rho + a_2 x_2^\rho + \cdots + a_n x_n^\rho)^{1/\rho}, \quad \rho \neq 0$$

Example 4.17 (Marginal rate of substitution) Given a utility function u, the marginal rate of substitution (example 4.10) between goods i and j is given by

$$\text{MRS}_{ij} = \frac{D_{x_i}u[\mathbf{x}_0]}{D_{x_j}u[\mathbf{x}_0]}$$

If g is a monotonic transformation, the function $v = g \circ u$ is another utility function representing the same preferences (exercise 2.37). Applying the chain rule, we have

$$\frac{D_{x_i}v[\mathbf{x}_0]}{D_{x_j}v[\mathbf{x}_0]} = \frac{g'[u(\mathbf{x}_0)]D_{x_i}u[\mathbf{x}_0]}{g'[u(\mathbf{x}_0)]D_{x_j}u[\mathbf{x}_0]} = \frac{D_{x_i}u[\mathbf{x}_0]}{D_{x_j}u[\mathbf{x}_0]} = \text{MRS}_{ij}$$

The marginal rate of substitution is invariant to monotonic transformations of the utility function.

4.3.2 Mean Value Theorem

Previously (section 4.1), we showed how the derivative approximates a differentiable function locally. Now, we show how the derivative can provide information about the function over the whole domain. Consider figure 4.6, which shows the graph of a differentiable function $f: \Re \to \Re$ and its derivative at the point x_0. Differentiability implies that

$$f(x_0 + x) \approx f(x_0) + Df[x_0](x)$$

for small x. Generally, the approximation deteriorates as x increases. However, for every x, there exists another linear approximation which is exact at x, that is there exists some point $\bar{x}$ between x_0 and $x_0 + x$ such that

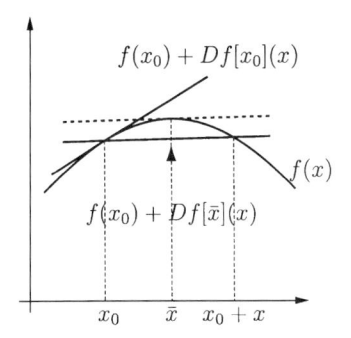

Figure 4.6
Illustrating the mean value theorem

$$f(x_0 + x) = f(x_0) + Df[\bar{x}](x)$$

This is the classical *mean value theorem*, which is readily generalized to functionals on normed linear spaces.

Theorem 4.1 (Mean value theorem) *Let $f: S \to \Re$ be a differentiable functional on a convex neighborhood S of $\mathbf{x}_0$. For every $\mathbf{x} \in S - \mathbf{x}_0$, there exists some $\bar{\mathbf{x}} \in (\mathbf{x}_0, \mathbf{x}_0 + \mathbf{x})$ such that*

$$f(\mathbf{x}_0 + \mathbf{x}) = f(\mathbf{x}_0) + Df[\bar{\mathbf{x}}](\mathbf{x}) \tag{9}$$

Proof For fixed $\mathbf{x}_0, \mathbf{x}$, the function $g: \Re \to X$ defined by $g(t) = \mathbf{x}_0 + t\mathbf{x}$ is differentiable with derivative $g'(t) = \mathbf{x}$ (exercise 4.6). We note that $g(0) = \mathbf{x}_0$ and $g(1) = \mathbf{x}_0 + \mathbf{x}$. The composite function $h = f \circ g: \Re \to \Re$ is differentiable (exercise 4.22) with derivative

$$h'(t) = Df[g(t)] \circ g'(t) = Df[g(t)](\mathbf{x})$$

By the classical mean value theorem (exercise 4.34), there exists an $\alpha \in (0, 1)$ such that

$$h(1) - h(0) = h'(\alpha) = Df[g(\alpha)](\mathbf{x})$$

Substituting in $h(0) = f(\mathbf{x}_0)$, $h(1) = f(\mathbf{x}_0 + \mathbf{x})$ and $\bar{\mathbf{x}} = g(\alpha)$ yields

$$f(\mathbf{x}_0 + \mathbf{x}) - f(\mathbf{x}_0) = Df[\bar{\mathbf{x}}](\mathbf{x}) \qquad \square$$

Exercise 4.34 (Classical mean value theorem)
Suppose that $f \in C[a, b]$ is differentiable on the open interval (a, b). Then there exists some $x \in (a, b)$ such that

$$f(b) - f(a) = f'[x](b - a)$$

[Hint: Apply Rolle's theorem (exercise 5.8) to the function

$$h(x) = f(x) - \frac{f(b) - f(a)}{b - a}(x - a)]$$

The mean value theorem is rightly seen as the cornerstone of differential calculus and a fertile source of useful propositions. We give some examples below. In subsequent sections we use it to derive Taylor's theorem on polynomial approximation, Young's theorem on symmetry of the Hessian and the fundamental inverse and implicit function theorems. Figure 4.7 illustrates the seminal role of the mean value theorem in the material of this chapter.

4.3 Properties of Differentiable Functions

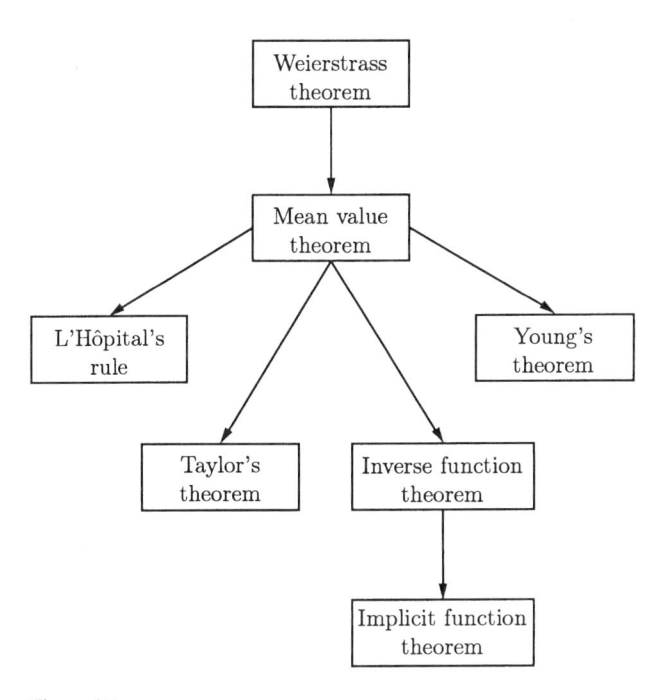

Figure 4.7
Theorems for smooth functions

Exercise 4.35 (Increasing functions)
A differentiable functional f on a convex set $S \subseteq \Re^n$ is increasing if and only $\nabla f(\mathbf{x}) \geq \mathbf{0}$ for every $\mathbf{x} \in S$, that is, if every partial derivative $D_{x_i} f[\mathbf{x}]$ is nonnegative.

Exercise 4.36 (Strictly increasing functions)
A differentiable functional f on a convex set $S \subseteq \Re^n$ is strictly increasing if $\nabla f(\mathbf{x}) > \mathbf{0}$ for every $\mathbf{x} \in X$, that is, if every partial derivative $D_{x_i} f[\mathbf{x}]$ is positive.

Example 4.18 It is worth noting that the converse of the previous result is not true. For example, $f(x) = x^3$ is strictly increasing on $\Re$ although $f'[0] = 0$.

Example 4.19 (Marginal utility) If u represents a preference ordering $\succsim$ on $X \subseteq \Re^n$, then u is increasing on X if and only if $\succsim$ is weakly monotonic (example 2.59). It is u strictly increasing if $\succsim$ is strongly monotonic. Consequently the marginal utility (example 4.9) of any good is nonnegative if and only if $\succsim$ weakly monotonic (exercise 4.35). Furthermore, if marginal utility is positive for all goods, this implies that $\succsim$ is strongly monotonic (exercise 4.36).

Exercise 4.37 (C^1 functions)
Let f be a functional on an open subset S of $\Re^n$. Then f is continuously differentiable (C^1) if and only if each of the partial derivatives $D_i f[\mathbf{x}]$ exists and is continuous on S.

We now present some extensions and alternative forms of the mean value theorem which are useful in certain applications. Theorem 4.1 applies only to functionals and has no immediate counterpart for more general functions. Furthermore, since the point of evaluation $\bar{\mathbf{x}}$ is unknown, the mean value theorem is often more usefully expressed as an inequality

$$|f(\mathbf{x}_1) - f(\mathbf{x}_0)| \leq \|Df(\bar{\mathbf{x}})\| \|\mathbf{x}_1 - \mathbf{x}_0\|$$

which follows immediately from (9). This alternative form of the mean value theorem generalizes to functions between normed linear spaces.

Corollary 4.1.1 (Mean value inequality) *Let $f: S \to Y$ be a differentiable function on an open convex set $S \subseteq X$. For every $\mathbf{x}_1, \mathbf{x}_2 \in S$,*

$$\|f(\mathbf{x}_1) - f(\mathbf{x}_2)\| \leq \sup_{\mathbf{x} \in [\mathbf{x}_1, \mathbf{x}_2]} \|Df[\mathbf{x}]\| \|\mathbf{x}_1 - \mathbf{x}_2\|$$

Proof The proof of corollary 4.1.1 is an insightful illustration of the use of the dual space in analysis. By the Hahn-Banach theorem (exercise 3.205), there exists a linear functional $\varphi \in Y^*$ such that $\|\varphi\| = 1$ and $\varphi(\mathbf{y}) = \|\mathbf{y}\|$. That is,

$$\varphi(f(\mathbf{x}_1) - f(\mathbf{x}_2)) = \|f(\mathbf{x}_1) - f(\mathbf{x}_2)\|$$

Then $\varphi \circ f: S \to \Re$ is a functional on S. Applying the mean value theorem (theorem 4.1), there exists some $\bar{\mathbf{x}} \in (\mathbf{x}_1, \mathbf{x}_2)$ such that

$$(\varphi \circ f)(\mathbf{x}_1) - (\varphi \circ f)(\mathbf{x}_2) = D(\varphi \circ f)[\bar{\mathbf{x}}](\mathbf{x}_1 - \mathbf{x}_2) \tag{10}$$

By the linearity of φ, the left-hand side of (10) is

$$\varphi \circ f(\mathbf{x}_1) - \varphi \circ f(\mathbf{x}_2) = \varphi(f(\mathbf{x}_1) - f(\mathbf{x}_2))$$
$$= \|f(\mathbf{x}_1) - f(\mathbf{x}_2)\|$$

4.3 Properties of Differentiable Functions

By the chain rule, the right-hand side becomes

$$D(\varphi \circ f)(\bar{\mathbf{x}})(\mathbf{x}_1 - \mathbf{x}_2) = D\varphi \circ Df[\bar{\mathbf{x}}](\mathbf{x}_1 - \mathbf{x}_2)$$
$$= \varphi \circ Df[\bar{\mathbf{x}}](\mathbf{x}_1 - \mathbf{x}_2)$$
$$\leq \|\varphi\| \, \|Df[\bar{\mathbf{x}}]\| \, \|\mathbf{x}_1 - \mathbf{x}_2\|$$
$$= \|Df[\bar{\mathbf{x}}]\| \, \|\mathbf{x}_1 - \mathbf{x}_2\|$$

since $\|\varphi\| = 1$. Substituting these relations into (10) gives the desired result, that is,

$$\|f(\mathbf{x}_2) - f(\mathbf{x}_2)\| \leq \|Df(\bar{\mathbf{x}})\| \, \|\mathbf{x}_1 - \mathbf{x}_2\| \leq \sup_{\mathbf{x} \in [\mathbf{x}_1, \mathbf{x}_2]} \|Df(\mathbf{x})\| \, \|\mathbf{x}_1 - \mathbf{x}_2\| \quad \square$$

If $\|Df\|$ is bounded on S, that is there exists a constant M such that $\|Df(\mathbf{x})\| < M$ for all $\mathbf{x} \in S$, corollary 4.1.1 implies that f is Lipschitz continuous on S, that is,

$$\|f(\mathbf{x}_1) - f(\mathbf{x}_2)\| \leq M\|\mathbf{x}_1 - \mathbf{x}_2\| \qquad \text{for every } \mathbf{x}_1, \mathbf{x}_2 \in S$$

Exercise 4.38 (Constant functions)
A differentiable function f on a convex set S is constant if and only if $Df[\mathbf{x}] = \mathbf{0}$ for every $\mathbf{x} \in S$.

Exercise 4.39 (Sequences of functions)
Let $f^n \colon S \to Y$ be a sequence of C^1 functions on an open set S, and define

$$f(\mathbf{x}) = \lim_{n \to \infty} f^n(\mathbf{x})$$

Suppose that the sequence of derivatives Df^n converges uniformly to a function $g \colon S \to BL(X, Y)$. Then f is differentiable with derivative $Df = g$.

Example 4.20 (Exponential function) The exponential function e^x (example 2.10) is the limit of a sequence of power series

$$e^x = \lim_{n \to \infty} f^n(x)$$

where

$$f^n(x) = 1 + \frac{x}{1} + \frac{x^2}{2!} + \frac{x^3}{3!} + \cdots + \frac{x^n}{n!}$$

Each partial function f^n is differentiable (exercise 4.13) with $Df^n(x) = f^{n-1}$. Since f^n converges uniformly on any compact set to e^x (exercise 2.8), so do the derivatives

$$\lim_{n \to \infty} Df^n[x] = \lim_{n \to \infty} f^{n-1}(x) = e^x$$

Using exercise 4.39, we conclude that the exponential function is differentiable with derivative

$$De^x = e^x$$

Example 4.21 Assume that $f: \Re \to \Re$ is differentiable with

$$f'(x) = f(x) \quad \text{for every } x \in \Re \tag{11}$$

Then f is exponential, that is, $f(x) = Ae^x$ for some $A \in \Re$. To see this, define

$$g(x) = \frac{f(x)}{e^x}$$

Now g is differentiable (exercise 4.30) with derivative

$$g'(x) = \frac{e^x f'(x) - f(x) e^x}{e^{2x}} = \frac{f'(x) - f(x)}{e^x}$$

Substituting (11), we have

$$g'(x) = \frac{f(x) - f(x)}{e^x} = 0 \quad \text{for every } x \in \Re$$

Therefore g is a constant function (exercise 4.38). That is, there exists $A \in \Re$ such that

$$g(x) = \frac{f(x)}{e^x} = A \quad \text{or} \quad f(x) = Ae^x$$

Equation (11) is the simplest example of a *differential equation* that relates a function to its derivative.

Exercise 4.40
Prove that $e^{x+y} = e^x e^y$ for every $x, y \in \Re$. [Hint: Consider $f(x) = e^{x+y}/e^y$.]

Exercise 4.41 (Constant elasticity)
The *elasticity* of a function $f: \Re \to \Re$ is defined to be

$$E(x) = x\frac{f'(x)}{f(x)}$$

In general, the elasticity varies with x. Show that the elasticity of a function is constant if and only if it is a power function, that is,

$$f(x) = Ax^a$$

Corollary 4.1.1 enables us to place bounds on the error in approximating the function f by its derivative anywhere in the domain. We give three useful variations. Exercise 4.43 will be used in the proof of theorem 4.4.

Exercise 4.42
Let $f: S \to Y$ be a differentiable function on an open convex set $S \subseteq X$. For every $\mathbf{x}_0, \mathbf{x}_1, \mathbf{x}_2 \in S$,

$$\|f(\mathbf{x}_1) - f(\mathbf{x}_2) - Df[\mathbf{x}_0](\mathbf{x}_1 - \mathbf{x}_2)\| \leq \sup_{\mathbf{x} \in [\mathbf{x}_1, \mathbf{x}_2]} \|Df[\mathbf{x}] - Df[\mathbf{x}_0]\| \|\mathbf{x}_1 - \mathbf{x}_2\|$$

[Hint: Apply corollary 4.1.1 to the function $g(\mathbf{x}) = f(\mathbf{x}) - Df[\mathbf{x}_0](\mathbf{x})$.]

Exercise 4.43 (Approximation lemma)
Let $f: X \to Y$ be C^1. For every $\mathbf{x}_0 \in X$ and $\varepsilon > 0$, there exists a neighborhood S containing $\mathbf{x}_0$ such that for every $\mathbf{x}_1, \mathbf{x}_2 \in S$,

$$\|f(\mathbf{x}_1) - f(\mathbf{x}_2) - Df[\mathbf{x}_0](\mathbf{x}_1 - \mathbf{x}_2)\| \leq \varepsilon \|\mathbf{x}_1 - \mathbf{x}_2\|$$

Exercise 4.44
Let $f: X \to Y$ be C^1. For every $\mathbf{x}_0 \in X$ and $\varepsilon > 0$, there exists a neighborhood S containing $\mathbf{x}_0$ such that for every $\mathbf{x}_1, \mathbf{x}_2 \in S$,

$$\|f(\mathbf{x}_1) - f(\mathbf{x}_2)\| \leq \|Df[\mathbf{x}_0] + \varepsilon\| \|\mathbf{x}_1 - \mathbf{x}_2\|$$

In fact we can go further and place upper and lower bounds on $f(\mathbf{x}_1) - f(\mathbf{x}_2)$, as in the following result, which is an interesting application of the separating hyperplane theorem.

Exercise 4.45 (Mean value inclusion)
Let $f: S \to Y$ be a differentiable function on an open convex set $S \subseteq X$. Then for every $\mathbf{x}_1, \mathbf{x}_2 \in S$,

$$f(\mathbf{x}_1) - f(\mathbf{x}_2) \in \overline{\text{conv}} \, A$$

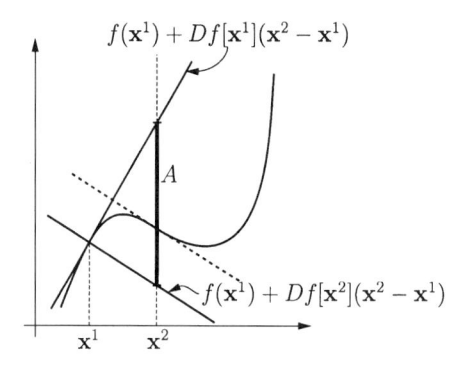

Figure 4.8
The mean value inclusion theorem

where $A = \{ \mathbf{y} \in Y : \mathbf{y} = Df[\bar{\mathbf{x}}](\mathbf{x}_1 - \mathbf{x}_2) \text{ for some } \bar{\mathbf{x}} \in [\mathbf{x}_1, \mathbf{x}_2]\}$. [Hint: Assume that $f(\mathbf{x}_1) - f(\mathbf{x}_2) \notin \overline{\text{conv}}\, A$, and apply proposition 3.14.]

Exercise 4.45 is illustrated in figure 4.8. The affine functions $f(\mathbf{x}_1) + Df[\mathbf{x}_1](\mathbf{x}_2 - \mathbf{x}_1)$ and $f(\mathbf{x}_1) + Df[\mathbf{x}_2](\mathbf{x}_2 - \mathbf{x}_1)$ bound the values of $f(\mathbf{x})$ between $\mathbf{x}_1$ and $\mathbf{x}_2$.

L'Hôpital's Rule

We sometimes need to evaluate the behavior of the ratio of two functionals close to a point where both are zero and their ratio is undefined. For example, if $c(y)$ denotes the total cost of production of output y (example 2.38), average cost is measured by $c(y)/y$. Provided there are no fixed costs, $c(y) = 0$ when $y = 0$ and their ratio is undefined. What can we say about the average cost of producing the first unit? *L'Hôpital's rule* provides an answer for differentiable functions, which we develop in the following exercises.

Exercise 4.46 (Cauchy mean value theorem)
Assume that f and g continuous functionals on $[a, b] \subseteq \Re$ that are differentiable on the open interval (a, b). Then there exists some $x \in (a, b)$ such that

$$(f(b) - f(a))g'(x) = (g(b) - g(a))f'(x)$$

Provided $g(a) \neq g(b)$ and $g'(x) \neq 0$, this can be written as

$$\frac{f'(x)}{g'(x)} = \frac{f(b)-f(a)}{g(b)-g(a)}$$

[Hint: Modify exercise 4.34.]

Exercise 4.47 (L'Hôpital's rule)
Suppose that f and g are functionals on $\Re$ such that

$$\lim_{x \to a} f(x) = \lim_{x \to a} g(x) = 0$$

If $\lim_{x \to a} f'(x)/g'(x)$ exists, then

$$\lim_{x \to a} \frac{f(x)}{g(x)} = \lim_{x \to a} \frac{f'(x)}{g'(x)}$$

Example 4.22 (CES function) The CES function (exercise 2.35)
$$f(\mathbf{x}) = (a_1 x_1^\rho + a_2 x_2^\rho + \cdots a_n x_n^\rho)^{1/\rho}$$
is not defined when $\rho = 0$. Using L'Hôpital's rule (exercise 4.47), we can show that the CES function tends to the Cobb-Douglas function as $\rho \to 0$. For simplicity, assume that $a_1 + a_2 + \cdots a_n = 1$ (see exercise 4.48). For any $\mathbf{x} > \mathbf{0}$, let $g: \Re \to \Re$ be defined by

$$g(\rho) = \log(a_1 x_1^\rho + a_2 x_2^\rho + \cdots a_n x_n^\rho)$$

Using the chain rule and exercise 4.28, we have

$$g'(\rho) = \frac{a_1 x_1^\rho \log x_1 + a_2 x_2^\rho \log x_2 + \cdots a_n x_n^\rho \log x_n}{a_1 x_1^\rho + a_2 x_2^\rho + \cdots a_n x_n^\rho}$$

and therefore

$$\lim_{\rho \to 0} g'(\rho) = \frac{a_1 \log x_1 + a_2 \log x_2 + \cdots a_n \log x_n}{a_1 + a_2 + \cdots + a_n}$$

$$= a_1 \log x_1 + a_2 \log x_2 + \cdots a_n \log x_n \tag{12}$$

assuming that $a_1 + a_2 + \cdots a_n = 1$. Now

$$\log f(\rho; \mathbf{x}) = \frac{g(\rho)}{\rho} = \frac{g(\rho)}{h(\rho)}$$

where h is the identity function on $\Re$, $h(\rho) = \rho$ and $h'(\rho) = 1$. Applying L'Hôpital's rule, we have

$$\lim_{\rho \to 0} \log f(\rho, \mathbf{x}) = \lim_{\rho \to 0} \frac{g(\rho)}{h(\rho)}$$

$$= \lim_{\rho \to 0} \frac{g'(\rho)}{h'(\rho)}$$

$$= a_1 \log x_1 + a_2 \log x_2 + \cdots a_n \log x_n$$

Since the exponential function is continuous,

$$\lim_{\rho \to 0} f(\rho, \mathbf{x}) = \lim_{\rho \to 0} \exp(\log f(\rho, \mathbf{x}))$$

$$= \lim_{\rho \to 0} \exp(a_1 \log x_1 + a_2 \log x_2 + \cdots a_n \log x_n)$$

$$= x_1^{a_1} x_2^{a_2} \ldots x_n^{a_n}$$

which is the Cobb-Douglas function (example 2.57).

Exercise 4.48
Recall that the CES function is homogeneous of degree one (exercise 3.163), while the Cobb-Douglas function is homogeneous of degree $a_1 + a_2 + \cdots + a_n$ (example 3.69). In the previous example we assumed that $a_1 + a_2 + \cdots + a_n = 1$. What is the limit of the CES function when $a_1 + a_2 + \cdots + a_n \neq 1$?

Exercise 4.49
Let $c(y)$ be the total cost of output y. Assume that c is differentiable and that there are no fixed costs ($c(y) = 0$). Show that the average cost of the first unit is equal to marginal cost.

The following variant of L'Hôpital's rule is also useful in evaluating behavior at infinity (example 4.23).

Exercise 4.50 (*L'Hôpital's rule for* ∞/∞)
Suppose that f and g are differentiable functionals on $\Re$ such that

$$\lim_{x \to \infty} f(x) = \lim_{x \to \infty} g(x) = \infty$$

while

$$\lim_{x \to \infty} \frac{f'(x)}{g'(x)} = k < \infty$$

Show that

1. For every $\varepsilon > 0$ there exists a_1 such that

$$\left|\frac{f(x) - f(a_1)}{g(x) - g(a_1)} - k\right| < \frac{\varepsilon}{2} \qquad \text{for every } x > a_1$$

2. Using the identity

$$\frac{f(x)}{g(x)} = \frac{f(x) - f(a)}{g(x) - g(a)} \times \frac{f(x)}{f(x) - f(a)} \times \frac{g(x) - g(a)}{g(x)}$$

deduce that there exists a such that

$$\left|\frac{f(x)}{g(x)} - k\right| < \varepsilon \qquad \text{for every } x > a$$

That is,

$$\lim_{x \to \infty} \frac{f(x)}{g(x)} = k = \lim_{x \to \infty} \frac{f'(x)}{g'(x)}$$

Example 4.23 (Logarithmic utility) The logarithmic utility function $u(x) = \log(x)$ is a common functional form in economic models with a single commodity. Although $u(x) \to \infty$ as $x \to \infty$, average utility $u(x)/x$ goes to zero, that is,

$$\lim_{x \to \infty} \frac{u(x)}{x} = \lim_{x \to \infty} \frac{\log(x)}{x} = 0$$

since (exercise 4.50)

$$\lim_{x \to \infty} \frac{u'(x)}{1} = \lim_{x \to \infty} \frac{1/x}{1} = 0$$

The limiting behavior of the log function can be compared with that of the exponential function (exercise 2.7), for which

$$\lim_{x \to \infty} \frac{e^x}{x} = \infty$$

4.4 Polynomial Approximation and Higher-Order Derivatives

The derivative provides a linear approximation to a differentiable function in the neighborhood of a point of interest. By virtue of the mean value theorem, it can also provide some information about the behavior of a function over a wider area of the domain. Frequently we require more exact information about the behavior of a function over a wide area. This can often be provided by approximating the function by a polynomial. In this section we show how the behavior of many smooth functions can be well approximated over an extended area by polynomials. These polynomials also provide additional information about the behavior of a function in the neighborhood of a point of interest. In particular, we can obtain information about the local curvature of the function, which is not evident from the derivative.

Example 4.24 (The exponential function) The derivative of the exponential function e^x is $f'[x_0] = e^{x_0}$ (example 4.13). Therefore the function

$$g(x) = e^{x_0} + e^{x_0} x$$

provides a linear approximation to the exponential function at the point x_0. Since $e^0 = 1$, the function

$$f_1(x) = 1 + x$$

gives a linear approximation to the exponential function $f(x) = e^x$ in the neighborhood of 0. As we can see from figure 4.9, this approximation deteriorates markedly even over the interval $[0, 1]$. The curvature of the exponential function implies that no linear function is going to provide a good approximation of the exponential function over an extended interval.

The approximation can be improved by adding an additional term x^2, as in

$$\hat{f}_2(x) = 1 + x + x^2$$

This quadratic approximation can be further improved by attenuating the influence of the new term x^2, yielding the quadratic polynomial

$$f_2(x) = 1 + x + \frac{x^2}{2}$$

as shown in figure 4.9. In fact we will show that f_2 is the best quadratic approximation to $f(x) = e^x$ in the neighborhood of the point 0.

If we extend the interval of interest to say $[0, 2]$, even the quadratic approximation f_2 deteriorates (figure 4.10). Accuracy of the approxima-

4.4 Polynomial Approximation and Higher-Order Derivatives

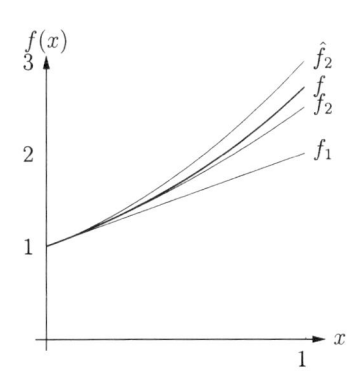

Figure 4.9
Approximating the exponential function

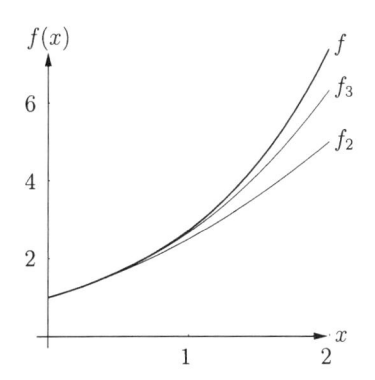

Figure 4.10
Adding another term extends the range of useful approximation

tion can be further improved by adding yet another term, yielding the cubic polynomial

$$f_3(x) = 1 + x + \frac{x^2}{2} + \frac{x^3}{6}$$

The power of the highest term in the polynomial is called the *order* of the polynomial. The cubic polynomial f_3 is a third-order polynomial.

This procedure of adding terms and increasing the order of the polynomial can be continued indefinitely. Each additional term improves the accuracy of the approximation in two ways:

- It improves the accuracy of the approximation in the neighborhood of the point 0.
- It extends the range over which the polynomial yields a useful approximation.

How can we find the best polynomial approximation to a given function? In section 4.4.3 we show that the best polynomial approximation involves the higher-order derivatives of the function. Therefore we must first investigate these higher-order derivatives.

4.4.1 Higher-Order Derivatives

The derivative Df of a differentiable function defines a function from X to $BL(X, Y)$. If the derivative is continuous ($f \in C^1$), we can ask whether the derivative $Df: X \to BL(X, Y)$ is itself differentiable at any point $\mathbf{x}$. In other words, does the (possibly nonlinear) map Df have a linear approximation at $\mathbf{x}$? If so, then the derivative $D(Df)[\mathbf{x}]$ is called the *second derivative* of f at $\mathbf{x}$. It is denoted $D^2 f[\mathbf{x}]$. The second derivative at a point $\mathbf{x}$ is a continuous linear map from X to $BL(X, Y)$, that is $D^2 f[\mathbf{x}] \in BL(X, BL(X, Y))$. If the derivative Df is differentiable at all $\mathbf{x} \in X$, then we say that f is twice differentiable, in which case the second derivative $D^2 f$ defines a function from X to $BL(X, BL(X, Y))$. If this function is continuous, that is, Df is continuously differentiable, we say that f is *twice continuously differentiable* or f is of class C^2. The set of all twice continuously differentiable functions between linear spaces X and Y is denoted $C^2(X, Y)$.

Although the second derivative function $D^2 f: X \to BL(X, BL(X, Y))$ looks complicated, it is still a mapping between normed linear spaces. We can continue in this fashion to define third, fourth, and still higher-order derivatives recursively as

$$D^n f[\mathbf{x}] = D(D^{n-1} f)[\mathbf{x}]$$

and obtain the corresponding classes of functions, $C^3, C^4, \ldots$. It is conventional to let C^0 denote the class of continuous functions. Clearly, the existence and continuity of a derivative of any order presupposes derivatives of lower orders so that these classes are nested

$$C^{n+1} \subset C^n \cdots \subset C^3 \subset C^2 \subset C^1 \subset C^0$$

If a function has continuous derivatives of all orders, that is f is of class

C^n for all $n = 1, 2, 3, \ldots$, we say that f is of class C^∞. A function f is *smooth* if it is sufficiently differentiable, that is f belongs to class C^n for given n, where n depends on the purpose at hand. For functionals on $\Re$, it is convenient to use the notation $f'[\mathbf{x}], f''[\mathbf{x}], f^{(3)}[\mathbf{x}], \ldots, f^{(r)}[\mathbf{x}], \ldots$ to denote successive derivatives.

Example 4.25 (Exponential function) The exponential function $f(x) = e^x$ is differentiable with derivative $f'[x] = e^x$. The derivative in turn is differentiable with derivative $f''[x] = e^x$. In fact the exponential function is differentiable to any order with $f^{(n)}[x] = e^x$. That is, the exponential function is infinitely differentiable and therefore smooth.

Example 4.26 (Polynomials) The nth-order polynomial

$$f_n(x) = a_0 + a_1 x + a_2 x^2 + \cdots + a_n x^n$$

is C^∞ with

$$f'_n[x] = a_1 + 2a_2 x^1 + \cdots + n a_n x^{n-1}$$
$$f''_n[x] = 2a_2 + 6a_3 x + \cdots + n(n-1) a_n x^{n-2}$$
$$f_n^{(n)}[x] = n! a_n$$
$$f_n^{(m)}[x] = 0, \quad m = n+1, n+2, \ldots$$

Remark 4.9 (Solow's convention) Typically economic models are formulated with general rather than specific functional forms, so that the degree of differentiability is a matter of assumption rather than fact. Usual practice is expressed by *Solow's convention*: "Every function is assumed to be differentiable one more time than we need it to be." For many purposes, it suffices if functions are twice continuously differentiable or C^2.

Exercise 4.51 (Composition)
If $f: X \to Y$ and $g: Y \to Z$ are C^n, then the composite function $g \circ f: X \to Z$ is also C^n.

4.4.2 Second-Order Partial Derivatives and the Hessian

When $X \subseteq \Re^n$, the partial derivative $D_{x_i} f[\mathbf{x}]$ of a functional f on X measures the rate of change of f parallel to the x_i axis. If $D_{x_i} f[\mathbf{x}]$ exists for all $\mathbf{x} \in X$, it defines another functional

$D_{x_i}f: X \to \Re$

If this is a continuous functional on X, we can consider in turn its partial derivatives $D_{x_j}(D_{x_i}f)[\mathbf{x}]$, which are called the *second-order partial derivative* of f. We will denote second-order partial derivatives by $D_{x_j x_i}f[\mathbf{x}]$, which we will sometimes abbreviate to $D_{ji}f[\mathbf{x}]$ when there is no ambiguity. The alternative notation

$$\frac{\partial^2 f(\mathbf{x})}{\partial x_i^2} \quad \text{and} \quad \frac{\partial^2 f(\mathbf{x})}{\partial x_j \partial x_i}$$

is also common. When $i \neq j$, $D_{x_j x_i}f[\mathbf{x}]$ is called the *cross partial derivative*.

Example 4.27 (Cobb-Douglas) The partial derivatives of the Cobb-Douglas function

$$f(x_1, x_2) = x_1^{a_1} x_2^{a_2}$$

are

$$D_1 f[\mathbf{x}] = a_1 x_1^{a_1 - 1} x_2^{a_2}$$
$$D_2 f[\mathbf{x}] = a_2 x_1^{a_1} x_2^{a_2 - 1}$$

The second-order partial derivatives are

$$D_{11}f[\mathbf{x}] = a_1(a_1 - 1)x_1^{a_1-2} x_2^{a_2}, \quad D_{21}f[\mathbf{x}] = a_1 a_2 x_1^{a_1-1} x_2^{a_2-1}$$
$$D_{12}f[\mathbf{x}] = a_1 a_2 x_1^{a_1-1} x_2^{a_2-1}, \quad D_{22}f[\mathbf{x}] = a_2(a_2 - 1)x_1^{a_1} x_2^{a_2-2}$$

Notice that the cross-partial derivatives are symmetric, that is,

$$D_{21}f[\mathbf{x}] = D_{12}f[\mathbf{x}]$$

We will see shortly that this is a general result.

Exercise 4.52
Compute the second-order partial derivatives of the quadratic function

$$f(x_1, x_2) = ax_1^2 + 2bx_1 x_2 + cx_2^2$$

Exercise 4.53 (C^2 functions)
Let f be a functional on an open subset S of $\Re^n$. Then f is twice continuously differentiable (C^2) if and only if each of the second-order partial derivatives $D_{ij}f[\mathbf{x}]$ exists and is continuous on S.

4.4 Polynomial Approximation and Higher-Order Derivatives

Example 4.28 (Cobb-Douglas) The Cobb-Douglas function

$$f(x_1, x_2) = x_1^{a_1} x_2^{a_2}$$

is C^2, since the second-order partial derivatives of the Cobb-Douglas function (example 4.27) are continuous. In fact the Cobb Douglas function is C^∞.

When f is a C^2 functional on $X \subseteq \Re^n$, the gradient

$$\nabla f(\mathbf{x}) = (D_{x_1} f[\mathbf{x}], D_{x_2} f[\mathbf{x}], \ldots, D_{x_n} f[\mathbf{x}])$$

is a differentiable function from $X \to \Re^n$. The second derivative of f is the derivative of the gradient, which can be represented by a matrix (the Jacobian of ∇f), the elements of which comprise the second-order partial derivatives of f. This matrix representing the second derivative of f is called the *Hessian* of f, denoted H_f:

$$H_f(\mathbf{x}) = \begin{pmatrix} D_{11} f[\mathbf{x}] & D_{12} f[\mathbf{x}] & \cdots & D_{1n} f[\mathbf{x}] \\ D_{21} f[\mathbf{x}] & D_{22} f[\mathbf{x}] & \cdots & D_{2n} f[\mathbf{x}] \\ \vdots & \vdots & \ddots & \vdots \\ D_{n1} f[\mathbf{x}] & D_{n2} f[\mathbf{x}] & \cdots & D_{nn} f[\mathbf{x}] \end{pmatrix}$$

Example 4.29 Using example 4.27, the Hessian of the Cobb-Douglas function

$$f(x_1, x_2) = x_1^{a_1} x_2^{a_2}$$

is

$$H(\mathbf{x}) = \begin{pmatrix} a_1(a_1 - 1) x_1^{a_1 - 2} x_2^{a_2} & a_1 a_2 x_1^{a_1 - 1} x_2^{a_2 - 1} \\ a_1 a_2 x_1^{a_1 - 1} x_2^{a_2 - 1} & a_2(a_2 - 1) x_1^{a_1} x_2^{a_2 - 2} \end{pmatrix}$$

which can be expressed more compactly as

$$H(\mathbf{x}) = \begin{pmatrix} \dfrac{a_1(a_1 - 1)}{x_1^2} & \dfrac{a_1 a_2}{x_1 x_2} \\ \dfrac{a_1 a_2}{x_1 x_2} & \dfrac{a_2(a_2 - 1)}{x_2^2} \end{pmatrix} f(\mathbf{x})$$

Exercise 4.54
Compute the Hessian of the quadratic function

$$f(x_1, x_2) = a x_1^2 + 2b x_1 x_2 + c x_2^2$$

Exercise 4.55
Compute the Hessian of the CES function

$$f(x_1, x_2) = (a_1 x_1^p + a_2 x_2^p)^{1/p}$$

Remark 4.10 (Second derivative as a bilinear function) We can use the theory developed in the previous chapter to provide an alternative representation of the second derivative applicable to more general spaces. The second derivative $D^2 f$ of a function $f \colon X \to Y$ is a linear mapping from the set X to the set $BL(X, Y)$. That is, $D^2 f$ is an element of the space $BL(X, BL(X, Y))$, which is equivalent (exercise 3.58) to $BiL(X \times X, Y)$, the set of all continuous bilinear maps from $X \times X \to Y$. In other words, the second derivative is a bilinear function from $X \times X \to Y$. Similarly higher-order derivatives can be identified with multilinear functions, with $D^k f$ mapping the k times product $X \times X \times \cdots \times X$ to Y. Moreover, when f is a functional $Y = \Re$ and X is finite dimensional, then the bilinear functional $D^2 f$ can be represented by a matrix (exercise 3.55). The Hessian is the matrix representation of $D^2 f$ with respect to the standard basis.

We have already remarked on the symmetry of the cross partial derivatives in the preceding examples. The order of differentiation does not matter. This somewhat counterintuitive property is exhibited by all C^2 functions, a result that is often referred to as Young's theorem. Yet another consequence of the mean value theorem, Young's theorem has some surprising results for economics (example 4.30).

Theorem 4.2 (Young's theorem) *If f is a C^2 functional on an open set $X \subseteq \Re^n$, its Hessian is symmetric, that is,*

$$D_{ij} f[\mathbf{x}] = D_{ji} f[\mathbf{x}] \quad \text{for every } i,j$$

for every $\mathbf{x} \in X$.

Proof To aid interpretation, we will assume that function f is a production function. The function $D_i f[\mathbf{x}]$ is the marginal product of input i. $D_{ji} f[\mathbf{x}]$ measures the rate at which the marginal product of input i changes with changes in the use of input j. Similarly $D_{ij} f[\mathbf{x}]$ measures the rate at which the marginal product of input j changes with changes in the use of input i. Young's theorem asserts that these two apparently different measures are equal if the production function is C^2.

4.4 Polynomial Approximation and Higher-Order Derivatives

As a partial derivative, $D_{ji}f[\mathbf{x}]$ involves changes in the use of only two commodities, with the quantities of all other commodities held constant. Without loss of generality, we can ignore the other components and assume that f is a function of only two variables (inputs) x_1 and x_2.

Consider two input bundles (x_1, x_2) and $(x_1 + dx_1, x_2 + dx_2)$, where dx_1 and dx_2 are small changes in the quantities of inputs 1 and 2 respectively. Define the function $\Delta f \colon \Re^2 \to \Re$ by

$$\Delta f(dx_1, y) = f(x_1 + dx_1, y) - f(x_1, y) \tag{13}$$

For each level of use of input 2, $\Delta f(y)$ measures the incremental product of an additional dx_1 of input 1. Next define the function $\Delta\Delta f \colon \Re^2 \to \Re$ by

$$\Delta\Delta f(dx_1, dx_2) = \Delta f(dx_1, x_2 + dx_2) - \Delta f(dx_2, x_2) \tag{14}$$

Expanding (14) using (13), we have

$$\Delta\Delta f(dx_1, dx_2) = (f(x_1 + dx_1, x_2 + dx_2) - f(x_1, x_2 + dx_2))$$
$$- (f(x_1 + dx_1, x_2) - f(x_1, x_2))$$

For small changes, the first term in parentheses approximates the marginal product of input 1 given the use of $x_2 + dx_2$ units of input 2. The second term in parentheses approximates the marginal product of input 1 given the use of only x_2 units of input 2. Their difference approximates the change in the marginal product of input 1 with a small change in use of input 2. Reordering the terms, we can rewrite the previous equation as

$$\Delta\Delta f(dx_1, dx_2) = (f(x_1 + dx_1, x_2 + dx_2) - f(x_1 + dx_1, x_2))$$
$$- (f(x_1, x_2 + dx_2) - f(x_1, x_2))$$

This expression also approximates the change in marginal product of input 2 with a small change in use of input 1. This is the economic content of Young's theorem. It remains to show that this approximation becomes exact for marginal changes.

Applying the mean value theorem (theorem 4.1) to (14) and using (13), we have

$$\Delta\Delta f(dx_1, dx_2) = (D_2 f(x_1 + dx_1, \bar{x}_2) - D_2 f(x_1, \bar{x}_2)) \, dx_2 \tag{15}$$

Since $D_2 f$ is differentiable (f is C^2), we can apply the mean value theorem to the first coordinate of (15) to give

$$\Delta\Delta f(dx_1, dx_2) = (D_{12}f(\bar{x}_1, \bar{x}_2) - D_{12}f(\bar{x}_1, \bar{x}_2))\, dx_1\, dx_2 \qquad (16)$$

where $(\bar{x}_1, \bar{x}_2)$ is a commodity bundle between (x_1, x_2) and $(x_1 + dx_1, x_2 + dx_2)$. Interchanging the roles of inputs 1 and 2 in the above derivation, there also exists a commodity bundle $(\tilde{x}_1, \tilde{x}_2)$ between (x_1, x_2) and $(x_1 + dx_1, x_2 + dx_2)$ at which

$$\Delta\Delta f(dx_1, dx_2) = (D_{21}f(\tilde{x}_1, \tilde{x}_2) - D_{12}f(\tilde{x}_1, \tilde{x}_2))\, dx_1\, dx_2 \qquad (17)$$

dx_1 and dx_2 were arbitrary use changes so that equations (16) and (17) apply for all $dx_1, dx_2 \in \Re$. Allowing dx_1, dx_2 to become arbitrarily small and invoking the continuity of $D_{12}f$ and $D_{21}f$, we have

$$D_{12}f(x_1, x_2) = \lim_{dx_i \to 0} \frac{\Delta\Delta f(dx_1, dx_2)}{dx_1 dx_2} = D_{21}f(x_1, x_2) \qquad \square$$

Remark 4.11 Although the preceding proof was couched in terms of a production function, the only property of f which was invoked was the fact that it was C^2 and the proof is completely general. It is worth observing that symmetry of the Hessian matrix at a particular point only requires continuity of the partial derivatives in the neighborhood of that point and not continuity everywhere.

The preceding discussion can be extended in a straightforward manner to higher orders of derivatives that arise occasionally in economics. For any C^r function ($r \geq 3$), the third- and higher-order derivatives $D^k f$ can be represented as a symmetric multilinear functions of partial derivatives of a corresponding order. This means that each nth-order partial derivative is independent of the order of differentiation.

Example 4.30 (Input demand functions) Suppose that a competitive firm produces a single output y using n inputs $(x_1, x_2, \ldots, x_n)$ according to the production function

$$y = f(x_1, x_2, \ldots, x_n)$$

Let p denote the output price and $\mathbf{w} = (w_1, w_2, \ldots, w_n)$ the input prices. The input demand functions $x_i(\mathbf{w}, p)$ specify the profit-maximizing demand for input i as a function of $\mathbf{w}$ and p. We will show in example 6.15

that the partial derivatives of the input demand function are proportional to the inverse of the Hessian matrix of the production function, that is

$$\begin{pmatrix} D_{w_1}x_1 & D_{w_2}x_1 & \cdots & D_{w_n}x_1 \\ D_{w_1}x_2 & D_{w_2}x_2 & \cdots & D_{w_n}x_2 \\ \vdots & \vdots & \ddots & \vdots \\ D_{w_1}x_n & D_{w_2}x_n & \cdots & D_{w_n}x_n \end{pmatrix} = \frac{1}{p} H_f^{-1}$$

By Young's theorem, the Hessian H_f is symmetric and so therefore is its inverse H_f^{-1}. This implies the surprising result that

$$D_{w_j}x_i(\mathbf{w}, p) = D_{w_i}x_j(\mathbf{w}, p)$$

That is, the change in demand for input i following a change in price of input j is equal to the change in demand for input j following a change in price of input i.

4.4.3 Taylor's Theorem

We return now to our primary goal of approximating an arbitrary smooth function by a polynomial, beginning with an example.

Example 4.31 (Best quadratic approximation) Suppose that we seek the best quadratic approximation to an arbitrary smooth function $f: \Re \to \Re$. That is, we seek to determine the coefficients a_0, a_1, a_2 in the quadratic polynomial

$$f_2(x) = a_0 + a_1 x + a_2 x^2$$

so that

$$f(x_0 + x) \approx a_0 + a_1 x + a_2 x^2$$

for small x. We note that f_2 is twice differentiable (example 4.26) with

$$f_2'(x) = a_1 + 2a_2 x \quad \text{and} \quad f_2''(x) = 2a_2$$

What do we mean by "best approximation." At minimum, we want f and its approximation to be tangential at x_0 ($x = 0$). This determines the first two coefficients of f_2, since

- $f_2(0) = f(x_0)$ implies that $a_0 = f(x_0)$
- $f_2'[0] = f'[x_0]$ implies that $a_1 = f'[x_0]$

That leaves a_2 to be determined. We could choose a_2 so that the functions f and f_2 have the same second derivative at $x = 0$ as well, which implies setting $a_2 = \frac{1}{2} f''[x_0]$. The quadratic polynomial

$$f_2(x) = f(x_0) + f'[x_0]x + \tfrac{1}{2} f''[x_0]x^2 \tag{18}$$

is the best quadratic approximation to f in the neighborhood of x_0 (exercise 4.58).

Alternatively, we could choose a_2 such that the approximation is exact at some other $x_1 \neq x_0$, that is, so that $f(x_1) = f_2(x_1 - x_0)$, which requires that $a_2 = \frac{1}{2} f''(\bar{x})$ for some $\bar{x}$ between x_0 and x_1 (exercise 4.56). The quadratic polynomial

$$\bar{f}_2(x) = f(x_0) + f'[x_0]x + \tfrac{1}{2} f''[\bar{x}]x^2 \tag{19}$$

is an approximation to f, which is exact at x_1.

This example illustrates the two fundamental results of polynomial approximation. For any smooth functional f on $\Re$, (18) is the best quadratic approximation to f in the neighborhood of x_0. Furthermore, in a generalization of the mean value theorem, for any specific $x_1 \neq x_0$, there exists a $\bar{x}$ between x_0 and x_1 such that the approximation (19) is exact at $x = x_1 - x_0$.

Exercise 4.56
Let f be a twice differentiable functional on some open interval $S \subseteq \Re$ containing x_0. For every $x_1 \in S$, there exists some $\bar{x}$ between x_0 and x_1 such that

$$f(x_1) = f(x_0) + f'[x_0]x + \tfrac{1}{2} f''[\bar{x}]x^2$$

[Hint: Consider the function $g(t) = f(t) + f'[t](x - t) + a_2(x - t)^2$.]

Exercise 4.56 is easily generalized to higher-order polynomials, which gives Taylor's theorem, which can be seen as a generalization of the mean value theorem to higher-level derivatives.

Exercise 4.57 (Taylor's theorem in $\Re$)
Let f be a $(n+1)$-times differentiable functional on some open interval $S \subseteq \Re$ containing x_0. For every $x \in S - x_0$, there exists some $\bar{x}$ between x_0 and $x_0 + x$ such that

$$f(x_0 + x) = f(x_0) + f'[x_0]x + \frac{1}{2}f''[x_0]x^2 + \frac{1}{3!}f^{(3)}[x_0]x^3$$
$$+ \cdots + \frac{1}{n!}f^{(n)}[x_0]x^n + \frac{1}{(n+1)!}f^{(n+1)}[\bar{x}]x^{n+1}$$

Example 4.32 (Exponential function) For the exponential function $f(x) = e^x$, $f^{(n)}[x] = e^x$, and therefore $f^{(n)}[0] = e^0 = 1$. Then, for every $n = 1, 2, \ldots$ and every $x \in \mathfrak{R}_+$, there exists some $\bar{x} \in (0, x)$ such that

$$e^x = 1 + x + \frac{1}{2!}x^2 + \frac{1}{3!}x^3 + \cdots + \frac{1}{n!}x^n + \varepsilon_n(x)$$

where

$$\varepsilon_n(x) = \frac{1}{(n+1)!}e^{\bar{x}}x^{n+1}$$

On $[0, 1]$, both x^{n+1} and $e^{\bar{x}}$ are increasing, so that the error is bounded by

$$\varepsilon_n(x) \leq \varepsilon_n(1) = \frac{1}{(n+1)!}e, \quad x \in [0, 1]$$

The error in approximating e^x by a polynomial over the interval $[0, 1]$ can be made as small as we like by choosing sufficiently many terms in the polynomial. For example, with five terms ($n = 5$), the approximation error is less than one percent over the interval $[0, 1]$. This confirms our discoveries in example 4.24.

Exercise 4.58
Let f be C^3 on some open interval $S \subseteq \mathfrak{R}$ containing x_0. For every $x \in S - x_0$,

$$f(x_0 + x) = f(x_0) + f'[x_0]x + \tfrac{1}{2}f''[x_0]x^2 + \varepsilon(x)$$

where

$$\lim_{x \to 0} \frac{\varepsilon(x)}{x^2} = 0$$

That is, the approximation error becomes very small as $x \to 0$. In this sense, (18) is the best quadratic approximation to f in a neighborhood of x_0.

Extending exercise 4.57 to functionals on multidimensional spaces gives us Taylor's theorem, which generalizes the mean value theorem (theorem 4.1) to higher-level derivatives.

Theorem 4.3 (Taylor's theorem) *Let f be a C^{n+1} functional on a convex neighborhood S of $\mathbf{x}_0$. For every $\mathbf{x} \in S - \mathbf{x}_0$, there exists some $\bar{\mathbf{x}} \in (\mathbf{x}_0, \mathbf{x}_0 + \mathbf{x})$ such that*

$$f(\mathbf{x}_0 + \mathbf{x}) = f(\mathbf{x}_0) + Df[\mathbf{x}_0]\mathbf{x} + \frac{1}{2!}D^2 f[\mathbf{x}_0](\mathbf{x},\mathbf{x}) + \frac{1}{3!}D^3 f[\mathbf{x}_0]\mathbf{x}^{(3)}$$

$$+ \cdots + \frac{1}{n!}D^n f[\mathbf{x}_0]\mathbf{x}^{(n)} + \frac{1}{(n+1)!}D^{n+1}f[\bar{\mathbf{x}}]\mathbf{x}^{(n+1)}$$

where $\mathbf{x}^{(k)}$ denotes the k-tuple $(\mathbf{x},\mathbf{x},\ldots,\mathbf{x})$.

Proof For fixed $\mathbf{x} \in S - \mathbf{x}_0$, define $g: \Re \to S$ by $g(t) = \mathbf{x}_0 + t\mathbf{x}$. The composite function $h = f \circ g: \Re \to \Re$ is C^{n+1} (exercise 4.59) with derivatives

$$h^{(k)}(t) = D^{(k)}f[g(t)](\mathbf{x})^{(k)} \qquad (20)$$

Applying the univariate Taylor's theorem (exercise 4.57), there exists an $\alpha \in (0,1)$ such that

$$h(1) = h(0) + h'[0] + \frac{1}{2}h''[0] + \frac{1}{3!}h^{(3)}[0] + \cdots + \frac{1}{n!}h^{(n)}[0] + \frac{1}{(n+1)!}h^{(n+1)}[\alpha]$$

Substituting $h(0) = f(\mathbf{x}_0)$, $h(1) = f(\mathbf{x}_0 + \mathbf{x})$, $\bar{\mathbf{x}} = g(\alpha)$, and (20) gives

$$f(\mathbf{x}_0 + \mathbf{x}) = f(\mathbf{x}_0) + Df[\mathbf{x}_0]\mathbf{x} + \frac{1}{2!}D^2 f[\mathbf{x}_0](\mathbf{x},\mathbf{x}) + \frac{1}{3!}D^3 f[\mathbf{x}_0]\mathbf{x}^{(3)}$$

$$+ \cdots + \frac{1}{n!}D^n f[\mathbf{x}_0]\mathbf{x}^{(n)} + \frac{1}{(n+1)!}D^{n+1}f[\bar{\mathbf{x}}]\mathbf{x}^{(n+1)}$$

as required. □

Exercise 4.59
Assume that f is a C^{n+1} functional on a convex set S. For fixed $\mathbf{x}_0 \in S$ and $\mathbf{x} \in S - \mathbf{x}_0$, define $g: \Re \to S$ by $g(t) = \mathbf{x}_0 + t\mathbf{x}$. Show that the composite function $h = f \circ g: \Re \to \Re$ is C^{n+1} with derivatives

$$h^{(k)}(t) = D^{(k)}f[g(t)](\mathbf{x})^{(k)}$$

4.4 Polynomial Approximation and Higher-Order Derivatives

Taylor's theorem states that *any* smooth functional can be approximated by an *n*th-order polynomial f_n of its derivatives

$$f_n(\mathbf{x}) = f(\mathbf{x}_0) + Df[\mathbf{x}_0](\mathbf{x}) + \frac{1}{2!}D^2f[\mathbf{x}_0](\mathbf{x})^{(2)} + \cdots + \frac{1}{n!}D^nf[\mathbf{x}_0]\mathbf{x}^{(n)} \quad (21)$$

where the error in approximating f by f_n is

$$\varepsilon(\mathbf{x}) = f(\mathbf{x}_0 + \mathbf{x}) - f_n(\mathbf{x}) = \frac{1}{(n+1)!}D^{n+1}f[\bar{\mathbf{x}}]\mathbf{x}^{(n+1)}$$

Analogous to the definition of differentiability in section 4.1, we can decompose the actual error $\varepsilon(\mathbf{x})$ into two components

$$\varepsilon(\mathbf{x}) = \eta(\mathbf{x})\|\mathbf{x}\|^n$$

where

$$\eta(\mathbf{x}) = \frac{\varepsilon(\mathbf{x})}{\|\mathbf{x}\|^n} = \frac{f(\mathbf{x}_0 + \mathbf{x}) - f_n(\mathbf{x})}{\|\mathbf{x}\|^n}$$

is the approximation error *relative to* $\|\mathbf{x}\|^n$. If the relative error $\eta(\mathbf{x}) \to 0$ as $\mathbf{x} \to 0$, approximation error $\eta(\mathbf{x})\|\mathbf{x}\|^n$ becomes negligible very quickly. This is a very strong form of convergence, which will be satisfied if $D^{n+1}f[\bar{\mathbf{x}}]$ is bounded on $[\mathbf{0}, \mathbf{x}]$. In particular, this will be satisfied if $f \in C^{n+1}$, which gives us the following corollary.

Corollary 4.3.1 (Taylor series approximation) *Let f be a C^{n+1} functional on a convex neighborhood S of $\mathbf{x}_0$. For every $\mathbf{x} \in S - \mathbf{x}_0$,*

$$f(\mathbf{x}_0 + \mathbf{x}) = f(\mathbf{x}_0) + Df[\mathbf{x}_0]\mathbf{x} + \frac{1}{2!}D^2f[\mathbf{x}_0](\mathbf{x}, \mathbf{x})$$

$$+ \cdots + \frac{1}{n!}D^nf[\mathbf{x}_0](\mathbf{x})^n + \eta(\mathbf{x})\|\mathbf{x}\|^n$$

and $\eta(\mathbf{x}) \to 0$ as $\mathbf{x} \to 0$.

Proof By Taylor's theorem (4.3),

$$\eta(\mathbf{x}) = \frac{\varepsilon(\mathbf{x})}{\|\mathbf{x}\|^n} = \frac{1}{(n+1)!}\left(\frac{\mathbf{x}}{\|\mathbf{x}\|}\right)^{(n)}D^{n+1}f[\bar{\mathbf{x}}](\mathbf{x})$$

for some $\bar{\mathbf{x}} \in (\mathbf{x}_0, \mathbf{x}_0 + \mathbf{x})$, and therefore

$$|\eta(\mathbf{x})| \leq \frac{1}{(n+1)!} D^{n+1} f[\bar{\mathbf{x}}](\mathbf{x})$$

If $f \in C^{n+1}$, then $D^{n+1}f[\bar{\mathbf{x}}]$ is bounded on $[\mathbf{x}_0, \mathbf{x}_0 + \mathbf{x}]$, and therefore continuous (exercise 3.30). Consequently, $|\eta(\mathbf{x})| \to 0$ as $\mathbf{x} \to \mathbf{0}$. ☐

Corollary 4.3.1 generalizes the definition of the derivative in (2) to higher orders of approximation. The polynomial (21) is called the nth-order *Taylor series* for f. As n increases, convergence of the relative error $\eta(\mathbf{x})$ becomes more demanding. In this sense, the accuracy of the approximation f by f_n close to $\mathbf{x}_0$ increases with n.

For some functions, the absolute error $\varepsilon(\mathbf{x})$ becomes negligible for all $\mathbf{x}$ as the number of terms increases, that is, for all $\mathbf{x} \in X$

$$\lim_{n \to \infty} f(\mathbf{x}_0 + \mathbf{x}) - f_n(\mathbf{x}) = 0$$

Such functions are called *analytic*.

In economics, first- and second-order approximation usually suffices, and we seldom have to resort to higher-order polynomials. In Euclidean space, the second-order Taylor series has a convenient representation in terms of the gradient and Hessian of the function.

Example 4.33 (Quadratic approximation in $\Re^n$) Suppose that f is a smooth functional on a convex neighborhood $S \subseteq \Re^n$ of $\mathbf{x}_0$. For every $\mathbf{x} \in S - \mathbf{x}_0$,

$$f(\mathbf{x}_0 + \mathbf{x}) = f(\mathbf{x}_0) + \nabla f(\mathbf{x}_0)^T \mathbf{x} + \tfrac{1}{2} \mathbf{x}^T H_f(\mathbf{x}_0) \mathbf{x} + \eta(\mathbf{x}) \|\mathbf{x}\|^2 \tag{22}$$

where $\eta(\mathbf{x}) \to 0$ as $\mathbf{x} \to \mathbf{0}$. (corollary 4.3.1). Furthermore (theorem 4.3) there exists $\bar{\mathbf{x}} \in (\mathbf{x}_0, \mathbf{x}_0 + \mathbf{x})$ such that

$$f(\mathbf{x}_0 + \mathbf{x}) = f(\mathbf{x}_0) + \nabla f(\mathbf{x}_0)^T \mathbf{x} + \tfrac{1}{2} \mathbf{x}^T H_f(\bar{\mathbf{x}}) \mathbf{x} \tag{23}$$

Both quadratic approximations (22) and (23) are useful in practice (see the proof of proposition 4.1).

Example 4.34 (Cobb-Douglas) Let us compute the best quadratic approximation to the Cobb-Douglas function

$$f(x_1, x_2) = x_1^{1/3} x_2^{2/3}$$

in the neighborhood of the point $(8, 8)$. From example 4.29 the Hessian of f is

4.4 Polynomial Approximation and Higher-Order Derivatives

$$H_f(\mathbf{x}) = \begin{pmatrix} \frac{\frac{1}{3}(\frac{1}{3}-1)}{x_1^2} & \frac{\frac{1}{3}(\frac{2}{3})}{x_1 x_2} \\ \frac{\frac{1}{3}(\frac{2}{3})}{x_1 x_2} & \frac{\frac{2}{3}(\frac{2}{3}-1)}{x_2^2} \end{pmatrix}$$

$$f(\mathbf{x}) = \frac{2}{9} \begin{pmatrix} \frac{-1}{x_1^2} & \frac{1}{x_1 x_2} \\ \frac{1}{x_1 x_2} & \frac{-1}{x_2^2} \end{pmatrix} f(\mathbf{x})$$

which evaluates to

$$H(8,8) = \frac{2}{72} \begin{pmatrix} -1 & 1 \\ 1 & -1 \end{pmatrix}$$

The gradient at $(8,8)$ is

$$\nabla f(8,8) = (\tfrac{1}{3}, \tfrac{2}{3})$$

Therefore the second-order Taylor series approximation of f at $\mathbf{x}_0 = (8,8)$ is

$$f(\mathbf{x}_0 + \mathbf{x}) \approx f(\mathbf{x}_0) + \nabla f(\mathbf{x}_0)^T \mathbf{x} + \frac{1}{2}\mathbf{x}^T H(\mathbf{x}_0) \mathbf{x}$$

$$= 8 + \left(\frac{1}{2}, \frac{2}{3}\right)\begin{pmatrix} x_1 \\ x_2 \end{pmatrix} + \frac{1}{2}(x_1, x_2)^T \frac{2}{72}\begin{pmatrix} -1 & 1 \\ 1 & -1 \end{pmatrix}\begin{pmatrix} x_1 \\ x_2 \end{pmatrix}$$

$$= 8 + \frac{1}{3}x_1 + \frac{2}{3}x_2 - \frac{1}{72}(x_1^2 - 2x_1 x_2 + x_2^2) \tag{24}$$

Table 4.5 compares the error of the quadratic approximation (24) to the Cobb-Douglas function with that of the linear (first-order) approximation

$$f(\mathbf{x}_0 + \mathbf{x}) = 8 + \tfrac{1}{3}x_1 + \tfrac{2}{3}x_2$$

developed in section 4.1. Columns 2 and 3 compare the actual error $\varepsilon(\mathbf{x})$ for linear and quadratic approximations. We see that the quadratic approximation is uniformly more accurate at all points evaluated in the table—it provides a better *global* approximation. Columns 4 and 5 compare the relative (squared) error $\varepsilon(\mathbf{x})/\|\mathbf{x}\|^2$. Declining relative (squared) error is

Table 4.5
Approximating the Cobb-Douglas function

	Actual error		Relative error	
$\mathbf{x}_0 + \mathbf{x}$	Linear	Quadratic	Linear	Quadratic
Around the unit circle:				
(9.0, 8.0)	−0.0130	0.0009	−0.0130	0.0009
(8.7, 8.7)	−0.0000	−0.0000	−0.0000	−0.0000
(8.0, 9.0)	−0.0132	0.0007	−0.0132	0.0007
(7.3, 8.7)	−0.0281	−0.0003	−0.0281	−0.0003
(7.0, 8.0)	−0.0149	−0.0011	−0.0149	−0.0011
(7.3, 7.3)	0.0000	0.0000	0.0000	0.0000
(8.0, 7.0)	−0.0147	−0.0008	−0.0147	−0.0008
(8.7, 7.3)	−0.0276	0.0002	−0.0276	0.0002
Around a smaller circle:				
(8.1, 8.0)	−0.0001	0.0000	−0.0138	0.0001
(8.1, 8.1)	0.0000	0.0000	0.0000	0.0000
(8.0, 8.1)	−0.0001	0.0000	−0.0138	0.0001
(7.9, 8.1)	−0.0003	−0.0000	−0.0278	−0.0000
(7.9, 8.0)	−0.0001	−0.0000	−0.0140	−0.0001
(7.9, 7.9)	0.0000	0.0000	0.0000	0.0000
(8.0, 7.9)	−0.0001	−0.0000	−0.0140	−0.0001
(8.1, 7.9)	−0.0003	0.0000	−0.0278	0.0000
Parallel to the x_1 axis:				
(4.0, 8.0)	−0.3171	−0.0948	−0.0198	−0.0059
(6.0, 8.0)	−0.0649	−0.0093	−0.0162	−0.0023
(7.0, 8.0)	−0.0149	−0.0011	−0.0149	−0.0011
(7.5, 8.0)	−0.0036	−0.0001	−0.0144	−0.0005
(7.9, 8.0)	−0.0001	−0.0000	−0.0140	−0.0001
(8.0, 8.0)	0.0000	0.0000	NIL	NIL
(8.1, 8.0)	−0.0001	0.0000	−0.0138	0.0001
(8.5, 8.0)	−0.0034	0.0001	−0.0134	0.0005
(9.0, 8.0)	−0.0130	0.0009	−0.0130	0.0009
(10.0, 8.0)	−0.0489	0.0066	−0.0122	0.0017
(12.0, 8.0)	−0.1756	0.0466	−0.0110	0.0029
Parallel to the x_2 axis:				
(8.0, 4.0)	−0.2936	−0.0714	−0.0184	−0.0045
(8.0, 6.0)	−0.0628	−0.0073	−0.0157	−0.0018
(8.0, 7.0)	−0.0147	−0.0008	−0.0147	−0.0008
(8.0, 7.5)	−0.0036	−0.0001	−0.0143	−0.0004
(8.0, 7.9)	−0.0001	−0.0000	−0.0140	−0.0001
(8.0, 8.0)	0.0000	0.0000	NIL	NIL
(8.0, 8.1)	−0.0001	0.0000	−0.0138	0.0001
(8.0, 8.5)	−0.0034	0.0001	−0.0135	0.0004
(8.0, 9.0)	−0.0132	0.0007	−0.0132	0.0007
(8.0, 10.0)	−0.0502	0.0054	−0.0125	0.0013
(8.0, 12.0)	−0.1837	0.0385	−0.0115	0.0024

stringent requirement, which is met by the quadratic approximation but is not met by the linear approximation. In this sense the quadratic approximation provides a better *local* approximation. Corollary 4.3.1 shows that this is not a peculiarity of this example, but is a property of all smooth functions. The second-order Taylor series is the best quadratic approximation to a smooth function.

Exercise 4.60
Compute the second-order Taylor series expansion of the quadratic function

$$f(x_1, x_2) = ax_1^2 + 2bx_1x_2 + cx_2^2$$

around the point $(0, 0)$.

In economics the theory of polynomial approximations has two fundamental applications. In the first, quadratic approximations are used to analyze the behavior of unspecified function in the neighborhood of a point of interest. These underlie the second-order conditions in the theory of optimization (section 5.2) and are also used to characterize convexity and concavity (section 4.6). In the second major application, polynomial approximations are used to generate specific functional forms for economic models. We discuss this briefly now.

Although much economic analysis is conducted using general (unspecified) functional forms, there eventually comes a time where a functional form must be specified, whether to sharpen results, overcome intractability or provide a simple model for easy analysis. Furthermore, whenever empirical estimation is involved, a functional form must be chosen. Computational restrictions have favored functional forms that are linear in the parameters. Most popular functional forms in economic analysis can be viewed as linear or quadratic approximations to an arbitrary function.

From example 4.33, the best quadratic approximation to an arbitrary smooth functional f at a point $\mathbf{x}_0 \in S \subseteq \Re^n$ is given by

$$f(\mathbf{x}_0 + \mathbf{x}) \approx f(\mathbf{x}_0) + \nabla f(\mathbf{x}_0)^T \mathbf{x} + \tfrac{1}{2}\mathbf{x}^T H(\mathbf{x}_0)\mathbf{x}$$

where

$$\nabla f[\mathbf{x}_0] = (D_1 f[\mathbf{x}_0], D_2 f[\mathbf{x}_0], \ldots, D_n f[\mathbf{x}_0])$$

Table 4.6
Common linear-in-parameters functional forms

General quadratic	$y = a_0 + \sum_i a_i x_i + \sum_i \sum_j a_{ij} x_i x_j$
Cobb-Douglas	$\log y = a_0 + \sum_i a_i \log x_i$
CES	$y^p = a_0 + \sum_i a_i x_i^p$
Translog	$\log y = a_0 + \sum_i a_i \log x_i + \sum_i \sum_j a_{ij} \log x_i \log x_j$
Generalized Leontief	$y = a_0 + \sum_i a_i \sqrt{x_i} + \sum_i \sum_j a_{ij} \sqrt{x_i} \sqrt{x_j}$

and

$$H(\mathbf{x}_0) = \begin{pmatrix} D_{11}f[\mathbf{x}_0] & D_{12}f[\mathbf{x}_0] & \cdots & D_{1n}f[\mathbf{x}_0] \\ D_{21}f[\mathbf{x}_0] & D_{22}f[\mathbf{x}_0] & \cdots & D_{2n}f[\mathbf{x}_0] \\ \vdots & \vdots & \ddots & \vdots \\ D_{n1}f[\mathbf{x}_0] & D_{n2}f[\mathbf{x}_0] & \cdots & D_{nn}f[\mathbf{x}_0] \end{pmatrix}$$

Letting

$$a_0 = f(\mathbf{x}_0), \quad a_i = D_i f[\mathbf{x}_0], \quad a_{ij} = D_{ij} f[\mathbf{x}_0], \quad i,j = 1, 2, \ldots, n$$

gives the general quadratic function

$$f(\mathbf{x}) = a_0 + \sum_i a_i x_i + \sum_i \sum_j a_{ij} x_i x_j \tag{25}$$

Most of the common functional forms in economics are instances of this general quadratic function (25) for appropriate transformations of the variables. For example, the Cobb-Douglas and CES functions are linear ($a_{ij} = 0$) in logarithms and powers of the variables respectively. These forms are summarized in table 4.6.

4.5 Systems of Nonlinear Equations

Many economic models involve a system of equations relating one set of variables $(x_1, x_2, \ldots, x_n)$ to another set of variables $(y_1, y_2, \ldots, y_m)$ as in

$$y_1 = f_1(x_1, x_2, \ldots, x_n)$$
$$y_2 = f_2(x_1, x_2, \ldots, x_n)$$
$$\vdots$$
$$y_m = f_m(x_1, x_2, \ldots, x_n)$$

The system can be viewed as a function $\mathbf{f}\colon X \to Y$, where X is a subset of $\Re^n$ and Y a subset of $\Re^m$. Solving the equations amounts to expressing $\mathbf{x}$ in terms of $\mathbf{y}$, that is, to inverting the function $\mathbf{f}$. Under what conditions does the inverse $\mathbf{f}^{-1}$ exist, and how do we find it?

We studied linear systems of equations in chapter 3, where we learned that a necessary and sufficient condition for a linear system of equations to have a unique solution is that the function $\mathbf{f}$ have full rank. Remarkably this extends locally to smooth functions. A necessary and sufficient condition for the system $\mathbf{y} = \mathbf{f}(\mathbf{x})$ to have a unique solution in the neighborhood of a point $\mathbf{x}_0$ is that the corresponding linear system $y = D\mathbf{f}[\mathbf{x}_0](\mathbf{x})$ have a unique solution. This requires that the derivative $D\mathbf{f}[\mathbf{x}_0]$ be invertible, or equivalently that $\det J_\mathbf{f}(\mathbf{x}_0) \neq 0$. This fundamental result is known as the inverse function theorem.

4.5.1 The Inverse Function Theorem

A linear function f is nonsingular if it is both one-to-one and onto. A function $f\colon X \to Y$ is *locally one-to-one* at $\mathbf{x}_0$ if there exists a neighborhood of $\mathbf{x}_0$ such that the restriction of f to S is one-to-one. f is *locally onto* at $\mathbf{x}_0$ if for any neighborhood S of $\mathbf{x}_0$ there is a neighborhood T of $f(\mathbf{x}_0)$ such that $T \subseteq f(S)$.

The significance of these definitions for systems of equations is as follows. Suppose that $\mathbf{y}_0 = f(\mathbf{x}_0)$ and that $\mathbf{y}_1$ is sufficiently close to $\mathbf{y}_0$. If f is locally onto, then there exists $\mathbf{x}_1$ such that $\mathbf{y}_1 = f(\mathbf{x}_1)$. The system of equations $\mathbf{y} = f(\mathbf{x})$ is solvable in the neighborhood of $\mathbf{x}_0$. If f is also locally one-to-one, there is at most one $\mathbf{x}_1$ near $\mathbf{x}_0$ with $\mathbf{y}_1 = f(\mathbf{x}_1)$, and the local solution is unique.

Exercise 4.61
Let f be C^1 and suppose that $Df[\mathbf{x}_0]$ is one-to-one. Then

1. f is locally one-to-one, that is, there exists a neighborhood S of $\mathbf{x}_0$ such that f is one-to-one on S

2. f has an inverse $f^{-1}\colon f(S) \to S$ that is continuous

3. f is locally onto

[Hint: Use exercises 3.36 and 4.43.]

Remark 4.12 (Locally onto) Exercise 4.61 has an analogue that can be stated as follows: Let f be C^1, and suppose that $Df[\mathbf{x}_0]$ is onto. Then f is locally onto (Lang 1969, pp. 193–94).

To express the inverse function theorem succinctly, we say that a smooth function f is a C^n diffeomorphism if it is nonsingular and f^{-1} is also C^n. A function $f: X \to Y$ is called a local C^n diffeomorphism at $\mathbf{x}_0$ if there exists neighborhoods S of $\mathbf{x}_0$ and T of $f(\mathbf{x}_0)$ such that $f: S \to T$ is a C^n diffeomorphism. The inverse function theorem shows that a C^n function is a local diffeomorphism at every point $\mathbf{x}_0$ at which the derivative is nonsingular. This implies that

- f has an inverse in a neighborhood of $f(\mathbf{x}_0)$
- the inverse f^{-1} is continuous
- the inverse f^{-1} is also C^n

Theorem 4.4 (Inverse function theorem) *Let $\mathbf{x}_0$ be a regular point of the C^n function $f: X \to X$. Then there exists a neighborhood S of $\mathbf{x}_0$ on which f is invertible. The inverse f^{-1} is C^n on $f(S)$ with derivative*

$$Df^{-1}[\mathbf{x}_0] = (Df[\mathbf{x}_0])^{-1}$$

Proof The derivative $Df[\mathbf{x}_0]$ is one-to-one. By exercise 4.61, there exists a neighborhood S of $\mathbf{x}_0$ on which f is one-to-one. Furthermore $T = f(S)$ is neighborhood of $f(\mathbf{x}_0)$ and f has a local inverse $f^{-1}: T \to S$ which is continuous. Therefore f satisfies the conditions of exercise 4.27. f^{-1} is differentiable $f(\mathbf{x}_0)$ with derivative

$$Df^{-1}[\mathbf{x}_0] = (Df[\mathbf{x}_0])^{-1}$$

Since Df is C^{n-1}, so is Df^{-1} (exercise 4.51). Therefore $f^{-1} \in C^n$. □

The power of the inverse function theorem is starkly apparent when X is finite dimensional. Then the derivative can be represented by the Jacobian, and it is nonsingular precisely when the determinant of the Jacobian is nonzero. Therefore the question of whether a smooth function is a diffeomorphism, at least in the neighborhood of some point $\mathbf{x}_0$, reduces to verifying that a single number $\det J_f(\mathbf{x}_0)$ is nonzero.

The following corollary has an important implication for systems of equations. If $\mathbf{y}$ is a regular value of the function f describing a system of equations $f(\mathbf{x}) = \mathbf{y}$, then the system has a finite number of solutions.

Corollary 4.4.1 *If S is a bounded, open subset of $\Re^n$ and $\mathbf{y}$ is a regular value $f: S \to Y$, then $f^{-1}(\mathbf{y})$ is finite.*

Proof If **y** is a regular value, every $\mathbf{x} \in f^{-1}(\mathbf{y})$ is a regular point. For every $\mathbf{x} \in f^{-1}(\mathbf{y})$, there exists a neighborhood $S_\mathbf{x}$ on which f is one-to-one (theorem 4.4). That is, $f(\mathbf{x}') \neq \mathbf{y}$ for every $\mathbf{x}' \in S_\mathbf{x} \setminus \{\mathbf{x}\}$. The collection $\{S_\mathbf{x} : \mathbf{x} \in f^{-1}(\mathbf{y})\}$ is an open cover for $f^{-1}(\mathbf{y})$. Moreover $f^{-1}(\mathbf{y})$ is closed and also bounded since $f^{-1}(\mathbf{y}) \subseteq S$. Therefore $f^{-1}(\mathbf{y})$ is compact. So it has a finite subcover. That is, there exists a finite number of points $\mathbf{x}_1, \mathbf{x}_2, \ldots, \mathbf{x}_n$ such that

$$f^{-1}(\mathbf{y}) \subseteq \bigcup_{i=1}^{n} S_{\mathbf{x}_i}$$

and within each $S_{\mathbf{x}_i}$, $\mathbf{x}_i$ is the only solution. So $f^{-1}(\mathbf{y}) = \{\mathbf{x}_1, \mathbf{x}_2, \ldots, \mathbf{x}_n\}$. □

Remark 4.13 (Sard's theorem) A remarkable theorem (whose proof is beyond the scope of this book) states that for any differentiable function $f: \Re^n \to \Re^n$, irregular values are rare, in the sense that the set of non-regular values has "measure zero" (Smith 1983). Corollary 4.4.1 states that in this sense, most systems of equations have a finite number of solutions.

It is important to appreciate that the inverse function theorem is only a local result—it ensures that a smooth function is one-to-one in the neighborhood of a regular value. Even if the function is regular (has a nonsingular derivative everywhere), we cannot conclude that the function is one-to-one throughout its domain. More stringent conditions on the Jacobian are required to ensure that the function is *globally univalent*. We give one result in exercise 4.62. Other sources are given in the references.

Exercise 4.62 (Global univalence)
Let $f: S \to \Re^n$ be a differentiable function on a convex set $S \subseteq \Re^n$. Suppose that the Jacobian $J_f(\mathbf{x})$ is positive (or negative) definite for all $\mathbf{x} \in S$. Then f is one-to-one. [Hint: Assume, to the contrary, that $f(\mathbf{x}_1) = f(\mathbf{x}_0)$ for some $\mathbf{x}_1 \neq \mathbf{x}_0$ and consider the function $h(t) = \mathbf{x}^T(f(g(t)) - f(\mathbf{x}_0))$ where $g(t) = \mathbf{x}_0 + t\mathbf{x}$ and $\mathbf{x} = \mathbf{x}_1 - \mathbf{x}_0$.]

4.5.2 The Implicit Function Theorem

To usefully apply the inverse function theorem to economic models, we need to extend it systems in which there are more are more variables than unknowns. For example, suppose that we have a system of n equations in

$n + m$ variables, comprising n dependent variables x_i and m parameters θ_k as follows:

$$\begin{aligned} f_1(x_1, x_2, \ldots, x_n; \theta_1, \theta_2, \ldots, \theta_m) &= 0 \\ f_2(x_1, x_2, \ldots, x_n; \theta_1, \theta_2, \ldots, \theta_m) &= 0 \\ &\vdots \\ f_n(x_1, x_2, \ldots, x_n; \theta_1, \theta_2, \ldots, \theta_m) &= 0 \end{aligned} \qquad (26)$$

We would like to solve for the dependent variable $\mathbf{x}$ in terms of the parameters $\boldsymbol{\theta}$. This system of equations $f(\mathbf{x}, \boldsymbol{\theta}) = \mathbf{0}$ can be viewed as a function $f: X \times \Theta \to Y$ where $X, Y \subset \Re^n$ and $\Theta \subset \Re^m$. In solving this system of equations, we are looking for the function $g: \Theta \to X$ that determines $\mathbf{x}$ in terms of $\boldsymbol{\theta}$.

Example 4.35 (IS-LM model) The IS-LM model (example 4.12)

$$y = C(y, T) + I(r) + G$$

$$L(y, r) = \frac{M}{P}$$

specifies the relationship between two dependent variables r, y and the three parameters G, T, M. Analysis of the model reduces to analysis of the mapping f defined by

$$f_1(r, y; G, T, M) = y - C(y, T) - I(r) - G = 0 \qquad (27)$$

$$f_2(r, y; G, T, M) = PL(y, r) - M = 0 \qquad (28)$$

Suppose that the functions f in (26) are linear. Then there exists matrices A and B such that

$$A\mathbf{x} + B\boldsymbol{\theta} = \mathbf{0}$$

Provided that A is nonsingular, the expression can be inverted to yield $\mathbf{x}$ in terms of $\boldsymbol{\theta}$

$$\mathbf{x} = -A^{-1}B\boldsymbol{\theta}$$

The implicit function theorem generalizes this to smooth functions.

Exercise 4.63
Suppose that all the functions in the IS-LM are linear, for example,

$$C(y, T) = C_0 + C_y(y - T)$$

$$I(r) = I_0 + I_r r$$

$$L(Y, r) = L_0 + L_r r + L_y y$$

Solve for r and y in terms of the parameters G, T, M and $C_0, I_0, L_0, C_y, I_r, L_r, L_y$.

Theorem 4.5 (Implicit function theorem) *Let* $\mathbf{f}: X \times \Theta \to Y$ *be a* C^n *function on an open neighborhood of* $(\mathbf{x}_0, \boldsymbol{\theta})$ *at which* $D_{\mathbf{x}} f[\mathbf{x}_0, \boldsymbol{\theta}_0]$ *is nonsingular and* $f(\mathbf{x}_0, \boldsymbol{\theta}_0) = \mathbf{0}$. *Then there exists a neighborhood* Ω *of* $\boldsymbol{\theta}_0$ *and a unique function* $g: \Omega \to X$ *such that*

$$\mathbf{x}_0 = g(\boldsymbol{\theta}_0) \quad \text{and} \quad f(g(\boldsymbol{\theta}), \boldsymbol{\theta}) = \mathbf{0} \quad \text{for every } \boldsymbol{\theta} \in \Omega$$

Furthermore g *is* C^n *on* Ω *with*

$$D_{\boldsymbol{\theta}} g(\boldsymbol{\theta}_0) = -\bigl(D_{\mathbf{x}} f[\mathbf{x}_0, \boldsymbol{\theta}_0]\bigr)^{-1} \circ D_{\boldsymbol{\theta}} f[\mathbf{x}_0, \boldsymbol{\theta}_0]$$

Proof Define the function $F: X \times \Theta \to Y \times \Theta$ by

$$F(\mathbf{x}, \boldsymbol{\theta}) = (f(\mathbf{x}, \boldsymbol{\theta}), \boldsymbol{\theta}) = (\mathbf{y}, \boldsymbol{\theta})$$

F is differentiable at $(\mathbf{x}_0, \boldsymbol{\theta}_0)$ and $DF[\mathbf{x}_0, \boldsymbol{\theta}_0]$ is nonsingular (exercise 4.64). By the inverse function theorem (theorem 4.4), F is invertible on a neighborhood U of $(\mathbf{x}_0, \boldsymbol{\theta}_0)$ with $F^{-1}: V \to X \times \Omega$ being C^n on $V = F(U)$ and $(\mathbf{x}, \boldsymbol{\theta}) = F^{-1}(\mathbf{y}, \boldsymbol{\theta})$ for every $(\mathbf{y}, \boldsymbol{\theta}) \in V$. The required function g is the restriction of F^{-1} to $\mathbf{y} = \mathbf{0}$.

Specifically, let $\Omega = \{\boldsymbol{\theta} : (\mathbf{0}, \boldsymbol{\theta}) \in V\}$ and define $g: \Omega \to X$ by $g(\boldsymbol{\theta}) = \mathbf{x}$ where $\mathbf{x}$ is the first component of $(\mathbf{x}, \boldsymbol{\theta}) = F^{-1}(\mathbf{0}, \boldsymbol{\theta})$. Then

$$\mathbf{x}_0 = g(\boldsymbol{\theta}_0) \quad \text{and} \quad f(g(\boldsymbol{\theta}), \boldsymbol{\theta}) = \mathbf{0} \quad \text{for every } \boldsymbol{\theta} \in \Omega$$

Since F^{-1} is C^n, so is its restriction g. Furthermore, since $f(g(\boldsymbol{\theta}), \boldsymbol{\theta}) = \mathbf{0}$ is constant for all $\boldsymbol{\theta} \in \Omega$, $D_{\boldsymbol{\theta}} f(g(\boldsymbol{\theta}), \boldsymbol{\theta}) = \mathbf{0}$ by exercise 4.7. Applying the chain rule (exercise 4.22), we have

$$D_{\mathbf{x}} f[\mathbf{x}_0, \boldsymbol{\theta}_0] \circ D_{\boldsymbol{\theta}} g[\boldsymbol{\theta}_0] + D_{\boldsymbol{\theta}} f[\mathbf{x}_0, \boldsymbol{\theta}_0] = \mathbf{0}$$

or

$$D_{\boldsymbol{\theta}} g(\boldsymbol{\theta}_0) = -\bigl(D_{\mathbf{x}} f[\mathbf{x}_0, \boldsymbol{\theta}_0]\bigr)^{-1} \circ D_{\boldsymbol{\theta}} f[\mathbf{x}_0, \boldsymbol{\theta}_0] \qquad \square$$

Exercise 4.64
Show that F is C^n with $DF[\mathbf{x}_0, \boldsymbol{\theta}_0]$ is nonsingular.

Example 4.36 (Slope of the IS curve) The IS curve is implicitly defined by the equation

$$f(r, y; G, T, M) = y - C(y, T) - I(r) - G = 0$$

Assume that the consumption function $C(y, T)$ and the investment function $I(r)$ are smooth and that $D_y f = 1 - D_y C \neq 0$. Then, in the neighborhood of any point (r_0, y_0) on the IS curve, there exists a function $r = g(y)$ that can be considered the equation of the IS curve. The slope of the IS curve is given by the derivative $D_y g$, which by the implicit function theorem is given by

$$D_y g = -\frac{D_y f}{D_r f} = -\frac{1 - D_y C}{-D_r I}$$

Economic considerations imply that both the numerator $(1 - D_y C)$ and the denominator $(-D_r I)$ are positive so that the fraction, preceded by a negative sign, is negative. Under normal circumstances the IS curve is negatively sloped.

Exercise 4.65
Under what circumstances, if any, could the IS curve be horizontal?

Exercise 4.66
Determine the slope of the LM curve. Under what conditions would the LM curve be vertical?

Example 4.37 (Indifference curves and the MRS) Suppose that a consumer's preferences $\succsim$ can be represented (example 2.58) by the utility function $u \colon X \to \Re$, where $X \subset \Re^n$ is the consumption set. The contours of u,

$$u^{-1}(c) = \{ \mathbf{x} \in X : u(\mathbf{x}) = c \}$$

represent the indifference classes of $\succsim$.

Choose any consumption bundle $\mathbf{x}_0$ at which the marginal utility of good n is positive, that is, $D_{x_n} u[\mathbf{x}_0] \neq 0$. Let $c = u(\mathbf{x}_0)$. Applying the implicit function theorem, there exists a neighborhood $S \subset X$ of $\mathbf{x}_0$ and a function φ such that

$$x_n = \varphi(x_1, x_2, \ldots, x_{n-1}) \quad \text{and} \quad u(x_1, x_2, \ldots, x_{n-1}, \varphi(x_1, x_2, \ldots, x_{n-1})) = c$$

for all $\mathbf{x} \in S$. φ defines an *indifference surface*. Furthermore φ is C^r with

$$D_{x_i}\varphi(x_1, x_2, \ldots, x_{n-1}) = \frac{D_{x_i}u[\mathbf{x}]}{D_{x_n}u[\mathbf{x}]}$$

which is the marginal rate of substitution between i and n (example 4.10).

In the familiar case with two goods, the graph of the function $x_2 = \varphi(x_1)$ is called an *indifference curve*, the slope of which is given by

$$\varphi'(x_1) = \frac{D_{x_1}u[\mathbf{x}]}{D_{x_2}u[\mathbf{x}]}$$

the marginal rate of substitution between the two goods.

We will use the implicit function theorem in chapter 5 to derive the Lagrange multiplier rule for optimization. We will use it again in chapter 6 to derive comparative statics for economic models.

4.6 Convex and Homogeneous Functions

In this section we explore some special classes of smooth functions, such as convex, quasiconcave, and homogeneous functions.

4.6.1 Convex Functions

We saw in the previous chapter that convexity places restrictions on the behavior of a function. In particular, we saw that convex functions are generally continuous. Similarly convex functions are generally smooth functions (see remark 4.14). Moreover the derivative of a convex function bounds the function from below so that

$$f(\mathbf{x}) \geq f(\mathbf{x}_0) + Df[\mathbf{x}_0](\mathbf{x} - \mathbf{x}_0) \tag{29}$$

for every $\mathbf{x}, \mathbf{x}_0 \in S$. Similarly, if f is concave,

$$f(\mathbf{x}) \leq f(\mathbf{x}_0) + Df[\mathbf{x}_0](\mathbf{x} - \mathbf{x}_0) \tag{30}$$

for every $\mathbf{x}, \mathbf{x}_0$ in S. These bounds are extremely useful in computations with convex and concave functions. In fact the converse is also true, providing a valuable characterization of convex functions.

Exercise 4.67
A differentiable functional f on an open, convex set S is convex if and only if

$$f(\mathbf{x}) \geq f(\mathbf{x}_0) + Df[\mathbf{x}_0](\mathbf{x} - \mathbf{x}_0)$$

for every $\mathbf{x}, \mathbf{x}_0 \in S$. f is strictly convex if and only if

$$f(\mathbf{x}) > f(\mathbf{x}_0) + Df[\mathbf{x}_0](\mathbf{x} - \mathbf{x}_0)$$

for every $\mathbf{x} \neq \mathbf{x}_0 \in S$.

Geometrically exercise 4.67 implies that a differentiable function is convex if and only if it lies above its tangent hyperplane (example 4.3) at every point in the domain.

Remark 4.14 (Differentiability of convex functions) Equation (29) shows that the derivative of a convex function is a subgradient. Exercise 3.181 showed that a convex function has a subgradient at every interior point $\mathbf{x}_0$. If the function is differentiable at $\mathbf{x}_0$, then this subgradient is unique.

The converse holds for Euclidean space. That is, if S is finite dimensional, and the convex function $f \in F(S)$ has a unique subgradient at $\mathbf{x}_0 \in \text{int } S$, then f is differentiable at $\mathbf{x}_0$ (Roberts and Varberg 1973, p. 115). It can be shown that the subgradient is unique for almost all $\mathbf{x}_0 \in \text{int } S$, and therefore that f is differentiable almost everywhere. Furthermore, if a convex function is differentiable on $S \subseteq \Re^n$, then Df is continuous on S, that is $f \in C^1[S]$ (Roberts and Varberg 1973, pp. 110-11).

The uniqueness of the subgradient of a differentiable function (remark 4.14) implies the following envelope theorem for convex functions in Euclidean space. If a convex function is bounded by a differentiable convex function, which it intersects at some point, then the former function must be differentiable at the point of intersection.

Exercise 4.68
Suppose that f and h are convex functionals on a convex set S in Euclidean space with f differentiable at $\mathbf{x}_0$ and

$$f(\mathbf{x}_0) = h(\mathbf{x}_0) \quad \text{and} \quad f(\mathbf{x}) \geq h(\mathbf{x}) \quad \text{for every } \mathbf{x} \in S \tag{31}$$

Then h is differentiable at $\mathbf{x}_0$, with $Dh[\mathbf{x}_0] = Df[\mathbf{x}_0]$.

When $X \subseteq \Re^n$, the derivative can be represented by the gradient, which provides the following useful alternative characterization of a convex function.

Exercise 4.69
A differentiable function f on a convex, open set S in $\Re^n$ is convex if and only if

$$(\nabla f(\mathbf{x}) - \nabla f(\mathbf{x}_0))^T (\mathbf{x} - \mathbf{x}_0) \geq 0 \tag{32}$$

for every $\mathbf{x}_0, \mathbf{x} \in S$. f is strictly convex if and only if

$$(\nabla f(\mathbf{x}) - \nabla f(\mathbf{x}_0))^T (\mathbf{x} - \mathbf{x}_0) > 0$$

for every $\mathbf{x} \neq \mathbf{x}_0 \in S$.

In the special case of functions on $\Re$, this has the following useful form.

Exercise 4.70
A differentiable function f on an open interval $S \subseteq \Re$ is (strictly) convex if and only if f' is (strictly) increasing.

Combined with exercises 4.35 and 4.36, this means that convex and concave functions on $\Re$ can be identified through their second derivatives.

Exercise 4.71
Let f be a twice differentiable function on an open interval $S \subseteq \Re$. Then

$$f \text{ is } \begin{Bmatrix} \text{convex} \\ \text{concave} \end{Bmatrix} \text{ if and only if } f''(x) \text{ is } \begin{Bmatrix} \geq 0 \\ \leq 0 \end{Bmatrix} \text{ for every } x \in S$$

$$f \text{ is strictly } \begin{Bmatrix} \text{convex} \\ \text{concave} \end{Bmatrix} \text{ if } f''(x) \text{ is } \begin{Bmatrix} > 0 \\ < 0 \end{Bmatrix} \text{ for every } x \in S$$

Example 4.38 (Power function) Earlier (example 3.120), we showed that the power function

$$f(x) = x^n, \quad n = 1, 2, \ldots$$

is convex on $\Re_+$. Now, we extend this characterization to the general power function $f: \Re_+ \to \Re$ (example 2.56) defined by

$$f(x) = x^a, \quad a \in \Re$$

with

$$f'(x) = ax^{a-1} \quad \text{and} \quad f''(x) = a(a-1)x^{a-2}$$

We observe that for every $x > 0$

$$f''(x) = \begin{cases} = 0 & \text{if } a = 0, 1 \\ < 0 & \text{if } 0 < a < 1 \\ > 0 & \text{otherwise} \end{cases}$$

We conclude that the general power function is strictly concave if $0 < a < 1$, strictly convex if $a < 0$ or $a > 1$ and convex when $a = 0$ or $a = 1$ (see figure 3.11).

Note that this characterization is limited to the nonnegative domain $\Re_+$. For example, x^3 is neither convex nor concave on $\Re$.

Exercise 4.72
What can we say about the concavity/convexity of the simple power functions $f(x) = x^n$, $n = 1, 2, \ldots$ over $\Re$.

Example 4.39 (Exponential function) The first and second derivatives of the exponential function (example 3.49) $f(x) = e^x$ are

$$f'(x) = e^x \quad \text{and} \quad f''(x) = e^x > 0 \quad \text{for every } x \in \Re$$

by exercise 2.6. Therefore the exponential function is strictly convex on $\Re$.

Example 4.40 (Log function) The first and second derivatives of the log function $f(x) = \log(x)$ are

$$f'(x) = \frac{1}{x} = x^{-1} \quad \text{and} \quad f''(x) = -\frac{1}{x^2} < 0 \quad \text{for every } x > 0$$

which implies that f is strictly concave on $\Re_{++}$.

Proposition 4.71 generalizes this characterization to $\Re^n$, providing an important link between this and the previous chapter.

Proposition 4.1 *Let f be twice differentiable on an open convex S in $\Re^n$, and let $H_f(\mathbf{x})$ denote the Hessian of f at $\mathbf{x}$. Then*

$$f \text{ is locally } \begin{Bmatrix} convex \\ concave \end{Bmatrix} \text{ at } \mathbf{x} \text{ if and only if } H_f(\mathbf{x}) \text{ is } \begin{Bmatrix} nonnegative \\ nonpositive \end{Bmatrix} \text{ definite}$$

Furthermore

$$f \text{ is strictly locally } \begin{Bmatrix} convex \\ concave \end{Bmatrix} \text{ at } \mathbf{x} \text{ if } H_f(\mathbf{x}) \text{ is } \begin{Bmatrix} positive \\ negative \end{Bmatrix} \text{ definite.}$$

4.6 Convex and Homogeneous Functions

Proof Let **x** be a point at which $H_f(\mathbf{x})$ is nonnegative definite. Since $f \in C^2$, there exists a convex neighborhood S of **x** such that $H_f(\bar{\mathbf{x}})$ is nonnegative definite for every $\bar{\mathbf{x}} \in S$. By Taylor's theorem (example 4.33), for every $\mathbf{x}_0, \mathbf{x}_1 \in S$ there exists $\bar{\mathbf{x}} \in S$ such that

$$f(\mathbf{x}_1) = f(\mathbf{x}_0) + \Delta f(\mathbf{x}_0)^T(\mathbf{x}_1 - \mathbf{x}_0) + (\mathbf{x}_1 - x_0)^T H_f(\bar{\mathbf{x}})(\mathbf{x}_1 - \mathbf{x}_0)$$

Since $H_f(\bar{\mathbf{x}})$ is nonnegative definite, $(\mathbf{x}_1 - x_0)^T H_f(\bar{\mathbf{x}})(\mathbf{x}_1 - \mathbf{x}_0) \geq 0$, and therefore

$$f(\mathbf{x}_1) \geq f(\mathbf{x}_0) + \nabla f(\mathbf{x}_0)^T(\mathbf{x}_1 - \mathbf{x}_0)$$

So f is convex on S, a convex neighborhood of **x** (exercise 4.67). That is, f is locally convex at **x**. If $H_f(\mathbf{x}_0)$ is positive definite, then the inequality is strict

$$f(\mathbf{x}_0 + \mathbf{x}) > f(\mathbf{x}_0) + \nabla f(\mathbf{x}_0)^T(\mathbf{x}_1 - \mathbf{x}_0)$$

and f is strictly locally convex. Similarly, if $H_f(\mathbf{x}_0)$ is nonpositive definite,

$$f(\mathbf{x}_0 + \mathbf{x}) \leq f(\mathbf{x}_0) + \nabla f(\mathbf{x}_0)^T(\mathbf{x}_1 - \mathbf{x}_0)$$

and therefore

$$-f(\mathbf{x}_0 + \mathbf{x}) \geq -f(\mathbf{x}_0) + \nabla(-f)(\mathbf{x}_0)^T(\mathbf{x}_1 - \mathbf{x}_0)$$

$-f$ is locally convex and so f is locally concave (exercise 3.124). If H is negative definite, the inequalities are strict.

Conversely, suppose that f is locally convex at $\mathbf{x}_0$. Then there exists a convex neighborhood S of $\mathbf{x}_0$ such that

$$f(\mathbf{x}) \geq f(\mathbf{x}_0) + \nabla f(\mathbf{x}_0)^T(\mathbf{x} - \mathbf{x}_0)$$

for every $\mathbf{x} \in S$ (exercise 4.67). Since S is open, for every $\mathbf{x} \in \Re^n$, there exists $t > 0$ such that $\mathbf{x}_0 + t\mathbf{x} \in S$, and therefore

$$f(\mathbf{x}_0 + t\mathbf{x}) \geq f(\mathbf{x}_0) + \nabla f(\mathbf{x}_0)^T t\mathbf{x} \tag{33}$$

By Taylor's theorem (example 4.33),

$$f(\mathbf{x}_0 + t\mathbf{x}) = f(\mathbf{x}_0) + \nabla f(\mathbf{x}_0)^T t\mathbf{x} + (t\mathbf{x})^T H(\mathbf{x}_0) t\mathbf{x} + \eta(t\mathbf{x})\|t\mathbf{x}\|^2 \tag{34}$$

with $\eta(t\mathbf{x}) \to 0$ as $t\mathbf{x} \to \mathbf{0}$. Together, (33) and (34) imply that

$$(t\mathbf{x})^T H(\mathbf{x}_0) t\mathbf{x} + \eta(t\mathbf{x})\|t\mathbf{x}\|^2 = t^2 \mathbf{x}^T H(\mathbf{x}_0)\mathbf{x} + t^2 \eta(t\mathbf{x})\|\mathbf{x}\|^2 \geq 0$$

Dividing by $t^2 > 0$,

$$\mathbf{x}^T H(\mathbf{x}_0)\mathbf{x} + \eta(t\mathbf{x})\|\mathbf{x}\|^2 \geq 0$$

Since $\eta(t\mathbf{x}) \to 0$ as $t \to 0$, we conclude that $\mathbf{x}^T H(\mathbf{x}_0)\mathbf{x} \geq 0$ for every $\mathbf{x}$. H is nonnegative definite. The proof for f concave is analogous with the inequalities reversed. □

Corollary 4.1 *Let f be twice differentiable on an open convex S in $\Re^n$. Then f is convex on S if and only if $H_f(\mathbf{x})$ is nonnegative definite for every $\mathbf{x} \in S$. f is concave on S if and only if $H_f(\mathbf{x})$ is nonpositive definite for every $\mathbf{x} \in S$.*

Proof Apply exercise 3.142. □

The following analogous characterization is very useful in identifying supermodular functions (section 2.2.2). A function is supermodular if and only if its Hessian has nonnegative off-diagonal elements.

Proposition 4.2 *Let f be a twice differentiable functional on an open convex lattice S in $\Re^n$. Then f is supermodular if and only if $D_{ij}^2 f[\mathbf{x}] \geq 0$ for every $i \neq j$ and $\mathbf{x} \in S$.*

Proof By exercise 2.59, f is supermodular if and only if

$$f(x_i + r, x_j + t; \mathbf{x}_{-ij}) - f(x_i + t, x_j; \mathbf{x}_{-ij}) \geq f(x_i, x_j + t; \mathbf{x}_{-ij}) - f(x_i, x_j; \mathbf{x}_{-ij}) \tag{35}$$

for every $r > 0$ and $t > 0$. Dividing by $t > 0$,

$$\frac{f(x_i + r, x_j + t; \mathbf{x}_{-ij}) - f(x_i + r, x_j; \mathbf{x}_{-ij})}{t} \geq \frac{f(x_i, x_j + t; \mathbf{x}_{-ij}) - f(x_i, x_j; \mathbf{x}_{-ij})}{t}$$

Therefore

$$D_{x_j} f[x_i + r, x_j; \mathbf{x}_{-ij}] = \lim_{t \to 0} \frac{f(x_i + r, x_j + t; \mathbf{x}_{-ij}) - f(x_i + r, x_j; \mathbf{x}_{-ij})}{t}$$

$$\geq \lim_{t \to 0} \frac{f(x_i, x_j + t; \mathbf{x}_{-ij}) - f(x_i, x_j; \mathbf{x}_{-ij})}{t}$$

$$= D_{x_j} f[x_i, x_j; \mathbf{x}_{-ij}]$$

that is,

$$D_{x_j}f[x_i + r, x_j; \mathbf{x}_{-ij}] - D_j f[x_i, x_j; \mathbf{x}_{-ij}] \geq 0$$

Dividing now by $r > 0$,

$$D_{ij}f[\mathbf{x}] = \lim_{r \to 0} \frac{D_j f[x_i + r, x_j; \mathbf{x}_{-ij}] - D_j f[x_i, x_j; \mathbf{x}_{-ij}]}{r} \geq 0$$

Conversely, if $D^2_{ij}f[\mathbf{x}] < 0$ for some i, j and $\mathbf{x}$, there exists some $r > 0$ and $t > 0$ violating (35). □

Example 4.41 Generalizing example 4.29, the off-diagonal elements of the Hessian of the Cobb-Douglas function

$$f(\mathbf{x}) = x_1^{a_1} x_2^{a_2} \ldots x_n^{a_n}, \qquad a_i > 0$$

are

$$D_{ij} = \frac{a_i a_j}{x_i x_j} f(\mathbf{x}) \geq 0 \qquad \text{for every } \mathbf{x} \in \Re^n_+$$

Therefore (proposition 4.2), the Cobb-Douglas function is supermodular on $\Re^n_+$, echoing our conclusion in example 2.68.

Quasiconcave Functions

A differentiable function is convex if and only if it lies above its tangent hyperplane (exercise 4.67), which forms a supporting hyperplane to the epigraph of a convex function. Similarly a differentiable function is quasiconcave if and only if its upper contour sets lie above the tangent to the contour. In other words, the derivative supports the upper contour set. This is formalized in the following exercise.

Exercise 4.73
A differentiable functional f on an open set $S \subseteq \Re^n$ is quasiconcave if and only if

$$f(\mathbf{x}) \geq f(\mathbf{x}_0) \Rightarrow \nabla f(\mathbf{x}_0)^T (\mathbf{x} - \mathbf{x}_0) \geq \mathbf{0} \qquad \text{for every } \mathbf{x}, \mathbf{x}_0 \text{ in } S \qquad (36)$$

Inequality (36) can be strengthened where f is regular.

Exercise 4.74
Suppose that a differentiable functional f on an open set $S \subseteq \Re^n$ is quasiconcave. At every regular point $\nabla f(\mathbf{x}_0) \neq \mathbf{0}$,

$$f(\mathbf{x}) > f(\mathbf{x}_0) \Rightarrow \nabla f(\mathbf{x}_0)^T (\mathbf{x} - \mathbf{x}_0) > 0 \qquad \text{for every } \mathbf{x}, \mathbf{x}_0 \text{ in } S \qquad (37)$$

A restricted form of quasiconcavity is useful in optimization (see section 5.4.3). A function is quasiconcave if it satisfies (37) at regular points of f. It is pseudoconcave if it satisfies (37) at all points of its domain. That is, a differentiable functional on an open convex set $S \in \Re^n$ is *pseudoconcave* if

$$f(\mathbf{x}) > f(\mathbf{x}_0) \Rightarrow \nabla f(\mathbf{x}_0)^T (\mathbf{x} - \mathbf{x}_0) > 0 \qquad \text{for every } \mathbf{x}, \mathbf{x}_0 \in S \qquad (38)$$

A function is pseudoconvex if $-f$ is pseudoconcave. Pseudoconcave functions have two advantages over quasiconcave functions—every local optimum is a global optimum, and there is an easier second derivative test for pseudoconcave functions. Nearly all quasiconcave functions that we encounter are in fact pseudoconcave.

Exercise 4.75 (Pseudoconvex functions)
A differentiable function $f : S \to \Re$ is pseudoconvex if

$$f(\mathbf{x}) < f(\mathbf{x}_0) \Rightarrow \nabla f(\mathbf{x}_0)^T (\mathbf{x} - \mathbf{x}_0) < 0 \qquad \text{for every } \mathbf{x}, \mathbf{x}_0 \text{ in } S$$

Exercise 4.76 (Pseudoconcave functions)
Show that

1. Every differentiable concave function is pseudoconcave.

2. Every pseudoconcave function is quasiconcave

3. Every regular quasiconcave function is pseudoconcave.

Example 4.42 The function $f(x) = x^3$ is quasiconcave on $\Re$. It is not pseudoconcave, since

$$f(1) > f(0) \quad \text{but} \quad \nabla f(0)(1-0) = 0$$

Example 4.43 The Cobb-Douglas function

$$f(\mathbf{x}) = x_1^{a_1} x_2^{a_2} \ldots x_n^{a_n}, \qquad a_i > 0$$

is pseudoconcave on $\Re_{++}^n$, since it is quasiconcave (example 3.59) and regular (example 4.16).

Exercise 4.77
Is the CES function

$$f(\mathbf{x}) = (\alpha_1 x_1^\rho + \alpha_2 x_2^\rho + \cdots \alpha_n x_n^\rho)^{1/\rho}, \qquad \alpha_i > 0, \rho \neq 0$$

pseudoconcave?

4.6.2 Homogeneous Functions

Recall that homogeneous functions behave like power functions along any ray through the origin (section 3.8). This structure is inherited by the derivatives of differentiable homogeneous functions.

Exercise 4.78
If a differentiable functional f is homogeneous of degree k, its partial derivatives are homogeneous of degree of $k - 1$.

Example 4.44 (Slope of a contour) If f is a differentiable functional on $S \subseteq \Re^2$, the slope of the contour at $\mathbf{x}_0$ (example 4.11) is given by

$$\text{slope}(\mathbf{x}_0) = -\frac{D_{x_1} f[\mathbf{x}_0]}{D_{x_2} f[\mathbf{x}_0]}$$

If f is homogeneous of degree k, its partial derivatives are homogeneous of degree $k - 1$, and

$$\text{slope}(t\mathbf{x}_0) = -\frac{D_{x_1} f(t\mathbf{x}_0)}{D_{x_2} f(t\mathbf{x}_0)} = -\frac{t^{k-1} D_{x_1} f[\mathbf{x}_0]}{t^{k-1} D_{x_2} f[\mathbf{x}_0]} = -\frac{D_{x_1} f[\mathbf{x}_0]}{D_{x_2} f[\mathbf{x}_0]} = \text{slope}(\mathbf{x}_0)$$

This means that the slope of the contours is constant along any ray through the origin.

The most important property of differentiable homogeneous functions is that the directional derivative of a homogeneous function along any ray through the origin is proportional to the value of the function, where the constant of proportionality is equal to the degree of homogeneity.

Exercise 4.79 (Directional derivative)
If f is homogeneous of degree k

$$\vec{D}_\mathbf{x} f(\mathbf{x}) = kf(\mathbf{x}) \tag{39}$$

[Hint: Use L'Hôpital's rule (exercise 4.47).]

This property is usually expressed in economics in an alternative form known as Euler's theorem. In fact this property is a necessary and sufficient condition for a differentiable function to be homogeneous.

Proposition 4.3 (Euler's theorem) *A differentiable functional $f: S \to \Re$ is homogeneous of degree k if and only if*

$$Df[\mathbf{x}](\mathbf{x}) = \sum_{i=1}^{n} D_{x_i} f[\mathbf{x}] x_i = kf(\mathbf{x}) \quad \text{for every } \mathbf{x} \in S \tag{40}$$

Proof Equation (40) can be derived from (39) by expressing the directional derivative in terms of partial derivatives (exercise 4.10). As an alternative, we derive (40) directly. If f is homogeneous of degree k,

$$f(t\mathbf{x}) = t^k f(\mathbf{x}) \quad \text{for every } \mathbf{x} \text{ and } t > 0$$

Differentiating with respect to t using the chain rule (exercise 4.22), we have

$$D_t f[t\mathbf{x}]\mathbf{x} = kt^{k-1} f(\mathbf{x})$$

Evaluating at $t = 1$ gives the desired result. The converse is exercise 4.80. □

Exercise 4.80
If f satisfies (40) for all $\mathbf{x}$, it is homogeneous of degree k. [Hint: Use exercise 4.38.]

Euler's theorem can be thought of as a multidimensional extension of the rule for the derivative of a power function (example 4.15). Two special cases are worth considering. When $n = 1$, equation (40) becomes

$$kf(x) = f'(x)x \Rightarrow f'(x) = k\frac{f(x)}{x}$$

which is precisely the rule for a power function (example 4.15). When $k = 1$, equation (40) becomes

$$Df[\mathbf{x}](\mathbf{x}) = \sum_{i=1}^{n} D_{x_i} f[\mathbf{x}] x_i = f(\mathbf{x}) \quad \text{for every } \mathbf{x} \in S$$

In this case the derivative is exact (rather than an approximation) along any ray through the origin. We give a sample of applications of Euler's theorem in economics.

Example 4.45 (Production with constant returns to scale) Suppose that a firm produces a single product using a constant returns to scale technology, so that its production function is linearly homogeneous. Recall that the partial derivatives of the production function are the marginal products of the respective inputs (example 4.5). By Euler's theorem, total output y is determined by the sum of the quantity of each input employed times its marginal product, that is,

$$y = f(\mathbf{x}) = \sum_{i=1}^{n} \mathrm{MP}_i x_i$$

Note that this is true for any production plan $(y, \mathbf{x})$.

We will show in the next chapter that if the firm is in a competitive market, profit is maximized if each input is paid the value of its marginal product, $p\mathrm{MP}_i$ where p is the price of the output. In this case Euler's theorem requires that factor payments exhaust the total revenue py,

$$py = \sum_{i=1}^{n} p\mathrm{MP}_i x_i$$

so that maximum profit will be precisely zero. This is a general characteristic of competitive firms with constant returns to scale.

Example 4.46 (Wicksell's law) Again, suppose that a firm produces a single product using a constant returns to scale technology f. Then the marginal products are homogeneous of degree 0 (exercise 4.78) and satisfy Euler's theorem

$$\sum_{i=1}^{n} D_{x_i} \mathrm{MP}_j x_i = \sum_{i=1}^{n} D_{x_i x_j}^2 f[\mathbf{x}] x_i = 0 \times \mathrm{MP}_j = 0$$

which implies that

$$D_{x_i x_i}^2 f(\mathbf{x}) = -\sum_{j \neq i} \frac{x_j}{x_j} D_{x_i x_j}^2 f(\mathbf{x})$$

With only two inputs, this reduces to

$$D_{x_i x_i}^2 f(\mathbf{x}) = -\frac{x_i}{x_j} D_{x_i x_j}^2 f(\mathbf{x})$$

The term on the left measures the way in which the marginal product of input i varies with the quantity of i; the term on the right measures the way in which the marginal product of i varies with quantity of j.

Wicksell's law holds that these two effects are equal in magnitude and opposite in sign.

Exercise 4.81 (Complementary inputs)
Two inputs are said to *complementary* if their cross-partial derivative $D^2_{x_i x_j} f(\mathbf{x})$ is positive, since this means that increasing the quantity of one input increases the marginal productivity of the other. Show that if a production function of two inputs is linearly homogeneous and quasi-concave, then the inputs are complementary. [Hint: Use proposition 4.1.]

Exercise 4.82 (Elasticity of scale)
In a generalization of the notion of elasticity of univariate functions (example 4.41), the elasticity the *elasticity of scale* of a functional f is defined by

$$E(\mathbf{x}) = \frac{t}{f(t\mathbf{x})} D_t f(t\mathbf{x}) \Big|_{t=1}$$

where the symbol $|_{t=1}$ means that the expression is evaluated at $t = 1$. In general, the elasticity of scale varies with $\mathbf{x}$. Show that the elasticity of scale is constant k if and only if f is homogeneous of degree k.

Exercise 4.83 (Regularity)
A differentiable functional f homogeneous of degree $k \neq 0$ is regular wherever $f(\mathbf{x}) \neq 0$.

Exercise 4.84
If f is C^2 and homogeneous of degree k with Hessian H, then

$$\mathbf{x}^T H_f(\mathbf{x}) \mathbf{x} = k(k-1) f(\mathbf{x})$$

Note that this equation holds only at the point $\mathbf{x}$ at which the Hessian is evaluated.

Example 4.47 (Homogeneity conditions) If the consumer's preferences are strictly convex, the optimal solution to the consumer's problem is a set of demand functions $x_i(\mathbf{p}, m)$, which are homogeneous of degree zero (example 3.62). Applying Euler's theorem to the demand for good i, we get

$$\sum_{i=1}^{n} p_j D_{p_j} x_i(p, m) + m D_m x_i(p, m) = 0$$

Dividing by $x_i(p,m)$ gives

$$\sum_{i=1}^{n} \frac{p_j}{x_i} D_{p_j} x_i(p,m) + \frac{m}{x_i} D_m x_i(p,m) = 0$$

which can be expressed as

$$\sum_{i=1}^{n} \varepsilon_{ij} + \eta_i = 0 \qquad (41)$$

where

$$\varepsilon_{ij}(\mathbf{p},m) = \frac{p_j}{x_i(\mathbf{p},m)} D_{p_j} x_i[\mathbf{p},m] \quad \text{and} \quad \eta_i(\mathbf{p},m) = \frac{m}{x_i(\mathbf{p},m)} D_m x_i[\mathbf{p},m]$$

are the price and income elasticities of demand respectively. Zero homogeneity of demand implies that the consumer's price and income elasticities for any commodity are related by (41).

Homothetic Functions

Recall that a function is homothetic if its contours are radial expansions of one another. Formally

$$f(\mathbf{x}_1) = f(\mathbf{x}_2) \Rightarrow f(t\mathbf{x}_1) = f(t\mathbf{x}_2) \qquad \text{for every } \mathbf{x}_1, \mathbf{x}_2 \in S \text{ and } t > 0$$

This implies that the slopes of the contours are constant along a ray. Exercise 4.44 demonstrated this for homogeneous functions. In the following exercise we extend this to homothetic functions.

Exercise 4.85
If $f: S \to \Re$, $S \subseteq \Re^n$ is strictly increasing, differentiable and homothetic, then for every i, j,

$$\frac{D_{x_i} f(t\mathbf{x})}{D_{x_j} f(t\mathbf{x})} = \frac{D_{x_i} f(\mathbf{x})}{D_{x_j} f(\mathbf{x})} \qquad \text{for every } \mathbf{x} \in S \text{ and } t > 0$$

Example 4.48 (Homothetic preferences) If consumer preferences are homothetic (example 3.172), then any utility function u representing the preferences is homothetic. Exercise 4.85 implies that the marginal rate of substitution (example 4.10) is constant along any ray through the origin. It depends only on the relative proportions of different goods, not on the absolute quantities consumed.

4.7 Notes

Excellent expositions of the top-down approach (remark 4.8) to smooth functions are given by Dieudonne (1960) and Spivak (1965). Chillingworth (1976) is less rigorous but very insightful. Lang (1969) is elegant but more difficult. Smith (1983) and Robert (1989) are also worthwhile consulting. Spivak (1965, pp. 44–45) discusses the variety of notation for partial derivatives. Exercise 4.9 is from Robert (1989).

The bottom-up approach is followed by Simon and Blume (1994) and Sydsaeter and Hammond (1995). Leading mathematical texts covering this material include Apostol (1974), Bartle (1976), and Rudin (1976). Spivak (1980) is an excellent source for elementary (single variable) calculus. Although old, Allen (1938) is an excellent source for economic applications.

Precise usage of the adjective "smooth" varies, with some authors restricting the class of smooth functions to C^∞ rather than C^n. Solow's convention (remark 4.9) is cited in Varian (1992, p. 487). The discussion of functional forms is based on Fuss et al. (1978). Our derivation of the inverse function theorem follows Bartle (1976) and Lang (1969). Global univalence is discussed by Gale and Nikaido (1965) and Nikaido (1968).

Differentiable convex functions are discussed by Bazaraa, Sherali, and Shetty (1993), Madden (1986), Roberts and Varberg (1973), and Rockafellar (1976). Madden (1986) also discusses homogeneous functions, as does Allen (1938).

5 Optimization

The fundamental economic hypothesis is that human beings behave as rational and self interested agents in the pursuit of their objectives, and that aggregate resources are limited. Economic behavior is therefore modeled as the solution of a constrained optimization problem.
—Bill Sharkey, 1995

5.1 Introduction

In earlier chapters we used optimization models as illustrations. Now optimization takes center stage and we utilize the tools we have developed to explore optimization models in some depth. The general constrained optimization problem was posed in example 2.30 as

$$\max_{\mathbf{x} \in G(\boldsymbol{\theta})} f(\mathbf{x}, \boldsymbol{\theta}) \tag{1}$$

The maximand $f(\mathbf{x}, \boldsymbol{\theta})$ is known as the *objective* function (example 2.26). Its value depends both upon the *choice* or *decision variables* $\mathbf{x}$ and the exogenous *parameters* $\boldsymbol{\theta}$. The choice of $\mathbf{x}$ is constrained to a *feasible set* $G(\boldsymbol{\theta}) \subseteq X$ that also depends on the parameters $\boldsymbol{\theta}$. Recall that (1) extends to minimization problems, since

$$\min_{\mathbf{x} \in G(\boldsymbol{\theta})} f(\mathbf{x}, \boldsymbol{\theta}) = \max_{\mathbf{x} \in G(\boldsymbol{\theta})} -f(\mathbf{x}, \boldsymbol{\theta})$$

Typically the feasible set $G(\boldsymbol{\theta})$ can be represented by a function $g \colon X \times \Theta \to Y$

$$G(\boldsymbol{\theta}) = \{\mathbf{x} \in X : \mathbf{g}(\mathbf{x}, \boldsymbol{\theta}) \leq \mathbf{0}\}$$

so that the constrained optimization problem becomes (example 2.40)

$$\max_{\mathbf{x} \in X} f(\mathbf{x}, \boldsymbol{\theta}) \quad \text{subject to} \quad \mathbf{g}(\mathbf{x}, \boldsymbol{\theta}) \leq \mathbf{0}$$

If $Y \subseteq \Re^m$, the function $\mathbf{g}$ can be decomposed (example 2.40) into m separate constraints (functionals) $g_j \colon X \times \theta \to \Re$,

$$g_1(\mathbf{x}, \boldsymbol{\theta}) \leq 0, \quad g_2(\mathbf{x}, \boldsymbol{\theta}) \leq 0, \quad \ldots, \quad g_m(\mathbf{x}, \boldsymbol{\theta}) \leq 0$$

Throughout this chapter we will assume that all spaces are finite dimensional, with $X \subseteq \Re^n$ and $Y \subseteq \Re^m$.

In this formulation the constraints on choice are of two types. The *functional constraints*, $g_j(\mathbf{x}, \boldsymbol{\theta}) \leq 0$, express an explicit functional relationship between the choice variables and the parameters. They can be used to derive explicit conditions to characterize the optimal choice, and to analyze the sensitivity of the optimal choice to changes in parameters. Feasible choices are also constrained to belong to the less-specified set X. The choice of whether to represent a particular constraint explicitly in the form $\mathbf{g}(\mathbf{x}, \theta) \leq \mathbf{0}$ or implicitly as in $\mathbf{x} \in X$ is largely an analytical choice depending on the purpose at hand. Sometimes it is productive to model a particular constraint explicitly. For other purposes it is more convenient to embody a constraint implicitly in the domain X. We illustrate with two important economic models.

Example 5.1 (The producer's problem) The simplest specification of the producer's problem is to choose a feasible production plan $\mathbf{y}$ in the production possibility set Y to maximize total profit

$$\max_{\mathbf{y} \in Y} \mathbf{p}^T \mathbf{y}$$

In this specification there are no explicit functional constraints.

In order to characterize the optimal solution, it is helpful to distinguish inputs and outputs. Where the firm produces a single output, we can represent the production possibility set by a production function $y = f(\mathbf{x})$ relating inputs $\mathbf{x}$ to output y (example 2.24). Letting p denote the price of the output and $\mathbf{w}$ the prices of the inputs, the producer's problem is to maximize profit $py - \mathbf{w}^T \mathbf{x}$ subject to the production constraint $y = f(\mathbf{x})$, that is,

$$\max_{\mathbf{x}} \ py - \mathbf{w}^T \mathbf{x}$$

subject to $\quad y = f(\mathbf{x})$

If the firm is competitive, the prices of inputs and outputs are taken as parameters. The producer's problem has a special structure in that the parameters p and $\mathbf{w}$ appear only in the objective function, not in the constraints.

Example 5.2 (The consumer's problem) The consumer's choice is constrained by her income m and the prices of goods and services $\mathbf{p}$. Representing her preferences by a utility function u (example 2.58), her problem is to choose that consumption bundle $\mathbf{x}$ in her budget set $X(\mathbf{p}, m)$ that maximizes her utility $u(\mathbf{x})$, that is,

$$\max_{\mathbf{x} \in X(\mathbf{p},m)} u(\mathbf{x})$$

The quantities of goods and services consumed $\mathbf{x}$ are the choice variables, while prices $\mathbf{p}$ and income m are parameters. Typically we isolate the effect of prices $\mathbf{p}$ and income m on feasible choice by specifying the budget set in terms of the budget constraint (example 1.113) so that the consumer's problem becomes

$$\max_{\mathbf{x} \in X} u(\mathbf{x})$$

subject to $\mathbf{p}^T \mathbf{x} \leq m$

The consumer's problem has a special structure in that the parameters $\mathbf{p}$ and m appear only in the constraint, not in the objective function.

Although physical and biological laws preclude negative consumption, this nonnegativity constraint is usually left implicit in the consumption set $X \subseteq \Re_+^n$. Sometimes a more complete analysis is required, in which case the nonnegativity requirement on consumption can be modeled by an explicit constraint $\mathbf{x} \geq \mathbf{0}$ (example 5.17).

A solution to the constrained optimization problem (1) is a feasible choice $\mathbf{x}^* \in G(\boldsymbol{\theta})$ at which the objective function $f(\mathbf{x}, \boldsymbol{\theta})$ attains a value at least as great as any other feasible choice, that is,

$$f(\mathbf{x}^*, \boldsymbol{\theta}) \geq f(\mathbf{x}, \boldsymbol{\theta}) \qquad \text{for every } \mathbf{x} \in G(\boldsymbol{\theta}) \tag{2}$$

We say that $\mathbf{x}^*$ maximizes $f(\mathbf{x}, \boldsymbol{\theta})$ over $G(\boldsymbol{\theta})$. Analysis of an economic model posed as a constrained maximization problem typically involves addressing four questions:

· Does the problem have a solution (*existence*)?

· What is the solution (*computation*)?

· How can we characterize the solution (*characterization*)?

· How does the solution vary with the parameters (*comparative statics* or sensitivity analysis)?

Sufficient conditions for existence of a solution were derived in chapter 2. Provided that f is continuous and $G(\boldsymbol{\theta})$ is compact, the Weierstrass theorem (theorem 2.2) guarantees the existence of a solution to a constrained optimization problem. It is usual practice in economics to construct the model in such a way as to ensure that these conditions are met and existence assured. When nonexistence is a problem, it can usually be traced to inadequacy in the specification of the model rather than an intrinsic feature of the problem (exercise 5.1).

Exercise 5.1
A small Pacific island holds the entire world stock K of a natural fertilizer. The market price p of the fertilizer varies inversely with the rate at which it is sold, that is,

$$p = p(x), \quad p'(x) < 0$$

where x denotes the number of tons of fertilizer mined and sold per year. Giving equal regard to present and future generations, the island government wishes to choose a rate of exploitation of their natural resource x which maximizes total revenue $f(x)$ from the resource where

$$f(x) = \begin{cases} 0, & x = 0 \\ Kp(x), & x > 0 \end{cases}$$

What is the optimal exploitation rate?

This chapter is devoted to answering the second and third questions, that is, the computation and characterization of optimal solutions. The fourth question, comparative statics, is the subject of chapter 6. Typically economic analysis is less concerned with computing optimal solutions to particular problems than with identifying and characterizing optimal solutions in terms of the ingredients of the model. This distinguishes economics from other disciplines focusing on constrained optimization, such as operations research, where a central concern is in the development and analysis of efficient computational procedures. The remainder of this section introduces additional vocabulary and summarizes some of the results from previous chapters.

A solution $\mathbf{x}^*$ to a constrained optimization problem satisfying (2) is known as a *global optimum*. A choice that is the best in its neighborhood is called a *local optimum*. Formally a choice $\mathbf{x}^* \in G(\boldsymbol{\theta})$ is a *local optimum* if there is a neighborhood S of $\mathbf{x}^*$ such that

$$f(\mathbf{x}^*, \boldsymbol{\theta}) \geq f(\mathbf{x}, \boldsymbol{\theta}) \quad \text{for every } \mathbf{x} \in S \cap G(\boldsymbol{\theta})$$

For example, each peak in a mountain range is a local maximum. Only the highest mountain is a global maximum. A global optimum is necessarily a local optimum, but not every local optimum is a global optimum.

Whether characterizing optima using differential calculus or computing optima with numerical techniques, our usual procedures at best identify only local optima. Since we usually seek global solutions to economic models, it is common to impose further conditions that ensure that any local optimum is also a global optimum (exercise 5.2).

Exercise 5.2 (Local-global)
In the constrained optimization problem

$$\max_{\mathbf{x} \in G(\boldsymbol{\theta})} f(\mathbf{x}, \boldsymbol{\theta})$$

suppose that f is concave and $G(\boldsymbol{\theta})$ convex. Then every local optimum is a global optimum.

Another distinction we need to note is that between strict and nonstrict optima. A point $\mathbf{x}^* \in G(\boldsymbol{\theta})$ is a *strict* local optimum if it is strictly better than all feasible points in a neighborhood S, that is,

$$f(\mathbf{x}^*, \boldsymbol{\theta}) > f(\mathbf{x}, \boldsymbol{\theta}) \quad \text{for every } \mathbf{x} \in S \cap G(\boldsymbol{\theta})$$

It is a *strict* global optimum if it is "simply the best," that is,

$$f(\mathbf{x}^*, \boldsymbol{\theta}) > f(\mathbf{x}, \boldsymbol{\theta}) \quad \text{for every } \mathbf{x} \in G(\boldsymbol{\theta})$$

If there are other feasible choices which are as good as $\mathbf{x}^*$, then $\mathbf{x}^*$ is a nonstrict or weak (local or global) optimum.

Quasiconcavity is not sufficient to ensure the equivalence between local and global optima (see exercise 5.2). However, strict quasiconcavity is sufficient and moreover every optimum is a strict optimum. Therefore, if a solution to the optimization problem exists, it is unique (corollary 3.1.1).

Exercise 5.3
In the constrained optimization problem

$$\max_{\mathbf{x} \in G(\boldsymbol{\theta})} f(\mathbf{x}, \boldsymbol{\theta})$$

suppose that f is strictly quasiconcave in $\mathbf{x}$ and $G(\boldsymbol{\theta})$ convex. Then every optimum is a strict global optimum.

There is one final distinction that we need to make. If the solution $\mathbf{x}^*$ of an optimization problem is an interior point (section 1.3) of the feasible set, then it is called an *interior* solution. A solution that lies on the boundary of the feasible set is called a *boundary* solution. This distinction is important, since tighter conditions can be developed for interior optima.

In characterizing optimal solutions to a constrained optimization problem, we develop *optimality* conditions which will identify a solution to (1). These optimality conditions are of two types, called *necessary* and *sufficient* conditions. *Necessary conditions* must be satisfied by every solution, and any choice that violates the necessary conditions cannot be an optimum. These must be distinguished from *sufficient conditions*, which guarantee that a point meeting the conditions is an optimum. Sufficient conditions do not preclude the possibility of other solutions to the problem which violate the sufficient conditions.

Example 5.3 Being a male is a necessary (but not a sufficient) condition for being a father. Having a son is a sufficient (but not a necessary) condition for being a father.

Example 5.4 Sufficient conditions to ensure that the existence of an optimal solution to the constrained optimization problem

$$\max_{\mathbf{x} \in G(\boldsymbol{\theta})} f(\mathbf{x}, \boldsymbol{\theta})$$

are that f is continuous and $G(\boldsymbol{\theta})$ compact. In addition sufficient conditions to ensure that the solution is unique are that $G(\boldsymbol{\theta})$ is convex and f is strictly concave. None of these conditions is necessary.

In this chapter we develop necessary and sufficient conditions to identify a solution to the general problem (1). We will proceed in stages, starting with intuitive solutions to familiar problems and building gradually to a general solution. In the course of this development, we will present four complementary approaches to the solution of the general constrained optimization problem. Each perspective offers a different insight into this fundamental tool of economic analysis.

For clarity of notation, we will suppress for most of this chapter the explicit dependence of the objective function and the constraints on the parameters $\boldsymbol{\theta}$, expressing the general constrained optimization problem as

$$\max_{\mathbf{x} \in X} f(\mathbf{x})$$

subject to $\quad \mathbf{g}(\mathbf{x}) \leq \mathbf{0}$

Accordingly G will denote the feasible set $G = \{\mathbf{x} \in X : \mathbf{g}(\mathbf{x}) \leq \mathbf{0}\}$. Explicit dependence parameters will be restored in the next chapter when we explore sensitivity of the optimal solution to the changes in the parameters. To utilize the tools of the previous chapter, we assume throughout that the functions f and $\mathbf{g}$ are twice continuously differentiable (C^2).

5.2 Unconstrained Optimization

We begin with the simple optimization problem

$$\max_{\mathbf{x} \in X} f(\mathbf{x}) \tag{3}$$

in which there are no functional constraints. Suppose that $\mathbf{x}^*$ is a local optimum of (3) in X so that there exists a neighborhood $S \subseteq X$ such that

$$f(\mathbf{x}^*) \geq f(\mathbf{x}) \qquad \text{for every } \mathbf{x} \in S \tag{4}$$

Provided that f is differentiable, we can use the derivative to approximate $f(\mathbf{x})$ in the neighborhood S, that is,

$$f(\mathbf{x}) \approx f(\mathbf{x}^*) + Df[\mathbf{x}^*](\mathbf{x} - \mathbf{x}^*)$$

If $\mathbf{x}^*$ is a local maximum, (4) implies that

$$f(\mathbf{x}^*) \geq f(\mathbf{x}^*) + Df[\mathbf{x}^*](\mathbf{x} - \mathbf{x}^*)$$

or

$$Df[\mathbf{x}^*](\mathbf{x} - \mathbf{x}^*) \leq 0 \qquad \text{for every } \mathbf{x} \in S$$

At a maximum it is impossible to move from $\mathbf{x}^*$ in any direction and increase the value of the objective function $f(\mathbf{x})$.

Furthermore, if $\mathbf{x}^*$ is an interior point of X, it is necessary that

$$Df[\mathbf{x}^*](\mathbf{x}^* - \mathbf{x}) = 0 \qquad \text{for every } \mathbf{x} \in S$$

since $Df[\mathbf{x}^*]$ is a *linear* function. That is the derivative $Df[\mathbf{x}^*]$ at $\mathbf{x}^*$ must be the zero function. Its matrix representation, the gradient, must be the *zero vector*

$$\nabla f(\mathbf{x}^*) = (D_{x_1}f[\mathbf{x}^*], D_{x_2}f[\mathbf{x}^*], \ldots, D_{x_n}f[\mathbf{x}^*]) = 0 \tag{5}$$

and all its components, the partial derivatives of f evaluated at $\mathbf{x}^*$, must be zero. This is a precise expression of the fact that, at a maximum, it is impossible to marginally change the value of any one of the variables x_i so as to increase the value of the function $f(\mathbf{x})$. Even more stringent, it implies that the directional derivative in any direction is zero. It is impossible to move from $\mathbf{x}^*$ in any direction and increase the value of the function $f(\mathbf{x})$. We say that f is *stationary* at $\mathbf{x}^*$ and $\mathbf{x}^*$ is *stationary point* of f.

In short, a local maximum must be either a *stationary point* of f or a *boundary point* of X (or both). Note again how we are using the linear approximation to an arbitrary function to reveal something about the underlying function. Since stationarity involves the first derivative, these are called the *first-order conditions*.

Proposition 5.1 (First-order conditions for a maximum) *If $\mathbf{x}^*$ is a local maximum of f in X, there exists a neighborhood S of $\mathbf{x}^*$ such that*

$$Df[\mathbf{x}^*](\mathbf{x} - \mathbf{x}^*) \leq 0 \quad \text{for every } \mathbf{x} \in S$$

Furthermore, if $\mathbf{x}^*$ is an interior point of X, then f is stationary at $\mathbf{x}^*$, that is,

$$\nabla f(\mathbf{x}^*) = \mathbf{0}$$

Exercise 5.4
Prove proposition 5.1 formally.

Example 5.5 (Nonnegative variables) Variables in economic problems are frequently restricted to be nonnegative, which creates the possibility of boundary solutions. Consider the single-variable optimization problem

$$\max_{x \geq 0} f(x)$$

where $x \in \Re$. Suppose that x^* maximizes $f(x)$ over $\Re_+$. Then x^* is either a stationary point or a boundary point or both (proposition 5.1). That is,

Either $x > 0$ and $f'(x^*) = 0$
or $x = 0$ and $f'(x^*) \leq 0$

These alternatives can be telescoped into the necessary condition

5.2 Unconstrained Optimization

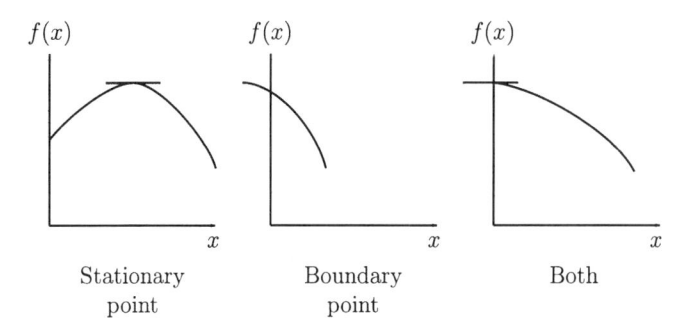

Figure 5.1
A maximum must be either a stationary point or a boundary point or both

$$f'(x^*) \leq 0, \quad x^* \geq 0, \quad x^* f'(x^*) = 0$$

The three cases are illustrated in figure 5.1.

We extend this result in the following corollary.

Corollary 5.1 (Nonnegative variables) *If $\mathbf{x}^*$ is a local maximum of f in $\Re_+^n$, then it is necessary that $\mathbf{x}^*$ satisfy*

$$\nabla f(\mathbf{x}^*) \leq \mathbf{0}, \quad \mathbf{x}^* \geq \mathbf{0}, \quad \nabla f(\mathbf{x}^*)^T \mathbf{x}^* = 0$$

which means that for every i,

$$D_{x_i} f[\mathbf{x}^*] \leq 0, \quad x_i^* \geq 0, \quad x_i^* D_{x_i} f[\mathbf{x}^*] = 0$$

Exercise 5.5
Prove corollary 5.1.

As we saw in the previous chapter, the derivative of a function at any point $\mathbf{x}$ defines the tangent hyperplane to the surface at that point. Stationarity (5) implies that the tangent hyperplane is horizontal at a stationary point $\mathbf{x}^*$, and so the graph of the function is flat at this point. Consider a physical analogy. We recognize the top of a hill by observing that we cannot climb any further, that all paths lead down. At the peak the gradient is flat. If we could identify the coordinates of all flat land in a given region, this list would necessarily include all mountain peaks. But it would also include many other points, including valley floors, lake bottoms, plains, cols, and so on. Further criteria are required to distinguish the mountain peaks from troughs and other points of zero gradient.

Example 5.6 (Saddle point) Consider the function

$$f(x_1, x_2) = x_1^2 - x_2^2$$

The first-order condition for stationarity is

$$\nabla f(\mathbf{x}) = (2x_1, -2x_2) = (0, 0)$$

In practice, the first-order condition is usually written as a system of partial differential equations

$$D_1 f(x) = 2x_1 = 0$$

$$D_2 f(x) = -2x_2 = 0$$

Since these equations have the unique solution $x_1 = x_2 = 0$, the function f has a unique stationary point $(0, 0)$. However, the stationary point $(0, 0)$ is neither a maximum nor a minimum of the function, since

$$-3 = f(1, 2) < f(0, 0) < f(2, 1) = 3$$

$(0, 0)$ is in fact a saddle point (section 3.7.4). The function is illustrated in figure 5.2.

To distinguish maxima from minima and other stationary points, we resort to a higher-order approximation. If $\mathbf{x}^*$ is a local maximum, then

$$f(\mathbf{x}^*) \geq f(\mathbf{x})$$

for every $\mathbf{x}$ in a neighborhood of $\mathbf{x}^*$. Previously we used a linear approximation to estimate $f(\mathbf{x})$. Assuming that f is C^2, $f(\mathbf{x})$ can be better approximated by the second-order Taylor series (example 4.33)

$$f(\mathbf{x}) \approx f(\mathbf{x}^*) + \nabla f(\mathbf{x}^*)^T \mathbf{dx} + \tfrac{1}{2} \mathbf{dx}^T H_f(\mathbf{x}^*) \mathbf{dx}$$

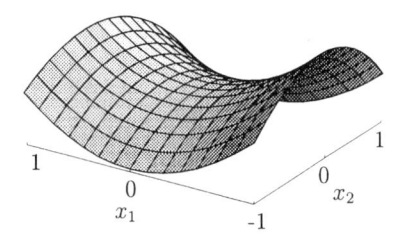

Figure 5.2
A saddle point

where $H_f(\mathbf{x}^*)$ is the Hessian of f at x^* and $\mathbf{dx} = \mathbf{x} - \mathbf{x}^*$. If $\mathbf{x}^*$ is a local maximum, then there exists a ball $B_r(\mathbf{x}^*)$ such that

$$f(\mathbf{x}^*) \geq f(\mathbf{x}^*) + \nabla f(\mathbf{x}^*)^T \mathbf{dx} + \tfrac{1}{2} \mathbf{dx}^T H_f(\mathbf{x}^*) \mathbf{dx}$$

or

$$\nabla f(\mathbf{x}^*)^T \mathbf{dx} + \tfrac{1}{2} \mathbf{dx}^T H_f(\mathbf{x}^*) \mathbf{dx} \leq 0$$

for every $\mathbf{dx} \in B_r(\mathbf{x}^*)$. To satisfy this inequality for all small $\mathbf{dx}$ requires that the first term be zero and the second term nonpositive. In other words, for a point $\mathbf{x}^*$ to be a local maximum of a function f, it is necessary that the gradient be zero and the Hessian be nonpositive definite at $\mathbf{x}^*$. We summarize these conditions in the following theorem.

Theorem 5.1 (Necessary conditions for interior maximum) For $\mathbf{x}^*$ to be an interior local maximum of $f(\mathbf{x})$ in X, it is necessary that

1. $\mathbf{x}^*$ be a stationary point of f, that is, $\nabla f(\mathbf{x}^*) = 0$, and
2. f be locally concave at $\mathbf{x}^*$, that is, $H_f(\mathbf{x}^*)$ is nonpositive definite

Theorem 5.1 states necessary conditions for an interior maximum—that is, these conditions must be satisfied at every local maximum. A slight strengthening gives sufficient conditions, but these are not necessary (example 5.7).

Corollary 5.1.1 (Sufficient conditions for interior maximum) If

1. $\mathbf{x}^*$ is a stationary point of f, that is, $\nabla f(\mathbf{x}^*) = 0$, and
2. f is locally strictly concave at $\mathbf{x}^*$, that is, $H_f(\mathbf{x}^*)$ is negative definite

then $\mathbf{x}^*$ is a strict local maximum of f.

Exercise 5.6
Prove corollary 5.1.1.

Example 5.7 The function $f(x) = -x^4$ has a strict global maximum at 0, since $f(0) = 0$ and $f(x) < 0$ for every $x \neq 0$. We note that 0 satisfies the necessary but not the sufficient conditions for a local maximum, since

$$f'(x) = -4x^3, \quad f'(0) = 0$$
$$f''(x) = -12x^2, \quad f''(0) = 0$$

Where f is concave, stationarity is both necessary and sufficient for an interior optimum.

Corollary 5.1.2 (Concave maximization) *Suppose that f is concave and $\mathbf{x}^*$ is an interior point of X. Then $\mathbf{x}^*$ is a global maximum of f on X if and only if $\nabla f(\mathbf{x}^*) = \mathbf{0}$.*

Proof Suppose that f is concave and $\nabla f(\mathbf{x}^*) = \mathbf{0}$. Then $\mathbf{x}^*$ satisfies the necessary conditions for a local optimum (theorem 5.1). Moreover, by Exercise 4.67,

$$f(\mathbf{x}) \leq f(\mathbf{x}^*) + \nabla f(\mathbf{x}^*)^T (\mathbf{x} - \mathbf{x}^*) = f(\mathbf{x}^*)$$

Therefore $\mathbf{x}^*$ is a global optimum. □

Exercise 5.7 (Interior minimum)
For $\mathbf{x}^*$ to be an interior minimum of $f(\mathbf{x})$, it is necessary that

1. $\mathbf{x}^*$ be a stationary point of f, that is, $\nabla f(\mathbf{x}^*) = 0$, and
2. f be locally convex at $\mathbf{x}^*$, that is, $H_f(\mathbf{x}^*)$ is nonnegative definite

If furthermore $H_f(\mathbf{x}^*)$ is positive definite, then $\mathbf{x}^*$ is a strict local minimum.

The following result was used to prove the mean value theorem (exercise 4.34) in chapter 4.

Exercise 5.8 (Rolle's theorem)
Suppose that $f \in C[a, b]$ is differentiable on (a, b). If $f(a) = f(b)$, then there exists $x \in (a, b)$ where $f'(x) = 0$.

Since the conditions of theorem 5.1 involve respectively the first and second derivatives of the objective function, they are called the *first-order* and *second-order* necessary conditions for an interior maximum. The usual technique for solving unconstrained optimization problems in economics is to use the first-order conditions to identify all the stationary points of the objective function. Normally this is done by solving the first-order conditions as a system of equations. This reduces enormously the number of potential candidates for an optimal solution. Second-order conditions are then used to distinguish the local maxima and minima from other stationary points. Having found all the local maxima, their values can be compared to find the global maximum. This technique is illustrated in the following examples and exercises.

5.2 Unconstrained Optimization

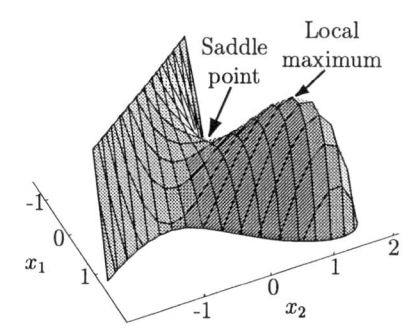

Figure 5.3
A local maximum which is not a global maximum

Example 5.8 Consider the problem

$$\max f(x_1, x_2) = 3x_1x_2 - x_1^3 - x_2^3$$

illustrated in figure 5.3. The first-order conditions for a maximum are

$$D_1 f = 3x_2 - 3x_1^2 = 0$$
$$D_2 f = 3x_1 - 3x_2^2 = 0$$

or

$$x_2 = x_1^2, \quad x_1 = x_2^2$$

These equations have two solutions: $x_1 = x_2 = 0$ and $x_1 = x_2 = 1$. Therefore $(0,0)$ and $(1,1)$ are the only stationary points of f.

The Hessian of f is

$$H(\mathbf{x}) = \begin{pmatrix} -6x_1 & 3 \\ 3 & -6x_2 \end{pmatrix}$$

At $(1,1)$ this evaluates to

$$H(\mathbf{x}) = \begin{pmatrix} -6 & 3 \\ 3 & -6 \end{pmatrix}$$

which is negative definite. Therefore, we conclude that the function f attains a local maximum at $(1,1)$ (corollary 5.1.1). At the other stationary point $(0,0)$, the Hessian evaluates to

$$H(\mathbf{x}) = \begin{pmatrix} 0 & 3 \\ 3 & 0 \end{pmatrix}$$

which is indefinite. $(0,0)$ is in fact a saddle point (section 3.7.4). Hence $\mathbf{x}^* = (1,1)$ is the unique local maximum of f, where the function attains the value $f(1,1) = 1$. Note, however, that $(1,1)$ is not a global maximum of f, since, for example, $f(-1,-1) = 5 > f(1,1)$. In fact f has no global maximum on $\Re^2$, since for any x_2, $f(x_1, x_2) \to \infty$ as $x_1 \to -\infty$.

Example 5.9 (Saddle point) We showed earlier (example 5.6) that the function

$$f(x_1, x_2) = x_1^2 - x_2^2$$

has a unique stationary point at $(0,0)$. However, the stationary point $(0,0)$ cannot be a maximum or minimum of f because $(0,0)$ does not satisfy the second-order necessary condition for a maximum or minimum. The Hessian of f

$$H = \begin{pmatrix} 2 & 0 \\ 0 & -2 \end{pmatrix}$$

is indefinite. For example, the quadratic form $(1,0)^T H (1,0) = 2$ while $(0,1)^T H (1,0) = -2$. We can conclude that $(0,0)$ is a *saddle point* of f, being simultaneously a minimum of f in the x_1 direction and a maximum of f in the x_2 direction.

Exercise 5.9
Solve the problem

$$\max_{x_1, x_2} f(x_1, x_2) = x_1 x_2 + 3x_2 - x_1^2 - x_2^2$$

Exercise 5.10
Show that $(0,0)$ is the only stationary point of the function

$$f(x_1, x_2) = x_1^2 + x_2^2$$

Is it a maximum, a minimum or neither?

A function that has a nonpositive definite Hessian at a point $\mathbf{x}^*$ is locally concave, that is, concave in a neighborhood of $\mathbf{x}^*$. Frequently, in economic analysis, conditions are imposed (e.g., strict concavity) that ensure that the function has a unique (global) maximum. In this case, should the function have only a single stationary point, the analyst can correctly deduce that this is the unique maximum, and the first-order conditions can be solved for this maximum. Without such an assumption, even if a function has a unique stationary point that is a local maximum, it is incorrect to infer that this is a global maximum (examples 5.8 and 5.10).

Example 5.10 Consider the function $f\colon \Re^2 \to \Re$ defined by

$$f(x_1, x_2) = 3x_1 e^{x_2} - x_1^3 - e^{3x_2}$$

Stationary points are defined by the first-order conditions

$$D_1 f(x) = 3e^{x_2} - 3x_1^2 = 0$$

$$D_2 f(x) = 3x_1 e^{x_2} - 3e^{3x_2} = 0$$

which can be reduced to

$$e^{x_2} - x_1^2 = 0$$

$$x_1 - e^{2x_2} = 0$$

These equations have the unique (real) solution $x_1 = 1$, $x_2 = 0$. Therefore $(1, 0)$ is the unique stationary point of f. The Hessian of f is

$$H = \begin{pmatrix} -6x_1 & 3e^{x_2} \\ 3e^{x_2} & -9e^{3x_2} \end{pmatrix}$$

which evaluates to

$$H = \begin{pmatrix} -6 & 3 \\ 3 & -9 \end{pmatrix}$$

at the point $(1, 0)$. This matrix is negative definite, which implies that f is locally concave at $(1, 0)$ and so attains a local maximum of $f(1, 0) = 1$ there. This is not a global maximum, since, for example, $f(-2, -2) = 7.18$.

Example 5.11 (The competitive firm) A competitive firm produces a single output y using n inputs $(x_1, x_2, \ldots, x_n)$ according to the production function

$$y = f(x_1, x_2, \ldots, x_n)$$

Let p denote the output price and $\mathbf{w} = (w_1, w_2, \ldots, w_n)$ the input prices. The firm's revenue is $pf(\mathbf{x})$ produced at a cost of $\mathbf{w}^T\mathbf{x}$, so that its net profit is given by

$$\Pi(\mathbf{x}; \mathbf{w}, p) = pf(\mathbf{x}) - \mathbf{w}^T\mathbf{x}$$

The firm's optimization problem is to choose $\mathbf{x}$ to maximize its profit, that is

$$\max_{\mathbf{x}} \Pi(\mathbf{x}; \mathbf{w}, p) = pf(\mathbf{x}) - \mathbf{w}^T\mathbf{x}$$

This implicitly determines the output level y according to the production function $y = f(\mathbf{x})$.

The first-order necessary condition for $\mathbf{x}^*$ to maximize profit is that $\Pi(\mathbf{x}; \mathbf{w}, p)$ be stationary at $\mathbf{x}^*$, that is,

$$D_{\mathbf{x}}\Pi[\mathbf{x}^*; \mathbf{w}, p] = pD_{\mathbf{x}}f[\mathbf{x}^*] - \mathbf{w} = \mathbf{0}$$

or

$$pD_{x_i}f[\mathbf{x}^*] = w_i \quad \text{for every } i = 1, 2, \ldots, n \tag{6}$$

$D_{x_i}f[\mathbf{x}^*]$ is the marginal product of input i in the production of y and $pD_{x_i}f[\mathbf{x}^*]$ is the value of the marginal product. Optimality (6) requires that the profit-maximizing firm should utilize each input until the value of its marginal product $pD_{x_i}f[\mathbf{x}^*]$ is equal to its cost w_i. This accords with common sense. If the value of an additional unit of input i ($pD_{x_i}f[\mathbf{x}^*]$) exceeds its cost (w_i), the firm would increase its profit by increasing the utilization of x_i. Conversely, if $pD_{x_i}f[\mathbf{x}^*] < w_i$, the firm would increase its profit by reducing the input of x_i (assuming that $x_i > 0$). Only if $pD_{x_i}f[\mathbf{x}^*] = w_i$ for every input i can the firm be maximizing its profit. The first-order necessary condition is the ubiquitous "marginal benefit equals marginal cost" familiar from elementary economics.

The second-order necessary condition for $\mathbf{x}^*$ to maximize profit is that

$$\mathbf{dx}^T H_{\Pi}(\mathbf{x}^*) \mathbf{dx} = \mathbf{dx}^T p H_f(\mathbf{x}^*) \mathbf{dx} \leq 0 \quad \text{for every } \mathbf{dx}$$

where $H_f(\mathbf{x}^*)$ is the Hessian of the production function f. That is, the Hessian matrix of the production function must be nonpositive definite at the optimal point. Note that the second-order necessary condition will be satisfied at every stationary point if the production function is concave.

5.2 Unconstrained Optimization

Exercise 5.11
A popular product called *pfillip*, a nonnarcotic stimulant, is produced by a competitive industry. Each firm in this industry uses the same production technology, given by the production function

$$y = k^{1/6} l^{1/3}$$

where y is the amount of pfillip produced, k is the amount of *kapitose* (a special chemical) and l is the amount of *legume* (a common vegetable) used in production. The current prices of kapitose and legume are $1 and $1/2 per unit respectively. Firms also incur fixed costs of $1/6. If the market price of pfillip is also $1, how profitable is the average firm?

Example 5.12 (Monopoly) Economic analysis of a monopolist usually focuses on its output decision. Presuming that production is undertaken efficiently (i.e., at minimum cost), the monopolist's technological constraints can be represented by the cost function $c(y)$ (example 2.38). Letting $R(y)$ denote the revenue from selling y units of output, the monopolist's net profit is

$$\Pi(y) = R(y) - c(y)$$

The first-order condition for maximizing profit is

$$D_y \Pi(y) = R'(y) - c'(y) = 0$$

or

$$R'(y) = c'(y)$$

which is the familiar optimality condition "marginal revenue equal to marginal cost."

The second-order condition necessary for a profit maximum is

$$D_y^2 \Pi(y) = R''(y) - c''(y) \leq 0$$

Sufficient conditions to ensure this are that marginal revenue is decreasing and marginal cost increasing with output.

Example 5.13 (Least squares regression) The standard linear regression model of econometrics postulates that some observed economic variable y is linearly related to a number of independent variables x_i together with some random error ε,

$$y = \beta_1 x_1 + \beta_2 x_2 + \cdots + \beta_n x_n + \varepsilon$$

although the precise coefficients $(\beta_1, \beta_2, \ldots, \beta_n)$ are unknown. The objective of an econometric analysis is to estimate the most likely values of the unknown coefficients $(\beta_1, \beta_2, \ldots, \beta_n)$, given some observations y_t of y and the corresponding independent variables $\mathbf{x}_t$. The least squares criterion chooses estimated coefficients $(\hat{\beta}_1, \hat{\beta}_2, \ldots, \hat{\beta}_n)$ so as to minimize the sum of the squared errors between the observed values y_t and the predicted values $\hat{\beta} \mathbf{x}_t$. The least squares estimates solve the following optimization problem

$$\min_{\hat{\beta}} \sum_t (y_t - \hat{\beta} \mathbf{x}_t)^2$$

The first-order conditions for a minimum are

$$D_{\hat{\beta}_i} \sum_t (y_t - \hat{\beta} \mathbf{x}_t)^2 = 2 \sum_t (y_t - \hat{\beta} \mathbf{x}_t)(-x_{it}) = 0, \qquad i = 1, 2 \ldots, n$$

which are known as the *normal equations*. Solving these equations for given data y_t and $\mathbf{x}_t$ gives the least squares estimators $\hat{\beta}$.

The following exercise shows that the first-order necessary condition is invariant to monotonic transformations of the objective function. This can be of considerable help in practice, where a monotonic transformation can make the solution of problem more tractable (e.g., see exercise 5.13).

Exercise 5.12 (Monotonic transformation)
Suppose that $h: \Re \to \Re$ is a monotonic transformation (example 2.60) of $f: X \to \Re$. Then $h \circ f$ has the same stationary points as f.

Exercise 5.13 (Maximum likelihood estimation)
Suppose that a random variable x is assumed to be normally distributed with (unknown) mean μ and variance σ^2 so that its probability density function is

$$f(x) = \frac{1}{\sqrt{2\pi}\sigma} \exp\left(-\frac{1}{2\sigma^2}(x-\mu)^2\right)$$

The probability (likelihood) of a sequence of independent observations $(x_1, x_2, \ldots, x_T)$ for given parameters μ and σ^2 is given by their joint distribution

5.2 Unconstrained Optimization

$$L(\mu, \sigma) = \prod_{t=1}^{T} f(x_t) = \frac{1}{(2\pi)^{T/2} \sigma^T} \exp\left(-\frac{1}{2\sigma^2} \sum_{t=1}^{T} (x_t - \mu)^2\right)$$

which is known as the *likelihood function*. The maximum likelihood estimators of μ and σ^2 are those values that maximize the likelihood of given set of observations $(x_1, x_2, \ldots, x_T)$. That is, they are the solution to the following maximization problem:

$$\max_{\mu, \sigma} L(\mu, \sigma) = \frac{1}{(2\pi)^{T/2} \sigma^T} \exp\left(-\frac{1}{2\sigma^2} \sum_{t=1}^{T} (x_t - \mu)^2\right)$$

Show that the maximum likelihood estimators are

$$\hat{\mu} = \bar{x} = \sum_{t=1}^{T} x_t, \quad \hat{\sigma}^2 = \frac{1}{T} \sum_{t=1}^{T} (x_t - \bar{x})^2$$

[Hint: It is convenient to maximize log L; see exercise 5.12.]

These examples and exercises illustrate the typical approach to optimization problems in economics. The first-order necessary conditions for a maximum are obtained and are used to chararacterize or identify the optimal solution. These conditions are then interpreted or implemented in terms of the original model. For example, in the model of a competitive firm, the first-order necessary conditions for a profit maximum (stationarity of the profit function) are interpreted in terms of the familiar marginality conditions, "value of marginal product equals factor price." In the least squares regression model, the first-order necessary conditions lead directly to the *normal equations*, which are used to define the least squares estimator. The second-order conditions play a secondary role. Rather than identifying optimal solutions, their role is to place bounds on the parameters of the model. For example, in a competitive firm, the second-order conditions require that the production function be locally concave. From this we infer that the competitive model is inappropriate unless the technology is such that the production function is locally concave. Often the underlying assumptions of the model are such as to ensure that the second-order conditions are satisfied everywhere, for example, that the production function is globally concave. The second-order conditions often play a crucial role in comparative statics (chapter 6).

Our derivation of the first-order conditions, by evaluating the effect of perturbations around a proposed solution, forms the foundation of all optimization theory. Constraints have the effect of restricting the set of feasible perturbations. It is no longer necessary that the fundamental inequality (4) be satisfied for all **x** in the neighborhood of **x*** but only for all neighboring **x**, which also satisfy the constraints **x** ∈ G. In the next two sections we show how this requirement can be translated into conditions on the functions *f* and **g** and how these conditions can be phrased in terms of the maximization of a new function called the *Lagrangean*, which is constructed from *f* and **g**.

In the next section (section 5.3) we consider a restricted class of problems in which all the functional constraints are binding (equalities). In this restricted context we develop the basic necessary conditions from four different perspectives, each of which contributes its own insight into the results. In following section (section 5.4) we generalize our results to encompass problems in which the constraints are inequalities.

5.3 Equality Constraints

Analysis of a constrained optimization problem is simplified if we assume that the functional constraints are equalities, so (1) becomes

$$\max_{\mathbf{x} \in X} f(\mathbf{x}) \tag{7}$$

subject to $\mathbf{g}(\mathbf{x}) = 0$

We present four complementary derivations of the basic necessary conditions for an optimal solution of (7). While the four approaches lead to the same result, each provides a different perspective contributing to a better understanding of this most fundamental tool of economic analysis.

5.3.1 The Perturbation Approach

The first approach to constrained optimization follows the perturbation procedure that we adopted for the unconstrained problem in the previous section. To introduce the idea, let us first apply the perturbation approach to a familiar problem, the consumer's problem assuming that the consumer spends all her income. In this case, the consumer's problem is

$$\max_{\mathbf{x} \in X} u(\mathbf{x})$$

subject to $\quad \mathbf{p}^T \mathbf{x} = m$

For simplicity, we assume that there are only two commodities ($X \subset \Re^2$).

Suppose that $\mathbf{x}^*$ is the consumer's most preferred choice. Then it is must be impossible to rearrange her expenditure and achieve a higher utility. In particular, it must be impossible to increase the purchases of x_1 at the expense of x_2 and achieve higher utility. To see what this implies, consider a small increase dx_1 in x_1 which costs $p_1 \, dx_1$. To satisfy the budget constraint, this must be offset by a corresponding reduction in $p_2 \, dx_2$ in expenditure on x_2 so that

$$p_1 \, dx_1 + p_2 \, dx_2 = 0$$

or

$$p_1 \, dx_1 = -p_2 \, dx_2 \tag{8}$$

That is, the increase in expenditure on x_1 must be exactly equal to the decrease in expenditure on x_2. If $\mathbf{x}^*$ is an optimal choice, this transfer of expenditure cannot generate an increase in utility, that is,

$$u(\mathbf{x}^*) \geq u(\mathbf{x}^* + \mathbf{dx})$$

where $\mathbf{dx} = (dx_1, dx_2)$. To evaluate the change in utility, we can use the linear approximation provided by the derivative, that is,

$$u(\mathbf{x}^* + \mathbf{dx}) \approx u(\mathbf{x}^*) + Du[\mathbf{x}^*](\mathbf{dx})$$

which can be decomposed into the separate effects of x_1 and x_2 (example 4.9),

$$u(\mathbf{x}^* + \mathbf{dx}) \approx u(\mathbf{x}^*) + D_{x_1} u[\mathbf{x}^*] \, dx_1 + D_{x_2} u[\mathbf{x}^*] \, dx_2$$

If $\mathbf{x}^*$ is optimal, this change in consumption cannot provide an increase in utility, that is, the change in utility must be nonpositive:

$$D_{x_1} u[\mathbf{x}^*] \, dx_1 + D_{x_2} u[\mathbf{x}^*] \, dx_2 \leq 0$$

or

$$D_{x_1} u[\mathbf{x}^*] \, dx_1 \leq -D_{x_2} u[\mathbf{x}^*] \, dx_2 \tag{9}$$

$D_{x_1}u[\mathbf{x}^*]$ is the marginal utility of good 1 at the consumption bundle $\mathbf{x}^*$. The left-hand side measures the increase in utility from consuming more x_1. If $\mathbf{x}^*$ is an optimal consumption bundle, this must be more than offset by the decrease in utility from consuming less of x_2, which is measured be the right-hand side. Dividing (9) by equation (8), we conclude that the optimal bundle $\mathbf{x}^*$ must satisfy the condition that

$$\frac{D_{x_1}u[\mathbf{x}^*]}{p_1} \leq \frac{D_{x_2}u[\mathbf{x}^*]}{p_2}$$

Similarly, by considering an increase in x_2 matched by a reduction in x_1, we conclude that

$$\frac{D_{x_1}u[\mathbf{x}^*]}{p_1} \geq \frac{D_{x_2}u[\mathbf{x}^*]}{p_2}$$

Together, these inequalities imply that

$$\frac{D_{x_1}u[\mathbf{x}^*]}{p_1} = \frac{D_{x_2}u[\mathbf{x}^*]}{p_2} \tag{10}$$

The ratio of marginal utility to the price of each good must be equal at the optimal choice $\mathbf{x}^*$. If this were not the case, it would be possible to rearrange the consumer's expenditure in such a way as to increase total utility. For example, if the left-hand side was greater than the right-hand side, one dollar of expenditure transferred from x_2 to x_1 would give greater satisfaction to the consumer.

Applying similar analysis to the general equality-constrained problem (7), suppose that $\mathbf{x}^*$ maximizes $f(\mathbf{x})$ subject to $\mathbf{g}(\mathbf{x}) = \mathbf{0}$. Consider a small change in $\mathbf{x}$ to $\mathbf{x}^* + \mathbf{dx}$. The change in $\mathbf{x}$ must satisfy

$$\mathbf{g}(\mathbf{x}^* + \mathbf{dx}) = \mathbf{0}$$

Provided that $\mathbf{x}^*$ is a regular point of $\mathbf{g}$, $\mathbf{dx}$ must satisfy the linear approximation

$$\mathbf{g}(\mathbf{x}^* + \mathbf{dx}) \approx \mathbf{g}(\mathbf{x}^*) + Dg[\mathbf{x}^*](\mathbf{dx}) = \mathbf{0}$$

or

$$Dg[\mathbf{x}^*](\mathbf{dx}) = \mathbf{0} \tag{11}$$

For all perturbations $\mathbf{dx}$ satisfying (11), we must have

5.3 Equality Constraints

$$f(\mathbf{x}^*) \geq f(\mathbf{x}^* + \mathbf{dx}) \approx f(\mathbf{x}^*) + Df[\mathbf{x}^*](\mathbf{dx})$$

This implies that a necessary condition for $\mathbf{x}^*$ to maximize $f(\mathbf{x})$ is

$$Df[\mathbf{x}^*](\mathbf{dx}) = 0 \text{ for all } \mathbf{dx} \text{ such that } D\mathbf{g}[\mathbf{x}^*](\mathbf{dx}) = 0 \tag{12}$$

In a constrained optimization problem, optimality does not require overall stationarity of the objective function, but stationarity with respect to a restricted class of perturbations (11). At the optimum $\mathbf{x}^*$ the objective function is stationary with respect to perturbations which continue to satisfy the constraint. For example, the optimal solution to the consumer's problem does not mean that there exist no preferred commodity bundles. Rather, optimality requires that any preferred bundles cost more than her available income. Her utility function is stationary with respect to affordable commodity bundles.

Condition (12) can be expressed alternatively as saying that there does not exist any perturbation $\mathbf{dx} \in \mathbb{R}^n$ such that

$$D\mathbf{g}[\mathbf{x}^*](\mathbf{dx}) = 0 \quad \text{and} \quad Df[\mathbf{x}^*](\mathbf{dx}) > 0$$

In other words, it is impossible to find any change from $\mathbf{x}^*$, which simultaneously satisfies the constraint and yields a higher value of the objective.

To derive useful criterion for identifying an optimal solution, we utilize the results of chapter 3. The derivative of the objective function $Df[\mathbf{x}^*]$ is a linear functional on X. The derivative of the constraint $D\mathbf{g}[\mathbf{x}^*]$ can be considered as a system of m linear functionals, $Dg_1[\mathbf{x}^*], Dg_2[\mathbf{x}^*], \ldots, Dg_m[\mathbf{x}^*]$ on X. By (12) these linear functionals satisfy the conditions of the Fredholm alterative (exercises 3.48 and 3.199), which implies that there exist constants λ_j such that

$$Df[\mathbf{x}^*] = \sum_{j=1}^{m} \lambda_j Dg_j[\mathbf{x}^*]$$

or alternatively that the gradient of the objective function is a linear combination of the gradient of constraints

$$\nabla f(\mathbf{x}^*) = \sum_{j=1}^{m} \lambda_j \nabla g_j(\mathbf{x}^*)$$

The constants λ_j are known as the *Lagrange multipliers*. This provides a useful criterion for identifying optimal solutions to equality constrained problems, which we summarize in the following theorem.

Theorem 5.2 (Lagrange multiplier theorem) *Suppose that* $\mathbf{x}^*$ *is a local optimum of*

$$\max_{\mathbf{x} \in X} f(\mathbf{x})$$

subject to $\quad \mathbf{g}(\mathbf{x}) = \mathbf{0}$

and a regular point of $\mathbf{g}$. *Then the gradient of* f *at* $\mathbf{x}^*$ *is a linear combination of the gradients of the constraints, that is there exist unique multipliers* $\lambda_1, \lambda_2, \ldots, \lambda_m$ *such that*

$$\nabla f(\mathbf{x}^*) = \sum_{j=1}^{m} \lambda_j \nabla g_j(\mathbf{x}^*) \tag{13}$$

The requirement that $\mathbf{x}^*$ be a regular point of $\mathbf{g}$, that is, $Dg[\mathbf{x}^*]$, has full rank, means that the gradients $\nabla g_1(\mathbf{x}^*), \nabla g_2(\mathbf{x}^*), \ldots, \nabla g_m(\mathbf{x}^*)$ are linearly independent. Known as the constraint qualification condition, it will be explored more fully in section 5.4.2.

Example 5.14 Consider the problem

$$\max_{x_1, x_2} f(\mathbf{x}) = x_1 x_2$$

subject to $\quad g(\mathbf{x}) = x_1 + x_2 = 1$

The gradient of the objective function is

$$\nabla f(\mathbf{x}) = (x_2, x_1)$$

while that of the constraint is

$$\nabla g(\mathbf{x}) = (1, 1)$$

A necessary condition for a solution is that these be proportional, that is,

$$\nabla f(\mathbf{x}) = (x_2, x_1) = \lambda(1, 1) = \nabla g(\mathbf{x})$$

which implies that

$$x_1 = x_2 = \lambda \tag{14}$$

The solution must also satisfy the constraint

$$x_1 + x_2 = 1 \tag{15}$$

Equations (14) and (15) can be solved to yield the solution

$$x_1 = x_2 = \lambda = \tfrac{1}{2}$$

Exercise 5.14
Solve

$$\max_{x_1, x_2} f(\mathbf{x}) = 1 - (x_1 - 1)^2 - (x_2 - 1)^2$$

subject to $\quad g(\mathbf{x}) = x_1^2 + x_2^2 = 1$

Example 5.15 (The consumer's problem) Returning to the consumer's problem

$$\max_{\mathbf{x} \in X} u(\mathbf{x})$$

subject to $\quad \mathbf{p}^T \mathbf{x} = m$

the gradient of the objective function

$$\nabla u(\mathbf{x}^*) = (D_{x_1} u[\mathbf{x}^*], D_{x_2} u[\mathbf{x}^*], \ldots, D_{x_n} u[\mathbf{x}^*])$$

lists the marginal utility of each good. The gradient of the budget constraint $\nabla g(\mathbf{x}^*) = (p_1, p_2, \ldots, p_n)$ lists the price of each good. A necessary condition for the $\mathbf{x}^*$ to be an optimal consumption bundle is that there exists a Lagrange multiplier λ such that these gradients are proportional, that is,

$$\nabla u(\mathbf{x}^*) = \lambda \mathbf{p} \quad \text{or} \quad D_{x_i} u[\mathbf{x}] = \lambda p_i \quad \text{for every } i = 1, 2, \ldots, n \tag{16}$$

At the optimal consumption bundle $\mathbf{x}^*$, the marginal utility of each good must be proportional to its price.

Equation (16) is known as the first-order condition for utility maximization. It can be rearranged to give

$$\mathrm{MRS}_{ij}(x^*) = \frac{D_{x_i} u[\mathbf{x}^*]}{D_{x_j} u[\mathbf{x}^*]} = \frac{p_i}{p_j} \tag{17}$$

Utility maximization requires that, at the optimal consumption bundle, marginal rate of substitution between any two goods is equal to their relative prices. With two goods the marginal rate of substitution measures the slope of the indifference curve through $\mathbf{x}^*$ (example 4.11). The price ratio measures the slope of the budget constraint. Equation (17) expresses the familiar condition that the optimal consumption bundle is found where the budget line is tangential to an indifference curve (figure 5.4).

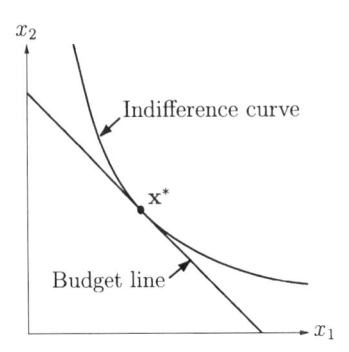

Example 5.16 (Cost minimization) Assume that the technology of a competitive firm producing a single output y can be represented by the production function $y = f(\mathbf{x})$. The minimum cost of producing output y is

$$\min_{\mathbf{x}} \mathbf{w}^T \mathbf{x}$$

subject to $f(\mathbf{x}) = y$

where $\mathbf{w}$ lists the prices of the inputs. This is equivalent to

$$\max_{\mathbf{x}} -\mathbf{w}^T \mathbf{x}$$

subject to $-f(\mathbf{x}) = -y$

The gradient of the objective function is the vector of factor prices $-\mathbf{w}$, while the gradient of the production constraint is $-\nabla f(\mathbf{x}^*)$. A necessary condition for minimizing cost is that these be proportional, that is,

$$-\mathbf{w} = -\lambda \nabla f(\mathbf{x}^*) \quad \text{or} \quad w_i = \lambda D_{x_i} f[\mathbf{x}^*] \quad \text{for every } i = 1, 2 \ldots, n$$

To minimize costs, inputs should be used in such proportions that their marginal products are proportional to the prices. This is known as the first-order condition for cost minimization.

For nonnegative variables, we have the following corollary to theorem 5.2.

5.3 Equality Constraints

Corollary 5.2.1 (Nonnegative variables) *Suppose that $\mathbf{x}^*$ is a local optimum of*

$$\max_{\mathbf{x} \geq \mathbf{0}} f(\mathbf{x})$$

subject to $\mathbf{g}(\mathbf{x}) = \mathbf{0}$

and a regular point of $\mathbf{g}$. *Then there exists multipliers $\lambda_1, \lambda_2, \ldots, \lambda_m$ such that*

$$\nabla f(\mathbf{x}^*) \leq \sum \lambda_j \nabla g_j(\mathbf{x}^*), \qquad \mathbf{x}^* \geq \mathbf{0}$$

$$\left(\nabla f(\mathbf{x}^*) - \sum \lambda_j \nabla g_j(\mathbf{x}^*)\right)^T \mathbf{x}^* = 0$$

Proof Suppose without loss of generality that the first k components of $\mathbf{x}^*$ are strictly positive while the remaining components are zero. That is

$$x_i^* > 0, \qquad i = 1, 2, \ldots, k$$
$$x_i^* = 0, \qquad i = k+1, k+2, \ldots, n$$

Clearly, $\mathbf{x}^*$ solves the problem

$$\max_{\mathbf{x} \geq \mathbf{0}} f(\mathbf{x})$$

subject to $\mathbf{g}(\mathbf{x}) = \mathbf{0}$

$$x_i = 0, \quad i = k+1, k+2, \ldots, n$$

By theorem 5.2, there exist multipliers $\lambda_1, \lambda_2, \ldots, \lambda_m$ and $\mu_{k+1}, \mu_{k+2}, \ldots, \mu_n$ such that

$$\nabla f(\mathbf{x}^*) = \sum_{j=1}^{m} \lambda_j \nabla g_j(\mathbf{x}^*) + \sum_{i=k+1}^{n} \mu_i \mathbf{e}_i$$

where $\mathbf{e}_i$ is the ith unit vector (example 1.79). Furthermore $\mu_i \geq 0$ for every i so that

$$\nabla f(\mathbf{x}^*) \leq \sum_{j=1}^{m} \lambda_j \nabla g_j(\mathbf{x}^*) \qquad (18)$$

and for every $i = 1, 2, \ldots, n$,

either $x_i^* = 0$

or $D_{x_i} f[\mathbf{x}^*] = \sum \lambda_j D_{x_i} g_j[\mathbf{x}^*]$

or both.

Given (18) and $\mathbf{x}^* \geq \mathbf{0}$, this is expressed concisely in the condition

$$\left(\nabla f(\mathbf{x}^*) - \sum \lambda_j \nabla g_j(\mathbf{x}^*) \right)^T \mathbf{x}^* = 0$$

Example 5.17 (The consumer's problem) Since negative consumption is impossible, the consumer's problem should properly be posed as

$$\max_{\mathbf{x} \geq \mathbf{0}} u(\mathbf{x})$$

subject to $\mathbf{p}^T \mathbf{x} = m$

The first-order condition for utility maximization is

$$\nabla u(\mathbf{x}^*) \leq \lambda \mathbf{p}, \qquad \mathbf{x}^* \geq \mathbf{0}$$

$$(\nabla u(\mathbf{x}^*) - \lambda p)^T \mathbf{x}^* = 0$$

This means for every good i,

$$D_i u[\mathbf{x}^*] \leq \lambda p_i \quad \text{and} \quad D_i u[\mathbf{x}^*] = \lambda p_i \quad \text{if } x_i^* > 0$$

A good will not be purchased ($x_i^* = 0$) if its marginal utility fails to exceed the critical level λp_i. Furthermore, for goods i, j that are purchased, the latter condition can be equivalently expressed as

$$\text{MRS}_{ij} = \frac{D_i u[\mathbf{x}^*]}{D_j u[\mathbf{x}^*]} = \frac{p_i}{p_j} \tag{19}$$

The consumer optimality conditions (marginal rate of substitution equal to price ratio) apply only to goods that are actually included in the optimal consumption bundle ($x_j > 0$). Typical practice is to focus on attention on the goods that are actually purchased (interior solutions) and to ignore the rest. The analyst must keep the possibility of boundary solutions in mind, and remain alert to the complications that might arise if a change in conditions causes the optimal solution to move away from a particular boundary. For instance, the price and income elasticities of a unpurchased good are zero. However, if prices and incomes change sufficiently to induce the purchase of a previously unpurchased product, its price and income elasticities will become nonzero. Caviar does not feature in the shopping baskets of the average consumer, and her price elasticity would be zero. A sufficient fall in the price of caviar might alter that substantially.

Exercise 5.15 (Quasi-linear preferences)
Analyze the consumer's problem where

$$u(\mathbf{x}) = x_1 + a \log x_2$$

ensuring that consumption is nonnegative. For simplicity, assume that $p_1 = 1$.

Sometimes nonnegativity constraints apply to a subset of the variables, for which we have the following variation of theorem 5.2 and corollary 5.2.1.

Exercise 5.16
Suppose that $(\mathbf{x}^*, \mathbf{y}^*)$ is a local optimum of

$$\max_{\mathbf{x}, \mathbf{y}} f(\mathbf{x}, \mathbf{y})$$

subject to $\mathbf{g}(\mathbf{x}, \mathbf{y}) = \mathbf{0}$ and $\mathbf{y} \geq \mathbf{0}$

and a regular point of $\mathbf{g}$. Then there exist multipliers $\lambda_1, \lambda_2, \ldots, \lambda_m$ such that

$$D_\mathbf{x} f[\mathbf{x}^*, \mathbf{y}^*] = \sum \lambda_j D_\mathbf{x} g_j[\mathbf{x}^*, \mathbf{y}^*] \quad \text{and} \quad D_\mathbf{y} f[\mathbf{x}^*, \mathbf{y}^*] \leq \sum \lambda_j D_\mathbf{y} g_j[\mathbf{x}^*, \mathbf{y}^*]$$

with

$$D_{y_i} f[\mathbf{x}^*, \mathbf{y}^*] = \sum \lambda_j D_{y_i} g_j[\mathbf{x}^*, \mathbf{y}^*] \quad \text{if } y_i > 0$$

5.3.2 The Geometric Approach

In example 5.15 we showed that the optimal choice of a consumer between two commodities was found where the budget line is tangent to an indifference curve. The tangency requirement extends to more general optimization problems. Consider the optimization problem depicted in figure 5.5. The objective is to locate the highest point on the concave surface while remaining on the curved constraint.

Analysis of such a problem is facilitated by considering the contours (section 2.1.4) of the objective function and the constraint respectively, as

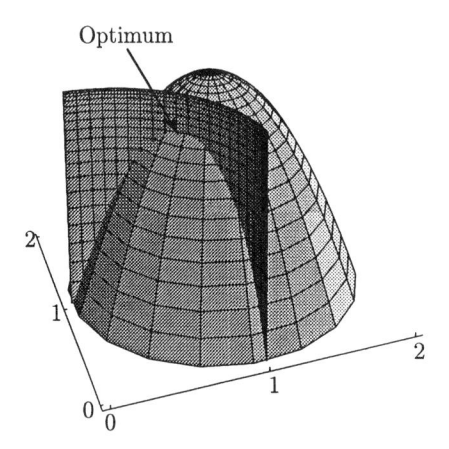

Figure 5.5
The optimum is the highest point that is common to the objective surface and the constraint

in figure 5.6. The contour $f(\mathbf{x}) = v_1$ shows the set of all $\mathbf{x}$ at which the objective function attains a given level v_1. The second contour $f(\mathbf{x}) = v_2$ contains points that achieve a higher level of the objective. Choice is constrained to lie on the curve $g(\mathbf{x}) = 0$.

Clearly, no point can be an optimal solution if it is possible to move along the constraint to a higher contour of the objective function. For example, $\hat{\mathbf{x}}$ cannot be an optimal solution, since it is possible to move along the constraint in the direction of $\mathbf{x}^*$ increasing the value of $f(\mathbf{x})$. Only at a point where the constraint curve is tangential to a contour of the objective function is it impossible to move to a higher contour while simultaneously satisfying the constraint. $\mathbf{x}^*$ is such a point, since it impossible to move along the constraint without reducing the value of $f(\mathbf{x})$. Tangency between a contour of the objective function and the constraint is a necessary condition for optimality. The point of tangency $\mathbf{x}^*$, which lies on the highest possible contour consistent with the constraint curve $g(\mathbf{x}) = 0$, is the optimal solution to this constrained maximization problem.

To show that tangency between the objective function and the constraint is equivalent to the first-order condition (13), we observe that the slope of the objective function at $\mathbf{x}^*$ is given by (example 4.11)

$$\text{Slope of } f = -\frac{D_{x_1} f[\mathbf{x}^*]}{D_{x_2} f[\mathbf{x}^*]}$$

5.3 Equality Constraints

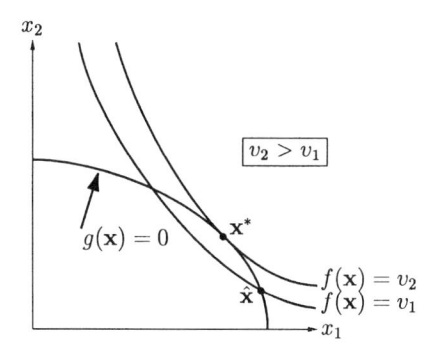

Figure 5.6
Tangency between the constraint and the objective function

Similarly, the slope of the constraint at $\mathbf{x}^*$ is

$$\text{Slope of } g = -\frac{D_{x_1} g[\mathbf{x}^*]}{D_{x_2} g[\mathbf{x}^*]}$$

Tangency requires that these slopes be equal, that is,

$$\frac{D_{x_1} f[\mathbf{x}^*]}{D_{x_2} f[\mathbf{x}^*]} = \frac{D_{x_1} g[\mathbf{x}^*]}{D_{x_2} g[\mathbf{x}^*]}$$

which can be rewritten as

$$\frac{D_{x_1} f[\mathbf{x}^*]}{D_{x_1} g[\mathbf{x}^*]} = \frac{D_{x_2} f[\mathbf{x}^*]}{D_{x_2} g[\mathbf{x}^*]} = \lambda \tag{20}$$

Letting λ denote the common value, (20) can be rewritten as two equations:

$$D_{x_1} f[x^*] = \lambda D_{x_1} g[\mathbf{x}^*]$$

$$D_{x_2} f[\mathbf{x}^*] = \lambda D_{x_2} g[\mathbf{x}^*]$$

or more succinctly

$$\nabla f(\mathbf{x}^*) = \lambda \nabla g(\mathbf{x}^*)$$

This is the basic first-order necessary condition for an optimum (theorem 5.2).

The same reasoning applies when there are more than two decision variables, as in

$$\max_{\mathbf{x} \in X} f(\mathbf{x})$$

subject to $\quad g(\mathbf{x}) = \mathbf{0}$

The constraint $g(\mathbf{x}) = 0$ defines a surface in X (section 4.2). A contour of the objective function passes through every point on the constraint surface. No point $\mathbf{x}$ on the constraint surface can be optimal if the contour of the objective function through $\mathbf{x}$ intersects the constraint surface. Therefore a necessary condition for a point $\mathbf{x}^*$ to be optimal is that the contour through $\mathbf{x}^*$ is tangential to the constraint surface at $\mathbf{x}^*$. Tangency of two surfaces requires that the gradient of the objective function $\nabla f(\mathbf{x}^*)$ be orthogonal to the constraint surface or alternatively that their respective normals be aligned, that is,

$$\nabla f(\mathbf{x}^*) = \lambda \nabla g(\mathbf{x}^*)$$

This again is the basic first-order necessary condition for an optimum.

With multiple constraints, the geometric perspective becomes harder to visualize. Provided that the constraints are differentiable at $\mathbf{x}^*$ and $D\mathbf{g}[\mathbf{x}^*]$ has full rank, the feasible set is a smooth surface in X to which $D\mathbf{g}[\mathbf{x}^*]$ defines the tangent plane at $\mathbf{x}^*$. Again, a point $\mathbf{x}^*$ cannot be optimal if the contour surface of the objective function through $\mathbf{x}^*$ intersects the constraint surface. Optimality requires that the gradient of the objective function be perpendicular (orthogonal) to the tangent plane of the constraint surface.

A case with two constraints is illustrated in figure 5.7. The feasible set is the curve G defined by the intersection of the two constraint surfaces $g_1(\mathbf{x}) = 0$ and $g_2(\mathbf{x}) = 0$. Any point on this curve is a feasible solution to the optimization problem. An optimal solution $\mathbf{x}^*$ requires that the gradient of the objective function at $\mathbf{x}^*$ be orthogonal to the curve at $\mathbf{x}^*$, since otherwise a better point could be found. This will only be the case if the gradient of f lies in the plane generated by the normals ∇g_1 and ∇g_2 to the constraint surfaces g_1 and g_2 respectively. That is, optimality requires that the gradient of the objective function at the optimal point $\mathbf{x}^*$ be a linear combination of the gradients of the constraints

$$\nabla f(\mathbf{x}^*) = \lambda_1 \nabla g_1(\mathbf{x}^*) + \lambda_2 \nabla g_2(\mathbf{x}^*)$$

where the weights λ_j are the Lagrange multipliers.

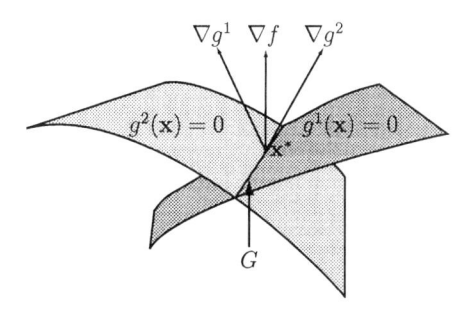

Figure 5.7
A problem with two constraints

Remark 5.1 The illustration in figure 5.6 showing two disjoint convex sets suggests a possible role for a separation theorem. This is indeed appropriate. Theorem 5.2 is an application of the Fredholm alternative, which we derived from the separating hyperplane theorem (exercise 3.199). Exercise 5.34 invites you to establish the converse, deriving the Farkas lemma from the necessary conditions for a constrained optimization problem. In section 5.4.5 we will apply the separating hyperplane theorem directly to obtain a stronger theorem for a particular class of problems. This underlines the close relationship between constrained optimization and the separation of convex sets.

5.3.3 The Implicit Function Theorem Approach

A straightforward approach to solving some constrained optimization problems is use the constraint to solve for some of the decision variables, converting the problem into unconstrained optimization problem. This technique is illustrated by the following example.

Example 5.18 (Designing a vat) Consider the problem of designing an open rectangular vat using a given quantity A of sheet metal so as to hold the maximum possible volume. It is seems intuitive that the base of the vat should be square, and this intuition is correct. But what is the optimal proportion of height to width? Should the vat be short and squat or alternatively tall and narrow? On this, our intuition is less clear. By formulating this as a problem of constrained maximization, the answer can be readily attained.

For simplicity, assume that the base of the vat is square and that all the sheet metal is utilized. There are two choice variables—width (w) and height (h). The volume of the vat is base times height

volume $= w^2 h$

while the area of sheet metal is base plus sides

area $= w^2 + 4wh$

The constrained maximization problem is

$$\max_{w,h} \; f(w,h) = w^2 h$$

subject to $\quad g(w,h) = w^2 + 4wh - A = 0$

Once the size of the base is determined, the height is implicitly determined by the available sheet metal. That is, w^2 is required for the base, leaving $A - w^2$ for the four walls, which implies that the height is limited to $(A - w^2)/4w$. That is, we can solve for h in terms of w from the constraint

$$h = \frac{A - w^2}{4w} \tag{21}$$

and substitute this into the objective function

$$\text{volume}(w) = w^2 \frac{A - w^2}{4w} = w \frac{A - w^2}{4} \tag{22}$$

obtaining an expression for the feasible volume in terms of w alone. This converts the constrained maximization problem into an equivalent unconstrained maximization problem that can be solved using the standard technique. The feasible volume (22) is maximized where

$$D_w \, \text{volume}[w] = w \frac{A - w^2}{4} - \frac{2w^2}{4} = 0$$

whose solution is $w = \sqrt{A/3}$. The corresponding h can be found by substituting in the constraint (21).

While this technique is conceptually straightforward, it is not always possible to solve the constraint explicitly. In many problems the con-

5.3 Equality Constraints

straint will be intractable or may not have an explicit functional form. However, provided that the constraint is differentiable, it is always possible to solve it approximately using the derivative. This is the essence of the implicit function theorem (theorem 4.5), which provides an alternative derivation of the Lagrange multiplier method. Using this approach, we use the implicit function theorem to solve the constraint locally, effectively converting a constrained maximization problem into an equivalent unconstrained problem using a linear approximation.

Applying the implicit function theorem to the general equality-constrained maximization problem

$$\max_{\mathbf{x} \in X} f(\mathbf{x}) \quad \text{subject to} \quad \mathbf{g}(\mathbf{x}) = \mathbf{0} \tag{23}$$

is an exercise in the manipulation of linear functions. Suppose that $\mathbf{x}^*$ solves (23). Provided that $D\mathbf{g}[\mathbf{x}^*]$ has full rank m, we can decompose $\mathbf{x} \in \mathfrak{R}^n$ into two subvectors $(\mathbf{x}_1, \mathbf{x}_2)$, where $\mathbf{x}_1 \in \mathfrak{R}^m$ and $\mathbf{x}_2 \in \mathfrak{R}^{n-m}$ such that $D_{\mathbf{x}_1}\mathbf{g}[\mathbf{x}^*]$ has rank m. Applying the implicit function theorem (theorem 4.5) to the constraint $\mathbf{g}(\mathbf{x}_1, \mathbf{x}_2) = \mathbf{0}$, there exists a differentiable function $\mathbf{h}: \mathfrak{R}^{n-m} \to \mathfrak{R}^m$ such that

$$\mathbf{x}_1 = \mathbf{h}(\mathbf{x}_2) \quad \text{and} \quad \mathbf{g}(\mathbf{h}(\mathbf{x}_2), \mathbf{x}_2) = \mathbf{0} \tag{24}$$

for all $\mathbf{x}_2$ in a neighborhood of $\mathbf{x}_2^*$ and

$$D_{\mathbf{x}_2}\mathbf{h}[\mathbf{x}_2^*] = -(D_{\mathbf{x}_1}\mathbf{g}[\mathbf{x}^*])^{-1} \circ D_{\mathbf{x}_2}\mathbf{g}[\mathbf{x}^*] \tag{25}$$

Substituting (24) into the objective function converts the constrained maximization problem (23) into an unconstrained maximization problem

$$\max_{\mathbf{x}_2} f(\mathbf{h}(\mathbf{x}_2), \mathbf{x}_2) \tag{26}$$

A necessary condition for $\mathbf{x}_2^*$ to solve (26) is that $f(\mathbf{h}(\mathbf{x}_2), \mathbf{x}_2)$ be stationary at $\mathbf{x}_2^*$ (theorem 5.1), that is $\mathbf{x}_2^*$ must satisfy the first-order condition

$$D_{\mathbf{x}_2} f(\mathbf{h}(\mathbf{x}_2), \mathbf{x}_2) = D_{\mathbf{x}_1} f \circ D_{\mathbf{x}_2}\mathbf{h} + D_{\mathbf{x}_2} f = 0$$

which we obtained by using the chain rule (exercise 4.22), suppressing function arguments for clarity. Substituting (25) yields

$$-D_{\mathbf{x}_1} f \circ (D_{\mathbf{x}_1}\mathbf{g})^{-1} \circ D_{\mathbf{x}_2}\mathbf{g} + D_{\mathbf{x}_2} f = 0 \tag{27}$$

Now $D_{\mathbf{x}_1}f$ is a linear functional on $\Re^m$. $D_{\mathbf{x}_1}\mathbf{g}$ comprises m linearly independent functionals on $\Re^m$, which span the dual space $(\Re^m)^*$. Therefore there exists $\lambda \in \Re^m$ such that

$$D_{\mathbf{x}_1}f = \lambda^T D_{\mathbf{x}_1}\mathbf{g} \tag{28}$$

Substituting into (27) yields

$$-\lambda^T D_{\mathbf{x}_1}\mathbf{g} \circ (D_{\mathbf{x}_1}\mathbf{g})^{-1} \circ D_{\mathbf{x}_2}\mathbf{g} + D_{\mathbf{x}_2}f = 0$$

or

$$D_{\mathbf{x}_2}f = \lambda^T D_{\mathbf{x}_2}\mathbf{g} \tag{29}$$

Combining (28) and (29) gives the necessary condition for a local optimum

$$D_{\mathbf{x}}f[\mathbf{x}^*] = \lambda^T D_{\mathbf{x}}\mathbf{g}[\mathbf{x}^*] \quad \text{or} \quad \nabla f(\mathbf{x}^*) = \sum_{j=1}^m \lambda_j \nabla g_j(\mathbf{x}^*) \tag{30}$$

Remark 5.2 The preceding derivation provides an alternative proof of the theorem 5.2, based on the implicit function theorem rather than the separating hyperplane theorem. Consequently its insight and motivation are quite distinct.

Exercise 5.17
Characterize the optimal solution of the general two-variable constrained maximization problem

$$\max_{x_1, x_2} f(x_1, x_2)$$

subject to $\quad g(x_1, x_2) = 0$

using the implicit function theorem to solve the constraint.

Exercise 5.18
The consumer maximization problem is one in which it is possible to solve the constraint explicitly, since the budget constraint is linear. Characterize the consumer's optimal choice using this method, and compare your derivation with that in example 5.15.

5.3.4 The Lagrangean

We have established that the first-order necessary condition for a constrained optimum is that the derivative of objective function be a linear combination of the derivatives of the constraints

$$Df[\mathbf{x}^*] = \sum_{j=1}^{m} \lambda_j Dg_j[\mathbf{x}^*]$$

This can be expressed alternatively as

$$Df[\mathbf{x}^*] - \sum_{j=1}^{m} \lambda_j Dg_j[\mathbf{x}^*] = \mathbf{0}$$

which is the necessary condition for a stationary value of the function

$$L(\mathbf{x}, \lambda) = f(\mathbf{x}) - \sum_{j=1}^{m} \lambda_j g_j(\mathbf{x})$$

$L(\mathbf{x}, \lambda)$ is called the *Lagrangean* and $\lambda = (\lambda_1, \lambda_2, \ldots, \lambda_m)$ the *Lagrange multipliers*.

The Lagrangean is constructed by taking a *linear combination* (weighted average) of the objective function f and the constraints g_j. Stationarity of the Lagrangean is necessary but not sufficient for a solution of the constrained optimization problem (7). The set of stationary points of the Lagrangean contains all local maxima (and minima) of f on the feasible set $G = \{\mathbf{x} \in X : \mathbf{g}(\mathbf{x}) = \mathbf{0}\}$, but it may contain other points as well. Other conditions are needed to distinguish maxima from minima and other critical points of the Lagrangean. Corollaries 5.2.2 and 5.2.3 give second-order necessary and sufficient conditions analogous to theorem 5.1 and corollary 5.1.1. For a local maximum, the Lagrangean must be locally concave, but only with respect to the subspace of feasible perturbations.

Corollary 5.2.2 (Necessary conditions for constrained maximum) Suppose that $\mathbf{x}^*$ is a local optimum of

$$\max_{\mathbf{x} \in X} f(\mathbf{x}) \quad \text{subject to} \quad \mathbf{g}(\mathbf{x}) = \mathbf{0} \tag{31}$$

and a regular point of $\mathbf{g}$. *Then there exist* $\lambda = (\lambda_1, \lambda_2, \ldots, \lambda_m)$ *such that*

1. $(\mathbf{x}^*, \lambda)$ *is a stationary point of the Lagrangean* $L(\mathbf{x}, \lambda) = f(\mathbf{x}) - \sum_{j=1}^{m} \lambda_j g_j(\mathbf{x})$, *that is*, $D_\mathbf{x} L[\mathbf{x}^*, \lambda] = \mathbf{0}$
2. $H_L(\mathbf{x}^*)$ *is nonpositive definite on the hyperplane tangent to* $\mathbf{g}$, *that is*,

$$\mathbf{x}^T H_L(\mathbf{x}^*) \mathbf{x} \leq 0 \text{ for every } \mathbf{x} \in T = \{\mathbf{x} \in X : D\mathbf{g}[\mathbf{x}^*](\mathbf{x}) = \mathbf{0}\}$$

Corollary 5.2.3 (Sufficient conditions for constrained maximum) Suppose there exists $\mathbf{x}^* \in X$ and $\lambda \in \Re^m$ such that

1. $(\mathbf{x}^*, \lambda)$ is a stationary point of the Lagrangean $L(\mathbf{x}, \lambda) = f(\mathbf{x}) - \sum_{j=1}^{m} \lambda_j g_j(\mathbf{x})$, that is, $D_\mathbf{x} L[\mathbf{x}^*, \lambda] = \mathbf{0}$
2. $H_L(\mathbf{x}^*)$ is negative definite on the hyperplane tangent to $\mathbf{g}$, that is,

$\mathbf{x}^T H_L(\mathbf{x}^*) \mathbf{x} < 0$ *for every nonzero* $\mathbf{x} \in T = \{\mathbf{x} \in X : D\mathbf{g}[\mathbf{x}^*](\mathbf{x}) = \mathbf{0}\}$

Then $\mathbf{x}^*$ *is a strict local maximum of* $\max_{\mathbf{x} \in X} f(\mathbf{x})$ *subject to* $\mathbf{g}(\mathbf{x}) = \mathbf{0}$.

Exercise 5.19
Prove corollary 5.2.3.

We will not prove the second part of corollary 5.2.2, since it will not be used elsewhere in the book (see Luenberger 1984, p. 226; Simon 1986, p. 85). Instead we develop below (corollary 5.2.4, section 5.4.3) some global conditions that are useful in practice.

Remark 5.3 (Form of the Lagrangean) At this point it seems immaterial whether we construct the Lagrangean by adding or subtracting the constraints from the objective function. If we write the Lagrangean as $L = f + \sum \lambda_j g_j$ rather than $L = f - \sum \lambda_j g_j$, we will arrive at the same stationary values $\mathbf{x}$, although the associated Lagrange multipliers will change sign. However, when we allow for inequality constraints $\mathbf{g}(\mathbf{x}) \leq \mathbf{0}$, the chosen form $f - \sum \lambda_j g_j$ ensures that the Lagrange multipliers are nonnegative, which is the appropriate sign for their interpretation as shadow prices of the constraints (proposition 5.2). Also, the form $f - \sum \lambda_j g_j$ allows the interpretation of the Lagrangean as the net benefit function in section 5.3.6.

In consulting other texts, you should be wary that some authors pose the general optimization problem as

max $f(\mathbf{x})$ subject to $\mathbf{g}(\mathbf{x}) \geq \mathbf{0}$

or

min $f(\mathbf{x})$ subject to $\mathbf{g}(\mathbf{x}) \leq \mathbf{0}$

in which cases the most appropriate form of the Lagrangean is $f + \sum \lambda_j g_j$.

Clearly, the second-order necessary conditions are always satisfied when the Lagrangean is concave, in which case stationarity is also sufficient for a global maximum.

Exercise 5.20
Suppose that $(\mathbf{x}^*, \lambda)$ is a stationary point of the Lagrangean

$$L(\mathbf{x}, \lambda) = f(\mathbf{x}) - \sum \lambda_j g_j(\mathbf{x})$$

and $L(\mathbf{x}, \lambda)$ is concave in $\mathbf{x}$. Then $\mathbf{x}^*$ is a global solution of the problem

$$\max_{\mathbf{x} \in X} f(\mathbf{x}) \quad \text{subject to} \quad \mathbf{g}(\mathbf{x}) = \mathbf{0}$$

In general, knowledge of the solution is required before concavity of the Lagrangean can be verified. However, in the common instance of concave objective function and affine constraint, the Lagrangean is always concave, providing the following useful analogue of corollary 5.1.2.

Corollary 5.2.4 (Affine constraint) *Suppose that f is concave and $\mathbf{g}$ is affine of full rank. Then $\mathbf{x}^*$ is a global solution of*

$$\max_{\mathbf{x} \in X} f(\mathbf{x}) \quad \text{subject to} \quad \mathbf{g}(\mathbf{x}) = \mathbf{0}$$

if and only if $\mathbf{x}^*$ is a stationary point of the Lagrangean

$$L(\mathbf{x}, \lambda) = f(\mathbf{x}) - \sum \lambda_j g_j(\mathbf{x})$$

satisfying the constraint $\mathbf{g}(\mathbf{x}) = \mathbf{0}$.

Proof Since $\mathbf{g}$ is affine, $\mathbf{g}(\mathbf{x}) = \mathbf{h}(\mathbf{x}) + \mathbf{y}$, with $\mathbf{h}$ linear of full rank (exercise 3.39). Therefore $D\mathbf{g} = \mathbf{h}$ and $\mathbf{g}$ is regular. Therefore, if $\mathbf{x}^*$ is a (local) optimum, it is necessary that the Lagrangean is stationary. Conversely, since an affine function is both concave and convex (exercise 3.130), the Lagrangean

$$L(\mathbf{x}, \lambda) = f(\mathbf{x}) - \sum \lambda_j g_j(\mathbf{x})$$

is concave irrespective of the sign of λ_j (exercise 3.131). By exercise 5.20, every stationary point is a global optimum. □

Remark 5.4 If we knew that the Lagrange multipliers were nonnegative, then corollary 5.2.4 could be generalized to convex constraints. In general, however, the sign of the Lagrange multipliers are not restricted, which is why corollary 5.2.4 is limited to an affine constraint function. We will provide more general sufficiency conditions in section 5.4.3.

It is sometimes said that the Lagrangean technique converts constrained maximization into unconstrained maximization. This is incorrect. Corollary 5.2.2 states that if $\mathbf{x}^*$ maximizes f on G, then $\mathbf{x}^*$ must be stationary point of the Lagrangean L. This does not imply that $\mathbf{x}^*$ maximizes L on X (example 5.19). However, $\mathbf{x}^*$ does maximize L on the feasible set G (exercise 5.21).

Example 5.19 The point $(1, 1)$ is the only optimum of the problem

$$\max_{x_1, x_2} x_1 x_2 \quad \text{subject to} \quad x_1 + x_2 = 2$$

At the optimum the Lagrange multiplier $\lambda^* = 1$ and the optimum $(1, 1)$ is in fact a saddle point of the Lagrangean

$$L(x_1, x_2, \lambda^*) = x_1 x_2 - (x_1 + x_2 - 2)$$

It maximizes the Lagrangean along the constraint $x_1 + x_2 = 2$.

Exercise 5.21
If $\mathbf{x}^*$ maximizes $f(\mathbf{x})$ on $G = \{\mathbf{x} \in X : \mathbf{g}(\mathbf{x}) = \mathbf{0}\}$, then $\mathbf{x}^*$ maximizes the Lagrangean $L = f(\mathbf{x}) - \sum \lambda_j g_j(\mathbf{x})$ on G.

Example 5.20 (Designing a vat) Consider again the vat design problem (example 5.18)

$$\max_{w, h} f(w, h) = w^2 h$$

subject to $\quad g(w, h) = w^2 + 4wh - A = 0$

The Lagrangean for this problem, formed by adjoining the constraint (area) to the objective function (volume), is

$$L(w, h) = w^2 h - \lambda(w^2 + 4wh - A) \tag{32}$$

where λ is the Lagrange multiplier. The solution requires that the Lagrangean be stationary, for which it is necessary that

$$D_w L(w, h) = 2wh - \lambda(2w + 4h) = 0$$

$$D_h L(w, h) = w^2 - \lambda 4w = 0$$

To solve these necessary conditions for the optimal solution, we rewrite these equations as

$$2wh = \lambda(2w + 4h) \tag{33}$$

$$w^2 = \lambda 4w \tag{34}$$

Divide (33) by (34),

$$\frac{2wh}{w^2} = \frac{\lambda(2w + 4h)}{\lambda 4w}$$

and cancel the common terms to give

$$\frac{2h}{w} = \frac{w + 2h}{2w} = \frac{1}{2} + \frac{h}{w}$$

which can be solved to yield

$$\frac{h}{w} = \frac{1}{2} \quad \text{or} \quad h = \frac{1}{2}w \tag{35}$$

We conclude that squat vat, whose height is half the width, is the optimal design.

Note that we derived the desired conclusion, namely $h = w/2$, without actually solving for h and w. This is a common feature of economic applications, where the desired answer is not a specific number but a rule or principle that applies to all problems of a given type. In this example the rule is that the height should be half the width. An analogous result in economics is that the consumer should allocate expenditure so that the marginal rate of substitution is equal to the ratio of relative prices. To find the actual dimensions of the optimal vat, we substitute (35) into the constraint, yielding

$$w = \sqrt{\frac{A}{3}}, \quad h = \sqrt{\frac{A}{12}} \tag{36}$$

Exercise 5.22
Show that the volume of the vat is maximized by devoting one-third of the material to the floor and the remaining two-thirds to the walls.

Exercise 5.23
Design a rectangular vat (open at the top) of 32 cubic meters capacity so as to minimize the required materials.

Example 5.21 (The consumer's problem) The Lagrangean for the consumer's problem (example 5.15) is

$$L(\mathbf{x}, \lambda) = u(\mathbf{x}) - \lambda(\mathbf{p}^T\mathbf{x} - m)$$

A necessary condition for an optimal solution is that the Lagrangean be stationary, that is,

$$D_\mathbf{x}L[\mathbf{x}^*, \lambda] = D_\mathbf{x}u[\mathbf{x}^*] - \lambda\mathbf{p} = 0$$

or

$$\nabla u(\mathbf{x}^*) = \lambda\mathbf{p}$$

which is the standard first-order condition (16) for maximizing utility. If the utility function is concave, the first-order condition is also sufficient, since the budget constraint is affine (corollary 5.2.4).

Exercise 5.24
Solve the problem

$$\min_{x_1, x_2, x_3} x_1^2 + x_2^2 + x_3^2$$

subject to $\quad 2x_1 - 3x_2 + 5x_3 = 19$

The Lagrangean converts the search for constrained extrema into an unconstrained search for stationary points, at the cost of introducing additional variables λ_j. With n decision variable and m constraints, the first-order conditions for stationarity of the Lagrangean yield a system of n equations in the n decision variables x_i and the m unknown multipliers λ_j. Together with the m constraints, we have $n + m$ equations in $n + m$ unknowns. In principle, these can be solved to yield the optimal solution $\mathbf{x}^*$ and λ^*. Since these equations are typically nonlinear, there is no unique general method for their solution and often a general solution is not tractable. Fortunately, in economic applications, an explicit solution is often not required. The desired results are obtained by manipulation of the first-order conditions, as we did in example 5.21.

In those cases where it is necessary, facility in solving the first-order conditions comes with practice and the accumulation of small number of successful techniques, which are illustrated in the following examples. Example 5.20 illustrated one technique, in which the Lagrange multiplier is eliminated from the first-order conditions. Example 5.22 also uses this technique.

Example 5.22 A consumer's preferences over two goods can be represented by the utility function

$$u(\mathbf{x}) = \sqrt{x_1 x_2}$$

Suppose that the consumer's income is $12, and the two goods are priced at $p_1 = 1$ and $p_2 = 2$ respectively. We wish to find the consumer's optimal consumption bundle. The Lagrangean is

$$L(\mathbf{x}, \lambda) = \sqrt{x_1 x_2} - \lambda(x_1 + 2x_2 - 12)$$

Necessary conditions for the Lagrangean to be stationary are

$$D_{x_1} L(\mathbf{x}, \lambda) = \sqrt{\frac{x_2}{x_1}} - \lambda = 0$$

$$D_{x_2} L(\mathbf{x}, \lambda) = \sqrt{\frac{x_1}{x_2}} - 2\lambda = 0$$

or

$$\sqrt{\frac{x_2}{x_1}} = \lambda$$

$$\sqrt{\frac{x_1}{x_2}} = 2\lambda$$

We can eliminate the unknown multiplier λ by dividing one equation into the other

$$\frac{\sqrt{x_2/x_1}}{\sqrt{x_1/x_2}} = \frac{\lambda}{2\lambda}$$

which simplifies to

$$x_1 = 2x_2$$

From the budget constraint, we conclude that

$$x_1^* = 6 \quad \text{and} \quad x_2^* = 3$$

Exercise 5.25
Generalize the preceding example to solve

$$\max_{\mathbf{x}} u(\mathbf{x}) = x_1^\alpha x_2^{1-\alpha}$$

subject to $\quad p_1 x_1 + p_2 x_2 = m$

Another useful technique uses the first-order conditions to express the decision variables in terms of the Lagrange multiplier. These are then substituted in the constraint, which is solved for an explicit value for the Lagrange multiplier. In turn, this yields an explicit solution for the decision variables. This technique is used in the following example.

Example 5.23 (Logarithmic utility function) A very common specific functional form is the log-linear utility function

$$u(\mathbf{x}) = \sum_{i=1}^{n} \alpha_i \log x_i$$

where without loss of generality we can assume that $\sum_{i=1}^{n} \alpha_i = 1$. The Lagrangean for the consumer's problem

$$\max_{\mathbf{x} \in X} u(\mathbf{x}) = \sum_{i=1}^{n} \alpha_i \log x_i \quad \text{subject to} \quad \sum_{i=1}^{n} p_i x_i = m$$

is

$$L(\mathbf{x}, \lambda) = \sum_{i=1}^{n} \alpha_i \log x_i - \lambda \left(\sum_{i=1}^{n} p_i x_i - m \right)$$

Stationarity requires that

$$D_{x_i} L(\mathbf{x}, \lambda) = \frac{\alpha_i}{x_i} - \lambda p_i = 0$$

or

$$\alpha_i = \lambda p_i x_i \tag{37}$$

Summing over all goods and substituting in the budget constraint

$$1 = \sum_{i=1}^{n} \alpha_i = \lambda \sum_{i=1}^{n} p_i x_i = \lambda m$$

which implies that

$$\lambda = \frac{1}{m}$$

Substituting in (37), we conclude that utility maximization requires that the consumer spends proportion α_i on good i, that is,

$$p_i x_i = \frac{\alpha_i}{m}$$

Solving, the optimal quantity of good i is

$$x_i = \alpha_i \frac{m}{p_i}$$

Exercise 5.26
Solve the general Cobb-Douglas utility maximization problem

$$\max_{\mathbf{x}} \ u(\mathbf{x}) = x_1^{\alpha_1} x_2^{\alpha_2} \ldots x_n^{\alpha_n}$$

subject to $\quad p_1 x_1 + p_2 x_2 + \cdots p_n x_n = m$

[Hint: Follow the technique in example 5.23.]

Exercise 5.27 (CES production function)
A common functional form in production theory is the CES (constant elasticity of substitution) function

$$y = f(\mathbf{x}) = (a_1 x_1^\rho + a_2 x_2^\rho)^{1/\rho}$$

In this case the competitive firm's cost minimization problem is

$$\min_{x_1, x_2} \ w_1 x_1 + w_2 x_2$$

subject to $\quad f(\mathbf{x}) = (a_1 x_1^\rho + a_2 x_2^\rho)^{1/\rho} = y$

To facilitate solution, it is convenient to assume that the inputs are denominated so that $a_1 = a_2 = 1$ and to rewrite the constraint so that the problem becomes

$$\min_{x_1, x_2} \ w_1 x_1 + w_2 x_2$$

subject to $\quad x_1^\rho + x_2^\rho = y^\rho$

Find the cost-minimizing input levels for given values of w_1, w_2, and y.

We have shown that the optimal solution of a constrained maximization problem is necessarily a stationary point of the Lagrangean. Not only does this provide a powerful aid to solution, the Lagrangean function and its associated Lagrange multipliers have an insightful economic interpretation, as we explore in the next section.

5.3.5 Shadow Prices and the Value Function

We now introduce a parameter into functional constraints. Specifically, we consider the family of optimization problems

$$\max_{\mathbf{x} \in X} f(\mathbf{x})$$

subject to $\mathbf{g}(\mathbf{x}) = \mathbf{c}$

and explore how the value attained varies with the parameter $\mathbf{c}$.

Suppose that for every $\mathbf{c}$, there is a unique optimal solution $\mathbf{x}^*(\mathbf{c})$. In general, the variation of $\mathbf{x}^*$ with $\mathbf{c}$ is complex. However, the variation of $f(\mathbf{x}^*)$ with $\mathbf{c}$ can be readily estimated. Let $\mathbf{dx}$ denote the change in $\mathbf{x}$ following a change $\mathbf{dc}$ in $\mathbf{c}$. Then the increase in $f(\mathbf{x})$ is approximated by the derivative

$$df = f(\mathbf{x}^* + \mathbf{dx}) - f(\mathbf{x}^*) \approx Df[\mathbf{x}^*](\mathbf{dx}) \tag{38}$$

To satisfy the constraints, $\mathbf{dx}$ must satisfy (to a linear approximation)

$$Dg_j[\mathbf{x}^*](\mathbf{dx}) = dc_j \quad \text{for every } j = 1, 2 \ldots m$$

But we know that $\mathbf{x}^*$ satisfies the first-order condition (12)

$$Df[\mathbf{x}^*] = \sum_{j=1}^{m} \lambda_j Dg_j[\mathbf{x}^*]$$

and therefore substituting into (38),

$$df = Df[x^*](\mathbf{dx}) = \sum_{j=1}^{m} \lambda_j Dg_j[\mathbf{x}^*](\mathbf{dx}) = \sum_{j=1}^{n} \lambda_j \, dc_j = \lambda^T \mathbf{dc}$$

The change in the optimal value of the objective function $f(\mathbf{x}^*)$ following a change in the level of the constraint $\mathbf{dc}$ is equal to $\lambda^T \mathbf{dc}$, where $\lambda = (\lambda_1, \lambda_2, \ldots, \lambda_m)$ is the vector of Lagrange multipliers. Each Lagrange multipliers measures the rate of change of the objective function with respect to the level of the corresponding constraint. This is the essential insight of the following proposition, which will be proved in chapter 6.

Proposition 5.2 (Shadow prices) *Let* $\mathbf{x}^*$ *be a strict local optimum for the equality-constrained maximization problem*

$$\max_{\mathbf{x} \in X} f(\mathbf{x})$$

subject to $\quad \mathbf{g}(\mathbf{x}) = \mathbf{c}$

If f and $\mathbf{g}$ *are* C^2 *and* $Dg[\mathbf{x}^*]$ *is of full rank, then the value function*

$$v(\mathbf{c}) = \sup\{f(\mathbf{x}) : \mathbf{g}(\mathbf{x}) = \mathbf{c}\}$$

is differentiable with $\nabla v(\mathbf{c}) = \lambda$, *where* $\lambda = (\lambda_1, \lambda_2, \ldots, \lambda_m)$ *are the Lagrange multipliers associated with* $\mathbf{x}^*$.

Example 5.24 (Marginal utility of income) For the utility maximising consumer, an increase in income m will lead to a change in consumption $\mathbf{x}^*$ and a corresponding increase in utility. Without restrictions on consumer preferences, the change in consumption $\mathbf{x}^*$ cannot be easily characterized. However, the change in $u(\mathbf{x}^*)$ can. The change in utility following a unit increase in income is measured by λ, the Lagrange multiplier. Hence λ is called the *marginal utility of income*. The basic necessary condition for utility maximization (10),

$$\frac{D_{x_1} u[\mathbf{x}^*]}{p_1} = \frac{D_{x_2} u[\mathbf{x}^*]}{p_2} = \lambda$$

can be interpreted as saying that, at the optimal consumption bundle $\mathbf{x}^*$, it is immaterial how a small increment in income is divided among the different goods. The change in utility is the same no matter how the increment in income is spent.

Example 5.25 (Marginal cost) In the cost minimization problem of the competitive firm, the Lagrange multiplier measures an important economic quantity. The level of the constraint is the desired level of output and the objective function is the cost of production. The Lagrange multiplier measures the additional cost of producing an additional unit of output, that is the *marginal cost*. The first-order conditions for a competitive firm (example 5.16) can be written as

544   Chapter 5   Optimization

$$\lambda = \text{marginal cost} = \frac{w_i}{D_{x_i} f[\mathbf{x}^*]} \quad \text{for every } i$$

If the input mix is chosen optimally, it is immaterial which inputs are used to produce a marginal increase in output.

Comparing the first-order conditions for the cost minimization problem of the competitive firm

$$\mathbf{w} = \lambda \nabla f(\mathbf{x}^*)$$

with those of the corresponding profit maximization problem (example 5.11)

$$p \nabla f(\mathbf{x}^*) = \mathbf{w}$$

establishes the elementary rule characterizing the profit maximizing level of output, $p = \lambda = $ marginal cost.

Example 5.26 (Vat design)   Recall the problem of designing a volume-maximizing vat (example 5.18). After completing the design, the engineer discovers a small quantity of additional sheet metal. She does not have to redesign the vat in order to determine the additional volume possible. Rather, it can be calculated simply by multiplying the additional area dA by the Lagrange multiplier λ, that is,

Additional volume $= \lambda \, dA$

From equation (34), we find that

$$\lambda = \frac{w}{4} \tag{39}$$

Substituting in (36) yields

$$\lambda = \sqrt{\frac{A}{48}} \tag{40}$$

This measures the *shadow price* of the available sheet metal in terms of terms of the area available. For example, if the available sheet metal $A = 12$ square meters, then the shadow price of sheet metal is

$$\lambda = \sqrt{\frac{12}{48}} = \frac{1}{2}$$

Any additional sheet metal is worth approximately one-half a cubic meter of volume.

Suppose that instead of discovering additional sheet metal, a trader arrives offering to sell sheet metal at a price q per square meter. Assume further that the volume of the vat is worth p per cubic meter. Then, purchasing additional sheet metal is worthwhile provided the additional volume $\lambda \, dA$ is worth more than the cost of the additional sheet metal $q \, dA$, that is,

value of volume $= p\lambda \, dA \geq q \, da =$ cost

or $p\lambda \geq q$. λ measures the shadow price of sheet metal in units of volume, and so $p\lambda$ measures the shadow price of sheet metal in dollars. w measures the market price of sheet metal. Purchasing additional sheet is worthwhile as long as its shadow price $p\lambda$ exceeds its market price.

One of the useful products of the Lagrangean approach to constrained optimization problems is the automatic generation of shadow prices, which measures the costliness of the constraint in terms of the objective function. These frequently have an immediate role in economic discussion. For example, the marginal utility of income and the marginal cost are central concepts in consumer theory and producer theory respectively. While the Lagrange multipliers express shadow prices in the units of the objective function (utility, cost, volume), these can often be related directly to market prices as in the previous example.

5.3.6 The Net Benefit Approach

The preceding discussion of shadow prices leads to yet another approach to the Lagrange multiplier method, a fundamentally economic approach to the problem. Again, we introduce the idea by means of an example.

Example 5.27 (Designing a vat) Let us assume that the person charged with designing the vat subcontracts the problem to another decision maker, remunerating the latter according the volume produced and charging her for the sheet metal used (and rewarding her for sheet metal left over). That is, the vat owner puts a price of λ on the sheet metal and instructs the designer to maximize the net profit, namely

$$\Pi(w,h) = f(w,h) - \lambda g(w,h) = w^2 h - \lambda(w^2 + 4wh - A) \tag{41}$$

Equation (41) is precisely the Lagrangean for the vat design problem (32), the first-order conditions for maximizing which are given by equations (33) and (34), which yield the rule

$$h = \tfrac{1}{2}w$$

Note that we achieve this result without specifying the price λ of the sheet metal. In choosing w and h to maximize net profit (41), the designer will always design a vat of the optimal proportions irrespective of the price of the sheet metal. However, the size of the vat will be governed by the price of the sheet metal. By adjusting the price, the owner can ensure that the designer uses all the available sheet metal.

Exercise 5.28
In the vat design problem, suppose that 48 square meters of sheet metal is available. Show that if the shadow price of sheet metal is 1, designing a vat to maximize the net profit function produces a vat of the optimal shape and exactly exhausts the available metal.

Example 5.27 gives the fundamental economic insight of the Lagrangean method. We impute a price to each constraint and maximize the net benefit function

$$L(\mathbf{x}, \lambda) = f(\mathbf{x}) - \sum \lambda_j g_j(\mathbf{x}) \qquad (42)$$

consisting of the objective function minus the imputed value of the constraints. The first-order conditions for unconstrained maximization of the net benefit (42) define the optimal value of $\mathbf{x}$ in terms of the shadow prices λ. Together with the constraints, this enables us to solve for both the optimal $\mathbf{x}$ and the optimal shadow prices.

Example 5.28 (The competitive firm again) The Lagrangean for the competitive firm's cost minimization problem is

$$L(\mathbf{x}, \lambda) = \mathbf{w}^T \mathbf{x} - \lambda(f(\mathbf{x}) - y)$$

which can be rewritten as

$$L(\mathbf{x}, \lambda) = -(\lambda f(\mathbf{x}) - \mathbf{w}^T \mathbf{x}) + \lambda y \qquad (43)$$

The first term (in brackets) corresponds to the firm's net profit when the output is valued at price λ. For fixed y, minimizing (43) is equivalent to maximizing the net profit

$$\Pi = \lambda f(\mathbf{x}) - \mathbf{w}^T \mathbf{x}$$

The profit-maximizing output level $f(\mathbf{x}^*)$ will vary with the shadow price λ. Setting the shadow price equal to the marginal cost of production at y will ensure that profit maximizing output level $f(\mathbf{x}^*)$ is equal to the required output level y.

Example 5.29 (Electricity generation) An electricity company operates n generating plants. At each plant i, it costs $c_i(x_i)$ to produce x_i units of electricity. If the company aims to meet electricity demand D at minimum cost, its optimization problem is

$$\min_{\mathbf{x}} \sum_{i=1}^{n} c_i(x_i) \quad \text{subject to} \quad \sum_{i=1}^{n} x_i = D$$

The Lagrangean is

$$L(\mathbf{x}, \lambda) = \sum_{i=1}^{n} c_i(x_i) - \lambda \left(\sum_{i=1}^{n} x_i - D \right)$$

which can be rewritten as

$$L(\mathbf{x}, \lambda) = \sum_{i=1}^{n} c_i(x_i) + \lambda \left(D - \sum_{i=1}^{n} x_i \right) \tag{44}$$

Suppose that the company has the option of purchasing electricity from outside suppliers at a price λ. Then the Lagrangean (44) represents the sum of the costs of producing electricity at its own plants and purchasing electricity from outside. For any arbitrary price λ, the company will choose an optimal mix of its own production and outside purchase to minimize the total costs.

The first-order conditions for a minimum of (44) are

$$D_{x_i} L[x, \lambda] = D_{x_i} c_i[x_i] - \lambda = 0 \quad \text{for every } i \tag{45}$$

Optimality requires that the company utilize each plant to the level at which its marginal cost is equal to the alternative price λ. As the price increases, the proportion of demand that it satisfies from its own resources will increase. At some price λ^* the company will be induced to fill total demand D from its own production. This is the shadow price which arises from the solution of (45) together with the constraint

$$\sum_{i=1}^{n} x_i = D$$

and is the marginal cost of producing the total demand D from its own plants.

Exercise 5.29
Show how the shadow price can be used to decentralize the running of the power company, leaving the production level at each plant to be determined locally.

5.3.7 Summary

We see that the Lagrangean approach to constrained maximization is not just a convenient mathematical trick for converting constrained to unconstrained problems (as it is sometimes presented). It embodies a fundamentally economic approach to constrained maximization, which proceeds by putting a price on the constrained resources and maximizing the net benefit. You will enhance your understanding of constrained maximization immensely by keeping this interpretation at the forefront of your mind.

To recap, we have derived the basic necessary conditions, stationarity of the Lagrangean, in four different ways. They are:

Perturbation If a decision $\mathbf{x}^*$ is an optimal solution, it must be impossible to change or perturb $\mathbf{x}^*$ in any way consistent with the constraints and increase the value of the objective function. Using linear approximations to define the set of perturbations consistent with the constraints and also to measure the impact on the objective function, we showed that the impossibility of finding a value-improving perturbation led directly to the basic necessary conditions.

Geometric For a two-dimensional problem with a single constraint, it is obvious that an optimal solution must occur at a point of tangency between the constraint curve and the contours of the objective function. In a multidimensional problem, this generalizes to the objective function being orthogonal to the tangent space of the constraint.

Implicit function theorem A natural way to attempt to solve a constrained maximization problem is to use the constraint to eliminate some of the decision variables, thus converting a constrained problem into an equivalent unconstrained problem. In general, this technique is not available because the constraints are not given explicit functional forms. However, if the functions are differentiable, we can use the implicit function theorem to solve the constraints locally. This is equivalent to using a linear approximation to solve the constraint, and yields the basic necessary conditions.

Net benefit An economist might approach a constrained maximization problem by putting a price on the constrained resources and attempting to maximize the net benefit. The net benefit function is precisely the Lagrangean. Maximizing the net benefit requires stationarity of the Lagrangean. The Lagrange multipliers are the appropriate shadow prices that lead to utilization of the available resources.

All four approaches lead to the same first-order necessary condition for the solution of the constrained maximization problem

max $f(\mathbf{x})$ subject to $\mathbf{g}(\mathbf{x}) = \mathbf{0}$

namely that the Lagrangean $L(\mathbf{x}, \lambda) = f(\mathbf{x}) - \sum \lambda_j g_j(\mathbf{x})$ be stationary, that is,

$$D_\mathbf{x} L(\mathbf{x}, \lambda) = D_\mathbf{x} f(\mathbf{x}) - \sum \lambda_j D_\mathbf{x} g_j(\mathbf{x}) = 0$$

This is one of the must frequently used results in mathematical economics.

5.4 Inequality Constraints

In this section we analyze a constrained optimization problem in which the functional constrains are inequalities, as in

max $f(\mathbf{x})$ subject to $\mathbf{g}(\mathbf{x}) \leq \mathbf{0}$

Superficially the generalization of our results to inequality constraints seems quite straightforward. Simply disregarding those constraints which are not binding, we can adapt our previous results to obtain necessary conditions for a local optimum (section 5.3). However, subtle difficulties arise. Optima often appear on the boundary of the feasible set, complicating the determination of the set of feasible perturbations to be used in characterizing local optima. The regularity condition becomes too stringent and difficult to verify, yet some *constraint qualification* is required to guard against invalid inferences (section 5.4.2). On the other hand, when the objective function f is concave and the functional constraints g_j are convex, a beautiful alternative derivation is obtainable that leads to stronger and more robust conclusions (section 5.4.5).

5.4.1 Necessary Conditions

Suppose that $\mathbf{x}^*$ is a local solution of the problem

$$\max_{\mathbf{x} \in X} f(\mathbf{x}) \tag{46}$$

subject to $\quad g_j(\mathbf{x}) \leq 0, \quad j = 1, 2, \ldots, m$

Constraint j is *binding* at $\mathbf{x}^*$ if $g_j(\mathbf{x}^*) = 0$. Otherwise, $g_j(\mathbf{x}^*) < 0$, and constraint j is *slack*. Let

$$B(\mathbf{x}^*) = \{j : g_j(\mathbf{x}^*) = 0\}$$

be the set of binding constraints at $\mathbf{x}^*$.

Disregarding for the time being the slack constraints, the optimal solution $\mathbf{x}^*$ to (46) must a fortiori solve the restricted equality constrained problem

$$\max_{\mathbf{x} \in X} f(\mathbf{x}) \tag{47}$$

subject to $\quad g_j(\mathbf{x}) = 0 \quad$ for every $j \in B(\mathbf{x}^*)$

Given that the binding constraints $B(\mathbf{x}^*)$ satisfy the regularity condition, we can apply theorem 5.2 to (47) to derive the first-order condition that the gradient of $f(\mathbf{x}^*)$ be a linear combination of the gradients of the binding constraints, that is,

$$\nabla f(\mathbf{x}^*) = \sum_{j \in B(\mathbf{x}^*)} \lambda_j \nabla g_j(\mathbf{x}^*)$$

Furthermore we can deduce the sign of the Lagrange multipliers. Let v denote the value function for the family of problems

5.4 Inequality Constraints

$$\max_{\mathbf{x} \in X} f(\mathbf{x}) \tag{48}$$

subject to $\quad g_j(\mathbf{x}) = c_j \quad$ for every $j \in B(\mathbf{x}^*)$

By the envelope theorem (proposition 5.2), $\lambda_j = D_{c_j} v(\mathbf{c})$. Because of the direction of the original inequalities $g_j(\mathbf{x}) \le 0$, increasing c_j implies relaxing the constraint and expanding the feasible set. This cannot reduce the attainable value, so the value function must be increasing in c_j, that is, $\lambda_j = D_{c_j} v(c) \ge 0$ for every $j \in B(\mathbf{x}^*)$. The Lagrange multipliers must be *nonnegative*.

Assigning zero multipliers to the slack constraints ($\lambda_j = 0$ for every $j \notin B(\mathbf{x}^*)$), we can extend the sum to all the constraints

$$\nabla f(\mathbf{x}^*) = \sum_{j \in B(\mathbf{x}^*)} \lambda_j \nabla g_j(\mathbf{x}^*) + \sum_{j \notin B(\mathbf{x}^*)} \lambda_j \nabla g_j(\mathbf{x}^*)$$

$$= \sum_{j=1}^{m} \lambda_j \nabla g_j(\mathbf{x}^*) \tag{49}$$

where

$$\lambda_j \ge 0 \quad \text{and} \quad \lambda_j = 0 \text{ if } g_j(\mathbf{x}^*) < 0 \quad \text{for every } j = 1, 2, \ldots, m \tag{50}$$

The first condition (49) is the same as (13) with the additional requirement that the Lagrange multipliers λ_j be nonnegative. The second condition (50) requires that for every constraint *either*

1. the constraint is binding ($g_j(\mathbf{x}^*) = 0$), *or*
2. the Lagrange multiplier is zero ($\lambda_j = 0$), *or*
3. both

This is known as the *complementary slackness* condition, and can be expressed concisely as

$$\lambda_j g_j(\mathbf{x}^*) = 0 \quad \text{for every } j \tag{51}$$

since the zero product requires that at least one of the terms are zero.

Finally, what about the regularity condition? The condition necessary to apply theorem 5.2 involves only to the binding constraints. We do not require that $\mathbf{x}^*$ be a regular point of $\mathbf{g}$ as a whole. It suffices if $\mathbf{x}^*$ is a regular point of the binding components of $\mathbf{g}$. To this end, we will say that binding constraints $B(\mathbf{x}^*)$ are *regular* at $\mathbf{x}^*$ if their gradients $\nabla g_j(\mathbf{x}^*)$, $j \in B(\mathbf{x}^*)$ at $\mathbf{x}^*$ are linearly independent.

We have established the fundamental theorem of nonlinear programming, usually attributed to Kuhn and Tucker. After stating the theorem, we will present an alternative proof based on the Farkas lemma. This leads to an insightful geometric characterization. Exercise 5.32 invites yet another derivation from earlier work.

Remark 5.5 (Mixed problems) It is worth noting that our derivation applies equally to mixed problems in which there is a mixture of equality and inequality constraints, provided the regularity condition is satisfied by the equality constraints and the binding constraints jointly. The Lagrange multipliers attached to binding inequality constraints are necessarily nonnegative, whereas those attached to equality constraints are unrestricted in sign.

Although mixed problems arise regularly in practice, it is customary to maintain the distinction in theoretical treatments, since it is constraint qualifications become complicated for mixed problems. Indeed, the distinction is rather artificial, since it is usually straightforward to transform from one to the other by rephrasing the problem.

Theorem 5.3 (Kuhn-Tucker theorem) *Suppose that $\mathbf{x}^*$ is a local optimum of*

$$\max_{\mathbf{x}} f(\mathbf{x}) \tag{52}$$

subject to $\mathbf{g}(\mathbf{x}) \leq \mathbf{0}$

and the binding constraints are regular at $\mathbf{x}^*$. Then there exist unique multipliers $\lambda_1, \lambda_2, \ldots, \lambda_m \geq 0$ such that

$$\nabla f(\mathbf{x}^*) = \sum_{j=1}^{m} \lambda_j \nabla g_j(\mathbf{x}^*) \quad \text{and} \quad \lambda_j g_j(\mathbf{x}^*) = 0, \quad j = 1, 2\ldots, m$$

Proof We give an alternative proof which follows the perturbation approach which we used to establish theorem 5.2. Suppose that $\mathbf{x}^*$ solves (52). That is, $\mathbf{x}^*$ maximizes $f(\mathbf{x})$ subject to $\mathbf{g}(\mathbf{x}) \leq \mathbf{0}$. If this is so, there must be no *feasible* perturbation which achieves a higher level of $f(\mathbf{x})$. Determining the set of feasible perturbations requires distinguishing between binding and slack constraints. Let $B(\mathbf{x}^*)$ denote the set of constraints that are binding at $\mathbf{x}^*$ and let $S(\mathbf{x}^*)$ be the set of slack constraints at $\mathbf{x}^*$. That is,

$g_j(\mathbf{x}^*) = 0 \quad$ for every $j \in B(\mathbf{x}^*)$

$g_j(\mathbf{x}^*) < 0 \quad$ for every $j \in S(\mathbf{x}^*)$

If $S(\mathbf{x}^*)$ is nonempty, then by continuity there exists an open set U about $\mathbf{x}^*$ such that $g_j(\mathbf{x}) < 0$ for every $\mathbf{x} \in U$ and $j \in S(\mathbf{x}^*)$. That is, the slack constraints remain slack in an neighborhood of $\mathbf{x}^*$, and therefore they place no restrictions on the feasible perturbations provided the perturbations are sufficiently small.

A perturbation $\mathbf{dx}$ is feasible provided $g_j(\mathbf{x}^* + \mathbf{dx}) \leq 0$ for every $j \in B(\mathbf{x}^*)$ or to a linear approximation (and provided the binding constraints are regular)

$$dg_j = g_j(\mathbf{x}^* + \mathbf{dx}) - g_j(\mathbf{x}^*) \approx Dg_j[\mathbf{x}^*]\mathbf{dx} \leq 0 \tag{53}$$

For all perturbations $\mathbf{dx}$ satisfying (53), we must have

$f(\mathbf{x}^*) \geq f(\mathbf{x}^* + \mathbf{dx}) \approx f(\mathbf{x}^*) + Df[\mathbf{x}^*](\mathbf{dx})$

This implies that a necessary condition for $\mathbf{x}^*$ to maximize $f(\mathbf{x})$ subject to $\mathbf{g}(\mathbf{x}) \leq \mathbf{0}$ is

$Df[\mathbf{x}^*](\mathbf{dx}) \leq 0$

for all $\mathbf{dx}$ such that

$$Dg_j[\mathbf{x}^*](\mathbf{dx}) \leq 0 \quad \text{for every } j \in B(\mathbf{x}^*) \tag{54}$$

which should be compared to the corresponding condition (12) for equality-constrained problems. Condition (54) can be expressed alternatively as saying that there does not exist any perturbation $\mathbf{dx} \in \mathbb{R}^n$ such that

$Dg[\mathbf{x}^*](\mathbf{dx}) \leq 0 \quad$ for every $j \in B(\mathbf{x}^*)$ and $Df[\mathbf{x}^*](\mathbf{dx}) > 0$

In other words, it is impossible to find any change from $\mathbf{x}^*$ which simultaneously continues to satisfy the binding constraints and improves on the objective criterion.

To derive a useful criterion, we again draw on chapter 3. The derivatives of the objective function $Df[\mathbf{x}^*]$ and the constraints $Dg_j[\mathbf{x}^*]$ are linear functionals. We can apply the Farkas lemma (proposition 3.18) to (54) to deduce that $Df[\mathbf{x}^*]$ in the conic hull of the $Dg_j[\mathbf{x}^*]$. That is, there exists nonnegative multipliers $\lambda_1, \lambda_2, \ldots, \lambda_m$ such that

$$Df[\mathbf{x}^*] = \sum_{j \in B(\mathbf{x}^*)} \lambda_j Dg_j[\mathbf{x}^*]$$

Regularity implies that the multipliers are unique. If we assign zero multipliers to the slack constraints, we can extend the sum to all the constraints

$$Df[\mathbf{x}^*] = \sum_{j \in B(\mathbf{x}^*)} \lambda_j Dg_j[\mathbf{x}^*] + \sum_{j \in S(\mathbf{x}^*)} \lambda_j Dg_j[\mathbf{x}^*] = \sum_{j=1}^{m} \lambda_j Dg_j[\mathbf{x}^*]$$

Representing the derivatives by their gradients gives the expression in the theorem. In addition every constraint j is either binding ($g_j(\mathbf{x}^*) = 0$) or slack ($g_j(\mathbf{x}^*) < 0$), with $\lambda_j = 0$ for the slack constraints. Since $g_j(\mathbf{x}^*) \leq 0$ and $\lambda_j \geq 0$, this is expressed concisely by the condition $\lambda_j g_j(\mathbf{x}^*) = 0$ for every j. □

In application, it is usually more convenient to express these conditions in terms of stationarity of the Lagrangean.

Corollary 5.3.1 (Kuhn-Tucker conditions) *Suppose that $\mathbf{x}^*$ is a local solution of*

$$\max_{\mathbf{x}} \ f(\mathbf{x}) \tag{55}$$

subject to $\mathbf{g}(\mathbf{x}) \leq \mathbf{0}$

and the binding constraints are regular at $\mathbf{x}^*$. Then there exist unique multipliers $\boldsymbol{\lambda} = (\lambda_1, \lambda_2, \ldots, \lambda_m) \geq 0$ such that the Lagrangean

$$L(\mathbf{x}, \boldsymbol{\lambda}) = f(\mathbf{x}) - \sum_{j=1}^{m} \lambda_j g_j(\mathbf{x})$$

is stationary at $(\mathbf{x}^*, \boldsymbol{\lambda})$, that is,

$$D_{\mathbf{x}} L[\mathbf{x}^*, \boldsymbol{\lambda}] = D_{\mathbf{x}} f[\mathbf{x}^*] - \sum_{j=1}^{m} \lambda_j D_{\mathbf{x}} g_j[\mathbf{x}^*] = \mathbf{0} \quad \text{and} \quad \lambda_j g_j(\mathbf{x}^*) = 0 \tag{56}$$

for every $j = 1, 2, \ldots, m$.

These necessary first-order conditions for optimality (56) are known as the *Kuhn-Tucker conditions*.

Example 5.30 To solve the problem

$$\max_{x_1, x_2} 6x_1 + 2x_1x_2 - 2x_1^2 - 2x_2^2$$

subject to $\quad x_1 + 2x_2 - 2 \leq 0$

$-x_1 + x_2^2 - 1 \leq 0$

we look for stationary values of the Lagrangean

$$L(x_1, x_2, \lambda_1, \lambda_2) = 6x_1 + 2x_1x_2 - 2x_1^2 - 2x_2^2 - \lambda_1(x_1 + 2x_2 - 2)$$
$$- \lambda_2(-x_1 + x_2^2 - 1)$$

that satisfy the complementary slackness conditions. The first-order conditions for stationarity are

$$D_{x_1} L = 6 + 2x_2 - 4x_1 - \lambda_1 + \lambda_2 = 0 \qquad (57)$$

$$D_{x_2} L = 2x_1 - 4x_2 - 2\lambda_1 - 2\lambda_2 x_2 = 0 \qquad (58)$$

while the complementary slackness conditions are

$$x_1 + 2x_2 - 2 \leq 0, \quad \lambda_1 \geq 0, \quad \lambda_1(x_1 + 2x_2 - 2) = 0$$

$$-x_1 + x_2^2 - 1 \leq 0, \quad \lambda_2 \geq 0, \quad \lambda_2(x_1 - x_2^2 - 1) = 0$$

This reduces the problem to one of finding a solution to this system of equations and inequalities. Typically some trial and error is involved.

Suppose, for a start, that $\lambda_1 > 0$ and $\lambda_2 > 0$ so that both constraints are binding. That is, suppose that

$$x_1 + 2x_2 = 2 \quad \text{and} \quad -x_1 + x_2^2 = 1$$

These equations have a unique solution $x_1 = 0$ and $x_2 = 1$. Substituting in (57) and (58) gives

$$8 - \lambda_1 + \lambda_2 = 0 \quad \text{and} \quad -4 - 2\lambda_1 - 2\lambda_2 = 0$$

These equations have a unique solution $\lambda_1 = 3$ and $\lambda_2 = -5$. Since $\lambda_2 < 0$, we conclude that this cannot be a local optimum. Therefore at least one of the constraints must be slack at the optimum.

Suppose next that the second constraint is slack so $\lambda_2 = 0$. The binding constraint is regular for all x_1 and x_2. The first-order conditions and the binding constraint constitute a systems of three equations in three unknowns, namely

$6 + 2x_2 - 4x_1 - \lambda_1 = 0$

$2x_1 - 4x_2 - 2\lambda_1 = 0$

$x_1 + 2x_2 = 2$

which have a unique solution $x_1 = 10/7$, $x_2 = 2/7$, $\lambda_1 = 6/7$ that satisfies the second inequality

$$-\frac{10}{7} + \left(\frac{2}{7}\right)^2 = -\frac{66}{49} < 1$$

This point satisfies the necessary conditions for a local optimum.

For completeness, we should really check the case in which the first constraint is slack and $\lambda_1 = 0$, yielding the system

$6 + 2x_2 - 4x_1 + \lambda_2 = 0$

$2x_1 - 4x_2 - 2\lambda_2 x_2 = 0$

$-x_1 + x_2^2 = 1$

This system has three solutions, but each solution has $\lambda_2 < 0$. Therefore there cannot be a solution with the first constraint slack.

We conclude that $x_1 = 10/7$, $x_2 = 2/7$ is the only possible solution of the problem. In fact this is the unique global solution.

Exercise 5.30
Solve the problem

max $x_1 x_2$

subject to $\quad x_1^2 + 2x_2^2 \leq 3$

$2x_1^2 + x_2^2 \leq 3$

Example 5.31 (The consumer's problem) Previously (example 5.15) we assumed that the consumer would spend all her income. Dispensing with this assumption, the consumer's problem is

5.4 Inequality Constraints

$$\max_{\mathbf{x} \in X} u(\mathbf{x})$$

subject to $\quad \mathbf{p}^T\mathbf{x} \le m$

The constraint $g(\mathbf{x}) = \mathbf{p}^T\mathbf{x} - m \le 0$ satisfies the regularity condition, since $\nabla g(\mathbf{x}) = \mathbf{p} \ne \mathbf{0}$. The Lagrangean is $L(\mathbf{x},\lambda) = u(\mathbf{x}) - \lambda(\mathbf{p}^T\mathbf{x} - m)$. If $\mathbf{x}^*$ is optimal, then it must satisfy the Kuhn-Tucker conditions

$$D_{\mathbf{x}}L[\mathbf{x}^*,\lambda] = Du[\mathbf{x}^*] - \lambda \mathbf{p} = \mathbf{0} \quad \text{and} \quad \lambda(\mathbf{p}^T\mathbf{x} - m) = 0 \tag{59}$$

Two cases must be distinguished.

Case 1. $\lambda > 0$ This implies that $\mathbf{p}^T\mathbf{x} = m$, the consumer spends all her income. Condition (59) implies that

$$D_{x_i}u[\mathbf{x}^*] = \lambda p_i \quad \text{for every } i$$

Case 2. $\lambda = 0$ This allows the possibility that the consumer does not spend all her income. Substituting $\lambda = 0$ in (59), we have

$$D_{x_i}u[\mathbf{x}^*] = 0 \quad \text{for every } i$$

At the optimal consumption bundle $\mathbf{x}^*$, the marginal utility of every good is zero. The consumer is satiated; that is, no additional consumption can increase satisfaction.

We conclude that at the optimal consumption bundle, either the consumer is satiated or she spends all her income.

Example 5.32 (Rate of return regulation) A regulated monopolist produces a single output y from two inputs: capital, k, and labor, l, utilizing the technology (production function) $y = f(k,l)$. The unit costs of labor and capital are w and r respectively. Demand for the firm's product is represented by the inverse demand function $p(y)$. The firm's profit is

$$\Pi(k,l) = p(y)y - rk - wl = p(f(k,l))f(k,l) - rk - wl$$

To facilitate analysis, it is convenient to define the revenue function

$$R(k,l) = p(f(k,l))f(k,l)$$

so that profit is given by

$$\Pi(k,l) = R(k,l) - rk - wl$$

The regulator imposes a ceiling $s > r$ on the firm's rate of return to capital, that is,

$$\frac{R(k,l) - wl}{k} \leq s$$

Therefore the regulated firm's optimization problem is to choose k and l so as to maximize its total profit consistent with the constraint, that is,

$$\max_{k,l} \Pi(k,l) = R(k,l) - rk - wl$$

subject to $\quad R(k,l) - wl - sk \leq 0$

We investigate the effect of the profit constraint on the firm's optimal production decision.

Provided that the constraint satisfies the regularity condition, there exists a nonnegative multiplier $\lambda \geq 0$ such that the Lagrangean

$$L(k,l,\lambda) = R(k,l) - rk - wl - \lambda(R(k,l) - wl - sk)$$

is stationary at the optimal solution. That is,

$$D_k L[k^*, l^*, \lambda] = D_k R[k^*, l^*] - r - \lambda D_k R[k^*, l^*] + \lambda s = 0$$

$$D_l L[k^*, l^*, \lambda] = (1-\lambda) D_l R[k^*, l^*] - (1-\lambda) w = 0$$

with

$$\lambda(R(k,l) - wl - sk) = 0$$

These first-order conditions can be rewritten as

$$(1-\lambda) D_k R[k^*, l^*] = (1-\lambda) r - \lambda(s-r) \qquad (60)$$

$$(1-\lambda) D_l R[k^*, l^*] = (1-\lambda) w$$

$$\lambda(R(k^*, l^*) - wl - sk) = 0$$

Given that $r > s$, (60) ensures that $\lambda \neq 1$. Therefore the first-order conditions can be further simplified to give

$$D_k R[k^*, l^*] = r - \frac{\lambda(s-r)}{1-\lambda}$$

$$D_l R[k^*, l^*] = w$$

$$\lambda(R[k^*, l^*] - wl - sk) = 0$$

5.4 Inequality Constraints

To verify that these conditions are necessary for an optimum, we have to confirm that the regularity condition is satisfied at (k^*, l^*), a step that is often overlooked. The constraint is

$$g(k, l) = R(k, l) - wl - sk \leq 0$$

with gradient

$$\nabla g(k, l) = (D_k R - s, D_l R - w)$$

Substituting from the first-order conditions, the gradient of the constraint at (k^*, l^*) is

$$\nabla g(k^*, l^*) = \left(\frac{1}{1-\lambda}, 0\right) \neq \mathbf{0}$$

Since the gradient is nonzero, any solution of the first-order conditions satisfies the regularity condition, and therefore the first-order conditions are necessary for a solution.

In characterizing the optimal solution, we distinguish two cases:

Case 1. $\lambda = 0$ The regulatory constraint is not (strictly) binding and the firm's first-order conditions for profit maximization reduce to those for an unregulated monopolist, namely produce where marginal revenue product is equal to factor cost:

$$D_k R[k^*, l^*] = r$$

$$D_l R[k^*, l^*] = w$$

Case 2. $\lambda > 0$ The regulatory constraint is binding. The first-order conditions are

$$D_k R[k^*, l^*] = r - \frac{\lambda(s-r)}{1-\lambda}$$

$$D_l R[k^*, l^*] = w$$

$$R(k, l) - wl = sk$$

Compared to the unregulated firm, the profit constraint drives a wedge between the marginal revenue product of capital and its price. Using the second-order condition for a maximum, it can be shown (Baumol and Klevorick 1970) that $0 < \lambda < 1$, and therefore that $D_k R(k, l) < r$ at the optimal solution. The firm uses capital inefficiently, in that its marginal revenue product is less than its opportunity cost.

Exercise 5.31
Develop the conclusion of the preceding example to show that the regulated firm does not produce at minimum cost (see example 5.16).

Exercise 5.32
Inequality constraints $g_j(\mathbf{x}) \le 0$ can be transformed into to equivalent equality constraints by the addition of slack variables $s_j \ge 0$,

$$g_j(\mathbf{x}) + s_j = 0$$

Use this transformation to provide an alternative derivation of theorem 5.3 from theorem 5.2. Disregard the regularity condition.

Exercise 5.33
An equality quality constraint $\mathbf{g}(\mathbf{x}) = \mathbf{0}$ can be represented by a pair of inequality constraints

$$\mathbf{g}(\mathbf{x}) \le \mathbf{0}, \quad \mathbf{g}(\mathbf{x}) \ge \mathbf{0}$$

Use this transformation to derive theorem 5.2 from theorem 5.3. Disregard the regularity condition.

Exercise 5.34
Use the Kuhn-Tucker conditions to prove the Farkas alternative (proposition 3.19). [Hint: Consider the problem maximize $\mathbf{c}^T\mathbf{x}$ subject to $A\mathbf{x} \le \mathbf{0}$.]

Nonnegative Variables

Suppose (as is often the case) that the decision variables are restricted to be nonnegative so that the optimization problem is

$$\max_{\mathbf{x} \ge 0} f(\mathbf{x})$$

subject to $\mathbf{g}(\mathbf{x}) \le \mathbf{0}$

which can be written in standard form as

$$\max_{\mathbf{x}} f(\mathbf{x}) \tag{61}$$

subject to $\mathbf{g}(\mathbf{x}) \le \mathbf{0}$ and $\mathbf{h}(\mathbf{x}) = -\mathbf{x} \le \mathbf{0}$

5.4 Inequality Constraints

Suppose that $\mathbf{x}^*$ is a local optimum of (61) and the binding constraints (including the nonnegativity constraint $\mathbf{h}$) are regular at $\mathbf{x}^*$. Applying the Kuhn-Tucker theorem (theorem 5.3), there exist $\lambda = (\lambda_1, \lambda_2, \ldots, \lambda_m) \geq \mathbf{0}$ and $\mu = (\mu_1, \mu_2, \ldots, \mu_n) \geq \mathbf{0}$ such that

$$\nabla f(\mathbf{x}^*) = \sum_{j=1}^{m} \lambda_j \nabla g_j(\mathbf{x}^*) + \mu \tag{62}$$

and

$$\lambda_j g_j(\mathbf{x}^*) = 0 \quad \text{and} \quad \mu_i h_i(\mathbf{x}^*) = -\mu_i x_i^* = 0 \tag{63}$$

for every $i = 1, 2, \ldots n$ and $j = 1, 2, \ldots m$. Since $\mu \geq \mathbf{0}$, a necessary condition for $\mathbf{x}^*$ to solve (61) is

$$\nabla f(\mathbf{x}^*) - \sum \lambda_j \nabla g_j(\mathbf{x}^*) \leq \mathbf{0} \tag{64}$$

Equation (62) can be solved for μ to yield

$$-\mu = \nabla f(\mathbf{x}^*) - \sum \lambda_j \nabla g_j(\mathbf{x}^*)$$

Substituting into (63) yields

$$\lambda_j g_j(\mathbf{x}^*) = 0 \quad \text{and} \quad \left(\nabla f(\mathbf{x}^*) - \sum \lambda_j \nabla g_j(\mathbf{x}^*)\right)^T \mathbf{x}^* = 0$$

which should be compared with corollary 5.2.1. Define

$$L(\mathbf{x}, \lambda) = f(\mathbf{x}) - \sum_{j=1}^{n} \lambda_j g_j(\mathbf{x})$$

which is the Lagrangean for the problem *ignoring* the nonnegativity constraints. Then (64) can be written as

$$D_{x_i} L[\mathbf{x}^*, \lambda] \leq 0 \quad \text{for every } i = 1, 2, \ldots, n$$

and the necessary conditions can be written compactly in a form that emphasizes the symmetry between decision variables $\mathbf{x}$ and Lagrange multipliers $\lambda = (\lambda_1, \lambda_2, \ldots, \lambda_m)$.

$$D_{x_i} L[\mathbf{x}^*, \lambda] \leq 0, \quad x_i^* \geq 0, \quad x_i^* D_{x_i} L[\mathbf{x}^*, \lambda] = 0, \quad i = 1, 2, \ldots, n$$

$$g_j(\mathbf{x}^*) \leq 0, \quad \lambda_j \geq 0, \quad \lambda_j g_j(\mathbf{x}^*) = 0, \quad j = 1, 2, \ldots, m$$

Optimality requires the following:

- For every decision variable x_i, the Lagrangean is stationary (with respect to x_i) or $x_i = 0$, or both.
- For every constraint, the constraint is binding or its shadow price is zero, or both.

To summarize, to deal compactly with nonnegativity constraints, we may omit them explicitly from the Lagrangean, and then we modify first-order conditions appropriately—from (56) to those immediately above. These stationarity conditions are also known as the *Kuhn-Tucker conditions*. We record this conclusion in the following corollary.

Corollary 5.3.2 (Nonnegative variables) *Suppose that $\mathbf{x}^*$ is a local optimum of*

$$\max_{\mathbf{x} \geq 0} f(\mathbf{x})$$

subject to $\mathbf{g}(\mathbf{x}) \leq \mathbf{0}$

and that the binding constraints are regular at $\mathbf{x}^*$. Then there exist unique multipliers $\lambda = (\lambda_1, \lambda_2, \ldots, \lambda_m)$ such that

$$D_{x_i} L[\mathbf{x}^*, \lambda] \leq 0 \quad x_i^* \geq 0 \quad x_i^* D_{x_i} L[\mathbf{x}^*, \lambda] = 0, \qquad i = 1, 2, \ldots, n$$

$$g_j(\mathbf{x}^*) \leq 0 \quad \lambda_j \geq 0 \quad \lambda_j g_j(\mathbf{x}^*) = 0, \qquad j = 1, 2, \ldots, m$$

Although the Kuhn-Tucker conditions are a compact representation of the first-order conditions for a solution of an optimization problem with nonnegative variables and inequality constraints, their solution is not always transparent. This is because they are a system of inequalities, and we cannot apply all the transformations which we can apply to equations. Normally only some of the variables x_i or λ_j will be nonzero at the optimal solution. Some trial and error is usually involved in identifying the nonzero variables. In economic problems, economic intuition may guide the selection. The typical procedure is illustrated in the following examples.

Example 5.33 Consider the problem

$$\max_{x_1 \geq 0, x_2 \geq 0} \log x_1 + \log(x_2 + 5)$$

subject to $\quad x_1 + x_2 - 4 \leq 0$

5.4 Inequality Constraints

The Langrangean is

$$L(x_1, x_2, \lambda) = \log x_1 + \log(x_2 + 5) - \lambda(x_1 + x_2 - 4)$$

and the Kuhn-Tucker conditions are

$$D_{x_1} L = \frac{1}{x_1} - \lambda \leq 0, \quad x_1 \geq 0, \quad x_1 \left(\frac{1}{x_1} - \lambda\right) = 0$$

$$D_{x_2} L = \frac{1}{x_2 + 5} - \lambda \leq 0, \quad x_2 \geq 0, \quad x_2 \left(\frac{1}{x_2 + 5} - \lambda\right) = 0$$

$$x_1 + x_2 \leq 4, \quad \lambda \geq 0, \quad \lambda(4 - x_1 - x_2) = 0$$

Assuming that the regularity condition is satisfied (exercise 5.35), these conditions must be satisfied at any local optimum. To find local optima, we have to solve this system of inequalities. This process usually involves some trial and error.

Let us begin with the assumption that both $x_1 > 0$ and $x_2 > 0$. This implies that

$$\frac{1}{x_1} - \lambda = 0 \quad \text{and} \quad \frac{1}{x_2 + 5} - \lambda = 0$$

or

$$x_1 = x_2 + 5$$

which is inconsistent with the constraints; that is, the system

$$x_1 = x_2 + 5$$

$$x_1 + x_2 \leq 4$$

$$x_1 \geq 0$$

$$x_2 \geq 0$$

has no solution. Therefore our trial hypothesis that $x_1 > 0$ and $x_2 > 0$ is invalid.

Next we consider the possibility that $x_1 > 0$ and $\lambda > 0$, with $x_2 = 0$. In this case the Kuhn-Tucker conditions reduce to

$$\frac{1}{x_1} - \lambda = 0$$

$$\frac{1}{5} - \lambda = 0$$

$$4 - x_1 = 0$$

These have an obvious solution $x_1 = 4$ and $\lambda = \frac{1}{4}$.

Finally, we consider the possibility that $x_2 > 0$ and $\lambda > 0$, with $x_1 = 0$. In this case the Kuhn-Tucker conditions reduce to

$$\frac{1}{0} \leq \lambda = 0$$

$$\frac{1}{x_2 + 5} - \lambda = 0$$

$$4 - x_2 = 0$$

This system implies that $x_2 = 4$ and

$$\lambda \geq 1, \quad \lambda = \tfrac{1}{9}$$

which is clearly inconsistent.

The point $x_1 = 4$, $x_2 = 0$ and $\lambda = \frac{1}{4}$ is the only point which satisfies the Kuhn-Tucker conditions. Subject to verifying that this point satisfies the regularity (exercise 5.35) and second-order conditions (example 5.46), we conclude that this is the solution of the problem.

Exercise 5.35
In the previous example, verify that the binding constraints are regular at $\mathbf{x}^* = (4, 0)$.

Exercise 5.36 *(The consumer's problem)*
Derive and interpret the Kuhn-Tucker conditions for the consumer's problem

$$\max_{\mathbf{x} \geq \mathbf{0}} u(\mathbf{x})$$

subject to $\quad \mathbf{p}^T \mathbf{x} \leq m$

constraining consumption to be nonnegative, while allowing the consumer to spend less than her income. We will show later (example 5.41) that the regularity condition is satisfied.

Example 5.34 (Upper and lower bounds) A special case that is often encountered is where the constraints take the form of upper and lower bounds, as in the problem

$$\max_{0 \leq \mathbf{x} \leq \mathbf{c}} f(\mathbf{x})$$

which can be written in standard from as

$$\max_{\mathbf{x} \geq 0} f(\mathbf{x})$$

subject to $g_i(\mathbf{x}) = x_i - c_i \leq 0, \quad i = 1, 2, \ldots, n$

Assuming that $\mathbf{c} > \mathbf{0}$, the binding constraints are regular for all $\mathbf{0} \leq \mathbf{x} \leq \mathbf{c}$. By corollary 5.3.2, it is necessary for a local optimum that there exist $\lambda = (\lambda_1, \lambda_2, \ldots, \lambda_n) \geq \mathbf{0}$ such that

$$D_{x_i} L[\mathbf{x}^*, \lambda] = D_{x_i} f[\mathbf{x}^*] - \lambda_i \leq 0, \quad x_i^* \geq 0, \quad x_i^* (D_{x_i} f[\mathbf{x}^*] - \lambda_i) = 0$$

$$g_i(\mathbf{x}^*) = x_i^* - c_i \leq 0, \quad \lambda_i \geq 0, \quad \lambda_i (x_i^* - c_i) = 0$$

for every $i = 1, 2, \ldots, n$, where L is the Lagrangean $L(\mathbf{x}, \lambda) = f(\mathbf{x}) - \sum_{i=1}^n \lambda_i g_i(\mathbf{x})$. Now for every i,

$$x_i^* > 0 \Rightarrow D_{x_i} f[\mathbf{x}^*] = \lambda_i \geq 0$$

and

$$x_i^* < c \Rightarrow \lambda_i = 0 \Rightarrow D_{x_i} f[\mathbf{x}^*] \leq 0$$

These first-order conditions can be expressed very compactly as

$$D_{x_i} f[\mathbf{x}^*] \geq 0 \quad \text{if } x_i^* > 0 \quad \text{and} \quad D_{x_i} f[\mathbf{x}^*] \leq 0 \quad \text{if } x_i^* < c_i$$

for every $i = 1, 2, \ldots, n$. These conditions together imply that $D_{x_i} f[\mathbf{x}^*] = 0$ for every i such that $0 < x_i^* < c_i$.

Example 5.35 (Production with capacity constraints) Consider a monopolist that supplies a market where total demand can be segmented into n different periods and arbitrage between periods is precluded. Let y_i denote the firm's sales in period i and $\mathbf{y} = (y_1, y_2, \ldots, y_n)$ the firm's sales plan. Total revenue $R(\mathbf{y})$ is function of sales in each period, which is equal to production (storage is precluded). This specification allows for interdependence between the demand and price in different periods. Similarly total cost $c(\mathbf{y})$ depends on production in each period, which cannot exceed total capacity Y, that is, $y_i \leq Y$. The firm's problem is to determine the level of production in each period y_i so as to maximize total profit. The firm's maximization problem is

$$\max_{0 \leq y_i \leq Y} \Pi(\mathbf{y}) = R(\mathbf{y}) - c(\mathbf{y})$$

Following example 5.34, the Kuhn-Tucker conditions for an optimal production plan $\mathbf{y}^*$ are

$$D_{y_i}\Pi[\mathbf{y}^*] \geq 0 \quad \text{if } y_i^* > 0 \quad \text{and} \quad D_{y_i}\Pi(\mathbf{y}^*) \leq 0 \quad \text{if } y_i^* < Y \tag{65}$$

for every period i.

To interpret these conditions, we recognize that periods fall into two categories: *peak periods* in which production is at capacity ($y_i = Y$) and *off-peak periods* in which production falls below capacity ($y_i < Y$). To simplify notation, let $R_{y_i} = D_{y_i}R[\mathbf{y}^*]$ denote the marginal revenue in period i evaluated at the optimal plan $\mathbf{y}^*$. Similarly let c_{y_i} denote the marginal cost in period i. In off-peak periods, (65) implies that

$$R_{y_i} \leq c_{y_i} \quad \text{with } R_{y_i} = c_{y_i} \text{ if } y_i^* > 0$$

The production schedule $\mathbf{y}$ cannot be optimal if in any off-peak period marginal revenue of an additional unit exceeds marginal cost. Furthermore it cannot be optimal if marginal revenue is less than marginal cost unless a lower output is not feasible ($y_i^* = 0$). In peak periods, the firm is constrained by the capacity from producing at the level at which marginal revenue equals marginal cost. In such periods, production is equal to capacity $y_i^* = Y$ and

$$R_{y_i} \geq c_{y_i}$$

Marginal revenue may exceed marginal cost, because it is impossible to increase production.

We can derive additional insight by extending the model, allowing capacity to be a choice variable as well. Then cost depends on output and capacity $c(\mathbf{y}, Y)$, and the monopolist's problem is

$$\max_{\mathbf{y} \geq 0, Y \geq 0} \Pi(\mathbf{y}, Y) = R(\mathbf{y}) - c(\mathbf{y}, Y)$$

subject to the capacity constraints $y_i \leq Y$ for very $i = 1, 2, \ldots, n$. Since the

5.4 Inequality Constraints

upper bound is a decision variable, this problem no longer fits the formulation of example 5.34 and we apply corollary 5.3.2. The Lagrangean is

$$L(\mathbf{y}, Y, \lambda) = R(\mathbf{y}) - c(\mathbf{y}, Y) - \sum_{i=1}^{n} \lambda_j (y_i - Y)$$

The Kuhn-Tucker conditions for an optimum require that for every period $i = 1, 2, \ldots, n$,

$$D_{y_i} L = R_{y_i} - c_{y_i} - \lambda_i \leq 0, \quad y_i^* \geq 0, \quad y_i^* (R_{y_i} - c_{y_i} - \lambda_i) = 0$$

$$y_i^* \leq Y, \quad \lambda_i \geq 0, \quad \lambda(y_i^* - Y) = 0$$

and that capacity be chosen such that

$$D_Y L = -c_Y + \sum_{i=1}^{n} \lambda_i \leq 0, \quad Y \geq 0, \quad Y\left(c_Y - \sum_{i=1}^{n} \lambda_i\right) = 0$$

where $c_Y = D_y c[\mathbf{y}^*, Y]$ is the marginal cost of additional capacity. The conditions for optimal production are the same as before (65), with the enhancement that λ_i measures the degree to which marginal revenue may exceed marginal cost in peak periods. Specifically,

$$R_{y_i} = c_{y_i} + \lambda_i$$

The margin λ_i between marginal revenue and marginal cost represents the shadow price of capacity in period i.

The final first-order condition determines the optimal capacity. It says that capacity should be chosen so that the marginal cost of additional capacity is just equal to the total shadow price of capacity in the peak periods

$$c_Y = \sum_{i=1}^{n} \lambda_i$$

An additional unit of capacity would cost c_Y. It would enable the production of an additional unit of electricity in each peak period, the net revenue from which is given by $\lambda_i = R_{y_i} - c_{y_i}$. Total net revenue of an additional unit of capacity would therefore be $\sum_{i=1}^{n} \lambda_i$, which at the optimal capacity, should be precisely equal to its marginal cost.

5.4.2 Constraint Qualification

Faced with a constrained optimization problem, economists rely on the first-order conditions for stationarity to select out possible candidates for the solution. This technique relies on the first-order conditions being necessary for a solution, which requires that the functional specification of the constraint set be an adequate representation of its geometry.

In equality-constrained problems it is natural to assume that the constraint set is a smooth hypersurface or *manifold*. This is ensured by the regularity condition that the gradients of the constraints be linearly independent. We employed the same regularity condition in deriving the Kuhn-Tucker theorem in the previous section. However, with inequality-constrained problems, corner solutions arise and the regularity condition becomes too stringent. We need to consider alternative constraint qualifications.

A more practical difficulty in an inequality-constrained problem is that the regularity condition will not necessarily apply throughout the constraint set. Verifying the regularity condition requires knowing the optimal solution, but the first-order conditions will not identify the optimal solution unless the condition is satisfied. To avoid this "Catch 22," we would like to find global constraint qualification conditions that ensure that the first-order conditions are necessary. To illustrate the problem, we begin with a traditional example due to Kuhn and Tucker.

Example 5.36 Consider the problem

$$\max_{x_1,x_2} \; x_1$$

subject to $\quad x_2 - (1-x_1)^3 \leq 0$

$-x_2 \leq 0$

The Lagrangean for this problem is

$$L(x_1, x_2, \lambda_1, \lambda_2) = x_1 - \lambda_1(x_2 - (1-x_1)^3) + \lambda_2 x_2$$

The first-order conditions are

$$D_1 L = 1 - 3\lambda_1(1-x_1)^2 = 0$$
$$D_2 L = -\lambda_1 + \lambda_2 = 0$$

5.4 Inequality Constraints

$$\lambda_1(x_2 - (1-x_1)^3) = 0$$

$$\lambda_2 x_2 = 0$$

which can be written as

$$3\lambda_1(1-x_1)^2 = 1$$

$$\lambda_1 = \lambda_2$$

$$\lambda_1(x_2 - (1-x_1)^3) = 0$$

$$\lambda_2 x_2 = 0$$

These conditions have no solution. The first condition implies that $\lambda_1 \neq 0$ so that

$$\lambda_1 = \lambda_2 > 0$$

This implies that both constraints are binding, that is,

$$x_2 - (1-x_1)^3) = 0$$

$$x_2 = 0$$

which in turn implies that $x_1 = 1$. However, $x_1 = 1$ violates the first condition

$$3\lambda_1(1-x_1)^2 = 1$$

Hence the first-order conditions have no solution, and the Lagrangean technique fails to solve the problem.

The feasible set is illustrated in figure 5.8. It is obvious that the optimal solution occurs at the extremity $(1, 0)$. The difficulty is that the feasible set has a cusp at this point, so the derivatives of the constraints do not adequately represents the set of feasible perturbations at this point. The following discussion is aimed at making this precise.

Remark 5.6 (Fritz John conditions) In the preceding example the Lagrangean was not stationary at the optimal solution. If we augment that Lagrangean by attaching a coefficient (λ_0) to the objective function, we can formally circumvent this problem. The augmented Lagrangean

$$L(x_1, x_2, \lambda_0, \lambda_1, \lambda_2) = \lambda_0 x_1 - \lambda_1(x_2 - (1-x_1)^3) + \lambda_2 x_2$$

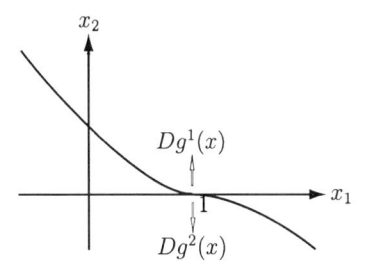

Figure 5.8
Kuhn and Tucker's example

is stationary at the optimal solution $(1, 0)$, that is,

$$D_1 L = \lambda_0 - 3\lambda_1 (1 - 1)^2 = 0$$

$$D_2 L = -\lambda_1 + \lambda_2 = 0$$

provided that $\lambda_0 = 0$. Stationarity of the augmented Lagrangean function together with the appropriate complementary slackness conditions are known as the Fritz John optimality conditions. These conditions are necessarily satisfied at any optimum without any constraint qualification. However, they provide no help in locating optima in cases that violate the constraint qualification condition. In such cases the Lagrange multiplier λ_0 is equal to zero, and the Fritz John conditions do not utilize any information relating to the gradient of the objective function. They merely state that there is a nonnegative and nontrivial linear combination of the gradients of the binding constraints that sum to zero. Constraint qualification can be seen as the search for conditions which ensure that $\lambda_0 \neq 0$.

The issue of constraint qualification involves the degree to which the functional constraints adequately describe the set of feasible perturbations from a given point. Let G be a nonempty set in $\Re^n$ and $\mathbf{x}^*$ a point in G. Another point $\mathbf{x} \neq \mathbf{0} \in \Re^n$ is a *feasible direction* at $\mathbf{x}^*$ if there exists some $\bar{\alpha} \in \Re$ such that $\mathbf{x}^* + \alpha \mathbf{x} \in G$ for every $0 < \alpha < \bar{\alpha}$. In other words, $\mathbf{x}$ is a feasible direction if a small perturbation of $\mathbf{x}^*$ in the direction $\mathbf{x}$ remains feasible. The set of all feasible directions at $\mathbf{x}^*$,

$$D(\mathbf{x}^*) = \{\mathbf{x} \in \Re^n : \mathbf{x}^* + \alpha \mathbf{x} \in G \text{ for every } \alpha \in [0, \bar{\alpha}] \text{ for some } \bar{\alpha} > 0\}$$

is called the *cone of feasible direction* in G at $\mathbf{x}^*$.

5.4 Inequality Constraints

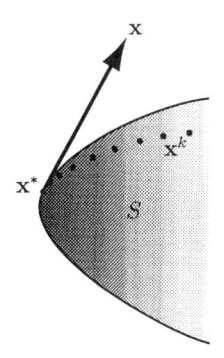

Figure 5.9
$\mathbf{x}^k$ converges to $\mathbf{x}^*$ from direction $\mathbf{x}$

Clearly, a point $\mathbf{x}^* \in G$ cannot be a local solution to

$$\max_{\mathbf{x} \in G} f(\mathbf{x})$$

if there exists any feasible direction in G at $\mathbf{x}^*$ in which $f(\mathbf{x})$ increases. To make this precise, let $H^+(\mathbf{x}^*)$ denote the set of all perturbations that increase the objective function from $\mathbf{x}^*$, that is,

$$H^+(\mathbf{x}^*) = \{\mathbf{x} \in \mathbb{R}^n : Df[\mathbf{x}^*](\mathbf{x}) > 0\}$$

$H^+(\mathbf{x}^*)$ is the open halfspace containing the upper contour set of the objective function. Then a necessary condition for $\mathbf{x}^*$ to be a local optimum is that $H^+(\mathbf{x}^*) \cap D(\mathbf{x}^*) = \emptyset$.

Exercise 5.37
Suppose that $\mathbf{x}^*$ is a local solution of $\max_{\mathbf{x} \in G} f(\mathbf{x})$. Then $H^+(\mathbf{x}^*) \cap D(\mathbf{x}^*) = \emptyset$.

Unfortunately, the set of feasible directions does not exhaust the set of relevant perturbations, and we need to consider a broader class. Suppose that $\mathbf{x} \in D(\mathbf{x}^*)$ with $\mathbf{x}^* + \alpha\mathbf{x} \in G$ for every $0 \leq \alpha \leq \bar{\alpha}$. Let $\alpha_k = 1/k$. Then the sequence $\mathbf{x}^k = \mathbf{x}^* + \alpha_k\mathbf{x} \in G$ converges to $\mathbf{x}^*$. Furthermore

$$\mathbf{x}^k - \mathbf{x}^* = \alpha_k \mathbf{x}$$

and therefore the sequence $(\mathbf{x}^k - \mathbf{x}^*)/\alpha_k$ converges trivially to $\mathbf{x}$. We say that $\mathbf{x}^k$ converges to $\mathbf{x}^*$ from the direction $\mathbf{x}$ (figure 5.9).

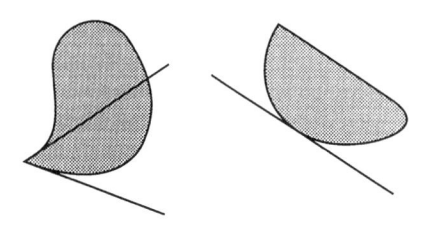

Figure 5.10
Examples of the cone of tangents

More generally, a feasible sequence $(\mathbf{x}^k) \subseteq G$ converges to $\mathbf{x}^*$ *from the direction* $\mathbf{x}$ if $\mathbf{x}^k \to \mathbf{x}^*$, and there exists a sequence $(\alpha_k) \subseteq \Re_+$ such that $(\mathbf{x}^k - \mathbf{x}^*)/\alpha_k$ converges to $\mathbf{x}$. A point $\mathbf{x} \in \Re^n$ is a *tangent to G at* $\mathbf{x}^*$ if there exists a feasible sequence $(\mathbf{x}^k)$ that converges to $\mathbf{x}^*$ from the direction $\mathbf{x}$. The set of all tangents to G at $\mathbf{x}^*$,

$$T(\mathbf{x}^*) = \left\{ \mathbf{x} \in \Re^n : \mathbf{x} = \lim_{k \to \infty} \frac{(\mathbf{x}^k - \mathbf{x}^*)}{\alpha_k} \text{ for some } \mathbf{x}^k \in G \text{ and } \alpha_k \in \Re_+ \right\}$$

is called the *cone of tangents* to G at $\mathbf{x}^*$.

Exercise 5.38
The cone $T(\mathbf{x}^*)$ of tangents to a set G at a point $\mathbf{x}^*$ is a nonempty closed cone. See figure 5.10.

Exercise 5.39
Show that $D(\mathbf{x}^*) \subseteq T(\mathbf{x}^*)$.

Clearly, $D(\mathbf{x}^*) \subseteq T(\mathbf{x}^*)$, and for some $\mathbf{x}^* \in G$ it may be proper subset. The significance of this is revealed by the following proposition. To adequately determine local optima, it does not suffice to consider only perturbations in $D(\mathbf{x}^*)$.

Proposition 5.3 (Basic necessary condition) Suppose that $\mathbf{x}^*$ is a local maximum of

$$\max_{\mathbf{x} \in G} f(\mathbf{x})$$

Then $H^+(\mathbf{x}^*) \cap T(\mathbf{x}^*) = \emptyset$.

Proof Let $\mathbf{x} \in T(\mathbf{x}^*)$ be a tangent to G at $\mathbf{x}^*$. That is, there exists a feasible sequence $(\mathbf{x}^k)$ converging to $\mathbf{x}^*$ and a sequence (α_k) of positive scalars such that $(\mathbf{x}^k - \mathbf{x}^*)/\alpha_k$ converges to $\mathbf{x}$. Since $\mathbf{x}^*$ is locally optimal, for large enough k

$$f(\mathbf{x}^*) \geq f(\mathbf{x}^k) \approx f(\mathbf{x}^*) + Df[\mathbf{x}^*](\mathbf{x}^* - \mathbf{x}^k)$$

so that

$$Df[\mathbf{x}^*](\mathbf{x}^k - \mathbf{x}^*) \approx f(\mathbf{x}^k) - f(\mathbf{x}^*) \leq \mathbf{0}$$

Dividing by α_k, we have

$$Df[\mathbf{x}^*]\left(\frac{\mathbf{x}^k - \mathbf{x}^*}{\alpha_k}\right) \leq 0$$

Letting $k \to \infty$ this implies that

$$Df[\mathbf{x}^*](\mathbf{x}) \leq 0$$

and so $\mathbf{x} \notin H^+(\mathbf{x}^*)$. Therefore $H^+(\mathbf{x}^*) \cap T(\mathbf{x}^*) = \emptyset$. □

This proposition suggests that the cone of tangents is the appropriate set of perturbations to be evaluated when considering the local optimality of a particular point. The issue of constraint qualification is the degree to this set of perturbations is adequately represented by the gradients of functional constraints.

Returning to the problem in which the feasible set G is defined by functional constraints, that is,

$$G = \{\mathbf{x} \in X : g_j(\mathbf{x}) \leq 0, j = 1, 2, \ldots, m\}$$

let $B(\mathbf{x}^*)$ denote the set of binding constraints at a point $\mathbf{x}^* \in G$. The system of inequalities

$$Dg_j[\mathbf{x}^*](\mathbf{dx}) \leq \mathbf{0}, \quad j \in B(\mathbf{x}^*) \tag{66}$$

is a linear approximation in the region of $\mathbf{x}^*$ to the system of inequalities

$$g_j(\mathbf{x}^*) \leq 0 \quad \text{for every } j$$

defining the feasible set G. The solution set of the linear system (66)

$$L(\mathbf{x}^*) = \{\mathbf{dx} \in X : Dg_j[\mathbf{x}^*](\mathbf{dx}) \leq 0, j \in B(\mathbf{x}^*)\}$$

is a closed convex polyhedral cone which is called the *linearizing cone* of G at $\mathbf{x}^*$. The cone of tangents $T(\mathbf{x}^*)$ is a subset of $L(\mathbf{x}^*)$ (exercise 5.40). The following example shows that it may be a proper subset.

Example 5.37 Consider the problem posed in example 5.36

$$\max_{x_1, x_2} x_1$$

subject to $x_2 - (1 - x_1)^3 \leq 0$

$-x_2 \leq 0$

The cone of tangents at $(1,0)$ is the half-line $T = \{(x_1, 0) \mid x_1 \leq 1\}$. The halfspace defined by the gradient of the first constraint at $(1, 0)$ is $\{(x_1, x_2) \mid x_2 \leq 0\}$, whereas that defined by the second constraint is $\{(x_1, x_2) \mid x_2 \geq 0\}$. The cone of locally constrained directions L is the intersection of these two halfspaces, namely the x_2 axis $L = \{(x_1, 0)\}$. In this problem the linearizing does not adequately restrict the set of feasible perturbations. This is why the Kuhn-Tucker conditions are not satisfied in this example.

Exercise 5.40
Show that $T(\mathbf{x}^*) \subseteq L(\mathbf{x}^*)$.

The Kuhn-Tucker first-order conditions are necessary for a local optimum at $\mathbf{x}^*$ *provided* that the linearizing cone $L(\mathbf{x}^*)$ is equal to the cone of tangents $T(\mathbf{x}^*)$, which is known as the Abadie constraint qualification condition. The Kuhn-Tucker conditions follow immediately from proposition 5.3 by a straightforward application of the Farkas lemma.

Proposition 5.4 (Abadie constraint qualification) Suppose that $\mathbf{x}^*$ is a local solution of

$$\max_{\mathbf{x} \in X} f(\mathbf{x}) \tag{67}$$

subject to $\mathbf{g}(\mathbf{x}) \leq \mathbf{0}$

and satisfies the Abadie constraint qualification $T(\mathbf{x}^*) = L(\mathbf{x}^*)$. Then there exist multipliers $\lambda_1, \lambda_2, \ldots, \lambda_m \geq 0$ such that

$$\nabla f(\mathbf{x}^*) = \sum_{j=1}^{m} \lambda_j \nabla g_j(\mathbf{x}^*) \quad \text{and} \quad \lambda_j g_j(\mathbf{x}^*) = 0, \quad j = 1, 2 \ldots, m$$

Proof By proposition 5.3, $H^+(\mathbf{x}^*) \cap T(\mathbf{x}^*) = \emptyset$, where

$$H^+(\mathbf{x}^*) = \{\mathbf{dx} : Df[\mathbf{x}^*](\mathbf{dx}) > 0\}$$

By the Abadie constraint qualification condition $T(\mathbf{x}^*) = L(\mathbf{x}^*)$, this implies that $H^+(\mathbf{x}^*) \cap L(\mathbf{x}^*) = \emptyset$, so the system

$$Df[\mathbf{x}^*](\mathbf{dx}) > 0$$

$$Dg_j[\mathbf{x}^*](\mathbf{dx}) \leq 0, \quad i \in B(\mathbf{x}^*)$$

has no solution. By the Farkas lemma (proposition 3.18), there exists nonnegative scalars λ_j such that

$$Df[\mathbf{x}^*] = \sum_{i \in B(\mathbf{x}^*)} \lambda_j Dg_j[\mathbf{x}^*]$$

Assigning zero multipliers to the slack constraints gives the desired result. □

Example 5.38 We have already seen how the Kuhn-Tucker example fails the Abadie constraint qualification condition (example 5.37). Suppose that we add an additional constraint $x_1 + x_2 \leq 1$ to this problem, which becomes

$$\max_{x_1, x_2} x_1$$

subject to $x_2 - (1 - x_1)^3 \leq 0$

$-x_2 \leq 0$

$x_1 + x_2 \leq 1$

The Lagrangean for this problem is

$$L(x_1, x_2, \lambda_1, \lambda_2, \lambda_3) = x_1 - \lambda_1(x_2 - (1 - x_1)^3) + \lambda_2 x_2 - \lambda_3(x_1 + x_2 - 1)$$

The first-order conditions are

$$D_1 L = 1 - 3\lambda_1(1 - x_1)^2 - \lambda_3 = 0$$

$$D_2 L = -\lambda_1 + \lambda_2 - \lambda_3 = 0$$

$$\lambda_1(x_2 - (1 - x_1)^3) = 0$$

$$\lambda_2 x_2 = 0$$

$$\lambda_3(x_1 + x_2 - 1) = 0$$

which can be written as

$$3\lambda_1(1-x_1)^2 = 1 - \lambda_3$$

$$\lambda_1 + \lambda_3 = \lambda_2$$

$$\lambda_1(x_2 - (1-x_1)^3) = 0$$

$$\lambda_2 x_2 = 0$$

$$\lambda_3(x_1 + x_2 - 1) = 0$$

These conditions are satisfied at $\mathbf{x}^* = (1,0)$ with $\lambda = (1,2,1)$. The additional constraint does not affect the feasible set or its cone of tangents T. However, it does alter the linearizing cone, which becomes

$$L(\mathbf{x}^*) = \{\mathbf{dx} : Dg_j[\mathbf{x}^*](\mathbf{dx}) \leq 0, i = 1, 2, 3\} = \{(x_1, 0) : x_1 \leq 1\} = T(\mathbf{x}^*)$$

With the additional constraint the linearizing cone $L(\mathbf{x}^*)$, adequately represents the set of appropriate perturbations $T(\mathbf{x}^*)$.

Unfortunately, the Abadie constraint qualification condition is difficult to verify. Consequently the optimization literature contains a variety of alternative constraint qualification conditions designed to ensure the necessity of the first-order conditions. In the following theorem we present a selection of constraint qualification conditions that are used in practice. Each implies the Abadie condition and hence ensures the necessity of the Kuhn-Tucker conditions for optimality.

First, we need to introduce some additional sets of perturbations. The linearizing cone

$$L(\mathbf{x}^*) = \{\mathbf{dx} \in X : Dg_j[\mathbf{x}^*](\mathbf{dx}) \leq \mathbf{0} \text{ for every } j \in B(\mathbf{x}^*)\}$$

is sometimes called *the cone of locally constrained directions*. Its interior is

$$L^0(\mathbf{x}^*) = \{\mathbf{dx} \in L : Dg_j[\mathbf{x}^*](\mathbf{dx}) < 0 \text{ for every } j \in B(\mathbf{x}^*)\}$$

which is known as the *cone of interior directions*. The *cone of semi-interior directions* L^1 is

$$L^1(\mathbf{x}^*) = \{\mathbf{dx} \in L : Dg_j[\mathbf{x}^*](\mathbf{dx}) < 0 \text{ for every } j \in B^N(\mathbf{x}^*)\}$$

where $B^N(\mathbf{x}^*)$ is the set of all nonconcave constraints binding at $\mathbf{x}^*$. Clearly, these cones are nested $L^0(\mathbf{x}^*) \subseteq L^1(\mathbf{x}^*) \subseteq L(\mathbf{x}^*)$. In fact all the cones of perturbations are nested, as specified in the following exercise. Each of the inclusions can be strict.

Exercise 5.41 (Cones of perturbations)
Show that

$$L^0(\mathbf{x}^*) \subseteq L^1(\mathbf{x}^*) \subseteq D(\mathbf{x}^*) \subseteq T(\mathbf{x}^*) \subseteq L(\mathbf{x}^*)$$

for every $\mathbf{x}^* \in G = \{\mathbf{x} : g_j(\mathbf{x}) \leq 0, j = 1, 2, \ldots, m\}$.

The following theorem establishes some practical constraint qualification conditions. The first two conditions are especially important in practice, since they provide global criteria which can be verified without knowing the optimal solution, eliminating the Catch 22 referred to above. The first criterion (concave CQ) applies in particular to linear constraints that are found in many economic models (example 5.2) and practical optimization problems (section 5.4.4). In economic models with nonlinear constraints, the second or third conditions are usually assumed (example 5.40). The regularity condition is the same as that used in section 5.3 and theorem 5.3. The Cottle and Arrow-Hurwicz-Uzawa conditions do not have practical application. They are vehicles for establishing the validity of the other four criteria.

Theorem 5.4 (Constraint qualification conditions) Suppose that $\mathbf{x}^*$ is a local solution of

$$\max_{\mathbf{x} \in X} f(\mathbf{x}) \quad \text{subject to} \quad g(\mathbf{x}) \leq \mathbf{0}$$

at which the binding constraints $B(\mathbf{x}^*)$ satisfy one of the following constraint qualification conditions.

Concave CQ g_j is concave for every $j \in B(\mathbf{x}^*)$

Pseudoconvex CQ g_j is pseudoconvex, and there exists $\hat{\mathbf{x}} \in X$ such that $g_j(\hat{\mathbf{x}}) < 0$ for every $j \in B(\mathbf{x}^*)$

Quasiconvex CQ g_j is quasiconvex, $\nabla g_j(\mathbf{x}^*) \neq \mathbf{0}$, and there exists $\hat{\mathbf{x}} \in X$ such that $g_j(\hat{\mathbf{x}}) < 0$ for every $j \in B(\mathbf{x}^*)$

Regularity The set $\{\nabla g_j(\mathbf{x}^*) : j \in B(\mathbf{x}^*)\}$ is linearly independent

Cottle CQ $L^0(\mathbf{x}^*) \neq \emptyset$

Arrow-Hurwicz-Uzawa (AHUCQ) $L^1(\mathbf{x}^*) \neq \emptyset$

Then $\mathbf{x}^*$ satisfies the Kuhn-Tucker conditions; that is, there exist multipliers $\lambda_1, \lambda_2, \ldots, \lambda_m \geq 0$ such that

$$\nabla f(\mathbf{x}^*) = \sum_{j=1}^{m} \lambda_j \nabla g_j(\mathbf{x}^*) \quad \text{and} \quad \lambda_j g_j(\mathbf{x}^*) = 0, \quad j = 1, 2 \ldots, m$$

The multipliers $\lambda_1, \lambda_2, \ldots, \lambda_m$ are unique if and only if $B(\mathbf{x}^*)$ satisfies the regularity condition; that is, $\{\nabla g_j(\mathbf{x}^*) : j \in B(\mathbf{x}^*)\}$ is linearly independent.

Proof Each of the constraint qualification conditions implies the Abadie constraint qualification, which by proposition 5.4 implies the necessity of the Kuhn-Tucker conditions. We prove that

pseudoconvex CQ $\Rightarrow$ Cottle $\Rightarrow$ Abadie

and leave the remaining implications for the exercises.

Assume that $\mathbf{g}$ satisfies the pseudoconvex CQ condition at $\mathbf{x}^*$. That is, for every $j \in B(\mathbf{x}^*)$, g_j is pseudoconvex, and there exists $\hat{\mathbf{x}}$ such that $g_j(\hat{\mathbf{x}}) < 0$ for every $j \in B(\mathbf{x}^*)$. Consider the perturbation $\mathbf{dx} = \hat{\mathbf{x}} - \mathbf{x}^*$. Since $g_j(\mathbf{x}^*) = 0$ and $g_j(\hat{\mathbf{x}}) < 0$, pseudoconvexity implies that (exercise 4.75)

$$g_j(\hat{\mathbf{x}}) < g_j(\mathbf{x}^*) \Rightarrow Dg_j[\mathbf{x}^*](\mathbf{dx}) < 0$$

for every binding constraint $j \in B(\mathbf{x}^*)$. Therefore $\mathbf{dx} \in L^0(\mathbf{x}^*) \neq \emptyset$, and $\mathbf{x}^*$ satisfies the Cottle constraint qualification condition.

For every j, let

$$S_j = \{\mathbf{dx} : Dg_j[\mathbf{x}^*](\mathbf{dx}) < 0\}$$

Each S_j is an open convex set (halfspace). If $\mathbf{x}^*$ satisfies the Cottle constraint qualification condition,

$$L^0(\mathbf{x}^*) = \bigcap_{j=1}^{m} S_j \neq \emptyset$$

By exercise 1.219,

$$\overline{L^0(\mathbf{x}^*)} = \bigcap_{j=1}^{m} \overline{S_j} = L(\mathbf{x}^*)$$

Since $L^0(\mathbf{x}^*) \subseteq T(\mathbf{x}^*) \subseteq L(\mathbf{x}^*)$ (exercise 5.41) and $T(\mathbf{x}^*)$ is closed, we have

$$L(\mathbf{x}^*) = \overline{L^0(\mathbf{x}^*)} \subseteq T(\mathbf{x}^*) \subseteq L(\mathbf{x}^*)$$

which implies that $T(\mathbf{x}^*) = L(\mathbf{x}^*)$, the Abadie constraint qualification condition.

Since $\lambda_j = 0$ for every $j \notin B(\mathbf{x}^*)$, the Kuhn-Tucker conditions imply that

$$\nabla f(\mathbf{x}^*) = \sum_{j \in B(\mathbf{x}^*)} \lambda_j \nabla g_j(\mathbf{x}^*)$$

If $\nabla g_j(\mathbf{x}^*)$, $j \in B(\mathbf{x}^*)$ are independent, then the λ_j are unique (exercise 1.137). Conversely, if there exist $\mu_1, \mu_2, \ldots, \mu_m$ such that with $\mu_j \neq \lambda_j$ for some j and

$$\nabla f(\mathbf{x}^*) = \sum_{j \in B(\mathbf{x}^*)} \mu_j \nabla g_j(\mathbf{x}^*)$$

then

$$\nabla f(\mathbf{x}^*) - \nabla f(\mathbf{x}^*) = \sum_{j \in B(\mathbf{x}^*)} (\lambda_j - \mu_j) \nabla g_j(\mathbf{x}^*) = \mathbf{0}$$

which implies that $\nabla g_j(\mathbf{x}^*)$, $j \in B(\mathbf{x}^*)$, are dependent (exercise 1.133). □

Example 5.39 The Kuhn-Tucker example (example 5.36)

$$\max_{x_1, x_2} x_1$$

subject to $\quad g_1(x_1, x_2) = x_2 - (1 - x_1)^3 \leq 0$

$g_2(x_1, x_2) = -x_2 \leq 0$

satisfies none of the constraint qualification conditions, since g_1 is neither concave nor pseudoconvex.

Example 5.40 (Slater constraint qualification) Economic models often assume the Slater constraint qualification condition, namely that g_j is convex and that there exists $\mathbf{x}$ such that $g_j(\mathbf{x}) < 0$ for every $j = 1, 2, \ldots, m$. Since every convex function is pseudoconvex (exercises 4.75 and 4.76), the Slater CQ implies the pseudoconvex CQ.

Exercise 5.42
Show

quasiconvex CQ $\Rightarrow$ Cottle

[Hint: See exercise 4.74.]

Exercise 5.43
Show that

regularity $\Rightarrow$ Cottle

[Hint: Use Gordan's theorem (exercise 3.239).]

Exercise 5.44
Show that

g_j concave $\Rightarrow$ AHUCQ $\Rightarrow$ Abadie

For nonnegative variables, we have the following corollary.

Corollary 5.4.1 (Constraint qualification with nonnegative variables)
Suppose that $\mathbf{x}^*$ *is a local solution of*

$$\max_{\mathbf{x} \geq 0} f(\mathbf{x}) \quad \text{subject to} \quad \mathbf{g}(\mathbf{x}) \leq \mathbf{0}$$

and the binding constraints $j \in B(\mathbf{x}^*)$ *satisfy any one of the following constraint qualification conditions:*

Concave CQ g_j *is concave for every* $j \in B(\mathbf{x}^*)$.

Pseudoconvex CQ g_j *is pseudoconvex, and there exists* $\hat{\mathbf{x}}$ *such that* $g_j(\hat{\mathbf{x}}) < 0$ *for every* $j \in B(\mathbf{x}^*)$.

Quasiconvex CQ g_j *is quasiconvex,* $\nabla g_j(\mathbf{x}^*) \neq \mathbf{0}$, *and there exists* $\hat{\mathbf{x}}$ *such that* $g_j(\hat{\mathbf{x}}) < 0$ *for every* $j \in B(\mathbf{x}^*)$.

Then $\mathbf{x}^*$ *satisfies the Kuhn-Tucker conditions; that is, there exist multipliers* $\lambda = (\lambda_1, \lambda_2, \ldots, \lambda_m)$ *such that*

$$D_{x_i} L[\mathbf{x}^*, \lambda] \leq 0, \quad x_i^* \geq 0, \quad x_i^* D_{x_i} L[\mathbf{x}^*, \lambda] = 0, \quad i = 1, 2, \ldots, n$$

$$g_j(\mathbf{x}^*) \leq 0, \quad \lambda_j \geq 0, \quad \lambda_j g_j(\mathbf{x}^*)) = 0, \quad j = 1, 2, \ldots, m$$

where L is the Lagrangean $L(\mathbf{x}, \lambda) = f(\mathbf{x}) - \sum_{j=1}^{m} \lambda_j g_j(\mathbf{x})$.

Proof The problem can be specified as

$$\max_{\mathbf{x}} f(\mathbf{x})$$

subject to $\mathbf{g}(\mathbf{x}) \leq \mathbf{0}$

$\mathbf{h}(\mathbf{x}) = -\mathbf{x} \leq \mathbf{0}$

We note that **h** is linear and is therefore both concave and convex. Further $D\mathbf{h}[\mathbf{x}] \neq \mathbf{0}$ for every **x**. Therefore, if **g** satisfies one of the three constraint qualification conditions, so does the combined constraint $(\mathbf{g}, \mathbf{h})$. By theorem 5.4, the Kuhn-Tucker conditions are necessary for a local optimum which, by corollary 5.3.2, can be expressed as above. □

Example 5.41 (The consumer's problem) The consumer's problem

$$\max_{\mathbf{x} \geq \mathbf{0}} u(\mathbf{x})$$

subject to $\quad \mathbf{p}^T \mathbf{x} \leq m$

has one functional constraint $g(\mathbf{x}) = \mathbf{p}^T \mathbf{x} \leq m$ and n inequality constraints $h_i(\mathbf{x}) = -x_i \leq 0$, the gradients of which are

$$\nabla g(\mathbf{x}) = \mathbf{p}, \quad \nabla h_i(\mathbf{x}) = \mathbf{e}_i, \quad i = 1, 2, \ldots, n$$

where $\mathbf{e}_i$ is the i unit vector (example 1.79). Provided that all prices are positive $\mathbf{p} > \mathbf{0}$, it is clear that these are linearly independent and the regularity condition of corollary 5.3.2 is always satisfied. However, it is easier to appeal directly to corollary 5.4.1, and observe that the budget constraint $g(\mathbf{x}) = \mathbf{p}^T \mathbf{x} \leq m$ is linear and therefore concave.

5.4.3 Sufficient Conditions

Theorem 5.4 and its corollary 5.4.1 provide practical criteria under which the Kuhn-Tucker conditions are necessary, ensuring that the solution of the problem

$$\max_{\mathbf{x} \in X} f(\mathbf{x}) \tag{68}$$

subject to $\quad \mathbf{g}(\mathbf{x}) \leq \mathbf{0}$

will be found among the solutions of the first-order conditions. However, not every solution of the Kuhn-Tucker conditions will necessarily be a local solution of the problem, let alone a global solution. As in section 5.2, discriminating among solutions of the first-order conditions requires second-order conditions. The specification of appropriate local conditions analogous to theorem 5.1 and corollary 5.1.1 is even more complicated than with equality constraints, since it is necessary to distinguish between binding and slack constraints. Consequently it is useful in economic analysis to seek global conditions that ensure the sufficiency of the first-order conditions, analogous to those in corollaries 5.1.2 and 5.2.4.

In essence, the Kuhn-Tucker conditions are sufficient for a global solution provided the objective function is pseudoconcave and the constraint functions are quasiconvex. Under these hypotheses, if a point $\mathbf{x}^*$ satisfies the Kuhn-Tucker conditions, then it is global solution of the problem. Since every concave function is pseudoconcave (exercise 4.76), this applies in particular to concave f.

Theorem 5.5 (Sufficient conditions for global optimum) *Suppose that f is pseudoconcave and $\mathbf{g}$ is quasiconvex, and let*

$$G = \{\mathbf{x} \in X : \mathbf{g}(\mathbf{x}) \leq \mathbf{0}\} \quad \text{and} \quad L(\mathbf{x}, \lambda) = f(\mathbf{x}) - \sum_{j=1}^{m} \lambda_j g_j(\mathbf{x})$$

Suppose that there exists $\mathbf{x}^* \in G$ and $\lambda = (\lambda_1, \lambda_2, \ldots, \lambda_m) \geq \mathbf{0}$, satisfying the Kuhn-Tucker conditions

$$\nabla f(\mathbf{x}^*) = \sum_{j=1}^{m} \lambda_j \nabla g_j(\mathbf{x}^*) \quad \text{and} \quad \lambda_j g_j(\mathbf{x}^*) = 0, \quad j = 1, 2 \ldots, m \qquad (69)$$

Then $f(\mathbf{x}^*) \geq f(\mathbf{x})$ for every $\mathbf{x} \in G$. That is, $\mathbf{x}^*$ solves

$$\max_{\mathbf{x} \in X} f(\mathbf{x}) \quad \text{subject to} \quad \mathbf{g}(\mathbf{x}) \leq \mathbf{0}$$

Proof Applying exercise 5.45, the first-order conditions (69) imply that

$$Df[\mathbf{x}^*](\mathbf{x} - \mathbf{x}^*) = \sum_{j=1}^{m} \lambda_j \nabla g_j (\mathbf{x} - \mathbf{x}^*) \leq 0 \qquad \text{for every } \mathbf{x} \in G$$

If f is pseudoconcave, this implies that

$$f(\mathbf{x}^*) \geq f(\mathbf{x}) \qquad \text{for every } \mathbf{x} \in G \qquad \square$$

Theorem 5.5 can be extended to quasiconcave objective functions, provided that the solution is not a critical point of the objective function.

Corollary 5.5.1 (Arrow-Enthoven theorem) *Suppose that f is quasiconcave and $\mathbf{g}$ is quasiconvex, and let*

$$G = \{\mathbf{x} \in X : \mathbf{g}(\mathbf{x}) \leq \mathbf{0}\} \quad \text{and} \quad L(\mathbf{x}, \lambda) = f(\mathbf{x}) - \sum_{j=1}^{m} \lambda_j g_j(\mathbf{x})$$

Suppose that there exists $\mathbf{x}^* \in G$ *and* $\lambda = (\lambda_1, \lambda_2, \ldots, \lambda_m) \geq \mathbf{0}$ *with* $Df[\mathbf{x}^*] \neq \mathbf{0}$ *satisfying the Kuhn-Tucker conditions*

$$\nabla f(\mathbf{x}^*) = \sum_{j=1}^{m} \lambda_j \nabla g_j(\mathbf{x}^*) \quad \text{and} \quad \lambda_j g_j(\mathbf{x}^*) = 0, \quad j = 1, 2 \ldots, m \quad (70)$$

Then $f(\mathbf{x}^*) \geq f(\mathbf{x})$ *for every* $\mathbf{x} \in G$.

Proof Since $Df[\mathbf{x}^*] \neq \mathbf{0}$, there exists $\mathbf{dx}$ such that $Df[\mathbf{x}^*](\mathbf{dx}) < 0$. Let $\mathbf{z} = \mathbf{x}^* + \mathbf{dx}$ so that $Df[\mathbf{x}^*](\mathbf{z} - \mathbf{x}^*) < 0$. Define $\mathbf{x}^*(t) = t\mathbf{z} + (1-t)\mathbf{x}^*$. For any $t \in (0, 1)$,

$$Df[\mathbf{x}^*](\mathbf{x}^*(t) - \mathbf{x}^*) = tDf[\mathbf{x}^*](\mathbf{z} - \mathbf{x}^*) < 0$$

Choose any $\mathbf{x} \in G$, and define $\mathbf{x}(t) = t\mathbf{z} + (1-t)\mathbf{x}$. The first-order conditions (70) imply that (exercise 5.45)

$$Df[\mathbf{x}^*](\mathbf{x}(t) - \mathbf{x}^*(t)) = (1-t)Df[\mathbf{x}^*](\mathbf{x} - \mathbf{x}^*)$$

$$= (1-t) \sum_{j=1}^{m} \lambda_j \nabla g_j(\mathbf{x} - \mathbf{x}^*) \leq 0$$

for any $t \in (0, 1)$. Adding these two inequalities, we have

$$Df[\mathbf{x}^*](\mathbf{x}(t) - \mathbf{x}^*) < 0$$

Since f is quasiconcave, this implies that (exercise 4.73)

$$f(\mathbf{x}^*) > f(\mathbf{x}(t)) \quad \text{for every } t \in (0, 1)$$

Letting $t \to 0$, we have $f(\mathbf{x}^*) \geq f(\mathbf{x})$ for every $\mathbf{x} \in G$. □

The extension to nonnegative variables is straightforward.

Corollary 5.5.2 (Nonnegative variables) *Suppose that f is quasiconcave and* $\mathbf{g}$ *is quasiconvex, and let*

$$G = \{\mathbf{x} \in \Re_+^n : \mathbf{g}(\mathbf{x}) \leq \mathbf{0},\} \quad \text{and} \quad L(\mathbf{x}, \lambda) = f(\mathbf{x}) - \sum_{j=1}^{m} \lambda_j g_j(\mathbf{x})$$

Suppose that there exists $\mathbf{x}^* \in G$ *and* $\lambda = (\lambda_1, \lambda_2, \ldots, \lambda_m) \geq \mathbf{0}$ *with* $Df[\mathbf{x}^*] \neq \mathbf{0}$ *satisfying the Kuhn-Tucker conditions*

$$D_{x_i}L[\mathbf{x}^*,\lambda] \leq 0, \quad x_i^* \geq 0, \quad x_i^* D_{x_i}L[\mathbf{x}^*,\lambda] = 0, \quad i=1,2,\ldots,n$$
$$g_j(\mathbf{x}^*) \leq 0, \quad \lambda_j \geq 0, \quad \lambda_j g_j(\mathbf{x}^*) = 0, \quad j=1,2,\ldots,m$$

Then $f(\mathbf{x}^*) \geq f(\mathbf{x})$ for every $\mathbf{x} \in G$.

Proof The first-order condition $D_{x_i}L[\mathbf{x}^*,\lambda] \leq 0$ for every i can be written

$$\nabla f(\mathbf{x}^*) - \sum_{j=1}^{m} \nabla \lambda_j g_j(\mathbf{x}^*) \leq \mathbf{0}$$

This implies that

$$\left(\nabla f(\mathbf{x}^*) - \sum_{j=1}^{m} \nabla \lambda_j g_j(\mathbf{x}^*) \right)^T \mathbf{x} \leq 0$$

for every $\mathbf{x} \in G$ (since $\mathbf{x} \geq 0$). The first-order conditions also require that

$$\left(\nabla f(\mathbf{x}^*) - \sum_{j=1}^{m} \lambda_j \nabla g_j(\mathbf{x}^*) \right)^T \mathbf{x}^* = 0$$

Subtracting

$$\left(\nabla f(\mathbf{x}^*) - \sum_{j=1}^{m} \lambda_j \nabla g_j(\mathbf{x}^*) \right)^T (\mathbf{x} - \mathbf{x}^*) \leq 0$$

or

$$\nabla f(\mathbf{x}^*)^T (\mathbf{x}-\mathbf{x}^*) \leq \sum_{j=1}^{m} \lambda_j \nabla g_j(\mathbf{x}^*)^T (\mathbf{x}-\mathbf{x}^*)$$

and therefore applying exercise 5.45, we have

$$\nabla f(\mathbf{x}^*)^T (\mathbf{x}-\mathbf{x}^*) \leq \sum_{j=1}^{m} \lambda_j \nabla g_j(\mathbf{x}^*)^T (\mathbf{x}-\mathbf{x}^*) \leq 0$$

for every $\mathbf{x} \in G$. The remainder of the proof is identical to that of corollary 5.5.1. □

Example 5.42 (Cost minimization) In full generality the competitive firm's cost minimization problem (example 5.16) is

$$\min_{\mathbf{x} \geq 0} \mathbf{w}^T \mathbf{x} \quad \text{subject to} \quad f(\mathbf{x}) \geq y$$

5.4 Inequality Constraints

The objective function is linear and hence concave. Quasiconcavity of the production function f is a natural assumption regarding the technology (example 1.163), and is less restrictive than assuming that f is fully concave. The cost minimization problem can be recast in the standard form as

$$\max_{\mathbf{x} \geq 0} -\mathbf{w}^T\mathbf{x} \quad \text{subject to} \quad -f(\mathbf{x}) \leq -y$$

with $-f$ quasiconvex. Let L denote the Lagrangean $L(\mathbf{x},\lambda) = -\mathbf{w}^T\mathbf{x} - \lambda(y - f(\mathbf{x}))$. By corollary 5.5.2, the Kuhn-Tucker conditions

$$D_{x_i}L[\mathbf{x}^*,\lambda] \leq 0, \quad x_i^* \geq 0, \quad x_i^* D_{x_i}L[\mathbf{x}^*,\lambda] = 0$$
$$-f(\mathbf{x}^*) + y \leq 0, \quad \lambda \geq 0, \quad \lambda(-f(\mathbf{x}^*) + y) = 0$$

for every $i = 1, 2, \ldots, n$ are sufficient for cost minimization. These require that $f(\mathbf{x}) \geq y$ and $f(\mathbf{x}) = y$ unless $\lambda = 0$. They also require that for each input i,

$$\lambda D_{x_i}f[\mathbf{x}^*] \leq w_i, \quad x_i \geq 0, \quad x_i(\lambda D_{x_i}f[\mathbf{x}^*] - w_i) = 0$$

No input is employed at a level at which the value of its marginal product $\lambda D_{x_i}f[\mathbf{x}]$ is greater than its cost w_i. Moreover, for each input i used in production $x_i > 0$, the value of its marginal product must be equal to its cost $\lambda D_{x_i}f[\mathbf{x}] = w_i$.

Theorem 5.5 and corollaries 5.5.1 and 5.5.2 establish the sufficiency of the first-order conditions provided the hypotheses are met. Theorem 5.4 and corollary 5.4.1 establishes the necessity of the first-order conditions provided an appropriate constraint qualification is met. In fact the Arrow-Enthoven constraint qualification condition is designed to match the hypotheses of the corollary 5.5.1. If the hypotheses of both theorem 5.4 and corollary 5.5.1 (or corollaries 5.4.1 and 5.5.2) are met, then the first-order conditions are both necessary and sufficient for a global optimum.

Corollary 5.5.3 (Necessary and sufficient conditions) *In the constrained optimization problem*

$$\max_{\mathbf{x} \in G} f(\mathbf{x}) \tag{71}$$

where $G\{\mathbf{x} \in X : g_j(\mathbf{x}) \leq 0, j = 1, 2, \ldots m\}$, suppose that

- f is quasiconcave and g_j are quasiconvex
- $\mathbf{x}^* \in G$ is a regular point of f
- $\mathbf{x}^*$ is a regular point of g_j for every $j \in B(\mathbf{x}^*)$
- there exists $\hat{\mathbf{x}}$ such that $g_j(\hat{\mathbf{x}}) < 0$ for every $j \in B(\mathbf{x}^*)$

where $B(\mathbf{x}^*)$ is the set of binding constraints at $\mathbf{x}^*$. Then $\mathbf{x}^*$ is a global solution of (71) if and only if there exist multipliers $\lambda_1, \lambda_2, \ldots, \lambda_m \geq 0$ such that

$$\nabla f(\mathbf{x}^*) = \sum_{j=1}^{m} \lambda_j \nabla g_j(\mathbf{x}^*) \quad \text{and} \quad \lambda_j g_j(\mathbf{x}^*) = 0, \qquad j = 1, 2 \ldots, m$$

Furthermore $\mathbf{x}^*$ is a global solution of

$$\max_{\mathbf{x} \geq 0} f(\mathbf{x}) \quad \text{subject to} \quad \mathbf{g}(\mathbf{x}) \leq \mathbf{0}$$

if and only if there exist multipliers $\lambda_1, \lambda_2, \ldots, \lambda_m$ such that

$$D_{x_i} L[\mathbf{x}^*, \lambda] \leq 0, \quad x_i^* \geq 0, \quad x_i^* D_{x_i} L[\mathbf{x}^*, \lambda] = 0, \qquad i = 1, 2, \ldots, n$$

$$g_j(\mathbf{x}^*) \leq 0, \quad \lambda_j \geq 0, \quad \lambda_j g_j(\mathbf{x}^*) = 0, \qquad j = 1, 2, \ldots, m$$

where $L(\mathbf{x}, \lambda) = f(\mathbf{x}) - \sum \lambda_j g_j(\mathbf{x})$ is the Lagrangean.

Proof By theorem 5.4 or corollary 5.4.1, the Kuhn-Tucker conditions are necessary. By corollary 5.5.1 or corollary 5.5.2, they are also sufficient. □

Example 5.43 (The consumer's problem) The consumer's problem is

$$\max_{\mathbf{x} \in X} u(\mathbf{x}) \quad \text{subject to} \quad \mathbf{p}^T \mathbf{x} \leq m$$

Assuming concavity of the utility function is too restrictive, while assuming quasiconcavity is natural, since it is equivalent to convexity of preferences (example 3.55). Nonsatiation implies that $Du[\mathbf{x}] \neq 0$ for every $\mathbf{x} \in X$. Under these assumptions, the first-order conditions

$$\nabla u(\mathbf{x}^*) = \lambda \mathbf{p}$$

are necessary and sufficient for utility maximization (corollary 5.5.3).

Exercise 5.45
Suppose that

$$G = \{\mathbf{x} \in X : g_j(\mathbf{x}) \leq 0, j = 1, 2, \ldots, m\}$$

with g_j is quasiconvex. Let $\mathbf{x}^* \in G$ and $\lambda \in \Re_+^m$ satisfy the complementary slackness conditions $\lambda_j g_j(\mathbf{x}^*) = 0$ for every $j = 1, 2, \ldots, m$. Then

$$\sum_j \lambda_j Dg_j[\mathbf{x}^*](\mathbf{x} - \mathbf{x}^*) \leq 0$$

for every $\mathbf{x} \in G$.

5.4.4 Linear Programming

A linear programming problem is a special case of the general constrained optimization problem

$$\max_{\mathbf{x} \geq 0} \; f(\mathbf{x}) \qquad (72)$$

subject to $\quad \mathbf{g}(\mathbf{x}) \leq \mathbf{0}$

in which both the objective function f and the constraint function $\mathbf{g}$ are linear. Linear programming is the most developed branch of optimization theory. It also the most important in practice. Since many real world systems are linear or can approximated linearly, linear programming provides an appropriate mathematical model for such systems. It would be hard to exaggerate the importance of linear programming in practical optimization, in applications such as production scheduling, transportation and distribution, inventory control, job assignment, capital budgeting, and portfolio management.

Example 5.44 Joel Franklin (1980) relates a visit to the headquarters of the Mobil Oil Corporation in New York in 1958. The purpose of his visit was to study Mobil's use of computers. In those days computers were rare and expensive, and Mobil's installation had cost millions of dollars. Franklin recognized the person in charge; they had been postdoctoral fellows together. Franklin asked his former colleague how long he thought it would take to pay off this investment. "We paid it off in about two weeks" was the surprise response. Elaborating, he explained that Mobil were able to make massive cost savings by optimizing production decision using linear programming, decisions that had previously been made heuristically. Franklin's anecdote highlights the enormous benefits that can accrue from optimizing recurrent decisions.

With linear objective and linear constraints, the standard linear programming problem (72) satisfies the hypotheses of both corollary 5.4.1 and theorem 5.5.1. Consequently the Kuhn-Tucker conditions are both necessary and sufficient for a global solution to a linear programming problem. The Kuhn-Tucker conditions for a linear programming problem determine a system of linear inequalities, whose every solution is a global optimum of the problem. Finding an optimal solution is a matter of finding a solution to this system of linear inequalities. However, for a problem with many variables or constraints, solving these inequalities could pose a formidable computational problem. A second reason for the popularity of linear programming, in practice, is the availability of a very efficient algorithm (the simplex algorithm) for solving the Kuhn-Tucker conditions for linear programs. The postwar conjunction of the availability of digital computers and the discovery of the simplex algorithm paved the way for successful industrial application as exemplified by Mobil's experience.

In the following example, we solve a simple linear programming problem to illustrate the general problem of solving the Kuhn-Tucker conditions. The example will also provide some insight into the practically important simplex algorithm.

Example 5.45 Suppose that a furniture maker can produce three products—bookcases, chairs, and desks. Each product requires machining, finishing, and some labor. The supply of these resources is limited. Unit profits and resource requirements are listed in the following table.

	Bookcases	Chairs	Desks	Capacity
Finishing	2	2	1	30
Labor	1	2	3	25
Machining	2	1	1	20
Net profit	3	1	3	

Letting x_b, x_c, and x_d stand for the output of bookcases, chairs, and desks respectively, we can model the production planning problem as follows:

5.4 Inequality Constraints

max $\Pi = 3x_b + x_c + 3x_d$

$2x_b + 2x_c + x_d \leq 30, \quad x_b + 2x_c + 3x_d \leq 25, \quad 2x_b + x_c + x_d \leq 20$

and

$x_a \geq 0, \quad x_b \geq 0, \quad x_c \geq 0$

Letting λ_f, λ_l, and λ_m be the Lagrange multipliers of the finishing, labor, and machining constraints respectively, the Lagrangean is

$$L(\mathbf{x}, \lambda) = 3x_b + x_c + 3x_d - \lambda_f(2x_b + 2x_c + x_d - 30)$$
$$- \lambda_l(x_b + 2x_c + 3x_d - 25) - \lambda_m(2x_b + x_c + x_d - 20)$$

The Kuhn-Tucker conditions are

$D_{x_b}L = 3 - 2\lambda_f - \lambda_l - 2\lambda_m \leq 0, \quad x_b \geq 0, \quad x_b D_{x_b}L = 0$

$D_{x_c}L = 1 - 2\lambda_f - 2\lambda_l - \lambda_m \leq 0, \quad x_c \geq 0, \quad x_c D_{x_c}L = 0$

$D_{x_d}L = 3 - \lambda_f - 3\lambda_l - \lambda_m \leq 0, \quad x_d \geq 0, \quad x_d D_{x_d}L = 0$

$2x_b + 2x_c + x_d \leq 30, \quad \lambda_f \geq 0, \quad \lambda_f(2x_b + 2x_c + x_d - 30) = 0$

$x_b + 2x_c + 3x_d \leq 25, \quad \lambda_l \geq 0, \quad \lambda_l(x_b + 2x_c + 3x_d - 25) = 0$

$2x_b + x_c + x_d \leq 20, \quad \lambda_m \geq 0, \quad \lambda_m(2x_b + x_c + x_d - 20) = 0$

and these characterize the optimal solution.

The essence of the problem of solving the first-order conditions is determining which constraints are binding and which are slack, and therefore which Lagrange multipliers are zero. We will show how the Lagrangean can aid in this choice; this is the essence of the simplex algorithm.

To start, suppose that we produce only desks. That is, let $x_c = x_d = 0$. The constraints become

$2x_b \leq 30, \quad x_b \leq 25, \quad 2x_b \leq 20$

The first two constraints are redundant, which implies that they must be slack and therefore $\lambda_f = \lambda_l = 0$. Complementary slackness requires that if $x_b > 0$,

$D_{x_b}L = 3 - 2\lambda_m = 0 \quad \text{or} \quad \lambda_m = \dfrac{3}{2}$

This in turn implies that the machining constraint $2x_b \le 20$ is binding. That is, $x_b = 10$. $x_b = 10$, $x_c = 0$, $x_d = 0$ is a feasible solution, since it satisfies the constraints. However, it cannot be an *optimal solution*, since the implied Lagrange multipliers ($\lambda_f = \lambda_l = 0$, $\lambda_m = 3/2$) do not satisfy the first-order conditions. Specifically,

$$D_{x_d} L = 3 - \lambda_f - 3\lambda_l - \lambda_m = \frac{3}{2} > 0$$

The implied shadow prices are not the optimal values.

Evaluating the Lagrangean at these shadow prices can lead us toward the optimal solution. At these shadow prices the Lagrangean is

$$L\left(\mathbf{x}, \left(0, 0, \frac{3}{2}\right)\right) = 3x_b + x_c + 3x_d - \frac{3}{20}(2x_b + x_c + x_d - 20)$$

$$= \left(3 - \frac{6}{2}\right)x_b + \left(1 - \frac{3}{2}\right)x_c + \left(3 - \frac{3}{2}\right)x_d - \frac{3}{2}(-20)$$

$$= 30 - \frac{1}{2}x_c + \frac{3}{2}x_d$$

This reveals that net profit at (10, 0, 0) is $30, but this is not optimal. At these shadow prices, profit would be increased by increasing x_d. (Profit would be decreased by increasing x_c.)

Taking this hint, let us now consider a production plan with both $x_b > 0$ and $x_d > 0$, leaving $x_c = 0$. The constraints become

$$2x_b + x_d \le 30, \quad x_b + 3x_d \le 25, \quad 2x_b + x_d \le 20$$

The first constraint is redundant (given the third constraint), which implies that $\lambda_f = 0$. Furthermore, if $x_b, x_d > 0$, complementary slackness requires that

$$D_{x_b} L = 3 - 2\lambda_f - \lambda_l - 2\lambda_m = 0$$

$$D_{x_d} L = 3 - \lambda_f - 3\lambda_l - \lambda_m = 0$$

which with $\lambda_f = 0$ implies that

$$\lambda_l + 2\lambda_m = 3, \quad 3\lambda_l + \lambda_m = 3$$

These equations have the unique solution $\lambda_l = 3/5$ and $\lambda_m = 6/5$. Substituting these shadow prices $\lambda^* = (0, 3/5, 6/5)$ into the Lagrangean, we get

5.4 Inequality Constraints

$$L(x, \lambda^*) = 3x_b + x_c + 3x_d - \frac{3}{5}(x_b + 2x_c + 3x_d - 25)$$

$$- \frac{6}{5}(2x_b + x_c + x_d - 20)$$

$$= \left(3 - \frac{3}{5} - \frac{12}{5}\right)x_b + \left(1 - \frac{6}{5} - \frac{6}{5}\right)x_c$$

$$+ \left(3 - \frac{9}{5} - \frac{6}{5}\right)x_d - 25\left(\frac{3}{5}\right) - 20\left(\frac{6}{5}\right)$$

$$= 39 - \frac{7}{5}x_c$$

This reveals that producing bookcases $x_b > 0$ and desks $x_d > 0$ but no chairs $x_c = 0$ will yield a profit of $39. The only possible perturbation involves producing chairs $x_c > 0$. Since the coefficient of x_c is negative, any marginal increase in x_c above 0 would decease profits. Because of linearity, any nonmarginal increase would also reduce profits. This establishes that an optimal plan produces no chairs.

It remains to find the optimal plan, the quantity of bookcases and desks. Since the labor and machining constraints are both binding, the equations

$$x_b + 3x_d = 25, \quad 2x_b + x_d = 20$$

can be solved to yield the optimal production of bookcases and desks. The optimal production plan is $x_b = 7$, $x_d = 6$.

Let us summarize the method adopted in example 5.45. Setting certain decision variables to zero yielded a basic feasible solution in which some of the constraints were slack. Using the first-order conditions, we derived the values of the Lagrange multiplier (resource prices) implied by this basic feasible solution. Evaluating the Lagrangean at these prices immediately revealed (1) whether or not this was an optimal solution and (2) if not, the direction of improvement. Making this improvement led to another basic feasible solution at which the procedure could be repeated. Since, at each stage, the value of the Lagrangean increases, this procedure must eventually lead us to the optimal solution. This is the basic simplex algorithm.

Exercise 5.46
Solve the preceding problem starting from the hypothesis that $x_c > 0$, $x_b = x_d = 0$. [Hint: If faced with a choice between $x_b > 0$ and $x_d > 0$, choose the latter.]

Exercise 5.47
What happens if you ignore the hint in the previous exercise?

5.4.5 Concave Programming

Concave programming is another special case of the general constrained optimization problem

$$\max_{\mathbf{x} \in X} f(\mathbf{x}) \tag{73}$$

subject to $\quad \mathbf{g}(\mathbf{x}) \leq \mathbf{0}$

in which the objective function f is concave and the constraint functions g_j are convex. For such problems an alternative derivation of the Kuhn-Tucker conditions is possible, providing yet another perspective on the Lagrangean method.

To achieve this perspective, we re-introduce a parameter $\mathbf{c}$ and consider the family of optimization problems

$$\max_{\mathbf{x} \in X} f(\mathbf{x})$$

subject to $\quad \mathbf{g}(\mathbf{x}) \leq \mathbf{c}$

of which (73) is the particular instance in which $\mathbf{c} = \mathbf{0}$. We direct our attention to what can be achieved, disregarding for the moment how it is achieved. That is, we suppress primary consideration of the decisions made ($\mathbf{x}$) and focus attention on the results achieved ($z = f(\mathbf{x})$) and the resources used ($\mathbf{c} = \mathbf{g}(\mathbf{x})$). Remarkably, answering the *what* question simultaneously answers the *how* question.

Regarding $\mathbf{c}$ as a parameter measuring the available resources, the value function

$$v(\mathbf{c}) = \max_{\mathbf{x} \in X} \{ f(\mathbf{x}) : g(\mathbf{x}) \leq \mathbf{c} \}$$

summarizes what can be achieved with different amounts of resources $\mathbf{c}$. The set of all attainable outcomes is given by its hypograph

5.4 Inequality Constraints

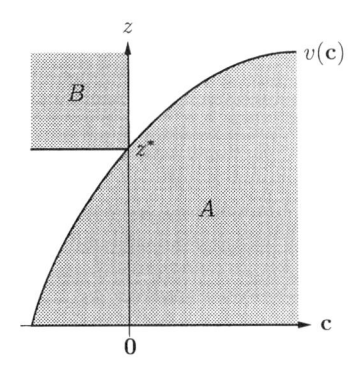

Figure 5.11
Concave programming

$$A = \{(\mathbf{c}, z) \in Y \times \Re : z \leq v(\mathbf{c})\}$$

Since the value function is concave (theorem 3.1), the attainable set A is convex (exercise 3.125).

Let $z^* = v(\mathbf{0})$; that is, z^* is the value attained by the optimal solution in the original problem (73). Define

$$B = \{(\mathbf{c}, z) \in Y \times \Re : \mathbf{c} \leq \mathbf{0}, z \geq z^*\}$$

B is a subset of the unattainable outcomes, comprising those outcomes that achieve at least the same level of the objective z^* with fewer resources $\mathbf{c} \leq \mathbf{0}$ or a higher level of the objective $z \geq z^*$ with the given resources $\mathbf{c} = \mathbf{0}$ (figure 5.11). B is a convex set with a nonempty interior (exercise 5.48) and A contains no interior points of B (exercise 5.49).

By the separating hyperplane theorem (corollary 3.2.1), there exists a linear functional $L \in (Y \times \Re)^*$ that separates A from B, that is,

$$L(\mathbf{c}, z) \leq L(\mathbf{0}, z^*) \quad \text{for every } (\mathbf{c}, z) \in A \tag{74}$$

$$L(\mathbf{c}, z) \geq L(\mathbf{0}, z^*) \quad \text{for every } (\mathbf{c}, z) \in B \tag{75}$$

Assuming that $Y \subseteq \Re^m$, there exists (exercise 5.50) $\alpha \geq 0$ and $\lambda \geq \mathbf{0}$ such that

$$L(\mathbf{c}, z) = \alpha z - \lambda^T \mathbf{c}$$

The Slater constraint qualification condition (example 5.4) ensures that $\alpha > 0$ (exercise 5.51), and we can without loss of generality select $\alpha = 1$.

Therefore (74) shows that $(\mathbf{0}, z^*)$ maximizes the linear functional $L(\mathbf{c}, z) = z - \lambda^T \mathbf{c}$ on A. That is,

$$z^* - \lambda^T \mathbf{0} \geq z - \lambda^T \mathbf{c} \quad \text{for every } (\mathbf{c}, z) \in A \tag{76}$$

By definition, for every $(\mathbf{c}, z) \in A$, there exists some $\mathbf{x} \in X$ such that $f(\mathbf{x}) = z$ and $g(\mathbf{x}) \leq \mathbf{c}$. Let $\mathbf{x}^*$ be such that $g(\mathbf{x}^*) \leq \mathbf{0}$ and $f(\mathbf{x}^*) = z^*$. Substituting in (76), we have

$$f(\mathbf{x}^*) \geq f(\mathbf{x}) - \lambda^T g(\mathbf{x}) \quad \text{for every } \mathbf{x} \in X \tag{77}$$

The right-hand side of this inequality is precisely the Lagrangean of the constrained optimization problem

$$\max_{\mathbf{x} \in X} f(\mathbf{x}) \quad \text{subject to} \quad \mathbf{g}(\mathbf{x}) \leq \mathbf{0}$$

We have shown that $\mathbf{x}^*$ maximizes the Lagrangean over the choice set X. Rearranging (77), we have

$$f(\mathbf{x}^*) - f(\mathbf{x}) \geq -\lambda^T \mathbf{g}(\mathbf{x})$$

for every $\mathbf{x} \in X$. In particular, $\mathbf{x}^* \in X$, and therefore

$$0 = f(\mathbf{x}^*) - f(\mathbf{x}^*) \geq -\lambda^T \mathbf{g}(\mathbf{x}^*)$$

that is, $\lambda^T g(\mathbf{x}^*) \geq 0$. But $\lambda \geq \mathbf{0}$ (exercise 5.50) and $\mathbf{g}(\mathbf{x}^*) \leq \mathbf{0}$, and therefore we must have

$$\lambda_j g_j(\mathbf{x}^*) = 0 \quad \text{for every } j = 1, 2, \ldots m \tag{78}$$

We have shown that, if $\mathbf{x}^*$ solves the problem

$$\max_{\mathbf{x}} f(\mathbf{x}) \quad \text{subject to} \quad \mathbf{g}(\mathbf{x}) \leq \mathbf{0}$$

there exist nonnegative Lagrange multipliers $\lambda = (\lambda_1, \lambda_2, \ldots, \lambda_m)$ such that

- $\mathbf{x}^*$ maximizes the Lagrangean function $L(\mathbf{x}, \lambda) = f(\mathbf{x}) - \sum_{j=1}^{m} \lambda_j g_j(\mathbf{x})$ on X
- $\mathbf{x}^*$ and λ satisfy the complementary slackness condition $\lambda_j g_j(\mathbf{x}^*) = 0$ for every $j = 1, 2, \ldots, m$

If f and $\mathbf{g}$ are differentiable and X is open, maximization requires that the Lagrangean be stationary at $\mathbf{x}^*$ (proposition 5.1), that is,

5.4 Inequality Constraints

$$D_\mathbf{x} L[\mathbf{x}^*, \lambda] = Df[\mathbf{x}^*] - \sum_{j=1}^{m} \lambda_j Dg_j[\mathbf{x}^*] = \mathbf{0}$$

or

$$\nabla f(\mathbf{x}^*) = \sum_{j=1}^{m} \lambda_j \nabla g_j(\mathbf{x}^*)$$

Furthermore, since the Lagrangean $L(\mathbf{x}, \lambda) = f(\mathbf{x}) - \sum_j \lambda_j g_j(\mathbf{x})$ is concave in $\mathbf{x}$, the first-order necessary conditions are sufficient (exercise 5.20)

We summarize these conclusions in the following theorem.

Theorem 5.6 (Concave programming) *Suppose that f is concave and g_j are convex and there exists an $\hat{\mathbf{x}} \in X$ for which $\mathbf{g}(\hat{\mathbf{x}}) < \mathbf{0}$. Then $\mathbf{x}^*$ is a global solution of*

$$\max_{\mathbf{x} \in X} f(\mathbf{x})$$

subject to $\mathbf{g}(\mathbf{x}) \leq \mathbf{0}$

if and only if there exist multipliers $\lambda = (\lambda_1, \lambda_2, \ldots, \lambda_m) \geq \mathbf{0}$ such that the Lagrangean $L(\mathbf{x}, \lambda) = f(\mathbf{x}) - \sum \lambda_j g_j(\mathbf{x})$ is maximized at $\mathbf{x}^*$ and $\lambda_j g_j(\mathbf{x}^*) = 0$ for every j. If f and $\mathbf{g}$ are differentiable, it is necessary and sufficient that $(\mathbf{x}^*, \lambda)$ is stationary point of L, that is $(\mathbf{x}^*, \lambda)$ satisfies the Kuhn-Tucker conditions

$$\nabla f(\mathbf{x}^*) = \sum_{j=1}^{m} \lambda_j \nabla g_j(\mathbf{x}^*) \quad \text{and} \quad \lambda_j g_j(\mathbf{x}^*) = 0, \quad j = 1, 2 \ldots, m$$

Exercise 5.48
Show that B is a convex set with a nonempty interior.

Exercise 5.49
Show that $A \cap \text{int } B = \emptyset$.

Exercise 5.50
Show that $L(\mathbf{c}, z) = \alpha z - \lambda^T \mathbf{c}$ with $\alpha \geq 0$ and $\lambda \geq \mathbf{0}$. [Hint: Use exercise 3.47 and apply (75) to the point $(\mathbf{c}, z^* + 1)$.]

Exercise 5.51
The constraint $\mathbf{g}$ satisfies the Slater constraint qualification condition if there exist $\hat{\mathbf{x}} \in X$ with $g(\hat{\mathbf{x}}) < \mathbf{0}$. Show that this implies that $\alpha > 0$.

Remark 5.7 (Normalization) Setting $\alpha = 1$ in defining the separating hyperplane is merely a normalization (see remark 3.2). The Lagrange multipliers α and λ constitute a system of shadow prices for the objective function and constraints (resources) respectively and the Lagrangean is a valuation function at these prices. Only relative prices matter for economic decisions and normalization corresponds to selecting the general price level. Fixing $\alpha = 1$ as the normalization amounts to selecting the value which is being maximized as the numéraire. For most optimization problems this seems an appropriate choice.

In other situations an alternative normalization might be useful. In the vat design problem (example 5.27) we measured the net benefit in terms of volume units, implicitly setting $\alpha = 1$. If volume could be assigned a value or market price, it might have been more useful to set α equal to that price, in which case the Lagrangean would have measured net revenue or profit. As another example, in a problem in which the objective is measured in a foreign currency, α might be set to the exchange rate. The point is that the optimal decision establishes only a system of relative prices; the general price level can be chosen arbitrarily to meet the needs of the analyst.

Remark 5.8 We cautioned earlier (section 5.3.4) that although a local optimum must be a stationary point of the Lagrangean, it does not necessarily maximize the Lagrangean over the set X. However, for concave problems (f concave, g_j convex), every stationary point is necessarily a maximum (exercise 5.20). Therefore, in concave programming, it is appropriate to say that solving of the constrained problem

$$\max_{\mathbf{x} \in X} f(\mathbf{x}) \quad \text{subject to} \quad \mathbf{g}(\mathbf{x}) \leq \mathbf{0}$$

is equivalent to unconstrained maximization of the Lagrangean $L(\mathbf{x}, \lambda) = f(\mathbf{x}) - \sum \lambda_j g_j(\mathbf{x})$ in X.

Example 5.46 In example 5.33 we showed that $(0, 4)$ is the only point satisfying the first-order necessary conditions for a local solution of the problem

$$\max_{x_1 \geq 0, x_2 \geq 0} \log x_1 + \log(x_2 + 5)$$

$$\text{subject to} \quad x_1 + x_2 - 4 \leq 0$$

Observing that the objective function f is concave (example 3.50 and exercise 3.131) and the constraints are affine and hence convex (exercise 3.130), we conclude that the first-order conditions are also sufficient. So $(0, 4)$ is a global solution.

Exercise 5.52 (Peak-load pricing)
Suppose that a public utility supplies a service, whose demand varies with the time of day. For simplicity, assume that demand in each period is independent of the price in other periods. The inverse demand function for each period is $p_i(y_i)$. Assume that marginal production costs c_i are constant, independent of capacity and independent across periods. Further assume that the marginal cost of capacity c_0 is constant. With these assumptions, the total cost function is

$$c(\mathbf{y}, Y) = \sum_{i=1}^{n} c_i y_i + c_0 Y$$

The objective is to determine outputs y_i (and hence prices p_i) and production capacity Y to maximize social welfare as measured by total consumer and producer surplus.

In any period i, total surplus is measured by the area between the demand and cost curves, that is,

$$S_i(\mathbf{y}, Y) = \int_0^{y_i} (p_i(\tau) - c_i) \, d\tau$$

So aggregate surplus is

$$S(\mathbf{y}, Y) = \sum_{i=1}^{n} \int_0^{y_i} (p_i(\tau) - c_i) \, d\tau - c_0 Y \tag{79}$$

The optimization problem is to choose nonnegative y_i and Y so as to maximize (79) subject to the constraints

$$y_i \leq Y, \quad i = 1, 2, \ldots, n$$

Show that it is optimal to price at marginal cost during off-peak periods, and extract a premium during peak periods, where the total premium is equal to the marginal cost of capacity c_0. Furthermore, under this pricing rule, the enterprise will break even. Note that

$$D_{y_i} S_i(\mathbf{y}, Y) = D_{y_i} \int_0^{y_i} (p_i(\tau) - c_i) \, d\tau = p_i(y_i) - c_i$$

5.5 Notes

There is an immense literature on optimization theory and practice. Most texts on mathematics for economists discuss optimization. Good treatments are given by Sydsaeter and Hammond (1995), Simon and Blume (1994), and Takayama (1994, 1985), listed in increasing order of rigor. There are also several texts on optimization written especially for economists. Dixit (1990) is a delightful and lucid introduction to optimization in economics written with a minimum of formalism. This is complemented by Sundaram (1996), which is a comprehensive and rigorous treatment of optimization in economics. Also recommended are Lambert (1985), Leonard and Long (1992), and Madden (1986). Simon (1986) is a concise survey of optimization in economics written for mathematicians.

Noneconomists often draw their examples from the physical world, and place relatively more emphasis on computation rather than characterization. However, there are insights to be gained from exploring this wider literature. Two personal favorites are Luenberger (1984) and Bazaraa et al. (1993). Other standard references include Mangasarian (1994) and Zangwill (1969). Chvatal (1983) is an excellent introduction to linear programming and the simplex algorithm. By analogy with linear programming, the general inequality-constrained optimization problem (section 5.4) is often call nonlinear programming. The problem with equality constraints (section 5.3) is sometimes called classical programming.

While all the examples presented in this chapter were finite dimensional, the results apply (with a little attention to topological niceties) to arbitrary Banach spaces, which is the appropriate setting for many optimization problems. A thorough coverage with many examples is given by Luenberger (1969).

The implicit function theorem approach is probably the most common textbook derivation of the Lagrange multiplier method. The reader who finds our coordinate-free approach intimidating will find a similar coordinate-based treatment in Beavis and Dobbs (1990, p. 38) and many similar texts.

The section on constraint qualification draws heavily on Abadie (1967) and Bazaraa et al. (1972, 1993). Theorem 5.5 is due to Mangasarian (1994), clarifying the classic theorem (corollary 5.5.1) of Arrow and Enthoven (1961). Our proof of the latter is adapted from Sundaram

(1996, p. 221). Takayama (1994, pp. 95–98, 615–619) discusses some other variations.

Example 5.1 is adapted from Whittle (1971). Exercise 5.11 is from Kreps (1990). Example 5.29 is adapted from Fryer and Greenman (1987). Examples 5.30 and 5.33 are adapted from Leonard and Long. The model of a regulated monopolist was proposed by Averch and Johnson (1962). Their analysis was clarified and extended in Baumol and Klevorick (1970). The peak-load pricing problem has a long tradition in economics. It was first formulated as an optimization problem by Williamson (1966). Exercise 5.52 is based on Littlechild (1970), who applied peak-load pricing to a telephone network.

6 Comparative Statics

One of the most important professional activities of economists is to carry out exercises in comparative statics: to estimate the consequences and the merits of changes in economic policy and in our economic environment.
—Herbert Scarf, 1994

Comparative statics is the name economists use to describe the analysis of way in which the solutions of an economic model change as the parameters and specification change. Most of the testable predictions and policy implications of economic models are generated by comparative static analysis.

Economic models fall into two classes: optimization models and equilibrium models. An optimization model can be represented as a constrained optimization problem

$$\max_{\mathbf{x} \in X} f(\mathbf{x}, \boldsymbol{\theta}) \quad \text{subject to} \quad \mathbf{x} \in G(\boldsymbol{\theta}) \tag{1}$$

while an equilibrium model can be expressed as a system of equations

$$f(\mathbf{x}, \boldsymbol{\theta}) = \mathbf{0} \tag{2}$$

Suppose that a change in policy or the context of a model can be summarized by a change in the parameters from $\boldsymbol{\theta}^1$ to $\boldsymbol{\theta}^2$. The objective of comparative static analysis is to assess the impact on the corrresponding solutions $\mathbf{x}^1$ and $\mathbf{x}^2$ to (1) or (2). A typical question might be: What is the impact on decision variable x_k of an increase in θ_i, holding all other θ_j constant? If we can determine that x_k must decrease, we can establish a qualitative proposition of the form

$$\theta_i^2 > \theta_i^1, \theta_{j \neq i} \text{ constant} \Rightarrow x_k^2 < x_k^1$$

As a further step, a quantitative analysis attempts to assess the magnitude of the change in x_k. The process is called comparative statics, since it *compares* equilibria before and after a change in a parameter. Comparative statics does not address the issue of the path by which economic variables move from one equilibrium solution to another. Neither does it consider the time taken for the adjustment. This is why it is called *comparative statics*.

One obvious approach to comparative static analysis would be to compute an explicit functional solution $\mathbf{x}(\boldsymbol{\theta})$ for (1) or (2), from which comparative statics would be derived immediately by substitution. Typically this procedure is not available to the economist since economic

Table 6.1
Maximum theorems

	Monotone maximum theorem (theorem 2.1)	Continuous maximum theorem (theorem 2.3)	Convex maximum theorem (theorem 3.10)	Smooth maximum theorem (theorem 6.1)
Objective function	Supermodular, increasing	Continuous	Concave	Smooth
Constraint correspondence	Weakly increasing	Continuous, compact-valued	Convex	Smooth, regular
Value function	Increasing	Continuous	Concave	Locally smooth
Solution correspondence	Increasing	Compact-valued, nonempty, uhc	Convex-valued	Locally smooth

models are usually formulated without specifying functional forms. Even when explicit functional forms are employed, derivation of an explicit solution to (1) or (2) may be computationally impractical. The skill of the economic analyst lies in deriving comparative static conclusions without explicit solution of the model. Economists employ several tools in this quest. This chapter explores the variety of tools which can be used for comparative static analysis in optimization and equilibrium models.

All comparative static analysis of optimization models is based on one or more of the maximum theorems, which are summarized in table 6.1. Crucially the continuous maximum theorem gives sufficient conditions for existence of an optimal solution, and ensures that the solution varies continuously with the parameters. The convex maximum theorem gives sufficient conditions for the optimal solution to be unique and value function concave. Analogously the monotone maximum theorem gives sufficient conditions for the optimal solution and value function to be increasing in the parameters. In section 6.1 we present yet another maximum theorem that shows that the optimal solution of a smooth model is a smooth function of the parameters. This leads to the *envelope theorem*, one of the most useful tools in the analysis of optimization models.

The envelope theorem is employed in section 6.2 to undertake comparative static analysis of the key models of microeconomic theory. This section also discusses two alternative approaches to the comparative statics of optimization models. Section 6.3 is devoted to the comparative statics of equilibrium models, where the basic tool is the implicit function theorem.

6.1 The Envelope Theorem

Comparative static analysis of optimization models is based upon the maximum theorems, which are summarized in table 6.1. Each theorem shows the extent to which the optimal solution and value function inherit the properties of the objective function and constraint correspondence. To apply standard differential analysis to the optimal solution, we need to ensure it is differentiable. Sufficient conditions for differentiability are provided by the smooth maximum theorem.

Theorem 6.1 (Smooth maximum theorem) *Suppose that* $\mathbf{x}_0$ *is a local maximum of*

$$\max_{\mathbf{x} \in G(\theta)} f(\mathbf{x}, \theta) \tag{3}$$

when $\theta = \theta_0$ *and*

- $G(\theta) = \{\mathbf{x} \in X : g_j(\mathbf{x}, \theta) \leq 0, j = 1, 2 \ldots, m\}$
- *f and* g_j *are* C^{n+1} *on* $X \times \Theta$, $n \geq 1, j = 1, 2, \ldots, m$
- *the binding constraints* $B(\mathbf{x}_0)$ *are regular*
- $\mathbf{x}_0$ *satisfies the sufficient conditions for a strict local maximum (corollary 5.2.3)*

Then there exists a neighborhood Ω *of* θ_0 *on which*

- *there exists a* C^n *function* $\mathbf{x}^* \colon \Omega \to X$ *such that* $\mathbf{x}^*(\theta)$ *solves (3) for every* $\theta \in \Omega$
- *the value function* $v(\theta) = \sup_{\mathbf{x} \in G(\theta)} f(\mathbf{x}, \theta)$ *is* C^n *on* Ω

Proof Assume first that all constraints are binding. Then (theorem 5.2) there exists λ such that $\mathbf{x}_0$ satisfies the first-order conditions

$$D_{\mathbf{x}} L[\mathbf{x}_0, \theta_0, \lambda] = \mathbf{0}$$

$$\mathbf{g}(\mathbf{x}_0, \theta) = \mathbf{0}$$

where L is the Lagrangean $L(\mathbf{x}, \theta, \lambda) = f(\mathbf{x}, \theta) - \sum \lambda_j g_j(\mathbf{x}, \theta)$. The Jacobian of this system is

$$J = \begin{pmatrix} H_L & J_{\mathbf{g}}^T \\ J_{\mathbf{g}} & 0 \end{pmatrix} \tag{4}$$

where H_L is the Hessian of the Lagrangean and $J_\mathbf{g}$ is the Jacobian of $\mathbf{g}$. Since $\mathbf{x}_0$ satisfies the conditions for a strict local maximum, J is nonsingular (exercise 6.1). By the implicit function theorem (theorem 4.5), there exist C^n functions $\mathbf{x}^*$ and λ^* such that $\mathbf{x}_0 = \mathbf{x}^*(\theta_0)$, $\lambda_0 = \lambda^*(\theta_0)$, and

$$D_\mathbf{x} L[\mathbf{x}^*(\theta), \theta, \lambda^*(\theta)] = \mathbf{0}$$

$$\mathbf{g}(\mathbf{x}^*(\theta), \theta) = \mathbf{0}$$

for every $\theta \in \Omega$. We can ensure that Ω is sufficiently small that $B(\mathbf{x}^*(\theta))$ is regular for all $\theta \in \Omega$ and also that $L(\mathbf{x}^*(\theta), \theta, \lambda^*(\theta))$ satisfies the second-order condition for a strict local maximum on Ω. Consequently $\mathbf{x}^*(\theta)$ solves (3) for every $\theta \in \Omega$ and the value function satisfies

$$v(\theta) = f(\mathbf{x}^*(\theta), \theta) \qquad \text{for every } \theta \in \Omega$$

By the chain rule (exercise 4.22), v is also C^n. If not all constraints are binding at $\mathbf{x}_0$, there exists a neighborhood in which they remain nonbinding, and we can choose Ω so that they remain nonbinding in Ω. □

Exercise 6.1
Show that the Jacobian J in (4) is nonsingular.

Having determined that v is differentiable, let us now compute its derivative. To simplify the notation, we will suppress the arguments of the derivatives. Let $f_\mathbf{x}$ denote $D_\mathbf{x} f[\mathbf{x}_0, \theta_0]$, the (partial) derivative of f with respect to $\mathbf{x}$ evaluated at $(\mathbf{x}_0, \theta_0)$. Similarly, let

$$f_\theta = D_\theta f[\mathbf{x}_0, \theta_0], \quad \mathbf{g}_\mathbf{x} = D_\mathbf{x} \mathbf{g}[\mathbf{x}_0, \theta_0], \quad \mathbf{g}_\theta = D_\theta \mathbf{g}[\mathbf{x}_0, \theta_0], \quad \mathbf{x}_\theta^* = D_\theta \mathbf{x}^*[\theta]$$

By theorem 6.1, there exists a neighborhood Ω around θ_0 and function $\mathbf{x}^*$ such that $v(\theta) = f(\mathbf{x}^*(\theta), \theta)$ for every $\theta \in \Omega$. By the chain rule,

$$D_\theta v[\theta] = f_\mathbf{x} \mathbf{x}_\theta^* + f_\theta \tag{5}$$

There are two channels by which a change in θ can affect $v(\theta) = f(\mathbf{x}^*(\theta), \theta)$. First, there is the direct effect f_θ since θ is an argument of f. Second, there is an indirect effect. A change in θ will also change $\mathbf{x}^*(\theta)$, which will in turn affect v. The first term $f_\mathbf{x} \mathbf{x}_\theta^*$ in (5) measures the indirect effect.

What do we know of the indirect effect? First, if $\mathbf{x}_0$ is optimal, it must satisfy the first-order conditions (theorem 5.3)

$$f_\mathbf{x} = \lambda_0^T \mathbf{g_x} \quad \text{and} \quad \lambda_0^T g(\mathbf{x}, \theta) = 0 \tag{6}$$

where λ_0 is the unique Lagrange multiplier associated with $\mathbf{x}_0$. Also $\mathbf{x}^*(\theta)$ satisfies the constraint $\mathbf{g}(\mathbf{x}^*(\theta), \theta) = \mathbf{0}$ for all $\theta \in \Omega$. Another application of the chain rule gives

$$\mathbf{g_x} \mathbf{x}_\theta^* + \mathbf{g}_\theta = \mathbf{0} \Rightarrow \lambda_0^T \mathbf{g_x} \mathbf{x}_\theta^* = -\lambda^T \mathbf{g}_\theta \tag{7}$$

Using (6) and (7), the indirect effect is $f_\mathbf{x} \mathbf{x}_\theta^* = \lambda_0^T g_\mathbf{x} \mathbf{x}_\theta^* = -\lambda^T \mathbf{g}_\theta$, and therefore

$$D_\theta v[\theta] = f_\theta - \lambda_0^T \mathbf{g}_\theta \tag{8}$$

Letting L denote the Lagrangean $L(\mathbf{x}, \theta, \lambda) = f(\mathbf{x}, \theta) - \lambda^T \mathbf{g}(\mathbf{x}, \theta)$, equation (8) implies that

$$D_\theta v[\theta] = D_\theta L[\mathbf{x}_0, \theta, \lambda_0]$$

This is the *envelope theorem*, which states that the derivative of the value function is equal to the partial derivative of the Lagrangean evaluated at the optimal solution $(\mathbf{x}_0, \lambda_0)$. In the special case

$$\max_{\mathbf{x} \in G} f(\mathbf{x}, \theta)$$

in which the feasible set G is independent of the parameters, $\mathbf{g}_\theta = \mathbf{0}$ and (8) becomes

$$D_\theta v[\theta] = f_\theta$$

The indirect effect is zero, and the only impact on v of a change in θ is the direct effect $\mathbf{f}_\theta$. We summarize this result in the following corollary.

Corollary 6.1.1 (Smooth envelope theorem) *Suppose that $\mathbf{x}_0$ is a local maximum of*

$$\max_{\mathbf{x} \in G(\theta)} f(\mathbf{x}, \theta) \tag{9}$$

when $\theta = \theta_0$ and

- $G(\theta) = \{\mathbf{x} \in X : g_j(\mathbf{x}, \theta) \leq 0, j = 1, 2 \ldots, m\}$
- *f and g_j are C^2 at $(\mathbf{x}_0, \theta_0), j = 1, 2, \ldots, m$*
- *the binding constraints $B(\mathbf{x}_0)$ are regular*

- $\mathbf{x}^0$ *satisfies the sufficient conditions for a strict local maximum*

Then the value function $v(\boldsymbol{\theta}) = \sup_{\mathbf{x} \in G(\boldsymbol{\theta})} f(\mathbf{x}, \boldsymbol{\theta})$ *is differentiable at* $\boldsymbol{\theta}_0$ *with*

$$D_\theta v[\boldsymbol{\theta}_0] = D_\theta L[\mathbf{x}_0, \boldsymbol{\theta}_0, \boldsymbol{\lambda}_0]$$

where L is the Lagrangean $L = f - \sum_{j=1}^m \lambda_j g_j$ *and* $\boldsymbol{\lambda}_0$ *is the (unique) Lagrange multiplier associated with* $\mathbf{x}_0$.

Example 6.1 The Lagrangean for the consumer's problem

$$\max_{\mathbf{x} \in X} u(\mathbf{x}) \quad \text{subject to} \quad \mathbf{p}^T \mathbf{x} = m$$

is $L(\mathbf{x}, \mathbf{p}, m, \lambda) = u(\mathbf{x}) - \lambda(\mathbf{p}^T\mathbf{x} - m)$. The envelope theorem states that

$$D_m v[\mathbf{p}, m] = D_m L[\mathbf{x}, \mathbf{p}, m, \lambda] = \lambda$$

which we have previously identified as the marginal utility of income. The envelope theorem fomalizes the observation made in example 5.24. At the optimum, it is immaterial how a small increment in income is divided among different commodities. The change in utility is the same no matter how the increment is spent.

Exercise 6.2
Prove proposition 5.2.

The envelope theorem is a key tool for comparative statics of optimization models. The derivation in corollary 6.2, based on the smooth maximum theorem, expresses the insight that the optimal solution is chosen so as to make the Lagrangean stationary with respect to feasible perturbations in $\mathbf{x}$. However, the assumptions required for corollary 6.1.1 are stringent. Where the feasible set is independent of the parameters, a more general result can be given.

Theorem 6.2 (Envelope theorem) Let $\mathbf{x}^*$ be the solution correspondence of the constrained optimization problem

$$\max_{\mathbf{x} \in G} f(\mathbf{x}, \boldsymbol{\theta})$$

in which $f: G \times \Theta \to \Re$ is continuous and G compact. Suppose that f is continuously differentiable in θ, that is, $D_\theta f[\mathbf{x}, \boldsymbol{\theta}]$ is continuous in $G \times \Theta$. Then the value function

6.1 The Envelope Theorem

$$v(\theta) = \sup_{x \in G} f(\mathbf{x}, \boldsymbol{\theta})$$

is differentiable wherever $\mathbf{x}^*$ *is single valued with* $D_\theta v[\theta] = D_\theta f[\mathbf{x}(\theta), \boldsymbol{\theta}]$.

Proof To simplify the proof, we assume that $\mathbf{x}^*$ is single valued for every $\theta \in \Theta$ (see Milgrom and Segal 2000 for the general case). Then $\mathbf{x}^*$ is continuous on Θ (theorem 2.3) and

$$v(\theta) = f(\mathbf{x}^*(\theta), \boldsymbol{\theta}) \qquad \text{for every } \theta \in \Theta$$

For any $\theta \neq \theta_0 \in \Theta$,

$$v(\theta) - v(\theta_0) = f(\mathbf{x}^*(\theta), \boldsymbol{\theta}) - f(\mathbf{x}^*(\theta_0), \boldsymbol{\theta}_0)$$
$$\geq f(\mathbf{x}^*(\theta_0), \boldsymbol{\theta}) - f(\mathbf{x}^*(\theta_0), \boldsymbol{\theta}_0)$$
$$= D_\theta f[\mathbf{x}^*(\theta_0), \boldsymbol{\theta}_0](\theta - \theta_0) + \eta(\theta)\|\theta - \theta_0\|$$

with $\eta(\theta) \to \mathbf{0}$ as $\theta \to \theta_0$. On the other hand, by the mean value theorem (theorem 4.1), there exist $\bar{\theta} \in (\theta, \theta_0)$ such that

$$v(\theta) - v(\theta_0) = f(\mathbf{x}^*(\theta), \boldsymbol{\theta}) - f(\mathbf{x}^*(\theta_0), \boldsymbol{\theta}_0)$$
$$\leq f(\mathbf{x}^*(\theta), \boldsymbol{\theta}) - f(\mathbf{x}^*(\theta), \boldsymbol{\theta}_0)$$
$$= D_\theta f[\mathbf{x}^*(\theta), \bar{\theta}](\theta - \theta_0)$$

Combining these inequalities, we have

$$D_\theta f[\mathbf{x}^*(\theta_0), \boldsymbol{\theta}_0](\theta - \theta_0) + \eta(\theta)\|\theta - \theta_0\| \leq v(\theta) - v(\theta_0)$$
$$\leq D_\theta f[\mathbf{x}^*(\theta), \bar{\theta}](\theta - \theta_0)$$

or

$$\frac{D_\theta f[\mathbf{x}^*(\theta_0), \boldsymbol{\theta}_0](\theta - \theta_0)}{\|\theta - \theta_0\|} + \eta(\theta) \leq \frac{v(\theta) - v(\theta_0)}{\|\theta - \theta_0\|} \leq \frac{D_\theta f[\mathbf{x}^*(\theta), \bar{\theta}](\theta - \theta_0)}{\|\theta - \theta_0\|}$$

Letting $\theta \to \theta_0$ yields

$$\lim_{\theta \to \theta_0} \frac{D_\theta f[\mathbf{x}^*(\theta_0), \boldsymbol{\theta}_0](\theta - \theta_0)}{\|\theta - \theta_0\|} + \lim_{\theta \to \theta_0} \eta(\theta)$$
$$\leq \lim_{\theta \to \theta_0} \frac{v(\theta) - v(\theta_0)}{\|\theta - \theta_0\|} \leq \lim_{\theta \to \theta_0} \frac{D_\theta f[\mathbf{x}^*(\theta), \bar{\theta}](\theta - \theta_0)}{\|\theta - \theta_0\|}$$

Since $\mathbf{x}^*$ and $D_\theta f$ are continuous,

$$\lim_{\theta \to \theta_0} \frac{D_\theta f[\mathbf{x}^*(\theta_0), \theta_0](\theta - \theta_0)}{\|\theta - \theta_0\|} \leq \lim_{\theta \to \theta_0} \frac{v(\theta) - v(\theta_0)}{\|\theta - \theta_0\|}$$

$$\leq \lim_{\theta \to \theta_0} \frac{D_\theta f[\mathbf{x}^*(\theta_0), \theta_0](\theta - \theta_0)}{\|\theta - \theta_0\|}$$

Therefore v is differentiable (exercise 4.3) and

$$Dv[\theta] = D_\theta f[\mathbf{x}^*(\theta), \theta]$$

where $D_\theta f[\mathbf{x}^*(\theta), \theta]$ denotes the partial derivative of f with respect to θ holding $\mathbf{x}$ constant at $\mathbf{x} = \mathbf{x}^*(\theta)$. □

Note that there is no requirement in theorem 6.2 that f is differentiable with respect to the decision variables $\mathbf{x}$, only with respect to the parameters. The practical importance of dispensing with differentiability with respect to $\mathbf{x}$ is that theorem 6.2 applies even when the feasible set is discrete. This is illustrated in figure 6.1, where the choice set $X = \{\mathbf{x}_1, \mathbf{x}_2, \mathbf{x}_3\}$ and there is a single parameter. Each curve represents the value of the objective function as a function of the parameter. Clearly, the optimal choice for each value of θ is that corresponding to the highest curve at that parameter value. Moreover the value function is the upper envelope of the objective functions $f(\mathbf{x}_i, \theta)$. We observe that the value function is differentiable wherever the the optimal choice is unique, and its derivative is equal to the derivative of the objective function at that point. This insight

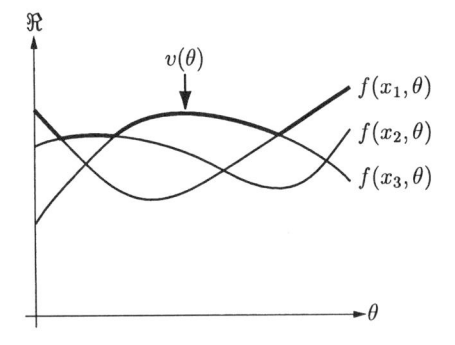

Figure 6.1
The value function is the upper envelope of the Lagrangean

extends to more general decision sets X, with the value function remaining the upper envelope of the objective function.

Example 6.2 (Optimal location) Suppose that demand for a product is distributed according to

$$f(x, \theta) = 1 - (x - \theta)^2$$

where $x \in [0, 1]$ is the location of the vendor, which can be interpreted as either physical location or location of the product in some notional "product space." θ is a parameter measuring the location of demand. For simplicity, assume that price is fixed at one and marginal costs are zero. Then the seller's optimization problem is

$$\max_{0 \le x \le 1} f(x, \theta) = 1 - (x - \theta)^2$$

The optimal solution is $x^*(\theta) = \theta$ and the value function is $v(\theta) = f(x^*(\theta), \theta) = 1$.

Suppose, however, that the seller is constrained to locate at one of the endpoints, that is, $x \in \{0, 1\}$. In this case the optimal solution is clearly to locate as close as possible to the demand

$$x^*(\theta) = \begin{cases} 0 & \text{if } \theta \le 0.5 \\ 1 & \text{if } \theta \ge 0.5 \end{cases}$$

and the value function is

$$v(\theta) = \max_{x \in \{0, 1\}} f(x, \theta)$$

$$= \max\{f(0, \theta), f(1, \theta)\} = \begin{cases} 1 - \theta^2 & \text{if } \theta \le 0.5 \\ 1 - (1 - \theta)^2 & \text{if } \theta \ge 0.5 \end{cases}$$

which is illustrated in figure 6.2. We observe that v is differentiable everywhere except at $\theta = 0.5$ (where x^* is multi-valued), with $D_\theta v[\theta] = D_\theta f[x^*(\theta), \theta]$.

6.2 Optimization Models

We now present three complementary approaches to the comparative static analysis of optimization models. The revealed preference approach (section 6.2.1) obtains limited conclusions with minimal assumptions. The

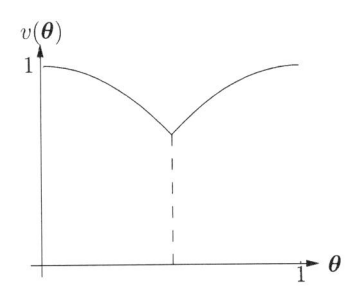

Figure 6.2
The value function is differentiable except at $\theta = 0.5$

value function approach (section 6.2.2) exploits the envelope theorem to deduce properties of the optimal solution from the known properties of the value function. The monotonicity approach (section 6.2.3) applies the monotone maximum theorem to derive strong qualitative conclusions in those models in which it is applicable.

6.2.1 Revealed Preference Approach

The revealed preference approach provides comparative static conclusions in the standard models of the consumer and producer, assuming nothing more than that the agent seeks to maximize utility or profit.

Remark 6.1 (Notation) To avoid a confusion of superscripts in this section, we will use $\mathbf{p} \cdot \mathbf{y}$ instead of our customary $\mathbf{p}^T \mathbf{y}$ to denote the inner product of $\mathbf{p}$ and $\mathbf{y}$. That is, if $\mathbf{p}, \mathbf{x}, \mathbf{y} \in \Re^n$,

$$\mathbf{p} \cdot \mathbf{y} = \sum_{i=1}^{n} p_i y_i \quad \text{and} \quad \mathbf{p} \cdot \mathbf{x} = \sum_{i=1}^{n} p_i x_i$$

Example 6.3 (Competitive firm) A competitive firm's optimization problem is to choose a feasible production plan $\mathbf{y} \in Y$ to maximize total profit

$$\max_{\mathbf{y} \in Y} \mathbf{p} \cdot \mathbf{y}$$

Consequently, if $\mathbf{y}^1$ maximizes profit when prices are $\mathbf{p}^1$, then

$$\mathbf{p}^1 \cdot \mathbf{y}^1 \geq \mathbf{p} \cdot \mathbf{y} \quad \text{for every } \mathbf{y} \in Y$$

Similarly, if $\mathbf{y}^2$ maximizes profit when prices are $\mathbf{p}^2$, then

$\mathbf{p}^2 \cdot \mathbf{y}^2 \geq \mathbf{p} \cdot \mathbf{y}$ for every $\mathbf{y} \in Y$

In particular,

$\mathbf{p}^1 \cdot \mathbf{y}^1 \geq \mathbf{p}^1 \cdot \mathbf{y}^2$ and $\mathbf{p}^2 \cdot \mathbf{y}^2 \geq \mathbf{p}^2 \cdot \mathbf{y}^1$

Adding these inequalities yields

$\mathbf{p}^1 \cdot \mathbf{y}^1 + \mathbf{p}^2 \cdot \mathbf{y}^2 \geq \mathbf{p}^1 \cdot \mathbf{y}^2 + \mathbf{p}^2 \cdot \mathbf{y}^1$

Rearranging, we have

$\mathbf{p}^2 \cdot (\mathbf{y}^2 - \mathbf{y}^1) \geq \mathbf{p}^1 \cdot (\mathbf{y}^2 - \mathbf{y}^1)$

Therefore

$$(\mathbf{p}^2 - \mathbf{p}^1) \cdot (\mathbf{y}^2 - \mathbf{y}^1) \geq 0 \quad \text{or} \quad \sum_{i=1}^{n}(p_i^1 - p_i^2)(y_i^1 - y_i^2) \geq 0 \tag{10}$$

If prices change from $\mathbf{p}^1$ to $\mathbf{p}^2$, the optimal production plan must change in such a way as to satisfy the inequality (10).

To understand the import of this inequality, consider a change in a single price p_i. That is, suppose that $p_j^2 = p_j^1$ for every $j \neq i$. Substituting in (10), this implies that

$$(p_i^2 - p_i^1)(y_i^2 - y_i^1) \geq 0 \quad \text{or} \quad p_i^2 > p_i^1 \Rightarrow y_i^2 \geq y_i^1 \tag{11}$$

Recall that y_i measures the net output of commodity i (example 1.7). Consequently, if i is an output, the quantity produced increases as its price increases (and fall as its price falls). If i is an input ($y_i^2 \leq 0$), (11) implies that the quantity of the input used falls as its price increases. In other words, it implies the elementary economic law that the supply curve of a competitive firm slopes upward, and its input demand curves slope downwards.

It is important to appreciate that (10) does not imply that other quantities $y_{j \neq i}$ do not change when p_i alone changes. In principle, all net outputs will change as the firm adjusts to a change in a single price. However, whatever the changes in the other commodities, the net output of good i must obey (11).

The term "revealed preference" stems from the observation that, in choosing $\mathbf{y}^1$ when the prices are $\mathbf{p}^1$, the decision maker reveals that she (weakly) prefers $\mathbf{y}^1$ to $\mathbf{y}^2$ given $\mathbf{p}^1$. Similarly her choice reveals that she prefers y^2 to y^1 given p^2. These revelations impose restriction (10) on y^1 and y^2. The beauty of this approach to the comparative statics is that the conclusions follow from nothing more than the assumption of maximizing behavior.

Example 6.4 (Expenditure minimization) An alternative way to approach the consumer's problem (example 5.1) is to consider the problem of minimizing the expenditure required to achieve a given level of utility

$$\min_{\mathbf{x} \in X} \mathbf{p} \cdot \mathbf{x} \quad \text{subject to} \quad u(\mathbf{x}) \geq u \tag{12}$$

The optimal solution of this problem specifies the optimal levels of consumption so as to achieve utility level u at minimum cost. If $\mathbf{x}^1$ and $\mathbf{x}^2$ represent optimal solutions when prices are $\mathbf{p}^1$ and $\mathbf{p}^2$ respectively, we conclude that

$$\mathbf{p}^1 \cdot \mathbf{x}^1 \leq \mathbf{p}^1 \cdot \mathbf{x}^2 \quad \text{and} \quad \mathbf{p}^2 \cdot \mathbf{x}^2 \leq \mathbf{p}^2 \cdot \mathbf{x}^1$$

Adding these inequalities, we have

$$\mathbf{p}^1 \cdot \mathbf{x}^1 + \mathbf{p}^2 \cdot \mathbf{x}^2 \leq \mathbf{p}^1 \cdot \mathbf{x}^2 + \mathbf{p}^2 \cdot \mathbf{x}^1$$

Rearranging yields

$$\mathbf{p}^2 \cdot (\mathbf{x}^2 - \mathbf{x}^1) \leq \mathbf{p}^1 \cdot (\mathbf{x}^2 - \mathbf{x}^1) \quad \text{or} \quad (\mathbf{p}^2 - \mathbf{p}^1) \cdot (\mathbf{x}^2 - \mathbf{x}^1) \leq 0 \tag{13}$$

For a single price change ($p_j^2 = p_j^1$ for every $j \neq i$), (13) implies that

$$(p_i^2 - p_i^1)(x_i^2 - x_i^1) \leq 0 \quad \text{or} \quad p_i^2 > p_i^1 \Rightarrow x_i^2 \leq x_i^1$$

Again, it is important to realise that (13) does not imply that other quantities do not change when p_i alone changes. It does imply that no matter what the other changes, the quantity of good i consumed falls when its price rises.

Note that the solution correspondence to the expenditure minimization problem (12) depends on prices $\mathbf{p}$ and utility u, and not on income. When it is single valued, the optimal solution function is known as the Hicksian or *compensated demand function*, and is often denoted $\mathbf{h}(\mathbf{p}, u)$ to distinguish it from the ordinary demand function $\mathbf{x}^*(\mathbf{p}, m)$. It specifies the consumer's behavior on the assumption that all prices changes are accompanied by compensating changes in income so as to leave the consumer's attainable utility unchanged.

6.2 Optimization Models

Reflecting on these two examples, we observe that the analysis depended on two characteristics of these models:

- The objective function is bilinear.
- The feasible set is independent of the parameters.

We formalize this observation in the following exercise.

Exercise 6.3
Suppose that f is bilinear and that

$$\mathbf{x}^1 \text{ solves } \max_{\mathbf{x} \in G} f(\mathbf{x}, \boldsymbol{\theta}^1) \text{ and } \mathbf{x}^2 \text{ solves } \max_{\mathbf{x} \in G} f(\mathbf{x}, \boldsymbol{\theta}^2)$$

Then

$$f(\mathbf{x}^1 - \mathbf{x}^2, \boldsymbol{\theta}^1 - \boldsymbol{\theta}^2) \geq 0$$

Examples 6.3 and 6.4 and exercise 6.3 do not exhaust the potential of the revealed preference approach. Other results can be derived with some ingenuity.

Example 6.5 (Monopoly) Suppose that the cost function of a monopolist changes from $c_1(y)$ to $c_2(y)$ in such a way that marginal cost is higher at every level of output

$$0 < c_1'(y) < c_2'(y) \quad \text{for every } y > 0$$

In other words, the marginal cost curve shifts upward. Then (exercise 6.4)

$$c_2(y_1) - c_2(y_2) \geq c_1(y_1) - c_1(y_2) \tag{14}$$

where y_1 and y_2 are the profit maximizing output levels when costs are c_1 and c_2 respectively. The *fundamental theorem of calculus* (Spivak 1980, p. 472) states: If $f \in C^1[a,b]$, then

$$f(b) - f(a) = \int_a^b f'(x)\,dx$$

Applying this theorem to both sides of (14), we deduce that

$$\int_{y_2}^{y_1} c_2'(y)\,dy \geq \int_{y_2}^{y_1} c_1'(y)\,dy$$

or

$$\int_{y_2}^{y_1} c_2'(y)\,dy - \int_{y_2}^{y_1} c_1'(y)\,dy = \int_{y_2}^{y_1} (c_2'(y) - c_1'(y))\,dy \geq 0$$

Since $c_2'(y) - c_1'(y) \geq 0$ for every y (by assumption), this implies that $y_2 \leq y_1$. Assuming the demand curve is downward sloping, this implies that $p_2 \geq p_1$. We have deduced that a monopolist's optimal price is increasing in marginal cost.

Exercise 6.4
Suppose that the cost function of a monopolist changes from $c_1(y)$ to $c_2(y)$ with

$$0 < c_1'(y) < c_2'(y) \qquad \text{for every } y > 0$$

Show that

$$c_2(y_1^*) - c_2(y_2^*) \geq c_1(y_1^*) - c_1(y_2^*) \tag{15}$$

where y_1^* and y_2^* are the profit maximizing output levels when costs are c_1 and c_2 respectively.

6.2.2 Value Function Approach

The value function approach to comparative statics exploits the special form of the value function in many economic models. It provides an especially elegant analysis of the core microeconomic models of the consumer and the producer.

Example 6.6 (Hotelling's lemma) The competitive producer's problem (example 5.1) is to choose a feasible production plan $\mathbf{y} \in Y$ to maximize total profit $\mathbf{p}^T\mathbf{y}$. Letting $f(\mathbf{y}, \mathbf{p}) = \mathbf{p}^T\mathbf{y}$ denote the objective function, the competitive firm solves

$$\max_{\mathbf{y} \in Y} f(\mathbf{y}, \mathbf{p})$$

Note that f is differentiable with $D_{\mathbf{p}}f[\mathbf{y}, \mathbf{p}] = \mathbf{y}$. Applying theorem 6.2, the profit function

$$\Pi(\mathbf{p}) = \sup_{\mathbf{y} \in Y} f(\mathbf{y}, \mathbf{p})$$

is differentiable wherever the supply correspondence $\mathbf{y}^*$ is single valued with

$$D_{\mathbf{p}}\Pi[\mathbf{p}] = D_{\mathbf{p}}f[\mathbf{y}^*(\mathbf{p}), \mathbf{p}] = \mathbf{y}^*(\mathbf{p}) \tag{16}$$

or

$$\mathbf{y}^*(\mathbf{p}) = \nabla\Pi(\mathbf{p})$$

In other words, whenever the optimal production plan $\mathbf{y}^*(\mathbf{p})$ is unique, it is equal to the gradient of the profit function at $\mathbf{p}$, an important result known as *Hotelling's lemma*.

Suppose that the supply correspondence is single valued for all $\mathbf{p}$; that is, $\mathbf{y}^*(\mathbf{p})$ is a function. Then the smooth maximum theorem (theorem 6.1) applies and $\mathbf{y}^*$ and v are smooth. The practical significance of Hotelling's lemma is that if we know the profit function, we can calculate the supply function by straightforward differentiation instead of solving a constrained optimization problem. But its theoretical significance is far more important. Hotelling's lemma enables us to deduce the properties of the supply function $\mathbf{y}^*$ from the already established properties of the profit function. In particular, we know that the profit function is convex (example 3.42).

From Hotelling's lemma (16), we deduce that the derivative of the supply function is equal to the second derivative of the profit function

$$D\mathbf{y}^*[\mathbf{p}] = D^2\Pi[\mathbf{p}]$$

or equivalently that the Jacobian of the supply function is equal to the Hessian of the profit function.

$$J_{\mathbf{y}^*}(\mathbf{p}) = H_{\Pi}(\mathbf{p})$$

Since Π is smooth and convex, its Hessian $H(\mathbf{p})$ is symmetric (theorem 4.2) and nonnegative definite (proposition 4.1) for all $\mathbf{p}$. Consequently the Jacobian of the supply function $J_{\mathbf{y}^*}$ is also symmetric and nonnegative definite. This implies that for all goods i and j,

$D_{p_i} y_i^*[\mathbf{p}] \geq 0$ \qquad Nonnegativity

$D_{p_i} y_j^*[\mathbf{p}] = D_{p_j} y_i^*[\mathbf{p}]$ \quad Symmetry

Nonnegativity is precisely the property we derived in example 6.3. Recalling that net outputs y_i are positive for outputs and negative for inputs, nonnegativity implies that the quantity of output supplied increases and the quantity of input demanded falls as its own price increases. *Symmetry* is new. It states the surprising proposition that the change in net output of good *i* following a change in the price of good *j* is precisely equal to the change in net output of good *j* following a change in the price of good *i*.

Exercise 6.5 (Single-output firm)
The preceding example is more familiar where the firm produces a single output and we distinguish inputs and outputs. Assume that a competitive firm produces a single output y from n inputs $\mathbf{x} = (x_1, x_2, \ldots, x_n)$ according to the production function $y = f(\mathbf{x})$ so as to maximize profit

$$\Pi(\mathbf{w}, p) = \max_{\mathbf{x}} pf(\mathbf{x}) - \mathbf{w}^T\mathbf{x}$$

Assume that there is a unique optimum for every p and $\mathbf{w}$. Show that the input demand $x_i^*(\mathbf{w}, p)$ and supply $y^*(\mathbf{w}, p)$ functions have the following properties:

$D_p y_i^*[\mathbf{w}, p] \geq 0$ Upward sloping supply

$D_{w_i} x_i^*[\mathbf{w}, p] \leq 0$ Downward sloping demand

$D_{w_j} x_i^*[\mathbf{w}, p] = D_{w_i} x_j^*[\mathbf{w}, p]$ Symmetry

$D_p x_i^*[\mathbf{w}, p] = -D_{w_i} y^*[\mathbf{w}, p]$ Reciprocity

Example 6.7 (Shephard's lemma) A cost-minimizing firm solves (example 5.16)

$$\min_{\mathbf{x} \geq 0} \mathbf{w}^T \mathbf{x} \quad \text{subject to} \quad f(\mathbf{x}) \geq y$$

which is equivalent to

$$\max_{\mathbf{x} \geq 0} -\mathbf{w}^T \mathbf{x} \quad \text{subject to} \quad -f(\mathbf{x}) \leq -y$$

Note that the feasible set is independent of the parameters $\mathbf{w}$, and theorem 6.2 applies. The cost function

$$c(\mathbf{w}, y) = \inf_{\mathbf{x} \geq 0} \mathbf{w}^T \mathbf{x} = -\sup_{\mathbf{x} \geq 0} -\mathbf{w}^T\mathbf{x}$$

is differentiable wherever $\mathbf{x}^*$ is single valued with

$$D_{\mathbf{w}} c[\mathbf{w}, y] = \mathbf{x}^*(\mathbf{w}, y) \tag{17}$$

Equation (17) is known as *Shephard's lemma*.

When it is single valued, $\mathbf{x}^*(\mathbf{w}, y)$ is called the *conditional demand function*, since it specifies the optimal demand conditional on producing a given level of output y. Practically, Shephard's lemma means that conditional demand function can be derived simply by differentiating the cost function. Theoretically, it means that the properties of the conditional demand function can be deduced from the known properties of the cost function. Provided that the production function is smooth, Theorem 6.1 guarantees the $\mathbf{x}^*$ and the cost function c are also smooth. From (17) we calculate that

$$D_\mathbf{w}\mathbf{x}^*[\mathbf{w}, y] = D^2_{\mathbf{ww}}c[\mathbf{w}, y] \tag{18}$$

For fixed y, the left-hand side of (18) can be represented by the Jacobian $J_{\mathbf{x}^*}(\mathbf{w})$ of $\mathbf{x}^*$ viewed as a function of $\mathbf{w}$ alone. Similarly the right-hand side of (18) can be represented by the Hessian $H_c(\mathbf{w})$ of c regarded as a function of $\mathbf{w}$ alone. (18) asserts that these two matrices are equal, that is,

$$J_{\mathbf{x}^*}(\mathbf{w}) = H_c(\mathbf{w})$$

We previously determined (example 3.126) that the cost function is concave in $\mathbf{w}$ so that $H_c(\mathbf{w})$ is nonpositive definite. It is also symmetric (theorem 4.2). Consequently, for all goods i and j,

$$D_{w_i}x_i^*[\mathbf{w}, y] \leq 0 \qquad \text{Negativity}$$
$$D_{w_i}x_j^*[\mathbf{w}, y] = D_{w_j}x_i^*[\mathbf{w}] \qquad \text{Symmetry}$$

Exercise 6.6 (Inferior inputs)
An input i is called *normal* its demand increases with output, that is, $D_y\mathbf{x}^*(\mathbf{w}, y) \geq 0$. Otherwise $(D_y\mathbf{x}^*(\mathbf{w}, y) < 0)$, i is called an *inferior input*. Show that an input i is normal if and only if marginal cost $D_yc[\mathbf{w}, y]$ is increasing in w_i.

Example 6.8 (Roys's identity) The consumer's problem is

$$\max_{\mathbf{x} \in X(\mathbf{p}, m)} u(\mathbf{x}) \tag{19}$$

where $X(\mathbf{p}, m) = \{\mathbf{x} \in X : \mathbf{p}^T\mathbf{x} - m \leq 0\}$ is the budget set. Theorem 6.2 cannot be applied, since the feasible set $X(\mathbf{p}, m)$ depends on the parameters $(\mathbf{p}$ and $m)$. Let $g(\mathbf{x}, \mathbf{p}, m) = \mathbf{p}^T\mathbf{x} - m$ so that $X(\mathbf{p}, m) = \{\mathbf{x} \in X : g(\mathbf{p}, m) \leq 0\}$. The Lagrangean is $u(\mathbf{x}) - \lambda(\mathbf{p}^T\mathbf{x} - m)$. Assuming that u is smooth and the demand correspondence $\mathbf{x}^*(\mathbf{p}, m)$ is single valued, we can apply corollary 6.1.1. The indirect utility function

$$v(\mathbf{p}, m) = \sup_{\mathbf{x} \in X(\mathbf{p}, m)} u(\mathbf{x})$$

is differentiable with

$$D_\mathbf{p} v[\mathbf{p}, m] = -\lambda D_\mathbf{p} g[\mathbf{x}^*, \mathbf{p}, m] = -\lambda \mathbf{x}^*$$

and

$$D_m v[\mathbf{p}, m] = -\lambda D_m g[\mathbf{x}^*, \mathbf{p}, m] = \lambda$$

where λ is the marginal utility of income (example 5.24). It follows that the demand function is

$$\mathbf{x}^*(\mathbf{p}, m) = -\frac{D_\mathbf{p} v[\mathbf{p}, m]}{\lambda} = -\frac{D_\mathbf{p} v[\mathbf{p}, m]}{D_m v[\mathbf{p}, m]} \tag{20}$$

Equation (20) is known as *Roy's identity*.

Example 6.9 (Slutsky equation) We have now encountered two different demand functions for the consumer. The solution $\mathbf{x}(\mathbf{p}, m)$ of (19) is known as the *ordinary demand function* to distinguish it from the *compensated demand function* $h(\mathbf{p}, u)$ defined in example 6.4. They differ in their independent variables but are related by the identity

$$\mathbf{x}(\mathbf{p}, m) \equiv \mathbf{h}(\mathbf{p}, v(\mathbf{p}, m)) \tag{21}$$

In other words, the ordinary and compensated demand functions intersect where utility $u = v(\mathbf{p}, m)$. Using Roy's identity, we can also show that their derivatives are also related. Differentiating (21) using the chain rule gives

$$D_m \mathbf{x}[\mathbf{p}, m] = D_u \mathbf{h}([\mathbf{p}, v(\mathbf{p}, m)] \circ D_m v[\mathbf{p}, m]$$
$$D_\mathbf{p} \mathbf{x}[\mathbf{p}, m] = D_\mathbf{p} \mathbf{h}[\mathbf{p}, v(\mathbf{p}, m)] + D_u \mathbf{h}[\mathbf{p}, v(\mathbf{p}, m)] \circ D_\mathbf{p} v[\mathbf{p}, m] \tag{22}$$

Substituting for $D_\mathbf{p} v[\mathbf{p}, m]$ using Roy's identity (20), we have

$$D_\mathbf{p} \mathbf{x}[\mathbf{p}, m] = D_\mathbf{p} \mathbf{h}[\mathbf{p}, m] + D_u \mathbf{h}([\mathbf{p}, m] \circ (-D_m v[\mathbf{p}, m] \mathbf{x}(\mathbf{p}, m))$$

and using (22),

$$D_\mathbf{p} \mathbf{x}[\mathbf{p}, m] = D_\mathbf{p} \mathbf{h}[\mathbf{p}, m] - D_m \mathbf{x}[\mathbf{p}, m] \mathbf{x}(\mathbf{p}, m)$$

which is known as the *Slutsky equation*. To understand the implication of this equation, consider the demand for good *i*. The Slutsky equation

$$D_{p_i} x_i[\mathbf{p}, m] = D_{p_i} h_i[\mathbf{p}, m] - x_i(\mathbf{p}, m) D_m x_i[\mathbf{p}, m] \tag{23}$$

decomposes the change in demand for good *i* (as its prices changes) into two terms—the substitution effect and the income effect. The *substitution effect* $D_{p_i} h_i[\mathbf{p}, m]$ measures the effect on demand of change in price, holding utility constant. We have already shown that the substitution effect $D_{p_i} h_i[\mathbf{p}, m]$ is always negative (example 6.4). Consequently the slope of the demand curve depends on the income effect.

An increase in p_i effectively reduces the consumer's real income by $p_i x_i$. The *income effect* $-x_i(\mathbf{p}, m) D_m x_i[\mathbf{p}, m]$ measures the effect on demand of the change in real income. $D_m x_i[\mathbf{p}, m]$ measures the responsiveness of demand to changes in income. If $D_m x_i[\mathbf{p}, m] \geq 0$, demand increases with income, and good *i* is called a *normal good*. In that case the income effect is also negative, and (23) implies that $D_{p_i} x_i[\mathbf{p}, m] \leq 0$. The Slutsky equations reveals that "normally" demand curves slope downward.

On the other hand, if $D_m x_i[\mathbf{p}, m] \leq 0$, good *i* is called an *inferior good*. For an inferior good, the income effect is positive. It is theoretically possible for a positive income effect to outweigh the negative substitution effect for an inferior good, in which case $D_{p_i} x_i[\mathbf{p}, m] \geq 0$ and the demand curve slopes upward. Such a good is called a *Giffen good*.

Exercise 6.7 (Aggregation conditions)
As the previous example demonstrated, utility maximization places no restrictions on the slopes of individual demand functions. However, the consumer's demand must always satisfy the budget constraint, which places certain restrictions on the system of demand functions as a whole. Show that for all goods *i*,

$\sum_{i=1}^{n} \alpha_i \eta_i = 1$ Engel aggregation condition

$\sum_{i=1}^{n} \alpha_i \varepsilon_{ij} = -\alpha_j$ Cournot aggregation condition

where $\alpha_i = p_i x_i / m$ is the budget share of good *i* and

$$\varepsilon_{ij}(\mathbf{p}, m) = \frac{p_j}{x_i(\mathbf{p}, m)} D_{p_j} x_i[\mathbf{p}, m] \quad \text{and} \quad \eta_i(\mathbf{p}, m) = \frac{m}{x_i(\mathbf{p}, m)} D_m x_i[\mathbf{p}, m]$$

are the price and income elasticities of demand respectively.

6.2.3 The Monotonicity Approach

The monotone maximum theorem (theorem 2.1) is an obvious tool for comparative static analysis in optimization models, since it generates precisely the type of conclusions we seek. Where applicable, it generates strong results with a minimum of extraneous assumptions. Unfortunately, its requirements are also strong, and it cannot supplant other tools in the economist's arsenal. It complements these other tools, generating additional conclusions in appropriate models and acting where other tools are unavailable. Often the model has to be transformed before the monotone maximum theorem can be applied. We give some examples. More examples can be found in the references cited in the notes.

The monotone maximum theorem is not directly applicable to many of the standard models of microeconomics, such as examples 5.1, 5.2, and 5.16, since their feasible sets are not lattices. However, it can be applied to the single-output firm, where the technology is represented by a production function. We generate stronger conclusions in the special case in which all inputs are complements in production.

Example 6.10 (Competitive firm) A firm produces a single product y using the technology

$$y = f(\mathbf{x})$$

The firm purchases inputs at fixed input prices $\mathbf{w}$ and sells its output at price p. Net revenue is

$$\mathrm{NR}(\mathbf{x}, p, \mathbf{w}) = pf(\mathbf{x}) - \sum_{i=1}^{n} w_i x_i$$

and the firm's objective is to maximize net revenue by appropriate choice of inputs, that is, to solve the optimization problem

$$\max_{\mathbf{x} \in \Re_+^n} \mathrm{NR}(\mathbf{x}, p, \mathbf{w})$$

The objective function is decreasing in factor prices $\mathbf{w}$. To conveniently apply the monotone maximum theorem, we recast the problem using the negative of the factor prices so that the objective function becomes

$$\mathrm{NR}(\mathbf{x}, p, -\mathbf{w}) = pf(\mathbf{x}) + \sum_{i=1}^{n} (-w_i) x_i$$

If all inputs are complements in production, the production function f is supermodular (example 2.70), and the net revenue function $\mathrm{NR}(\mathbf{x}, p, -\mathbf{w})$ is (exercise 6.8)

- supermodular in $\mathbf{x}$
- displays strictly increasing differences in $(\mathbf{x}, -\mathbf{w})$

Applying the monotone maximum theorem (corollary 2.1.2), the firm's input demand correspondences

$$x_i^*(\mathbf{w}) = \arg \max_{\mathbf{x} \in \Re_+^n} \left(pf(\mathbf{x}) + \sum_{i=1}^n (-w_i) x_i \right)$$

are always increasing in $-\mathbf{w}$ (always decreasing in $\mathbf{w}$). That is, provided the inputs are complements in production, an increase in input price will never lead to an increase in demand for any input.

This conclusion should be distinguished from our earlier conclusion that input demand functions are declining in their own price (example 6.3 and exercise 6.5). Exercise 6.5 assesses the impact of a change in the price of one factor holding all other prices constant. It does not predict the effect of changes in one factor price on the demand for other factors. It is valid irrespective of the technology. In contrast, the monotone maximum theorem predicts the impact of changes in one price on the demand for all factors, and in fact accounts for simultaneous increase in many factor prices, but requires a specific class of technology (complementary inputs).

Complementarity also implies that profit (Π) and output y^* are also decreasing functions of $\mathbf{w}$. However, complementarity is not necessary for these conclusions. Π is always decreasing in $\mathbf{w}$, while y^* is decreasing in $\mathbf{w}$ provided that there are no inferior inputs (exercise 6.6).

Exercise 6.8
In preceding example show that the net revenue function

$$\mathrm{NR}(\mathbf{x}, p, -\mathbf{w}) = pf(\mathbf{x}) + \sum_{i=1}^n (-w_i) x_i$$

is supermodular in $\mathbf{x}$ and displays strictly increasing differences in $(\mathbf{x}, -\mathbf{w})$ provided that the production function f is supermodular.

In the next example, we replicate the analysis of example 6.5 using the monotonicity approach.

Example 6.11 Suppose that a monopolist's cost function $c(y, \theta)$ displays increasing differences in a parameter θ; that is, $c(y^2, \theta) - c(y^1, \theta)$ for $y^2 > y^1$ is increasing in θ. This implies increasing marginal costs. Suppose also that demand $f(p)$ is decreasing in p. Then net revenue

$$\text{NR}(\mathbf{p}, \theta) = pf(p) - c(f(p), \theta)$$

- is supermodular in p (exercise 2.49)
- displays increasing differences in (p, θ) (exercise 6.9)

By the monotone maximum theorem (theorem 2.1), the profit-maximizing price is increasing in θ and therefore output $f(p)$ is decreasing in p.

Compared to example 6.5, the preceding analysis requires an additional assumption that demand is decreasing in price. Example 6.5 made no assumption regarding demand. On the other hand, example 6.5 required that marginal cost be continuous, which implicitly assumes that output is continuously variable. In contrast, this analysis makes no such assumption. Output and price can be discrete.

Exercise 6.9
Show that the objective function in the preceding example displays increasing differences in (p, θ).

Exercise 6.10
It is characteristic of microchip production technology that a proportion of output is defective. Consider a small producer for whom the price of good chips p is fixed. Suppose that proportion $1 - \theta$ of the firm's chips are defective and cannot be sold. Let $c(y)$ denote the firm's total cost function where y is the number of chips (including defectives) produced. Suppose that with experience, the yield of good chips θ increases. How does this affect the firm's production y? Does the firm compensate for the increased yield by reducing production, or does it take advantage of the higher yield by increasing production?

6.3 Equilibrium Models

The implicit function theorem (theorem 4.5) provides the mathematical foundation for most comparative static analysis of equilibrium models.

6.3 Equilibrium Models

For any solution $\mathbf{x}_0$ to a system of equations

$$\mathbf{f}(\mathbf{x}, \boldsymbol{\theta}_0) = \mathbf{0}$$

the implicit function theorem asserts the existence of a function $\mathbf{g}: \Omega \to X$ on a neighborhood Ω of $\boldsymbol{\theta}_0$ such that

$$\mathbf{x}_0 = g(\boldsymbol{\theta}_0) \quad \text{and} \quad f(g(\boldsymbol{\theta}), \boldsymbol{\theta}) = \mathbf{0} \quad \text{for every } \boldsymbol{\theta} \in \Omega$$

provided that $D_\mathbf{x} f[\mathbf{x}_0, \boldsymbol{\theta}_0]$ is nonsingular. Further $\mathbf{g}$ is differentiable with derivative

$$D_\theta \mathbf{g}[\boldsymbol{\theta}_0] = -(D_\mathbf{x} \mathbf{f}[\mathbf{x}_0, \boldsymbol{\theta}_0])^{-1} \circ D_\theta \mathbf{f}[\mathbf{x}_0, \boldsymbol{\theta}_0]$$

Although it is usually intractable to compute the explicit form of $\mathbf{g}$, its derivative provides a linear approximation that can be used to infer the behavior of $\mathbf{g}$ and hence of the equilibrium in the neighborhood of $\boldsymbol{\theta}_0$. We illustrate with some familiar examples.

Example 6.12 (A market model) The comparative statics of a simple market model of a single commodity is familiar from elementary economics. We will show how this can be formalized. Assume that demand q for some commodity is inversely related to the price p according the demand function

$$q = d(p)$$

where $D_p d < 0$. Supply is positively related to price, but is also dependent on some exogenous parameter θ (the "weather")

$$q = s(p, \theta)$$

with $D_p s > 0$ and $D_\theta s > 0$. We assume that both d and s are C^1.

Naturally equilibrium in this market pertains where demand equals supply. We will write the equilibrium relationship as

$$e(p, \theta) = d(p) - s(p, \theta) = 0$$

where $e(p, \theta)$ is called the excess demand function. Note that $D_p e = D_p d - D_p s$ and $D_\theta e = -D_\theta s$ (exercise 4.21). How does the equilibrium price depend upon θ (the "weather")?

Let (p_0, q_0, θ_0) denote an initial equilibrium. From the implicit function theorem, we know that there exists a function g relating the equilibrium market price p to the weather θ,

$p = g(\theta)$

in a neighborhood Ω of θ_0. Clearly, without specifying d and s, it is impossible to compute g explicitly. However, we can compute its derivative, which is given by

$$D_p g(\theta_0) = -(D_p e(p_0, \theta_0))^{-1} \circ D_\theta e(p_0, \theta_0) = -\frac{-D_\theta s}{D_p d - D_p s}$$

Armed with estimates regarding the magnitude of $D_p d$, $D_p s$, and $D_\theta s$, we could then assess the sensitivity of the market price to changes in weather. Even without any additional information, we can at least determine that the market price is negatively related to the weather. By assumption, $D_p d < 0$, $D_p s > 0$ and $D_\theta s > 0$, so $D_\theta g(\theta_0) < 0$. This confirms our economic intuition. Favourable weather increases supply, decreasing the market price. We can also assess the relationship between the weather and the equilibrium quantity. Substituting in the demand function (or the supply function) $q = d(g(\theta))$ and applying the chain rule

$$\frac{dq}{d\theta} = D_p d \circ D_\theta g > 0$$

The quantity traded is positively related to the weather.

Example 6.13 (The IS-LM model)
The IS-LM model (example 4.35) is specified by two equations

$$f_1(r, y; G, T, M) = y - C(y, T) - I(r) - G = 0$$
$$f_2(r, y; G, T, M) = PL(y, r) - M = 0$$

We assume that

$$0 < C_y < 1, \quad C_T < 0, \quad I_r < 0, \quad L_y > 0, \quad L_r < 0 \tag{24}$$

where C_y denotes the partial derivative of C with respect to y. The Jacobian of f is

$$J_f = \begin{pmatrix} D_r f_1 & D_y f_1 \\ D_r f_2 & D_y f_2 \end{pmatrix} = \begin{pmatrix} -I_r & 1 - C_y \\ L_r & L_y \end{pmatrix}$$

which is nonsingular, since its determinant

$$\Delta = -I_r L_y - (1 - C_y) L_r > 0$$

6.3 Equilibrium Models

is positive given the assumptions (24) regarding the slopes of the constituent functions.

Let $\mathbf{x} = (r, y) \in X$ and $\theta = (G, M, T) \in \Theta$. In a neighborhood $\Omega \subseteq \Theta$ around any equilibrium, there exists a function $\mathbf{g}: \Omega \to X$ specifying r and y as functions of G, M and T. Although the functional form of $\mathbf{g}$ is impossible to deduce without further specifying the model, we can derive some properties of $\mathbf{g}$. Specifically,

$$D_\theta \mathbf{g}[\theta_0] = -(D_\mathbf{x} f[\mathbf{x}_0, \theta_0])^{-1} \circ D_\theta f[\mathbf{x}_0, \theta_0]$$

or

$$J_g = J_f^{-1} K_f$$

where

$$K_f = \begin{pmatrix} D_G f_1 & D_M f_1 & D_T f_1 \\ D_G f_2 & D_M f_2 & D_T f_2 \end{pmatrix} = \begin{pmatrix} -1 & 0 & -C_T \\ 0 & -\dfrac{1}{P} & 0 \end{pmatrix}$$

is the matrix representing $D_\theta f$. The inverse of J_f is (exercise 3.104)

$$J_f^{-1} = \frac{1}{\Delta} \begin{pmatrix} L_y & -(1 - C_y) \\ -L_r & -I_r \end{pmatrix}$$

and therefore

$$J_g = J_f^{-1} K_f = -\frac{1}{\Delta} \begin{pmatrix} L_y & -(1 - C_y) \\ -L_r & -I_r \end{pmatrix} \begin{pmatrix} -1 & 0 & -C_T \\ 0 & -\dfrac{1}{P} & 0 \end{pmatrix}$$

$$= \frac{1}{\Delta} \begin{pmatrix} L_y & -\dfrac{1 - C_y}{P} & L_y \times C_T \\ -L_r & -\dfrac{I_r}{P} & -L_r \times C_T \end{pmatrix}$$

Using the alternative notation, we have shown that

$$\begin{pmatrix} \dfrac{\partial r}{\partial G} & \dfrac{\partial r}{\partial M} & \dfrac{\partial r}{\partial T} \\ \dfrac{\partial y}{\partial G} & \dfrac{\partial y}{\partial M} & \dfrac{\partial y}{\partial T} \end{pmatrix} = \frac{1}{\Delta} \begin{pmatrix} L_y & -\dfrac{1 - C_y}{P} & L_y \times C_T \\ -L_r & -\dfrac{I_r}{P} & -L_r \times C_T \end{pmatrix}$$

This summarizes the comparative statics of the IS-LM model. For example, we conclude that government spending is expansionary, since

$$\frac{\partial y}{\partial G} = -\frac{L_r}{\Delta} > 0$$

is positive given that $L_r < 0$ and $\Delta > 0$. The magnitude $\partial y/\partial G$ is known as the government-spending multiplier. The impact of a marginal increase in government spending matched by an equal change in taxes

$$\frac{\partial y}{\partial G} + \frac{\partial y}{\partial T} = -\frac{L_r(1 + C_T)}{\Delta} > 0$$

which is also positive. This magnitude is known as the balanced budget multiplier.

In practice, we often simplify the derivation of comparative statics by linearly approximating the original model, and analysing the behavior of the linear model, presuming that the behavior of the linear model will match that of the original nonlinear model. The justification for this presumption remains the implicit function theorem; the analysis is indeed equivalent and the conclusions identical (review the remarks proceeding theorem 4.5). The modified procedure is often easier to conduct and to follow. We illustrate this procedure in the IS-LM model. The reader should carefully compare the following example with example 6.13.

Example 6.14 (IS-LM model again) Recall that the original IS-LM model is described by two (nonlinear) equations

$$y = C(y, T) + I(r) + G$$

$$L(y, r) = \frac{M}{P}$$

For small changes in the income and the interest rate, the change in the demand for money can be approximated linearly by

$$dL = L_r\, dr + L_y\, dy$$

Treating all the functions in the model similarly, we can derive the following linearization of IS-LM model (exercise 4.63):

6.3 Equilibrium Models

$$dy = C_y\, dy + C_T\, dT + I_r\, dr + dG$$

$$L_r\, dr + L_y\, dy = \frac{1}{P}\, dM$$

Collecting the endogenous variables (dr, dy) on the left-hand side and the exogenous variable on the right-hand side, we have

$$-I_r\, dr + dy - C_y\, dy = C_T\, dT + dG$$

$$L_r\, dr + L_y\, dy = \frac{1}{P}\, dM$$

or in matrix form

$$\begin{pmatrix} -I_r & 1 - C_y \\ L_r & L_y \end{pmatrix} \begin{pmatrix} dr \\ dy \end{pmatrix} = \begin{pmatrix} dG + C_T\, dT \\ \frac{1}{P}\, dM \end{pmatrix}$$

Letting $dM = dT = 0$, the system reduces to

$$\begin{pmatrix} -I_r & 1 - C_y \\ L_r & L_y \end{pmatrix} \begin{pmatrix} dr \\ dy \end{pmatrix} = \begin{pmatrix} -dG \\ 0 \end{pmatrix}$$

which can be solved by Cramer's rule (exercise 3.103) to give

$$dy = \frac{\begin{vmatrix} -I_r & dG \\ L_r & 0 \end{vmatrix}}{\begin{vmatrix} -I_r & 1 - C_y \\ L_r & L_y \end{vmatrix}} = -\frac{L_r}{\Delta}\, dG$$

where

$$\Delta = -I_r L_y - (1 - C_y) L_r > 0$$

is the determinant of the Jacobian of f as in the previous example. We deduce that

$$\frac{\partial y}{\partial G} = -\frac{L_r}{\Delta} > 0$$

which is identical to the conclusion we derived more formally in the previous example.

Similarly, letting $dG = dT$ and $dM = 0$, we have

$$dy = \frac{\begin{vmatrix} -I_r & dG + C_T\,dT \\ L_r & 0 \end{vmatrix}}{\begin{vmatrix} -I_r & 1 - C_y \\ L_r & L_y \end{vmatrix}} = -\frac{L_r}{\Delta}(dG + C_T\,dT)$$

and the balanced budget multiplier is

$$\left.\frac{\partial y}{\partial G}\right|_{dG=DT} = -\frac{L_r(1 + C_T)}{\Delta} > 0$$

Since the optimal solution to an optimization model must satisfy a system of equations—the first-order conditions (theorems 5.1, 5.2, and 5.3)—every optimization model contains an equilibrium model. Consequently the tools of equilibrium analysis can also be applied to optimization models. In particular, the implicit function can be applied directly to the first-order conditions. We illustrate with an example.

Example 6.15 (The competitive firm) A competitive firm produces a single output y using n inputs $\mathbf{x} = (x_1, x_2, \ldots, x_n)$ according to the production function

$$y = f(x_1, x_2, \ldots, x_n)$$

Let p denote the output price and $\mathbf{w} = (w_1, w_2, \ldots, w_n)$ the input prices. The objective of the firm is to choose input levels and mix $\mathbf{x}$ to maximize profit $pf(\mathbf{x}) - \mathbf{w}^T\mathbf{x}$. Profit is maximized where (example 5.11)

$$Q(\mathbf{x}; \mathbf{w}, p) = pD_\mathbf{x}f - \mathbf{w} = 0 \tag{25}$$

The second-order condition for a strict local maximum is that

$$D_\mathbf{x}Q[\mathbf{x}, \mathbf{w}, p] = pD_{\mathbf{xx}}^2 f < 0$$

that is, f must be strictly locally concave. The first-order conditions (25) implicitly determine the optimal level of inputs for any combination of output price p and input prices $\mathbf{w}$. Assuming that f is C^2 and the optimal solution is unique, there exists a function $\mathbf{x}^*(\mathbf{p}, m)$ (*input demand function*) determining optimal demand as a function of $\mathbf{w}$ and p. Furthermore the input demand function $\mathbf{x}^*$ is C^1 with derivatives

$$D_\mathbf{w}\mathbf{x}^*(\mathbf{w}, p) = -(D_\mathbf{x}Q)^{-1} \circ D_\mathbf{w}Q = (pD_{\mathbf{ww}}^2 f)^{-1} \circ I_X \tag{26}$$

and

$$D_p \mathbf{x}^*(\mathbf{w}, p) = -(D_\mathbf{x} Q)^{-1} \circ D_p Q = -(pD_{\mathbf{xx}}^2 f)^{-1} \circ D_\mathbf{x} f$$

Regarding $\mathbf{x}^*$ as a function of $\mathbf{w}$ alone, equation (26) can be written

$$J_{\mathbf{x}^*}(\mathbf{w}) = \frac{1}{p} H_f^{-1}(\mathbf{x}^*)$$

The second-order condition requires that f be locally strictly concave so that H_f is negative definite (proposition 4.1). This implies identical conclusions to exercise 6.5.

Exercise 6.11
Suppose that there are only two inputs. They are complementary if $D^2 f_{x_1 x_2} > 0$. Show that $D_{w_1} x_2 < 0$ if the factors are complementary and $D_{w_1} x_2 \geq 0$ otherwise. Note that this is special case of example 6.10.

With explicit constraints, it can be more straightforward to linearize the first-order conditions before deducing the comparative statics, as we did in example 6.14. Again, we emphasize that the theoretical foundation for this analysis remains the implicit function theorem. The difference is merely a matter of computation.

Example 6.16 (Cost minimization) A firm produces a single output y from inputs $\mathbf{x}$ using technology $f(\mathbf{x})$. Assuming that the firm produces efficiently (that is, at minimum cost), we want to deduce the effect of changes in factor prices $\mathbf{w}$ on employment of factors $\mathbf{x}$. The first-order conditions for cost minimization are (example 5.16)

$$f(\mathbf{x}) = y$$

$$\mathbf{w} - \lambda D_\mathbf{x} f[\mathbf{x}] = 0$$

Suppose that $\mathbf{w}$ changes to $\mathbf{w} + \mathbf{dw}$. This will lead to consequent changes

$$\mathbf{x} \to \mathbf{x} + \mathbf{dx} \quad \text{and} \quad \lambda \to \lambda + d\lambda$$

which must again satisfy the first-order conditions

$$f(\mathbf{x} + \mathbf{dx}) = y$$

$$\mathbf{w} + \mathbf{dw} - (\lambda + d\lambda) D_\mathbf{x} f[\mathbf{x} + \mathbf{dx}] = 0$$

For small changes we can approximate $f(\mathbf{x} + \mathbf{dx})$ and $D_\mathbf{x} f(\mathbf{x} + \mathbf{dx})$ using the first and second derivatives of f:

$$f(\mathbf{x} + \mathbf{dx}) \approx f(\mathbf{x}) + D_\mathbf{x} f[\mathbf{x}](\mathbf{dx})$$

$$D_\mathbf{x} f[\mathbf{x} + \mathbf{dx}] \approx D_\mathbf{x} f[\mathbf{x}] + D_{\mathbf{xx}}^2 f[\mathbf{x}](\mathbf{dx})$$

Substituting and canceling common terms gives

$$D_\mathbf{x} f[\mathbf{x}](\mathbf{dx}) = 0$$

$$\mathbf{dw} - D_\mathbf{x} f[\mathbf{x}] \, d\lambda - \lambda D_{\mathbf{xx}}^2 f[\mathbf{x}](\mathbf{dx}) - D_{\mathbf{xx}}^2 f[\mathbf{x}](d\lambda \, \mathbf{dx}) = \mathbf{0}$$

Representing the first and second derivatives by the gradient and Hessian respectively, and recognizing that $d\lambda \, \mathbf{dx}$ is negligible for small changes, this system can be rewritten as

$$-\nabla f(\mathbf{x}) \, \mathbf{dx} = 0$$

$$-\nabla f(\mathbf{x})^T d\lambda - \lambda H_f(\mathbf{x}) \, \mathbf{dx} = -\mathbf{dw}$$

or in matrix form

$$\begin{pmatrix} 0 & -\nabla f(\mathbf{x}) \\ -\nabla f(\mathbf{x})^T & -\lambda H_f(\mathbf{x}) \end{pmatrix} \begin{pmatrix} d\lambda \\ \mathbf{dx} \end{pmatrix} = \begin{pmatrix} 0 \\ -\mathbf{dw} \end{pmatrix} \tag{27}$$

The matrix on the left is known as the *bordered Hessian* of f. Provided that f is locally strictly concave, $H_f(\mathbf{x})$ is negative definite (proposition 4.1) and symmetric (4.2). Consequently the bordered Hessian is nonsingular (exercise 6.1) and symmetric, and system (27) can be solved to give

$$\begin{pmatrix} d\lambda \\ \mathbf{dx} \end{pmatrix} = \begin{pmatrix} 0 & -\nabla f(\mathbf{x}) \\ -\nabla f(\mathbf{x})^T & -\lambda H_f(\mathbf{x}) \end{pmatrix}^{-1} \begin{pmatrix} 0 \\ -\mathbf{dw} \end{pmatrix}$$

It can be shown (Takayama 1985, p. 163) that the properties of H_f imply that

$$D_{w_i} x_i^*(w) \leq 0 \quad \text{and} \quad D_{w_i} x_j^*(w) = D_{w_j} x_i^*(w) \quad \text{for every } i \text{ and } j$$

which are identical to the conclusions obtained by the value function approach in example 6.7.

6.3 Equilibrium Models

With two inputs $\mathbf{x} = (x_1, x_2)$ the system (27) is

$$\begin{pmatrix} 0 & -f_1 & -f_2 \\ -f_1 & -\lambda f_{11} & -\lambda f_{12} \\ -f_2 & -\lambda f_{21} & -\lambda f_{22} \end{pmatrix} \begin{pmatrix} d\lambda \\ dx_1 \\ dx_2 \end{pmatrix} = \begin{pmatrix} 0 \\ -dw_1 \\ -dw_2 \end{pmatrix}$$

which can be solved by Cramer's rule (exercise 3.103) to give

$$dx_1 = \frac{\begin{vmatrix} 0 & 0 & -f_2 \\ -f_1 & 0 & -\lambda f_{12} \\ -f_2 & -dw_2 & -\lambda f_{22} \end{vmatrix}}{|\bar{H}|} = -\frac{f_1 f_2}{|\bar{H}|} dw_2$$

So

$$D_{w_2} x_1 = \frac{\partial x_1}{\partial w_2} = -\frac{f_1 f_2}{|\bar{H}|} > 0$$

since $\bar{H}$, the determinant of the bordered Hessian, is negative by the second-order condition.

One advantage of the implicit function approach to comparative statics is its flexibility. In the following example we use it to provide an alternative analysis of a question raised in exercise 6.12.

Example 6.17 It is characteristic of microchip production technology that a proportion of output is defective. Consider a small producer for whom the price of good chips p is fixed. Suppose that proportion $1 - \theta$ of the firm's chips are defective and cannot be sold. Let $c(y)$ denote the firm's total cost function where y is the number of chips (including defectives) produced. Suppose that with experience, the yield of good chips θ increases. How does this affect the firm's production y? Does the firm compensate for the increased yield by reducing production, or does it celebrate by increasing production?

The firm's optimization problem is

$$\max_y \; \theta p y - c(y)$$

The first-order and second-order conditions for profit maximization are

$$Q(y, \theta, p) = \theta p - c'(y) = 0 \quad \text{and} \quad D_y Q[y, \theta, p] = -c''(y) < 0$$

The second-order condition requires increasing marginal cost. Assuming that c is C^2, the first-order condition implicitly defines a function $y(\theta)$. Differentiating the first-order condition with respect to θ, we deduce that

$$p = c''(y)D_\theta y \quad \text{or} \quad D_\theta y = \frac{p}{c''(y)}$$

which is positive by the second-order condition. An increase in yield θ is analogous to an increase in product price p, inducing an increase in output y.

6.4 Conclusion

In this chapter we have introduced a variety of tools for comparative static analysis. No single tool is equal to every occasion, and each has a role to play. For standard analysis of the core microeconomic models of the consumer and the producer, it is hard to beat the elegance of the revealed preference and value function approaches. The implicit function theorem approach is the principal tool for the analysis of equilibrium models. It also provides flexibility to deal with modifications of the standard optimization models (example 6.17). It remains the most important tool in the research literature. The monotonicity approach, based on the monotone maximum theorem, is a relatively new development in analytical technique, and its potential is yet to be fully explored. Its key advantage is that it dispenses with irrelevant assumptions, focusing attention on the conditions necessary for particular relations to hold and generating more robust conclusions.

6.5 Notes

Silberberg (1990) gives a thorough account of the traditional methods of comparative statics in economic analysis. Mas-Colell et al. (1995) and Varian (1992) are good expositions of the value function approach in the theory of the consumer and the producer. Lambert (1985) is also insightful. Takayama (1985, pp. 161–166) discusses the implicit function approach to comparative statics in optimization models.

6.5 Notes

The monotonicity approach to comparative statics was introduced by Topkis (1978). Its application to economics is still in its infancy. Milgrom and Shannon (1994) is a good overview. Milgrom (1994) discusses applications to optimization models, Milgrom and Roberts (1994) applications to equilibrium models, and Milgrom and Roberts (1990) applications in game theory. Topkis (1995) applies the techniques to a variety of models of the firm. Athey et al. (2001) provide a comprehensive introduction to the monotonocity approach.

Theorem 6.2 is due to Milgrom and Segal (2000). A similar result is in Sah and Zhao (1998), from whom example 6.2 was adapted. Example 6.5 is from Tirole (1988).

References

Abadie, J. 1967. On the Kuhn-Tucker theorem. In *Nonlinear Programming*, edited by J. Abadie. Amsterdam: North-Holland, pp. 21–36.

Allen, R. G. D. 1938. *Mathematical Analysis for Economists*. London: Macmillan.

Apostol, T. 1974. *Mathematical Analysis*. Reading, MA: Addison-Wesley.

Arrow, K. J. 1963. *Social Choice and Individual Values*, 2nd ed. New Haven: Yale University Press.

Arrow, K. J., and A. C. Enthoven. 1961. Quasi-concave programming. *Econometrica* 29: 779–800.

Arrow, K. J., and F. H. Hahn. 1971. *General Competitive Analysis*. San Francisco: Holden Day.

Athey, S., P. Milgrom, and J. Roberts. 2001. *Robust Comparative Statics*. Princeton, NJ: Princeton University Press (forthcoming).

Averch, H., and L. L. Johnson. 1962. Behavior of a Firm under regulatory Constraint. *American Economic Review* 52: 1053–69.

Barten, A. P., and V. Böhm. 1982. Consumer theory. In *Handbook of Mathematical Economics*, vol. 2, edited by K. J. Arrow and M. D. Intrilligator. Amsterdam: North-Holland, pp. 381–429.

Bartle, R. G. 1976. *The Elements of Real Analysis*, 2nd ed. New York: Wiley.

Baumol, W. J., and A. K. Klevorick. 1970. Input choices and rate-of-return regulation: An overview of the discussion. *Bell Journal of Economics* 1: 162–90.

Bazaraa, M. S., J. J. Goode, and C. M. Shetty. 1972. Constraint qualifications revisited. *Management Science* 18: 567–73.

Bazaraa, M. S., H. D. Sherali, and C. M. Shetty. 1993. *Nonlinear Programming: Theory and Algorithms*, 2nd ed. New York: Wiley.

Beardon, A. F., and G. B. Mehta. 1994. The utility theorems of Wold, Debreu and Arrow-Hahn. *Econometrica* 62: 181–86.

Beavis, B., and I. Dobbs. 1990. *Optimization and Stability for Economic Analysis*. Cambridge: Cambridge University Press.

Berge, C. 1963. *Topological Spaces, Including a Treatment of Multi-valued Functions, Vector Spaces and Convexity*. New York: Macmillan.

Binmore, K. G. 1981. *The Foundations of Analysis: A Straightforward Introduction: Topological Ideas*, vol. 2. Cambridge: Cambridge University Press.

Birkhoff, G. 1973. *Lattice Theory*, 3rd ed. Providence, RI: American Mathematical Society.

Border, K. C. 1985. *Fixed Point Theorems with Applications in Economics and Game Theory*. Cambridge: Cambridge University Press.

Bosch, R. A., and J. A. Smith. 1998. Separating hyperplanes and the authorship of the disputed Federalist Papers. *American Mathematical Monthly* 105: 601–608.

Braunschweiger, C. C., and H. E. Clark. 1962. An extension of the Farkas theorem. *American Mathematical Monthly* 69: 271–77.

Carter, M., and P. Walker. 1996. The nucleolus strikes back. *Decision Sciences* 27: 123–36.

Chillingworth, D. R. J. 1976. *Differential Topology with a View to Applications*. London: Pitman.

Chvátal, V. 1983. *Linear Programming*. San Francisco: Freeman.

Debreu, G. 1954. Representation of a preference ordering by a numerical function. In *Decision Processes*, edited by R. Thrall, C. Coombs, and R. Davis. New York: Wiley. Reprinted on pp. 105–110 of *Mathematical Economics: Twenty Papers of Gerard Debreu*, Econometric Society Monographs no. 4. Cambridge: Cambridge University Press, 1983.

Debreu, G. 1959. *Theory of Value: An Axiomatic Analysis of Economic Equilibrium.* New Haven: Yale University Press.

Debreu, G. 1964. Continuity properties of Paretian utility. *International Economic Review* 5: 285–93. Reprinted on pp. 163–172 of *Mathematical Economics: Twenty Papers of Gerard Debreu*, Econometric Society Monographs no. 4. Cambridge: Cambridge University Press, 1983.

Debreu, G. 1982. Existence of competitive equilibrium. In *Handbook of Mathematical Economics*, vol. 2, edited by K. J. Arrow and M. D. Intrilligator. Amsterdam: North-Holland, pp. 697–743.

Debreu, G. 1991. The mathematization of economic theory. *American Economic Review* 81: 1–7.

Diamond, P. A. 1982. Aggregate demand management in search equilibrium. *Journal of Political Economy* 90: 881–94.

Dieudonne, J. 1960. *Foundations of Modern Analysis.* New York: Academic Press.

Diewert, W. E. 1982. Duality approaches to microeconomic theory. In *Handbook of Mathematical Economics*, vol. 2, edited by K. J. Arrow and M. D. Intrilligator. Amsterdam: North-Holland, pp. 535–99.

Diewert, W. E., M. Avriel, and I. Zang. 1981. Nine kinds of quasiconcavity and concavity. *Journal of Economic Theory* 25: 397–420.

Dixit, A. 1990. *Optimization in Economic Theory*, 2nd ed. London: Oxford University Press.

Duffie, D. 1992. *Dynamic Asset Pricing Theory.* Princeton: Princeton University Press.

Ellickson, B. 1993. *Competitive Equilibrium: Theory and Applications.* New York: Cambridge University Press.

Fan, K. 1956. On systems of linear inequalities. In *Linear Inequalities and Related Systems*, edited by H. W. Kuhn and A. W. Tucker. Princeton: Princeton University Press, pp. 99–156.

Franklin, J. 1980. *Methods of Mathematical Economics: Linear and Nonlinear Programming; Fixed Point Theorems.* New York: Springer-Verlag.

Fréchet, M. 1953. Emile Borel, initiator of the theory of psychological games and its application. *Econometrica* 21: 95–96.

Fryer, M. J., and J. V. Greenman. 1987. *Optimisation Theory: Applications in OR and Economics.* London: Edward Arnold.

Fudenberg, D., and J. Tirole. 1991. *Game Theory.* Cambridge: MIT Press.

Fuss, M., D. McFadden, and Y. Mundlak. 1978. A survey of functional forms in the economic analysis of production. In *Production Economics: A Dual Approach to Theory and Applications*, vol. 1, edited by M. Fuss and D. McFadden. Amsterdam: North-Holland, pp. 219–68.

Gale, D. 1960. *The Theory of Linear Economic Models.* New York: McGraw-Hill.

Gale, D., and H. Nikaido. 1965. The Jacobian matrix and the global univalence of mappings. *Mathematische Annalen* 159: 81–93. Reprinted on pp. 68–80 of *Readings in Mathematical Economics*, vol. 1, edited by P. Newman. Baltimore, MD: Johns Hopkins Press, 1968.

Gately, D. 1974. Sharing the gains from regional cooperation: A game theoretic application to planning investment in electric power. *International Economic Review* 15: 195–208.

Good, R. A. 1959. Systems of linear relations. *SIAM Review* 1: 1–31.

Green, J., and W. P. Heller. 1981. Mathematical analysis and convexity with applications to economics. In *Handbook of Mathematical Economics*, vol. 1, edited by K. J. Arrow and M. D. Intrilligator. Amsterdam: North-Holland, pp. 15–52.

Greenberg, H. J., and W. P. Pierskalla. 1971. A review of quasi-convex functions. *Operations Research* 19: 1553–70.

Hall, R. E. 1972. Turnover in the labor force. *Brookings Papers on Economic Activity* 3: 709ff.

Halmos, P. R. 1960. *Naive Set Theory*. Princeton: Van Nostrand.

Halmos, P. R. 1974. *Finite-Dimensional Vector Spaces*. New York: Springer-Verlag.

Hammond, P. J. 1976. Equity, Arrow's conditions, and Rawls' difference principle. *Econometrica* 44: 793–804.

Hildenbrand, W., and A. Kirman. 1976. *Introduction to Equilibrium Analysis: Variations on Themes by Edgeworth and Walras*, 1st ed. Amsterdam: North-Holland.

Holmes, R. B. 1975. *Geometric Functional Analysis and Its Applications*. New York: Springer-Verlag.

Janich, K. 1994. *Linear Algebra*. New York: Springer-Verlag.

Kakutani, S. 1941. A generalization of Brouwer's fixed point theorem. *Duke Mathematical Journal* 8: 457–59. Reprinted on pp. 33–35 of *Readings in Mathematical Economics*, vol. 1, edited by P. Newman. Baltimore, MD: Johns Hopkins Press, 1968.

Karlin, S. 1959. *Mathematical Methods and Theory in Games, Programming, and Economics*, vol. 1. Reading, MA: Addison-Wesley.

Kelley, J. L. 1975. *General Topology*. New York: Springer-Verlag. (Originally published by Van Norstrand in 1955.)

Klein, E. 1973. *Mathematical Methods in Theoretical Economics: Topological and Vector Space Foundations of Equilibrium Analysis*. New York: Academic Press.

Koopmans, T. C. 1957. *Three Essays on the State of Economic Science*. New York: McGraw-Hill.

Kreps, D. 1990. *A Course in Microeconomic Theory*. Princeton: Princeton University Press.

Lambert, P. J. 1985. *Advanced Mathematics for Economists: Static and Dynamic Optimization*. Oxford: Basil Blackwell.

Lang, S. 1969. *Real Analysis*. Reading, MA: Addison-Wesley.

Leonard, D., and N. V. Long. 1992. *Optimal Control Theory and Static Optimization in Economics*. Cambridge: Cambridge University Press.

Littlechild, S. C. 1970. Peak-load pricing of telephone calls. *Bell Journal of Economics* 1: 191–210.

Luenberger, D. 1969. *Optimization by Vector Space Methods*. New York: Wiley.

Luenberger, D. 1984. *Introduction to Linear and Nonlinear Programming*, 2nd ed. Reading, MA: Addison-Wesley.

Luenberger, D. 1995. *Microeconomic Theory*. New York: McGraw-Hill.

Luenberger, D. G. 1997. *Investment Science*. New York: Oxford University Press.

Madden, P. 1986. *Concavity and Optimization in Microeconomics*. Oxford: Basil Blackwell.

Magill, M., and M. Quinzii. 1996. *Theory of Incomplete Markets*. Cambridge: MIT Press.

Mangasarian, O. L. 1994. *Nonlinear Programming*. Philadelphia: SIAM. (Originally published by McGraw-Hill, 1969.)

Maor, E. 1994. *E: The Story of a Number*. Princeton: Princeton University Press.

Mas-Colell, A., M. D. Whinston, and J. R. Green. 1995. *Microeconomic Theory*. New York: Oxford University Press.

Milgrom, P. 1994. Comparing optima: Do simplifying assumptions affect conclusions? *Journal of Political Economy* 102: 607–15.

Milgrom, P., and J. Roberts. 1990. Rationalizability, learning, and equilibrium in games with strategic complements. *Econometrica* 58: 1255–77.

Milgrom, P., and J. Roberts. 1994. Comparing equilibria. *American Economic Review* 84: 441–59.

Milgrom, P., and I. Segal. 2000. *Envelope Theorems for Arbitrary Choice Sets*. Mimeo. Department of Economics, Stanford University.

Milgrom, P., and C. Shannon. 1994. Monotone comparative statics. *Econometrica* 62: 157–80.

Moulin, H. E. 1986. *Game Theory for the Social Sciences*, 2nd ed. New York: New York University Press.

Nikaido, H. 1968. *Convex Structures and Economic Theory*. New York: Academic Press.

Osborne, M., and A. Rubinstein. 1994. *A Course in Game Theory*. Cambridge: MIT Press.

Ostaszewski, A. 1990. *Advanced Mathematical Methods*. Cambridge: Cambridge University Press.

Panik, M. J. 1993. *Fundamentals of Convex Analysis: Duality, Separation, Representation and Resolution*. Dordrecht: Kluwer.

Pryce, J. D. 1973. *Basic Methods of Linear Functional Analysis*. London: Hutchinson University Library.

Rawls, J. 1971. *A Theory of Justice*. Cambridge: Harvard University Press.

Robert, A. 1989. *Advanced Calculus for Users*. Amsterdam: North-Holland.

Roberts, A. W., and D. E. Varberg. 1973. *Convex Functions*. New York: Academic Press.

Rockafellar, R. T. 1970. *Convex Analysis*. Princeton: Princeton University Press.

Rudin, W. 1976. *Principles of Mathematical Analysis*, 3rd ed. New York: McGraw-Hill.

Sah, R., and J. Zhao. 1998. Some envelope theorems for integer and discrete choice variables. *International Economic Review* 39: 623–34.

Sargent, T. 1987. *Dynamic Macroeconomic Theory*. Cambridge: Harvard University Press.

Schmeidler, D. 1969. The nucleolus of a characteristic function game. *SIAM Journal of Applied Mathematics* 17: 1163–70.

Sen, A. K. 1970a. *Collective Choice and Social Welfare*. San Francisco: Holden-Day.

Sen, A. K. 1970b. The impossibility of a Paretian liberal. *Journal of Political Economy* 78: 152–57.

Sen, A. 1995. Rationality and social choice. *American Economic Review* 85: 1–24.

Shapiro, C. 1989. Theories of oligopoly behavior. In *Handbook of Industrial Organization*, vol. 1, edited by R. Schmalensee and R. D. Willig. Amsterdam: North-Holland, pp. 329–414.

Shapley, L. 1971–72. Cores of convex games. *International Journal of Game Theory* 1: 11–26.

Shubik, M. 1982. *Game Theory in the Social Sciences*. Cambridge: MIT Press.

Silberberg, E. 1990. *The Structure of Economics: A Mathematical Analysis*, 2nd ed. New York: McGraw-Hill.

Simmons, G. F. 1963. *Topology and Modern Analysis*. New York: McGraw-Hill.

Simon, C. P. 1986. Scalar and vector maximization: Calculus techniques with economic applications. In *Studies in Mathematical Economics* edited by S. Reiter. Washington, DC: Mathematical Association of America, pp. 62–159.

Simon, C. P., and L. Blume. 1994. *Mathematics for Economists*. New York: Norton.

Smith, K. T. 1983. *Primer of Modern Analysis.* New York: Springer-Verlag.

Smith, L. 1998. *Linear Algebra,* 3rd ed. New York: Springer-Verlag.

Spivak, M. 1965. *Calculus on Manifolds.* Menlo Park, CA: Benjamin.

Spivak, M. 1980. *Calculus,* 2nd ed. Berkeley, CA: Publish or Perish, Inc.

Starr, R. 1969. Quasi-equilibria in markets with non-convex preferences. *Econometrica* 37: 25–38.

Starr, R. 1997. *General Equilibrium Theory: An Introduction.* Cambridge: Cambridge University Press.

Stoer, J., and C. H. R. Witzgall. 1970. *Convexity and Optimization in Finite Dimensions.* Berlin: Springer-Verlag.

Stokey, N. L., and R. E. Lucas. 1989. *Recursive Methods in Economic Dynamics.* Cambridge: Harvard University Press.

Sundaram, R. K. 1996. *A First Course in Optimization Theory.* New York: Cambridge University Press.

Sydsaeter, K., and P. J. Hammond. 1995. *Mathematics for Economic Analysis.* Englewood Cliffs, NJ: Prentice Hall.

Takayama, A. 1985. *Mathematical Economics,* 2nd ed. Cambridge: Cambridge University Press.

Takayama, A. 1994. *Analytical Methods in Economics.* Hemel Hempstead, Hertfordshire: Harvester Wheatsheaf.

Theil, H. 1971. *Principles of Econometrics.* New York: Wiley.

Tirole, J. 1988. *The Theory of Industrial Organization.* Cambridge: MIT Press.

Topkis, D. M. 1978. Minimizing a submodular function on a lattice. *Operations Research* 26: 305–21.

Topkis, D. 1995. Comparative statics of the firm. *Journal of Economic Theory* 67: 370–401.

Uzawa, H. 1962. Walras' existence theorem and Brouwer's fixed point theorem. *Economic Studies Quarterly* 8: 59–62.

Varian, H. 1987. The arbitrage principle in financial economics. *Journal of Economic Perspectives* 1: 55–72.

Varian, H. 1992. *Microeconomic Analysis,* 3rd ed. New York: Norton.

von Neumann, J. 1953. Communication of the Borel notes. *Econometrica* 21: 124–25.

Whittle, P. 1971. *Optimization under Constraints: Theory and Applications of Nonlinear Programming.* New York: Wiley-Interscience.

Whittle, P. 1992. *Probability via Expectation,* 3rd ed. New York: Springer-Verlag.

Williamson, O. E. 1966. Peak-load pricing and optimal capacity under indivisibility constraints. *American Economic Review* 56: 810–27.

Zangwill, W. I. 1969. *Nonlinear Programming: A Unified Approach.* Englewood-Cliffs, NJ: Prentice-Hall.

Zeidler, E. 1986. *Nonlinear Functional Analysis and Its Applications: Fixed Point Theorems,* vol. 1. New York: Springer-Verlag.

Zhou, L. 1993. A simple proof of the Shapley-Folkman theorem. *Economic Theory* 3: 371–72.

Zhou, L. 1994. The set of Nash equilibria of a supermodular game is a complete lattice. *Games and Economic Behavior* 7: 295–300.

Index of Symbols

Sets and Spaces

$\emptyset$	2
l_∞	117
$\mathfrak{N}$	2
$\mathfrak{R}$	2
$\mathfrak{R}^n$	6
$\mathfrak{R}^n_+$	6
$\mathfrak{R}^n_{++}$	6
$\mathfrak{R}^*$	29
X^*	280

Set Operations

$\in$	2
$\subseteq$	2
$\subset$	2
$\subsetneq$	2
$\supseteq$	2
$\mathscr{P}(S)$	2
$\|S\|$	2
$\cap$	4
$\cup$	4
S^c	4
$S \setminus T$	4
$X \times Y$	5
$\mathbf{x}$	6
$\mathbf{x}_{-i}$	7
$(x, \mathbf{x}_{-i})$	7

Relations

$\geq$	17, 22
$>$	17, 22
$\ngeq$	22
$\gg$	22
$\gtrsim$	13, 16, 26
$\succ$	16
$\nsucceq$	22
$\sim$	16
$\gtrsim_S$	30
$\gtrsim(a)$	18

$\sim(a)$	14
$[a, b]$	18
Pareto	33
$\vee$	26, 28
$\wedge$	26, 28
inf S	24
sup S	24

Metric Spaces

$\rho(x, y)$	45
$B_r(x)$	49
b(S)	5
$d(S)$	48
int S	50
lim	58
$\overline{S}$	50
(x^n)	58
$x^n \to x$	58

Linear Spaces

aff S	86
cone S	106
conv S	92
dim S	79
$\mathbf{e}_i$	77
ext(S)	96
lin S	72
ri S	128
Δ^m	99
$S^\perp$	292
S^*	382
S^{**}	382
$\|\mathbf{x}\|$	115
$[\mathbf{x}, \mathbf{y}]$	88

Functions and Correspondences

A^T	265
$B(X)$	155
$C(X)$	216

Index of Symbols

det A	297	$H^+(\mathbf{x}^*)$	571
epi f	154	$L(\mathbf{x}^*)$	573
$f(S)$	149	$L(\mathbf{x},\theta)$	533
$f^{-1}(S)$	149	$T(\mathbf{x}^*)$	572
$F(X)$	155		
$g \circ f$	150		
graph(f)	146		

Economic Models and Games

graph(φ)	182	$B_i(\mathbf{s})$	180
$H_f(c)$	284	core	39
hypo f	154	$\mathbf{e}_S$	282
kernel f	269	$E(x)$	453, 494
$\mathbf{x}^T\mathbf{y}$	290	Eff(Y)	53
$\mathbf{x}^T A \mathbf{x}$	303	$\mathscr{G}^N$	67
$X \to Y$	145	Nu	42
$X \rightrightarrows Y$	177	$v(\theta)$	160
$\succsim_f(a)$	154	$V(y)$	9
$\varphi^+(T)$	222	$w(S)$	39
$\varphi^-(T)$	222	$W(S)$	37
		$\mathbf{x}$	139

Smooth Functions

$X(\mathbf{p},m)$	130

$\approx$	423
C^1	427
C^n	460
C^∞	461
Df	425
$D^n f$	460
$D_{x_i} f$	430
$\vec{D}_\mathbf{x} f$	431
$d\mathbf{x}$	427
J_f	438
H_f	463
$(\mathbf{x})^n$	471
∇f	432

Optimization

arg max	182
$B(\mathbf{x}^*)$	550
$D(\mathbf{x}^*)$	570
$G(\theta)$	161

General Index

Action profile, 6, 43–45
Acyclical relation, 17
Adjoint, 295, 300
Admissible labeling, 110–11, 246–47
Affine
  combination, 86
  function, 276–77
  hull, 86
  set, 83–86, 309, 314
Affine dependence, 87
Always increasing, 198
Analytic, 472
Antisymmetric relation, 13
Arrow-Enthoven theorem, 582
Arrow's impossibility theorem, 35–36
Ascending, 195, 209
Ascoli's theorem, 221
Asymmetric relation, 13
Axiom of choice, 26

Banach fixed point theorem, 238–41
Banach space, 121
Bellman's equation, 165, 168–69
Balanced, 407–409
Barycentric coordinates, 87–88, 99
Basis, 77–81, 292
Best element, 20, 135
Best response correspondence, 180–81
Bilinear function, 287–90, 464
Binary relation. *See* Relation
Binding constraint, 550
Bolzano-Weierstrass theorem, 65
Boundary
  optimum, 502
  point, 50, 56, 125–26
  relative, 129
  set, 50–53, 128–29
Bounded functional, 155
Bounded. *See also* Totally bounded
  linear function, 273–74
  sequence, 59–60
  set, 48, 53, 61–62
Brouwer fixed point theorem, 245–51
Budget set, 131–32, 179–80

Cantor intersection theorem, 61
Carathéodory's theorem, 94, 108–10, 126–27
Cauchy sequence, 59
Cauchy mean value theorem, 454–55
Cauchy-Schwartz inequality, 290–91
CES function, 191, 200, 215, 339–40, 352, 355–56, 446, 455–56, 490, 541
Chain, 24
Chain rule, 441–42, 461

Characteristic function, 156, 171, 179
Closed
  ball, 51, 124–25
  correspondence, 182–83
  graph, 212
  interval, 18
  set, 50–53, 56, 60–61
Closed graph theorem, 275–76
Closed-valued, 179
Closure
  point, 50
  of a correspondence, 185
  of a set, 50
Cobb-Douglas function, 190–91, 199–200, 215, 342, 352–53, 355, 446, 463, 490, 539–41
Coalition, 3, 407–409
Coalitional game, 36–42, 68–69, 73, 78, 179
  balanced, 409
  convex, 200
  core, 39, 140, 380–81, 406–15
  essential, 39
  imputation, 101
  nucleolus, 41–42
  potential function, 268–69
  Shapley value, 266–69, 271
  superadditive, 106
Column space, 310
Compact
  function, 258
  metric space, 61–65
  set, 61–62, 121
Compact-valued, 179
Comparative statics, 601–602
Competitive equilibrium, 139–41, 251–54, 385–86
Complement of a set, 4
Complementary inputs, 203, 494
Complementary slackness, 395–96, 551
Complete
  lattice, 28
  metric space, 59–60
  order or relation, 13
Composition, 150, 184–85, 187, 212, 269, 333–34
Concave
  function, 328–36, 355 (*see also* Convex function)
  maximum theorem, 343–45
  programming, 592–95
Concavifiable, 340–41, 358
Cone, 104, 382–83
Conic hull, 106–108
Connected metric space, 52, 56

Constant function, 149, 425, 451
Constant returns to scale, 71–72, 104–106, 161, 354, 379, 492–93
Constraint qualification, 520, 568–81
  conditions, 577
Consumption set, 7–8
Continuous
  function 210–16 (*see also* Uniformly continuous)
  preference relation, 132–35
Continuously differentiable, 427, 460–62
Contours, 150, 436–37, 491
Contraction, 218–20, 238–41
Convergence, 58–60, 61, 118, 124
  pointwise, 151
  uniform, 151–52
Convex
  combination, 92
  cone, 104–106, 314
  correspondence, 182–83
  game, 200
  hull, 92, 126–27, 185
  maximum theorem, 342
  preference relation, 136–37, 339
  set, 88–95, 125–26, 293–94, 314
  technology, 89, 90, 105, 338–39
Convex function, 324–36, 483–88
  continuity, 334–36
  differentiability, 484
Convex-valued, 179
Cramer's rule, 310–11
Core, 39, 140, 380–81, 406–15
Correspondence, 177–85
  always increasing, 198
  ascending, 195
  closed-valued, 179
  closed, 182–83
  closed versus uhc, 226–27
  compact-valued, 179
  composition, 184–85
  continuous, 224
  convex-valued, 179
  convex, 182–83
  descending, 195
  domain, 182
  hemicontinuous, 223–27
  graph, 182
  monotone, 195–98
  product of, 185, 198, 229
  range, 182
  sum of, 185
  versus function, 178
Cost function, 163, 208–209, 331, 343, 345–46, 354, 358

Countable, 156
Cover, 62–64
Critical point, 440

Decision variable, 159, 497
Decreasing. *See* Monotone
Definite
  functional, 155
  matrix, 304
  quadratic form, 304
Demand
  correspondence, 180, 231, 345
  function, 354, 612, 616–19
DeMorgan's laws, 4
Dense, 53
Dependent variable, 145, 148–49
Derivative, 424–26
  second, 460
Determinant, 296–99, 302
Diameter of a set, 48
Differentiable, 424–25, 427, 433
Differential
  equation, 452
  total, 427
Diffeomorphism, 444, 478
Dimension
  affine set, 85
  convex set, 94
  finite, 79
  linear space, 79–80
  simplex, 99
Directional derivative, 430–32
Distance function, 159
Distribution function, 158
Domain, 12, 145, 182
Dual space, 280–84, 377
Duality, 284, 377–79, 382–84
Dynamical system, 158
Dynamic programming, 164–69, 241–44

Eigenvalue, 299–302
Elasticity, 452–53
  of scale, 494
Empty set, 2
Envelope theorem, 603–609
Epigraph, 154, 326, 355
Equicontinuous, 220–21, 259
Equivalence
  class, 14
  relation, 14–15
Euclidean
  metric, 47
  norm, 116–17
  space, 116–17, 291

Euler's theorem, 491–92
Exchange economy, 138–39, 251–54, 385–86, 414–15
Existence
 competitive equilibrium, 241–54
 eigenvalues, 300–301
 extreme point, 291, 374
 maximal element, 20
 Nash equilibrium, 255–57
 optimal choice, 135
 retraction, 294
 utility function, 192–93
Expected value, 278
Exponential function, 152–53, 189, 332–33, 443
Extended real numbers, 29, 154
Extreme point, 96–97, 125–26, 291, 374

Face, 96–98
Fan's condition, 292, 409
Farkas lemma, 383–84, 389–93
Feasible set, 162, 208, 497
Finite
 cover, 62–64
 dimension, 79
 intersection property, 64
First theorem of welfare economics, 141
First-order conditions, 504, 508
Fixed point, 149, 185, 296
Fixed point theorems
 Banach, 238–41
 Brouwer, 245–51
 Kakutani, 254–55
 Schauder, 257–59
 Tarksi, 233–35
Fredholm alternative, 283–84, 369–70, 392–93
Free disposal, 9, 54
Function, 145–50. *See also* Functional
 affine, 276–77
 analytic, 472
 bilinear, 287–90
 compact, 258
 composition, 150
 concave, 328–36, 355
 continuous, 210–16
 convex, 324–36, 483–88
 differentiable, 424–25, 427, 433
 domain, 145
 homogeneous, 351–55, 491–92
 homothetic, 356–58, 495
 inverse, 150
 linear, 263–76, 425
 Lipschitz, 218

 locally convex, 332, 336
 monotone, 186–94
 pseudoconcave, 490
 quasiconcave, 336–42
 onto, 146, 150
 one-to-one, 146, 150
 range, 145
 regular, 440
 space, 151
 supermodular, 198–201, 334, 388
 uniformly continuous, 217–18
Functional, 154
 bounded, 155
 constraints, 173, 497–98
 continuous, 213–16
 definite, 155
 equation, 168
 form, 190, 475–76
 linear, 277–80, 284–87, 294
 semicontinuous, 216–17

Game. *See* Coalitional game; Strategic game
Geometric series, 119
Global
 optimum, 500–501
 univalence, 479
Gordan's theorem, 393–98
Gradient, 432–33
Graph, 146, 148–49, 182
Greatest lower bound, 21

Hahn-Banach theorem, 369, 371–73
Halfspace, 358
Hemicontinuous, 223–27
Hessian, 463–66, 486–88
Hilbert space, 291, 294
Homeomorphism, 211
Homogeneous
 function, 351–55, 491–92
 system of equations, 308
Homogenization, 109
Homothetic
 function, 356–58, 495
 preferences, 356, 495
 technology, 358
Hotelling's lemma, 614–15
Hyperplane, 85, 284–87, 313–14, 358–59
Hypograph, 154, 329–30

Identity
 function, 149, 187
 matrix, 296
 operator, 296

Image, 145, 149, 269
Implicit function theorem, 479–81, 531–32, 622–23
Inclusion, 2
Increasing. *See* Monotone
Increasing differences, 201–203, 331
Independent variable, 145, 148–49
Indifference
  class, 15
  surface, 436–37, 482–83
Indirect utility function, 160–61, 209, 339, 348–49, 354
Induction, 188–89
Infimum (inf), 24
Inner product, 290–93
Input requirement set, 9, 90, 183–84, 378–79
Input-output model, 307–308, 319–20
Interior
  of a set, 50–51, 53
  point, 50–51, 56
  relative, 55–56, 128–30
  optimum, 502, 507
Intermediate value theorem, 216
Intersection, 4–5, 53, 90, 106, 125, 198
Interval, 18, 56, 88
Inverse
  function, 150, 270
  image, 149–50, 210–23
  matrix, 273
Inverse function rule, 444
Inverse function theorem, 477–78
IS-LM model, 439–40, 482, 624–28
Isoquants, 436–37

Jacobian, 438–39
Jensen's inequality, 327
Join, 26

Kakutani fixed point theorem, 254–55
Kernel, 269
K-K-M theorem, 250–51
Krein-Milman theorem, 375
Kuhn-Tucker
  conditions, 554–55, 562
  theorem, 552–54

Lagrange multiplier theorem, 520
Lagrangean, 532–34
Lattice, 26–31
Least upper bound, 21
Lebesgue number, 63

Leontief
  matrix, 320
  production function, 379
Lexicographic order, 22, 134, 192–93
lhc, 223–27
L'Hôpital's rule, 454–57
Limit, 58–60
Linear
  approximation, 417–24
  combination, 72
  dependence, 76, 122–23
  equations, 306–14, 384
  functional, 277–80, 284–87, 294
  hull, 72, 75
  inequalities, 306–308, 314–19, 380, 388–89
  operator, 295–96, 299–302
  ordering, 24
  programming, 317–18
  space, 66–82 (*see also* Normed linear space)
  subspace, 72–76
Linear function, 263–76
  bounded, 273–74
  continuous, 273–74
  derivative, 425
  nonsingular, 270
Linearly homogeneous, 353, 355
Lipschitz function, 218
Local optimum, 500, 511
Locally
  convex, 332, 336
  one-to-one, 477
  onto, 477
Log function, 189–90, 333, 444
Lower
  bound, 21
  contour set, 18, 154, 178–79, 214, 338

Manifold, 440
Map, 145
Markov chain, 249, 320–23
Marginal
  cost, 543–44
  product, 430, 434
  rate of substitution, 435–37, 446–47, 482–83
  utility, 435, 449–50, 543
Matrix, 264, 272–73
  Hessian, 463–66, 486–88
  Jacobian, 438–39
  Leontief, 320
  transition, 321
Maximal element, 20

Maximum theorems, 602
  continuous, 229–31
  convex, 342
  monotone, 205–207
  smooth, 603–604
Mean value theorem, 447–51
Meet, 26
Metric, 45–46, 156
Metric space, 45–66. *See also* Normed linear space
  compact, 61–65
  complete, 59–60
  connected, 52, 56
Michael selection theorem, 229
Minimax theorem, 349–51, 398–406
Minkowski's theorem, 377
Minimum. *See* Maximum
Mixed strategy, 102–103, 128–29
Modulus, 218
Monotone
  correspondence, 195–98
  function, 186–94, 433, 449
  maximum theorem, 205–207
  operator, 194
  preferences, 131, 191, 449–50
  selection, 197–98
  sequence, 59–60
  technology, 9
Monotonic transformation, 191–92, 514
Monotonicity. *See* Monotone
Motzkin's theorem, 397

n-tuple, 6
Nash equilibrium, 43–44, 181, 185–86, 255–57, 403
Natural order on $\mathscr{R}^n$, 22
Necessary conditions, 502, 507, 533, 550–55, 560–62, 585–86, 595
Neighborhood, 50
Net output, 8
Nested intersection theorem, 64
No-arbitrage theorem, 387–88
Nondecreasing, 187
Nonhomogeneous. *See* Homogeneous
Nonlinear equations, 476–83
Nonnegative orthant, 7
Nonnegativity constraints, 504–505, 523, 525, 560–62
Nonsatiation, 132, 138, 252
Nonsingular, 270, 296, 299, 305–306
No-retraction theorem, 250
Norm, 115, 156, 212, 274, 290–91
Normal
  of a hyperplane, 284
  space, 54

Normalization, 103, 286–87, 596
Normed linear space, 114–30
Nucleolus, 41–42, 136, 137
Nullity, 269–70, 271

Objective function, 160
Oligopoly, 170, 204, 257
One-to-one, 146, 150, 270, 271, 477
Onto, 146, 150, 270, 477
Open
  ball, 49, 120–21
  cover, 62
  interval, 18
  mapping, 212
  set, 50–53, 55
Open mapping theorem, 274–75
Operator, 145, 295–96
Optimal economic growth, 165–68, 244–45, 346–47
Order
  relation, 16–23
  topology, 134
Ordered set, 16
Ordering, 32
Orthogonal, 292
Orthonormal, 292

Parameter, 159, 497
Pareto
  efficient or optimal, 33–34, 139–41
  order, 33–34, 139–40
Partial
  derivative, 429–33, 437–38, 461–62
  order, 23–26
Partition, 14, 110, 407
Payoff function, 170, 193
Perturbation function. *See* Value function
Pointwise convergence, 151
Polyhedral set, 380–81
Polynomial, 68, 151
  approximation, 457–60, 467–75
Polytope, 98, 128, 381
Poset, 23–26
Power function, 147, 188, 190, 325, 337, 351, 443–44
Power set, 2
Preference relation, 13, 32–33, 130–38, 339, 356
Preimage, 149–50, 179
Preorder, 13
Present value, 120
Primal space, 280–81, 377
Principal axis theorem, 304
Principle of optimality, 169
Probability, 157

Product
  of correspondences, 185, 198, 229
  of functions, 188, 199, 214, 341
  order, 21–22
  rule, 442
  of sets, 5–7, 91
Production
  function, 159, 330, 338–39
  possibility set, 8, 91, 183–84, 331–32
Profit function, 161, 325, 353–54
Proper
  coalition, 3
  subset, 2
Pseudoconcave, 490
Pure strategy, 102–103

Quadratic
  approximation, 467–68, 472–75
  form, 302–306
Quasi-order, 13
Quasiconcave, 336–42, 489
  programming, 582–86
Quasiconvex. See Quasiconcave

Random variable, 157
Range
  of correspondence, 182
  of function, 145
  of relation, 12
Rank, 269–70, 271
Reflexive
  normed linear space, 295
  relation, 12
Regular function, 440, 478–79, 494
Regularity, 520, 551–52, 577
Relation, binary, 10–14, 146
  acyclical, 17
  antisymmetric, 13
  asymmetric, 13
  complete, 13
  continuous, 132–35
  equivalence, 14–15
  order, 16–23
  preference, 13, 32–33, 130–38
  reflexive, 12
  transitive, 12
Relative
  complement, 4
  interior point, 56, 128
  topology, 55–56, 128–30
Retraction, 249, 294
Riesz representation theorem, 294
Rolle's theorem, 508
Roy's identity, 617–18

Sard's theorem, 479
Saddle point, 349–50, 509–10
Sample space, 3
Schauder's theorem, 257–59
Second derivative, 460
Second theorem of welfare economics, 361–63, 385–86
Second-order conditions, 508
Selection, 186, 197–98, 229
Semicontinuous, 216–17
Semidefinite, 304
Separating hyperplane theorem, 358–60
  with nonnegative normal, 384–87
  strong, 366–69
Separation theorems, 55, 121, 358–71
Sequence, 57–62, 156
  Cauchy, 59
  monotone, 59–60
  of functions, 451
Series, 119
Set, 1
  countable, 156
  feasible, 162, 208
  polyhedral, 380–81
Shadow price, 542–45
Shapley-Folkman theorem, 93–94, 318–19, 375–77
Shapley value, 266–69, 271
Shephard's lemma, 616–17
Simplex, 98–103, 110–14
Simplicial partition, 110
Single crossing condition, 204–205
Singular, 270, 299. See also Nonsingular
Slack constraint, 550
Slater condition, 579
Slutsky equation, 618–19
Slack variables, 316, 560
Smooth function, 428, 461
Solow's convention, 461
Social choice, 34–36
Solution correspondence, 182
Space, 1
  action, 6, 42–45
  Banach, 121
  dual, 280–84
  Euclidean, 291
  function, 151
  Hilbert, 291, 294
  inner product, 290–92
  linear, 66–82
  metric, 45–66
  mixed strategy, 128
  normal, 54
  normed linear, 114–30

sample, 3
topological, 54
Span, 72, 77
Sperner's lemma, 110–14, 246–48
Spectral theorem, 301–302
Standard
 basis, 77–78
 simplex, 99
Stationary point, 504
Stochastic process. *See* Markov chain
Stiemke's theorem, 394–95
Strategic game, 3, 42–45
 Nash equilibrium, 43–44, 181, 185–86, 255–57, 403
 rationalizabilty, 181
 repeated, 57
 supermodular, 194, 207–208, 236–38
 zero sum, 193–94, 398–406
Strict or strong order, 16. *See also* Pareto order
Strong separation, 366–69
Strong set order, 30–31
Strong separation, 366–69
Subgradient, 364
Sublattice, 28
Submodular. *See* Supermodular
Subset, 2
Subspace
 of linear space, 72–76
 of metric space, 47
Sufficient conditions, 502, 507, 534, 581–86, 595
Sum
 of correspondences, 185
 of sets, 75–76, 91, 106, 118
Superadditive game, 194
Supermodular
 function, 198–201, 334, 488
 game, 194, 207–208, 236–38
Supporting hyperplane, 358–60
Supremum (sup), 24
Symmetric
 linear operator, 300–302
 relation, 12

Tangent hyperplane, 427
Tarksi fixed point theorem, 233–35
Taylor series, 471–72
Taylor's theorem, 468–72
Theorems of the alternative, 388–98
Topology, 54. *See also* Order topology; Relative topology
Total differential, 427
Totally bounded, 62
Transitive relation, 12

Transition matrix, 321
Transpose, 265–66
Triangle inequality, 45–46, 115
Tucker's theorem, 396–97
Tychonoff's theorem, 65

uhc, 223–27
Unanimity game, 40–41, 73, 76, 78, 268
Uniform
 convergence, 151–52
 continuity, 217–18
Union, 4–5, 53
Uniqueness
 Nash equilibrium, 257
 nucleolus, 137
 optimal choice, 136, 137, 345–46
 Shapley value, 271
Unit
 ball, 49–50, 120–21, 124–25, 250
 simplex, 99
 sphere, 52
 vector, 77
Univalent, 146, 479
Upper
 bound, 21
 contour set, 18, 154, 178–79, 337–38
Upper contour set, 154, 214
Utility function, 191–93, 212, 339
Uzawa equivalence theorem, 254

Value
 function, 160–61, 543 (*see also* Maximum theorems)
 of a game, 171 (*see also* Shapley value)
Vector space. *See* Linear space
Vertex, 96
Von Neumann's alternative theorem, 395–96

Walras's law, 252–53
Weak order, 16, 32–33. *See also* Pareto order
Weierstrass theorem, 215
Wicksell's law, 493–94

Young's theorem, 464–66

Zero-sum game, 402
Zorn's lemma, 26